Friederici
Lacerta
Deneb
Lactea
Albireo
Anser
Vulpecula
Scheat
Sagitta
Delphinus
Pegasus
Equuleus
Markab
Enif
Aquila
Aquarius
Antinous
Ecliptica
Capricornus
Piscis Notius
Globus
Fomalhaut
Grus
Microscopium
Indus
P9-AGL-235

URANOMETRIA 2000.0

Volume 1 - The Northern Hemisphere to -6°

URANOMETRIA 2000.0

Volume 1 – The Northern Hemisphere to −6°

Wil Tirion
Barry Rappaport
George Lovi

Published by:

Willmann–Bell, Inc.
P.O. Box 35025
Richmond, Virginia 23235 ☎ (804)
United States of America 320-7016

First Published August 1987
Reprinted:
October 1987
December 1988
June 1989
July 1991

Printed in the United States of America

Library of Congress Cataloging-in-Publication Data

Tirion, Wil
Uranometria 2000.0.
Includes Bibliographies.
Contents: v. 1. The Northern Hemisphere to −6° –
v. 2. The Southern Hemisphere to +6°.
1. Stars–Atlases. I. Rappaport, Barry, 1960 –
II. Lovi, George. III. Title
QB65.T65 1988 534.8′9 87-14769
ISBN 0-943396-14-X (v. 1)
ISBN 0-943396-15-8 (v. 2)

The front and back endpapers to this volume are full-scale reverse-printed reproductions of the equatorial zones from the hemispheric maps as published by Johann Elert Bode in his Uranographia *of 1801. Bode's atlas was published almost midway between the nearly 400 years separating Bayer's* Uranometria *of 1603 and* URANOMETRIA 2000.0 *and it is generally regarded as representing the pinnacle of oldtime pictorial star atlases. U.S. Naval Observatory, Washington, D.C.*

FOREWORD

As I hold this splendid atlas in hand, filled with admiration for such a magnificent cartographic achievement, I have to wonder what some of our illustrious astronomical forebears might think of it.

Claudius Ptolemy might be heard to mutter under his breath "too many stars," just as Emperor Joseph II complained that Mozart's *Marriage of Figaro* contained too many notes! The second-century Alexandrian was accustomed to display the stars on a globe—yellow spots against a dark background, he tells us in his *Almagest*—with dark lines connecting the stars to remind him of their mythological configurations. He described the locations of his 1028 stars with respect to these constellation figures, somewhat differently from his predecessors. Thus, for instance, those stars that Hipparchus places "on the shoulders of Virgo" he describes as "on her sides". How far is the present atlas from delineating the heavenly scenery with a thousand naked-eye stars!

Tycho Brahe, the great Danish observer, would undoubtedly be jealous and more than a little perplexed. By the end of the sixteenth century he managed to record 777 stars with an accuracy better than 1 arcminute, which is pushing the naked eye to its very limit of resolution. When it came time to draw up a star catalog, he hastily added another 223 so that the sheer bulk of Ptolemy's listing could not outrival his own. Tycho of course realized that there were several times as many stars actually visible to the unaided eye than those recorded in the catalogs, but the 332,556 in the Tirion-Rappaport-Lovi atlas would have been beyond the comprehension of anyone in the pre-telescopic age.

Galileo Galilei would nod in enthusiastic acclamation. "It's exactly as I said in my *Sidereus nuncius* more than three hundred years ago. With the aid of my telescope I settled the question of the Milky Way, which had vexed philosophers for all those centuries, and I declared that the galaxy is nothing but a congeries of innumerable stars." Galileo had, in fact, managed in 1610 to get approximately to the 9.5-magnitude limit of this atlas, but only in a few rare cases, and he made no attempt to map the entire sky.

John Flamsteed, appointed the first Astronomer Royal in 1675, would certainly appreciate this accomplishment. He never lived to see his own great atlas completed. As one of the first positional astronomers to exploit the power of the telescope, he recorded thousands of positions, but his three-volume catalog of observations finally appeared posthumously. As for the accompanying atlas, only the chart of Orion was finished at the time of his death in 1719; the continuing efforts of his co-worker Abraham Sharp and the artist Sir James Thornhill at last completed the volume a decade later. The speed with which the present charts were finished would surely, in the modern idiom, have boggled Flamsteed's mind.

William Herschel, discoverer of the planet Uranus and one of the greatest observers of all time, would be delighted, though hardly astonished, by this work. He had personally counted thousands of stars during his comprehensive "star gages" when he was probing the structure of the Milky Way, and eventually he catalogued 2500 nebulae found by sweeping the sky with the giant reflectors he had constructed. Herschel was never an atlas maker, but naturally he was a user, though he never had one this good; it would have been of inestimable value in the making of his several catalogs of double stars. And his sister Caroline would undoubtedly have profited from such charts while she was discovering her eight comets.

I'm not at all sure how James Challis, Plumian Professor of Astronomy at the University of Cambridge, would react to this new atlas. Had he had it in hand on July 29, 1846, he would probably have been the observer-discoverer of the planet Neptune. It would have meant exchanging infamy for glory. Challis, having access to the best refractor in England at the time, had agreed with the Astronomer Royal (George Biddell Airy) to search for a new planet predicted by young John Couch Adams. In the absence of a suitable chart, Challis was left with the task of tediously recording all the stars in the area, with the idea of going back later to check if one had moved. Armed with the Tirion-Rappaport-Lovi atlas, he could have found the interloper within an hour. Instead, he was beaten to the mark on September 23/24, 1846 by J.G. Galle in Berlin. Galle, who did have a suitable chart, promptly found the new planet on the first night he looked for it.

My ruminations cannot pass as history, but in their fantasy they hint both at the progress in astronomical map-making and at the powerful uses for such a marvelous tool as the URANOMETRIA 2000.0. As the title page of Copernicus' *De revolutionibus* proclaimed,

Igitur eme, lege, fruere.
"Therefore buy, read, profit!"

May 1987

Owen Gingerich
Harvard-Smithsonian
Center for Astrophysics

Contents

INTRODUCTION

Long before the invention of the telescope the stars were thought to be countless. Genesis 15:5 states, "Look up at the sky and count the stars, if you can." But how "countless" are the naked-eye stars? Assuming that stellar magnitude 6.5 is the average naked-eye limit, there are about 9,100. Other factors limit what we can see at one time and place. For example, we see, at best, only one celestial hemisphere at a time. Even a casual scanning of the skies reveals that the stars are not evenly distributed, and as we approach the horizon atmospheric extinction further limits what we see. Therefore, the number of "countless" visible stars probably does not exceed 2,500—a large number to be sure, but not staggering.

What happens when we use a telescope, even a modest one by today's standards? Under clear, dark skies the popular 8-inch Schmidt-Cassegrain telescopes can reach to about 14th magnitude, or almost 14 million stars! In the last 20 years thousands of these telescopes have been purchased by eager amateurs interested in exploring the wonders of the night skies. A major reason for the success of these instruments is their portability, since most modern-day amateurs do not have the luxury of an observatory or even dark skies at home. The movement to portability and light grasp has continued in recent years with the advent of the Dobsonian—a telescope noted for its stability and easily constructed mount—particularly for apertures over 12 inches. However, modern-day atlases have lagged behind this movement to expanded coverage and portability.

A comprehensive review of both old and new atlases will show the almost universal tendency of celestial map makers to create physically large maps, often on individual, unbound sheets. Maps like these must be used on a table, something not often carried out of doors. How did this happen? As George Lovi reveals in the following section of this atlas, most of these works have been designed by individuals who had the convenience of a well appointed observatory which provided protection from wind and dew as well as adequate space to lay out a large atlas.

The only easy-to-use atlases are certain "amateur" works such as *Norton's Star Atlas*—which, despite its nearly century old cartographic style, remains in print with a loyal following among observers. There is another reason, often overlooked by the user, for the cartographer to make large maps and hence fewer of them: A good set of maps will have overlap. Fewer maps require less overlap and hence less work. For example this atlas plots more than 332,000 stars and 10,300 deep-sky objects with generous overlap; therefore as many as 100,000 were drawn and labeled twice!

In 1979, I purchased a used copy of *Webbs Atlas of the Stars*, bound in book form and comprising 126 charts of about 8 by 10 inches. It plotted stars to 9th magnitude and extended to $-23°$. Now all but forgotten, H.B. Webb was a member of the AAVSO Chart Committee from 1934 to 1944. Although his atlas was intended for the variable star observer all the Messier objects were plotted, and because of this and the limiting magnitude of 9, I found it a convenient aid to navigating through the "fog of stars" seen even with a modest telescope. And here, in a single volume of manageable size, was a guide to the sky that I could easily use in the field. In the years since first finding this atlas I have not seen another for sale but have found other happy owners—all unwilling to part with theirs for the very same reasons I found mine so useful.

As an interesting aside, Wil Tirion thought that *Webbs Atlas of the Stars* looked familiar. Turning to his collection of atlases he found that the *Beyer-Graff Stern Atlas 1855.0* was apparently traced by Webb, since the scale, number of stars, and errors are exactly the same in each. Webb just changed the epoch (1920) and redivided the sky to accommodate the book format. In turn, the Stern Atlas is largely based upon the *Bonner Durchmusterung*, which is also one of the *Durchmusterung* atlases upon which URANOMETRIA 2000.0 is based—we have come full circle!

Since I am both a publisher and amateur astronomer it naturally follows that my experience with the *Webbs Atlas* and my observations of how others used it would suggest to me that there might be enough interest in a new work designed along these lines but reflecting more modern mapping techniques.

While it is one thing to recognize a need it is quite another to design a product to meet that need. Accordingly, in 1983, I sought out two acknowledged experts: George Lovi and Wil Tirion. George Lovi has been a popularizer of astronomy for over 25 years, a *Sky and Telescope* magazine columnist, and a planetarian. One of his particular long-time interests has been sky mapping. Wil Tirion, on the other hand, has become, in the last decade, the acknowledged master of modern-day celestial cartography. Both thought that the time was right for a new atlas with many stellar and nonstellar objects.

A sample chart was drafted from which it was determined that about 160 charts would be required to cover the sky. This decided, Wil Tirion had to set the project aside for nearly two years because of the backlog of prior commitments. To make sure that we developed the best possible specifications for URANOMETRIA 2000.0, George Lovi and I began a series of meetings at the U.S. Naval Observatory where we could, in one place, inspect most of the significant atlases since 1600. Brenda Corbin, Librarian, was of invaluable assistance on numerous occasions spanning the nearly three years of this effort. George shared the results of his research at the Naval Observatory with Wil Tirion and Barry

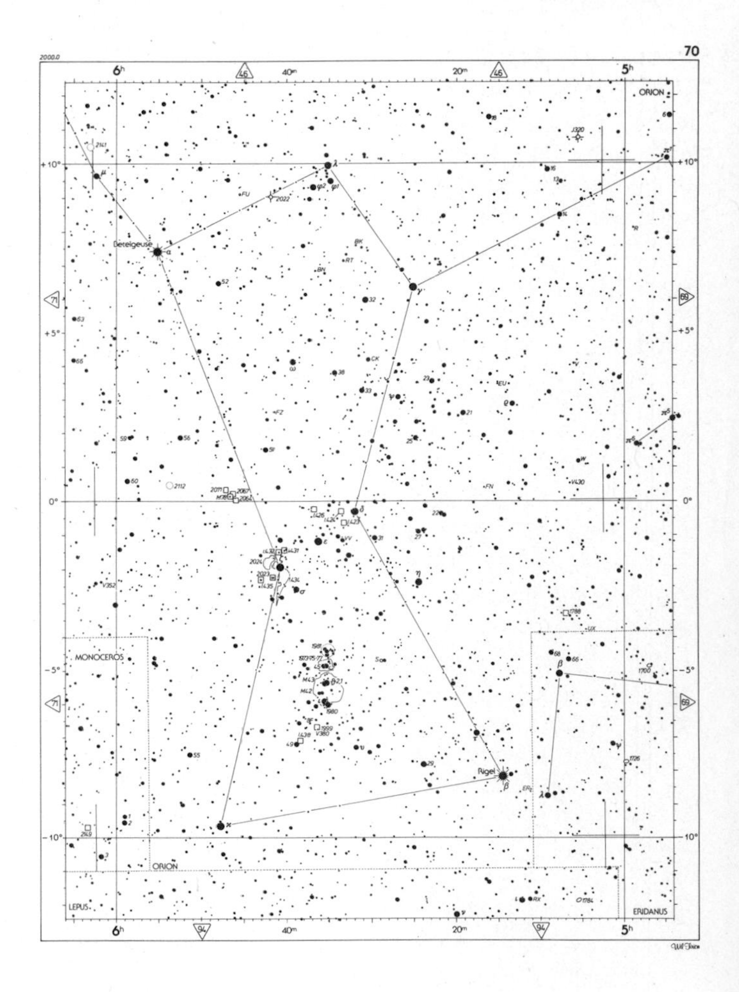

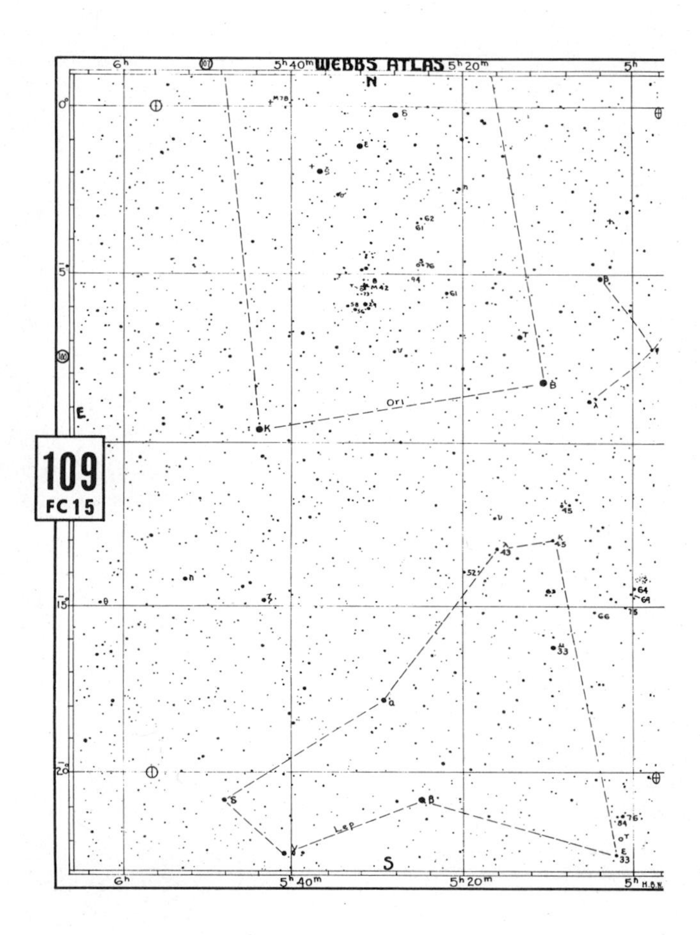

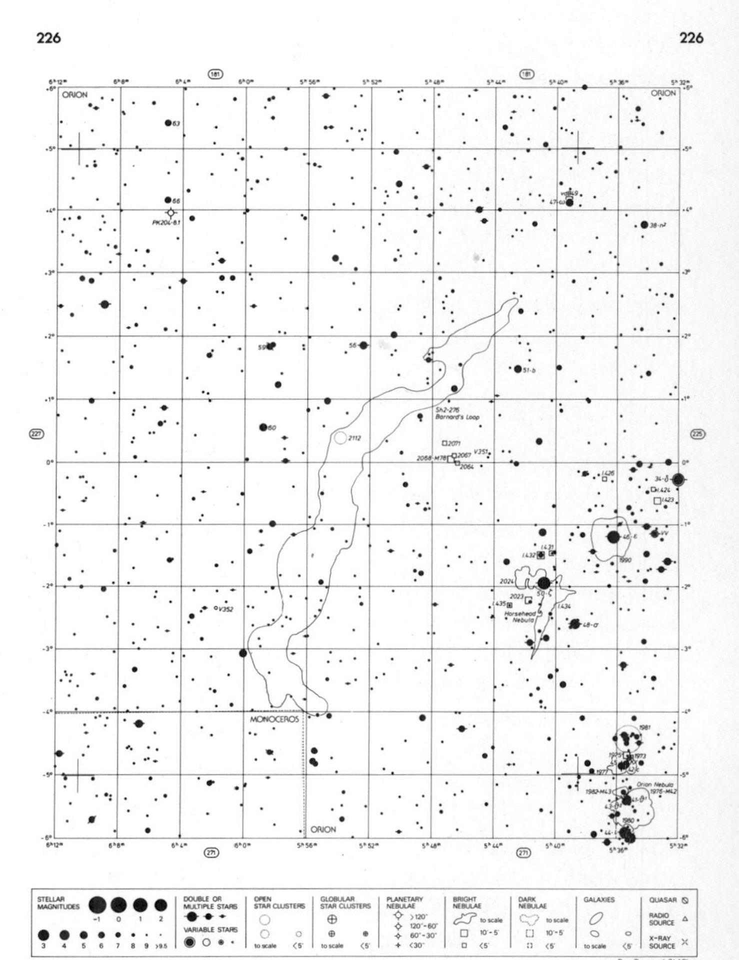

Fig. 1. How did URANOMETRIA 2000.0 *evolve? Here three charts of the area around Orion (reduced to 45%) demonstrate* URANOMETRIA 2000.0*'s development from a rather simple small-scale atlas to one of relatively large size and complexity. Immediately above is Chart 109 from* Webbs Atlas of the Stars *which was published in the 1940's. The chart shown at the upper right was executed by Wil Tirion as a prototype. If the atlas had been drawn like this a total of about 160 charts would have been required to cover the entire sky. At right is Chart 226 from the finished work. Notice that this final chart looks much less cluttered but in fact shows and labels much more than the other two because it it covers less sky area on the same page size. Moving from one chart to an adjacent one is facilitated by the adjacent chart numbers found around the outside border. With a little practice one can rapidly move over large segments of the sky yet retain the advantage of a large scale atlas to magnify a particular area of interest.*

Rappaport, and all three decided the final specifications for URANOMETRIA 2000.0. George also used this material to write "Uranography Yesterday and Today," a section of this atlas intended to familiarize the reader with the rich history of celestial cartography and provide an appreciation of why we chose to title this work URANOMETRIA 2000.0.

The final author, Barry Rappaport, was introduced to me by Richard Berry, Editor-in-Chief of *Astronomy* magazine. Rappaport was another person interested in celestial cartography. Upon graduating from Case Western Reserve University with a degree in astronomy, Rappaport joined the staff at the Jet Propulsion Laboratory. He had been encouraged while at Case by Dr. C. Bruce Stephenson to explore computer plotting of star atlases. He was assisted in this work by Dr. Charles Seitz of the California Institute of Technology, who provided access to computers and plotters to create an atlas based upon the *Cape Photographic Durchmusterung* star catalog. This atlas, of which only 50 copies were printed, extends from −18° to the south pole and plots 455,000 stars on 113 22-by-22 inch charts. He has also plotted an unpublished version of the *Cordoba Durchmusterung* with 600,000 stars. At the time this atlas goes to press Barry is on an educational leave of absence from JPL to do graduate work in astronomy at New Mexico State University.

In 1985 it became possible to assemble a full-sky catalog to 9.5 magnitude based upon the highly respected *Bonner Durchmusterung* (BD), *Südliche Bonner Durchmusterung* (SBD), and *Cordoba Durchmusterung* (CoD). A long-term project to convert these catalogs to computer readable form was just then being completed by Dr. Wayne H. Warren, Jr. of the National Space Science Data Center and Dr. Carlos Jaschek of Centre de Donneés Stellaires at the Observatoire de Strasbourg.

With this new data base the atlas grew from about 160 charts to the final 473. It also became possible to create, in addition to a highly accurate atlas, a companion star catalog which will be published as a three-volume set. URANOMETRIA 2000.0 STAR CATALOG will, in addition to providing the standard magnitude and positional data, cross-reference the *Durchmusturung* stars with those in a number of other recognized catalogs.

The Data Base

The Revised New General Catalogue of Nonstellar Objects (RNGC) by Sulentic and Tifft formed the machine readable nonstellar data base for URANOMETRIA 2000.0. This catalogue is also available as a book. To augment the RNGC, additional deep sky objects have been plotted by Wil Tirion. Many of these additional nonstellar objects accessible to the observer are listed in the excellent *Sky Catalogue 2000.0, Volume 2: Double Stars, Variable Stars and Nonstellar Objects* by Hirshfeld and Sinnott. To facilitate cross referencing between these two catalogues and this atlas we have generally adhered to the abbreviations used in the catalogues.

The BD, SBD, and CoD star catalogs provided the basic stellar data for this atlas. These works record about 1,100,000 stars, and in some areas of the sky reach magnitude 10 to 11. The BD and SBD had a working limit of magnitude 9.5, but later studies have shown that various systematic errors caused star brightnesses to be underestimated by about one-half magnitude toward the faint end. To achieve a relatively uniform limiting magnitude, all BD and SBD stars brighter than 9.3 were combined with 9.5 magnitude or brighter stars from the CoD. The result was a data base of 332,556 stars.

One major departure from earlier comprehensive atlases is our decision to show stellar magnitudes in whole-magnitude steps (magnitude bins) on the charts; for example, any star that falls between 4.50 and 5.49 is shown as 5th magnitude. This not only makes it easier to differentiate the various magnitudes on the charts by eye (and to compare them with the scale on each chart) but in practice astronomers still speak of 4th magnitude stars, 7th-magnitude stars, and so on. For exact values one should refer to standard catalogues or listings.

URANOMETRIA 2000.0's stellar data base was then augmented to include the 50 nearest stars and the 25 greatest proper motion stars no matter how faint (stars beyond 9.5 have their magnitudes noted). All the stars' positions were then precessed to equinox 2000.0. Analysis of the resulting data has shown the error in precession to epoch 2000.0 is less than the resolution of the charts. Specifically, an error of one second (of time) exists at zone 70°, while at zone 80° the error increases to approximately 30 seconds (time) for precession of 125 years (1875.0 to 2000.0). The overall working resolution of the atlas is estimated to be approximately 1′ at 85°, 30″ at 60°, and 10″ at 10°.

Since many catalogs use epoch 1950.0, we indicate on each chart how its coordinate grid could be shifted to read 1950 positions. Note the thin crosses close to each corner. Near each cross, an intersection of the coordinate grid is slightly darkened. If the grid were moved so the darker intersections fell on the crosses it would show 1950 positions.

In order to correct for stars with large proper motion the data were matched to a Luyten (NLTT) star data base to identify stars with proper motions greater than 0″.2 a year. A final positional and magnitude test was to compare URANOMETRIA 2000.0's data base to that of the *Yale Bright Star Catalogue* (YBS), 4th edition. Where differences were found the YBS data were used.

While the *Durchmusterung* star catalogs have been in use for years and are reliable, a small number of errors and omissions remain. To further reduce these, URANOMETRIA 2000.0's plots were compared to *The True Visual Magnitude Photographic Star Atlas*. When differences existed other atlases and catalogs were consulted to resolve the discrepancy; these are listed in the Reference Section.

Chart Arrangement and Labeling

Table 1 describes the general arrangement of URANOMETRIA 2000.0's charts, which were computer plotted at twice the final size of 1.85 cm = 1° per degree of declination. Stars were drawn as open circles, their relative sizes determined by their visual magnitudes rounded to the nearest whole number. The stars were plotted as open circles so as not to hide fainter companion stars. These circles were then filled by hand and where a bright star could hide a fainter neighbor, a ring of white space was left around the smaller disk. Working with the computer plots, Wil Tirion provided stellar no-

TABLE 1 URANOMETRIA 2000.0 CHART ARRANGEMENT					
Declination Range	Number of Charts	Vol. No.	R.A. Range of Chart Main	Overlap	Total
North Pole	2	I	12^h00^m	1^h00^m	13^h00^m
+85° to +72°	12	I	2^h00^m	20^m	2^h20^m
+73° to +60°	20	I	1^h12^m	8^m	1^h20^m
+62° to +49°	24	I	1^h00^m	4^m	1^h04^m
+51° to +38°	30	I	48^m	8^m	56^m
+40° to +27°	36	I	40^m	20^m	56^m
+29° to +16°	45	I	32^m	8^m	40^m
+18° to +05°	45	I	32^m	8^m	40^m
+06° to −06°	45	I, II	32^m	8^m	40^m
−05° to −18°	45	II	32^m	8^m	40^m
−16° to −29°	45	II	32^m	8^m	40^m
−27° to −40°	36	II	40^m	20^m	56^m
−38° to −51°	30	II	48^m	8^m	56^m
−49° to −62°	24	II	1^h00^m	4^m	1^h04^m
−60° to −73°	20	II	1^h12^m	8^m	1^h20^m
−72° to −85°	12	II	2^h00^m	20^m	2^h20^m
South Pole	2	II	12^h00^m	1^h00^m	13^h00^m
Total Charts	473				

tation based upon the following criteria: Star brightnesses are shown in whole magnitudes by tapered circles from −1 through 9.5. All first magnitude stars and other well known stars are labeled with their proper names (Castor, Polaris, etc.). The brighter stars are labeled first by Flamsteed number, then Bayer letter (if one exists). Double or multiple stars are shown with a line protruding from opposite sides of standard-sized dots. When these stars are separated by less than 1′ they are plotted with a dot size based on integrated magnitude (rounded to whole magnitudes). With separations greater than 1′ the components are plotted independently.

Variable stars are denoted with a concentric circle-and-dot symbol showing approximate maximum (outer) and minimum (inner) circle. For stars whose minimum brightness is below the atlas's 9.5 limit an open circle stands alone. Novae and supernovae whose maxima exceeded 9.5 are indicated as 9.5 magnitude variables with year of outburst preceded by an N or SN.

The 50 nearest stars are shown with their labels (no matter how faint) as catalogued by Batten in the Royal Astronomical Society of Canada's *1985 Observer's Handbook*. If these stars do not have a Flamsteed number or Bayer letter, their popular name, e.g. Barnard's Star, Ross 154, or catalog number (BD+36°2147) is shown; if fainter than 9.5 magnitude, their magnitude is added: Proxima ($11^{m}.1$).

The 25 greatest-proper-motion stars as catalogued in *Burnham's Celestial Handbook* have been plotted (no matter how faint). Labeling follows the procedure described for the nearest stars.

Constellation boundaries are those of E. Delporte endorsed, in 1930, by the International Astronomical Union (*Report of Commission 3*). Constellation names are placed near a border for easy identification. The Ecliptic is shown with a dashed line, labeled (once per chart) "Ecliptic," and calibrated each 1° in longitude. The Galactic Equator is shown with a dashed-dotted line, labeled (once per chart) "Galactic Equator," calibrated each 1° in galactic longitude. Also labeled are the North and South Galactic Poles and the North and South Ecliptic Poles. For added convenience, along the margins of each chart are numbers in ovals indicating the chart that covers that adjoining area.

In general, deep sky objects are labeled by NGC number (without prefix) or IC number (with prefix "I") and in the Messier catalogue, by a number preceded by an "M." Deep sky objects may also be labeled by popular names. Objects catalogued only in other works generally are identified by the catalogue or by an abbreviation of the authors' names. The abbreviations used are given in brackets at the end of each listing in the Reference Section. Distinctive symbols are used for 9 types of objects:

1. **Globular Clusters** are shown as a continuous-line open circle with an internal cross. Clusters larger than 5′ are drawn to scale.
2. **Open Clusters** are shown as a dotted open circle. Not included are scattered groups larger than 1°. Objects larger than 5′ but smaller than 1° are drawn to scale.
3. **Planetary Nebulae** are shown as open circles with four protruding lines in 4 selected sizes: Greater than 120″, 120″ to 60″, 60″ to 30″ and less than 30″.
4. **Bright Diffuse Nebulae** (selected) are drawn to scale with a solid outline if larger than 10′. Objects smaller than 10′ are drawn as solid-line square boxes in two sizes (10′ to 5′ and less than 5′).
5. **Dark Nebulae** (selected) are drawn to scale with a dotted line if larger than 10′. Objects smaller than 10′ are drawn as dotted square boxes in two sizes (10′ to 5′ and less than 5′).
6. **Galaxies** are drawn as open ovals, with objects greater than 5′ to scale. Included are members of the Local Group of galaxies not in the RNGC which are designated with their IC or UGC number, or if none, by their popular name (e.g. Sculptor System or UMi Dwarf galaxy).
7. **Radio Sources** are shown as open triangles when the source is invisible or below the chart's limit. Where the visible source is plotted, as in the case of a bright star (e.g. δ Cas) or a galaxy, the special symbol does not appear. In order of preference they are identified by 3C number, 4C number and then PKS without prefix formed by concatenating the hours and minutes of 1950 R.A. with declination truncated to tenths of a degree.
8. **X-ray Sources** are shown with an open-centered "X" when the source is invisible or below the chart's limit. Where the visible source is plotted the special symbol does not appear. Objects are generally labeled by common name (e.g., M82, NGC 1851, LMC X-2 [Large Magellanic Cloud], or γ Cas).
9. **Quasars** (Quasi-stellar objects) are identified by an open circle with a superimposed diagonal line. In order of preference they are labeled: 3C number, 4C number, and then coordinate designation as established in *Sky Catalogue 2000.0, Volume 2,* formed by concatenating the hours and minutes of 1950 R.A. with declination truncated to tenths of a degree.

Acknowledgments

Initially this atlas was to be hand drawn in about a year or two. When Barry Rappaport brought the capability to manipulate large data bases and then plot the results, it grew almost three times larger. While the computer increased overall plotting accuracy it did not create the atlas without significant human intervention. Many of the BD zones had been transferred to computer files but some zones remained, which had to be entered and verified by Barry. This done he spent hundreds of hours tending the plotter.

The first computer plotted atlas was the SAO. It required the services of six people working six months to apply standard labels to about 13,200 nonstellar objects. The stellar labeling was minimal (see Fig. 26 in the following section "Uranography Yesterday and Today"). URANOMETRIA 2000.0 plots about 10,300 nonstellar objects which are fully labeled with many of the symbols adjusted to size; the stars (of which there are about one-third more) are also labeled and many other refinements have been added—all by the hand of Wil Tirion or his young assistant Raymond de Visser—and all of this in less than two years. If this were not enough, Wil constantly had to refer to catalogs and other atlases to resolve the problems and inconsistencies associated with data derived from diverse sources.

Assistance has been received from many people. Jack W. Sulentic and William G. Tifft provided permission to use the RNGC tapes, Stephen J. Edberg offered suggestions on specifications, C. Bruce Stephenson advised on atlas mapping and gave stellar catalog information. Charles Seitz helped with large capacity computers and support equipment. Wayne H. Warren Jr. and Carlos Jaschek provided stellar catalogue data and advice on its use. Brenda Corbin, Librarian and our guide through the vast Naval Observatory Library, as well as Sandra Kitt, Librarian at the Richard S. Perkin Memorial Library at the American Museum–Hayden Planetarium, New York City, made their facilities available and offered assistance to George Lovi. Owen Gingerich and Deborah J. Warner critically read and commented on the text. Raymond de Visser, young friend and neighbor of Wil Tirion, wanted to help in some way—so each day, after school, he filled in star dots—more than 100,000 in this volume alone and in the process learned the sky "pretty well" according to Wil. Raymond is 12 years old.

The charts in this third printing of Volume I have been corrected based upon an errata developed with the assistance of the following persons: H.G. Adams, Erwin de Blok, Murray Cragin, Michael Ferrio, Steve Gottlieb, James Lucyk, Randy Lutz, Chris Quinnert, Alan M. MacRobert, Ronald Ravneberg, John Shannon, and D. Michael Walter. Brent A. Archinal, Harold G. Corwin, Jr. and Brian Skiff generously provided unpublished lists of corrections to the (R)NGC.

The errata to the first two printings of Volume I was published in the first printing of Volume II.[1]

No matter how hard one strives for perfection one can approach, but never quite achieve it. Astronomical data are not homogeneous and must always be used with a certain amount of respect for their diverse sources. While the authors have not knowingly permitted errors, no doubt errors continue to exist and some will be found—with more than a half-million hand entries—to be self-inflicted. Please let us know about them so that we may make corrections in subsequent printings.

Finally, the product of large projects tends to be reduced along the way to the lowest common denominator, but this need not be so. Here writ large for all to see is vivid testimony that three very talented people, separated by both a vast ocean and continent, can combine their strengths to depict the night sky as never before. I am indeed honored to have had the opportunity to chronicle their behind-the-scenes efforts in creating URANOMETRIA 2000.0. Each night that I use this atlas to sweep away my ignorance of God's universe I will remember Wil Tirion, Barry Rappaport and George Lovi and marvel anew at what they have wrought.

PERRY WILLMANN REMAKLUS
Publisher, Willmann-Bell, Inc.

[1]Individual copies are available from Willmann-Bell; please include a stamped, self-addressed business size envelope with your request.

References

STARS

Argelander, F.W.A., Bonner Sternverzeichniss, Sec 1-3. *Astron. Beob. Sternwarte Königl.* Rhein. Friedrich-Wilhelms-Univ. Bonn, Vols. 3, 4, 5. 1859–1862. [BD]

Batten, A.H., "The Nearest Stars," *Observers Handbook 1986,* Toronto, 1986: Royal Astronomical Society of Canada.

Bečvář, A., *Atlas of the Heavens—II: Catalogue 1950.0*, 4th edition, Prague and Cambridge, MA., 1964: Czechoslovak Academy of Sciences.

Bečvář, A., *Atlas Australis 1950.0*, 2nd edition, Prague and Cambridge, MA., 1976: Czechoslovak Academy of Sciences.

Bečvář, A., *Atlas Borealis 1950.0*, 2nd edition, Prague and Cambridge, MA., 1978: Czechoslovak Academy of Sciences.

Bečvář, A., *Atlas Eclipticalis*, 2nd edition, Prague and Cambridge, MA., 1974: Czechoslovak Academy of Sciences.

Delporte, E. *Délimitation Scientifique des Constellations*, Cambridge, 1930: Cambridge University Press.

Hirshfeld, A., R.W. Sinnott, *Sky Catalogue 2000.0, Vol. 1., Stars to Magnitude 8.0*, Cambridge, MA. 1982: Sky Publishing Corp.

Hoffleit, D., *The Bright Star Catalogue,* 4th revised edition, New Haven, CT., 1982: Yale University Observatory. [YBS]

Kholopov, P.N., *General Catalogue of Variable Stars,* 4th edition, Vols 1 and 2, Moscow, 1985: Astronomical Council of the USSR Academy of Sciences.

Kukarkin, B.V., *et al., General Catalogue of Variable Stars,* 3rd edition, Moscow, 1969–70; 1st, 2nd and 3rd Supplements, Moscow, 1971, 1974, 1976: Astronomical Council of the USSR Academy of Sciences.

Kukarkin, B.V., *et al., New Catalogue of Suspected Variable Stars,* Moscow, 1982: Astronomical Council of the USSR Academy of Sciences.

Luyten, W.J., *New Less Than Two Tenths Catalogue,* Minneapolis, MN. 1979, 1980: University of Minnesota. [NLTT]

Papadopoulos, C., *True Visual Magnitude Photographic Star Atlas*, Vol. 1.—Southern Stars, Vol 2.—Equatorial Stars, Oxford, 1979: Pergamon Press Ltd.

Papadopoulos, C., C. Scovil, *True Visual Magnitude Photographic Star Atlas*, Vol. 3.—Northern Stars, Oxford, 1980: Pergamon Press Ltd.

Perrine, C.D., Cordoba Durchmusterung, Part V. *Resultados Obs. Nacional Argentino*, vol. 21, −62° to −90°. 1932. [CoD].

Pickering, E.C., *Scale of the Bonn Durchmusterung,* Harvard College Observatory Annals, **72,** 6, Cambridge, MA, 1913.

Pickering, E.C., *Scale of the Cordoba Durchmusterung*, Harvard College Observatory Annals, **72,** 7, Cambridge, MA, 1913.

Schönfeld, E., Bonner Sternverzeichniss, Sec. 4. *Astron. Beob. Sternwarte Königl.* Rhein. Friedrich-Wilhelms-Univ. Bonn, vol. 8. 1886. [SBD]

Smithsonian Institution, *Smithsonian Astrophysical Observatory Star Catalog* Washington, D.C., 1966, 1971: Smithsonian Institution. [SAO]

Scovil, C.E., *The AAVSO Variable Star Atlas*, Cambridge, MA, 1980: Sky Publishing Corporation.

Thome, J.M., Cordoba Durchmusterung, Parts I-IV. *Resultados Obs. Nacional Argentino*, vol 16, −22° to −32°; vol. 17, −32° to −42°; vol. 18, −42° to −52°; vol. 21, −52° to −62°. 1892–1914. [CoD]

General Stellar and Nonstellar Objects

Burnham, Jr., R., *Burnham's Celestial Handbook,* New York, 1978: Dover Publications, Inc.

Hirshfeld, A., R.W. Sinnott, *Sky Catalogue 2000.0, Vol 2., Double Stars, Variable Stars and Nonstellar Objects*, Cambridge, MA, 1985: Sky Publishing Corp.

Neckel, Th., H. Vehrenberg, *Atlas of Galactic Nebulae*, Düsseldorf, vols. 1 and 2, 1985 and 1986: Treugesell-Verlag Dr. Vehrenberg KG.

Sulentic, J.W., and W.G. Tifft, *The Revised New General Catalogue of Nonstellar Astronomical Objects* Tucson, AZ, 1973, 1980: University of Arizona Press.

TABLE 2
GLOBULAR AND OPEN CLUSTER DESIGNATIONS

A–Antalova	K–King
Bar–Barkhatova	Lo–Loden
Bas–Basel	Ly–Lynga
Be–Berkley	Mrk–Markarian
Bi–Biurakan	Mel–Melotte
Bl–Blanco	Pi–Pismis
Bo–Bochum	Ro–Roslund
Cr–Collinder	Ru–Ruprecht
Cz–Czernik	Sh–Sher
Do–Dolidze	Ste–Stephenson
DoDz–Dolidze/ Dzimselejsvili	St–Stock
Fei–Feinstein	Tom–Tombaugh
Fr–Frolov	Tr–Trumpler
Haf–Haffner	Up–Upgren
H–Harvard	vdB–van den Bergh–Waterloo
Ho–Hogg	We–Westerlund
Isk–Iskudarian	

Globular and Open Clusters

Ruprecht, J., B. Baláz, R.E. White, *Catalogue of Star Clusters and Associations,* Supplement 1: Part A (Introduction), Part B1 (New Data for Open Clusters), Part B2 (New Data for Associations, Globular Clusters and Extragalactic Objects, Budapest, 1981: Akadémiai Kiadó. [See Table 2 for author(s) symbol]

van den Bergh, S., and G.L. Hagen, "UBV Photometry of Star Clusters in the Magellanic Clouds," *Astronomical Journal*, **73,** 569, 1968. [vdB-Ha]

Bright Nebulae

Cederblad, S., "Catalogue of Bright Diffuse Galactic Nebulae," *Meddelanden* fran Lunds Astronomiska Observatorium, Ser. 2, **12**, No. 119, 1946. [Ced]

Gum, C.S., "A Survey of Southern H II Regions," *Memoirs* of the Royal Astronomical Society, **67**, 21, 1955 [Gum]

Lynds, B.T., "Catalogue of Bright Nebulae," *Astrophysical Journal Supplement Series*, **12,** 163, 1965. [LBN]

Minkowski, R., "New Emission Nebulae," *Publications* of the Astronomical Society of the Pacific, **58**, 305, 1946. [Mi]

Rogers, A.W., C.T. Campbell, and J.B. Whiteoak, "A Catalogue of Hα-Emission Regions in the Southern Milky Way," *Monthly Notices* of the Royal Astronomical Society, **121**, 103, 1960. [RCW]

Sharpless, S. "A Catalogue of H II Regions," *Astrophysical Journal Supplement Series*, **4**, 257, 1959. [Sh2]

van den Bergh, S., "A Study of Reflection Nebulae." *Astronomical Journal*, **71**, 990, 1966. [vdB]

van den Bergh, S., and W. Herbst, "Catalogue of Southern Stars Embedded in Nebulosity," *Astronomical Journal*, **80**, 212, 1975. [vdBH]

Planetary Nebulae

Perek, L., and L. Kohoutek, *Catalogue of Galactic Planetary Nebulae,* Prague, 1967: Academia Publishing House of the Czecholosovak Academy of Sciences. [PK]

Dark Nebulae

Barnard, E.E., "Catalogue of 349 Dark Objects in the Sky," *A Photographic Atlas of Selected Regions of the Milky Way* Washington, D.C., 1927: Carnegie Institution. [B]

Bernes, C., "A Catalogue of Bright Nebulosities in Opaque Dust Clouds," *Astronomy and Astrophysics Supplement Series*, **29,** 65, 1977. [Be]

Lynds, B.T., "Catalogue of Dark Nebulae," *Astrophysical Journal Supplement Series,* **7,** 1, 1962. [LDN]

Sandqvist, Aa., "More Southern Dark Dust Clouds," *Astronomy and Astrophysics,* **57,** 467, 1977 [Sa]

Sandqvist, Aa., and K.P. Lindroos, "Interstellar Formaldehyde in Southern Dark Dust Clouds," *Astronomy and Astrophysics* **53,** 179, 1976. [SL]

Galaxies

Abell, G.O. "The Distribution of Rich Clusters of Galaxies," *Astrophysical Journal Supplement Series,* **3,** 211, 1958. [A]

Corwin, Jr. H.G., A. de Vaucouleurs, and G. de Vaucouleurs, *Southern Galaxy Catalogue,* Austin, TX, 1985: University of Texas Monographs in Astronomy No. 4.

de Vaucouleurs, G., A. de Vaucouleurs, and H.G. Corwin, Jr., *Second Reference Catalogue of Bright Galaxies,* Austin, TX, 1976: University of Texas Press. [RC2]

Nilson, P.N., *Uppsala General Catalogue of Galaxies,* Uppsala, 1973: Uppsala Astronomical Observatory. [UGC]

Nilson, P.N., *Catalogue of Selected Non-UGC Galaxies,* Uppsala, 1974: Uppsala Astronomical Observatory. [UGCA]

van den Bergh, S., "Luminosity Classifications of Dwarf Galaxies," *Astronomical Journal*, **71,** 922, 1966. [D or DDO]

Zwicky, F., *Catalogue of Galaxies and Clusters of Galaxies* Pasadena, CA, 1961–68, 6v.: California Institute of Technology. [ZWG]

RADIO SOURCES

Bennett, A.S., "The Revised 3C Catalogue of Radio Sources," *Memoirs* of the Royal Astronomical Society, **68**, 163, 1962. [3C or 3CR]

Boulton, J.G., and A.J. Shimmins, "The Parkes 2700 MHz Survey (fifth Part): Catalogue for the Declination Zone −35° to −45°," *Australian Journal of Physics*, Astrophysical Supplement, No. 30, 1, 1973. [PKS]

Boulton, J.G., A.J. Shimmins, J.V. Wall, and P.W. Butler, "The Parkes 2700 MHz Survey (Seventh, Eighth, Ninth and Tenth Parts)," *Australian Journal of Physics,* Astrophysical Supplement, No. 34, 1, 1975. [PKS]

Ekers, J.A., "The Parkes Catalogue of Radio Sources," *Australian Journal of Physics,* Astrophysical Suppl., No. 7, 1, 1969. [PKS]

Gower, J.F.R., P.F. Scott, and D. Wills, "A Survey of Radio Sources in the Declination Ranges −07° to 20° and 40° to 80°," *Memoirs* of the Royal Astr. Soc., **71,** 49, 1967. [4C part 2]

Pilkington, J.D.H., and P.F. Scott, "A Survey of Radio Sources Between Declinations 20° and 40°," *Memoirs* of the Royal Astronomical Society, **69,** 183, 1965. [4C, part 1]

Shimmins, A.J., "The Parkes 2700 MHz Survey: Catalogue for 03^h, 1^h, 19^h, and 23^h Zone, Declinations −33° to −75°," *Australian Journal of Physics,* Astrophysical Supplement, No. 21, 1, 1971. [PKS]

Shimmins, A.J., and J.G. Boulton, "The Parkes 2700 MHz Survey (Fourth Part): Catalogue for the South Polar Cap Zone, declination −75° to −90°," *Australian Journal of Physics,* Astrophysical Supplement, No. 26, 1, 1972. [PKS]

Shimmins, A.J., and H. Spinard, and E.O. Smith, "The Parkes 2700 MHz Survey (Sixth Part): Catalogue for the Declination Zone −30° to −35°," *Australian Journal of Physics* Astrophysical Supplement, No. 32, 1, 1974. [PKS]

Wall, J.V., A.J. Shimmins, and J.K. Merkelijn, "The Parkes 2700 MHz Survey: Catalogues for the ±4° Declination Zone and for the Selected Regions," *Australian Journal of Physics,* Astrophysical Supplement, No. 19, 1, 1971. [PKS]

X-RAY SOURCES

Amnuel, P.R., O.H. Guseinov, and Sh. Yu. Rakhamimov, "A Catalog of X-ray Sources," *Astrophysical Journal Supplement Series,* **41,** 327, 1979.

Amnuel, P.R., O.H. Guseinov, and Sh. Yu. Rakhamimov, "Second Catalogue of X-ray Sources," *Astrophysics and Space Science,* **82,** 3, 1982.

Bradt, H.V., R.E. Doxsey, and J.G. Jernigan, "Positions and Identifications of Galactic X-ray Sources, " *Advances in Space Exploration*, **3,** 3, 1979.

Bradt, H.V., and J.E. McClintock, "The Optical Counterparts of Compact galactic X-ray Sources," *Annual Review of Astronomy and Astrophysics* **21,** 13, 1983.

Manchester, R.N., and J.H. Taylor, "Observed and Derived Parameters for 330 Pulsars," *Astronomical Journal,* **86,** 1953, 1981.

QUASI-STELLAR OBJECTS

Craine, E.R., *A Handbook of Quasi-stellar and BL Lacertae Objects,* Tucson, AZ., 1977: Pachart.

Hewitt, A., and G. Burbidge, "A Revised Optical Catalogue of Quasi-stellar Objects," *Astrophysical Journal Supplement Series,* **43,** 57, 1980; **46,** 113, 1981.

Véron-Cetty, M.-P., and P. Véron, "A Catalogue of Quasars and Active Nuclei," European Southern Observatory *Report,* No. 1, 1984.

In 1603, nearly 400 years ago, Johann Bayer published his epochal Uranometria. *The above title page is from the 1655 edition—30 years after the author's death. Three earlier editions preceded it and another was to come in 1661. The title page reads, in part, as follows: "Uranometria, containing charts of all the constellations, drawn by a new method and engraved on copper plates." The pedestal on the left is inscribed: "To Atlas, the earliest teacher of astronomy" while the one on the right reads "To Hercules, the earliest student of astronomy." The atlas contained 51 star maps and was engraved by Alexander Mair. Using Tycho Brahe's star catalogue of 1592 as its primary "data base" it set, for its era, an unprecedented, highly-advanced scientific, graphic, and artistic standard for star charts.*

URANOGRAPHY

YESTERDAY AND TODAY

To the ancient Greeks, Urania was the Muse of the Heavens. There was also Uranus, the god of the celestial realm, for whom a major planet was named. Both derive from the Greek, root for sky, *urano*. Accordingly, there is a word, *uranography,* derived from the same source, which refers to celestial cartography just as geography refers to terrestrial mapping. However, the word uranography is little known today.

Fig. 1. This 1687 allegorical engraving from Johannes Hevelius' Firmamentum Sobiescianum sive Uranographia *depicts Urania, the Muse of the Heavens, with a stellar tiara and holding the Sun and Moon. She is surrounded by nymphs depicting the five bright planets. U.S. Naval Observatory, Washington, D.C.*

What does the uranographer actually map? A geographer delineates the actual physical form of the surface of our home planet, but charting the heavens is not that simple. What we regard as the sky, or firmament, is but a view of the universe from the vantage point of our Earth, whose members—Sun, Moon, planets, stars, clusters, nebulae, galaxies, and much else—each occupy a "position" in the sky that merely represents its *direction* in space from us. The universe is an enormous, mind-boggling three-dimensional entity, not the surface of a sphere.

Yet to chart it we must picture it to be just that. Astronomers make use of a gigantic imaginary *celestial sphere* which is infinitely large—larger than the entire universe. Everything within it, which is to say everything in the universe, has a definite position on this celestial sphere that is the result of a line from the Earth out to the celestial body in question and extended still further onto the inner surface of this imaginary—but mathematically real—sphere.

Like the Earth, this immense hollow globe has an imaginary grid system similar to longitude and latitude; the celestial version is called, respectively, right ascension and declination. This heavenly grid, which is most commonly used today for uranography, is called the *equatorial system* because it has a celestial equator resulting from extending the Earth's equator out to the sky; the same is done with the north and south poles, which now become the north and south celestial poles. (For uranographical purposes we can use the terms "celestial sphere" and "sky" interchangeably.)

Another basic system of celestial coordinates is the *ecliptic system*, little used today but the primary scheme during ancient, Medieval, and early-Renaissance times in the Western World. Its main reference circle is the ecliptic, which is

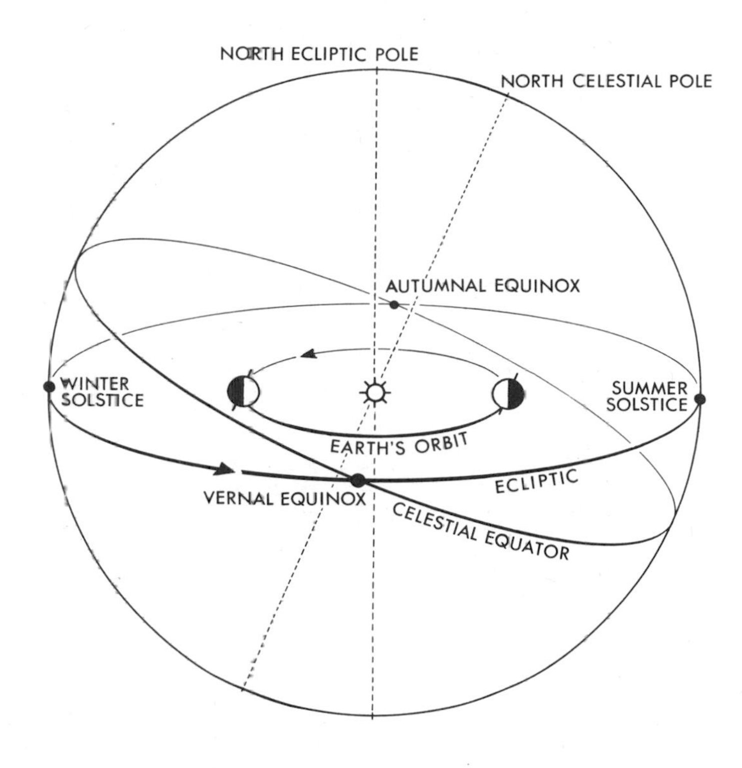

Fig. 2. The Earth's orbital motion and its rotation define, respectively, the ecliptic and equatorial coordinate systems. The celestial sphere upon which these coordinates are projected is infinitely large, so the Earth is in effect always in its center. Diagram by Wil Tirion.

Fig. 3. Depicting the heavens has taken various forms down through the ages. Here three examples illustrate these diverse methods. (a) Farnese Globe is a marble sculpture in the National Museum in Naples, Italy, believed to antedate 70 B.C. It shows Atlas holding up, for a change, the sky (celestial sphere) instead of the Earth. Clearly recognizable on the closeup of the globe are high relief carvings of the ancient constellations. Like all celestial globes, the Farnese Globe shows the sky mirror-reverse with east to the right and west to the left; this is the opposite of directions in the sky when we look up at it. This mirror-reverse mode was also unnecessarily adopted on some flat star maps until after the Renaissance. One interesting departure from usual uranographical practice for its era is that the Farnese Globe employs equatorial rather than ecliptic coordinates as its primary system. (b) The astrolabe was a useful, versatile kind of adjustable star map or flattened celestial globe that was commonly used during Medieval times to help locate celestial objects and solve problems in practical astronomy and navigation. This Arabic example, front and back, from the collection of the Smithsonian Institution, dates from 1281 and reminds us of the major role Mideast people played in keeping astronomy and other learning alive during that millenium-long mental and cultural sabbatical called the Middle or the Dark Ages. (c) A sixteenth century geocentric armillary sphere that now resides in the Science Museum, London.

the apparent annual path of the Sun in the sky, yet more fundamentally the extension of the plane of the Earth's orbit out to the celestial sphere. A major reason this was of primary importance in earlier eras is that the Moon and planets stay close to the ecliptic line in the sky since the tilts of their orbits are not greatly different from that of the Earth. The ecliptic system originally had fundamental astrological importance, and the constellations through which this coordinate line passes comprise the zodiac. There are also north and south ecliptic poles 90° above and below the ecliptic, and the other reference circles in this system are called celestial longitude and latitude. Since they do not correspond to longitude and latitude on Earth, these terms sometimes cause confusion for the astronomical neophyte.

Those who wish to know more about celestial coordinates should consult any of the numerous high-quality introductory astronomy texts currently available. Here we have but rudimentarily defined the equatorial and ecliptic systems because of their uranographical importance.

The ancients regarded the heavens as an intrinsic physical entity, either as an actual "ceiling" or dome over the Earth, and later as a sphere surrounding it, which represented the origin of the celestial sphere concept. This was responsible for some of the earliest attempts to map the sky with celestial globes. On their surfaces were delineated the imaginary constellation figures, most often pictures or carvings of the people, creatures, or objects the constellations represent with the actual stars frequently omitted. Not many early celestial globes have survived; a leading example is the beautiful and well-preserved Farnese Globe in Naples, Italy, a marble sculpture antedating 70 B.C. which shows Atlas holding a celestial globe over his head. The constellation figures are shown in bas-relief without any stars, although some astronomical historians believe that this sculpture originally did have stars on it.

Another form of early celestial globe is the armillary sphere, which could be called a "skeleton of the sky." It was an open framework showing only the principal coordinate circles and thus served as an early type of analog computer with which to solve astronomical, and often astrological, problems.

A derivative of the armillary sphere was the astrolabe, which utilized the mathematically elegant stereographic projection to flatten the sphere while preserving some of the latter's geometrical properties (such as angles) on a flat surface. The astrolabe, which was often a highly ornate and attractive artifact, could solve many of the problems that the armillary sphere was used for, but without the latter's bulk. It was therefore a standby of the astronomer and navigator for many centuries through the Renaissance.

A major disadvantage of the celestial globe, aside from its unwieldiness, was the way it showed the sky backwards, or mirror-reverse. This gave rise to the need for celestial charts which could be compared to the actual face of the sky as we see it. Not very many of these have come down to us from ancient and Medieval times; some of the better examples were made in the Orient, particularly in China. Some of the early Western star charts also, like the celestial globes, showed only constellation figures without any stars. A leading example of such a work is the planisphere ("flat sphere") of the Medieval monk Geruvigus, which is believed to date just before 1000 A.D.

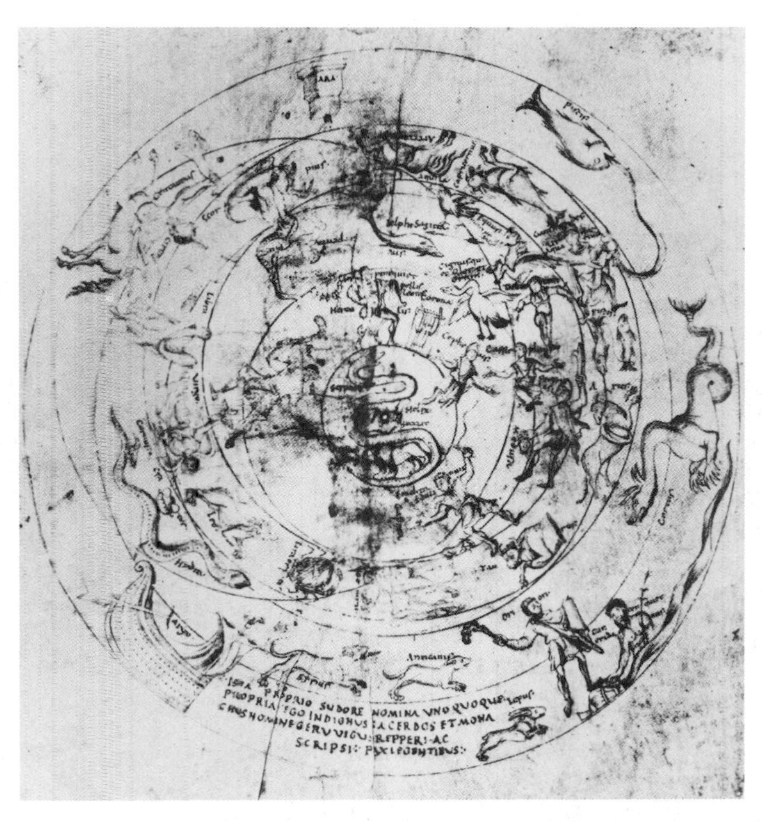

Fig. 4. Over a thousand years have elapsed since the monk Geruvigus drew this planisphere in Harleian MS. 647 which now resides in the British Museum library. Notice that the constellation figures are drawn without stars.

The first "modern" set of star charts, and without a doubt the most famous of the old ones, is the *Uranometria* atlas of the Bavarian lawyer and astronomy aficionado Johann Bayer (1572–1625), whose first edition appeared in 1603. This work was to uranography what the Gutenberg Bible was to printing. For its era, it set an unprecedented, highly-advanced scientific, graphic, and artistic standard for star charts. Bayer was the first to utilize and plot anything like accurate star positions; those in his atlas were based largely on the unprecedentedly precise—for that era—observations by the Danish astronomer Tycho Brahe (1546–1601). Despite being made in the pre-telescopic era with instruments equipped with only naked-eye sights, many of Tycho's positions were accurate to virtually one arc-minute. And in Bayer's *Uranometria* the accuracy with which most of the constellation patterns are recreated equals—and even exceeds—that of some present-day popular star charts.

Like other Western star maps of his era and earlier, Bayer used the ecliptic system for constructing his charts. Relatively few coordinate circles are shown, primarily vertical longitude circles 30° apart delineating the signs of the zodiac all the way to the ecliptic poles. The margins of the main maps are calibrated for every degree and the straight-line map projection used (which has been given the name trapezoidal projection) enables star positions to be read off with the aid of a simple straightedge.

Bayer's monumental work is also famous for initiating the still-used convention of labeling the brighter naked-eye stars in a constellation with lower-case Greek letters, which first appeared on his main atlas charts. In referring to a

star by this system, the Greek letter and genitive (possessive case) form of the constellation's Latin name is used; thus, the brightest star in our sky, Sirius, is designated α Canis Majoris, which means the Alpha star in the constellation Canis Major (the Great Dog).

It has long been widely believed that Bayer's sequence of Greek letters in a particular constellation is according to the apparent magnitudes (brightnesses as we see them from Earth) of the stars, starting with the most luminous one being α (Alpha), the second brightest being β (Beta), and so on down the alphabet. This is only very roughly so, with a great number of exceptions. There is ample evidence that Bayer used multiple criteria for his order of Greek letters in each constellation: not only the stars' order of magnitude, but their location in the constellation pattern or picture was a major regular consideration. A familiar example is the way he lettered the seven prominent Big Dipper stars in Ursa Major (the Great Bear) from west to east along the "flow" of this stellar asterism. Another convention Bayer seems to have sometimes followed is that if two stars are both in the same magnitude class, the northernmost one is assigned the prior letter in the alphabet. Thus, in Orion the star Betelgeuse is Alpha, and Rigel—which despite being brighter was still placed by Bayer in the first-magnitude class—has been made Beta because it is to the south of Betelgeuse. A totally specious conclusion has sometimes been advanced, often by popular writers, that discrepancies between the Greek-letter order in a constellation and their order of brightnesses represent changes in the stars' luminosities since Bayer's day. At that time magnitude estimates, made by the eye alone, were often quite rough and crude.

In constellations with many naked-eye stars, where the 24 letters of the Greek alphabet were exhausted, Bayer supplemented the sequence with Roman letters, both lower and upper case. His general star-labeling system has since his time been expanded to far-southerly constellations around the south celestial pole (and invisible from Europe) by the French astronomer Nicolas Louis de Lacaille (1713–62) who during a mid-18th-century expedition at the Cape of Good Hope compiled a catalogue of southern stars and created a group of new southern constellations named after inanimate objects. However, this astronomer assigned Greek letters to stars far fainter than Bayer would have, such as Omicron Octantis, of magnitude 7.2 and not even visible without optical aid. Unfortunately, this matter of stellar letter labels has since Bayer's time suffered from the "too many cooks" syndrome where other astronomers decided to get in on the act and assigned new letters to a great many previously unlabeled northern stars as well, which often created confusion because of labeling discrepancies from one source to another. Even today a great many stars—particularly brighter naked-eye ones—retain a multiplicity of aliases largely as a result of this.

Fig. 5. On the facing page is the famous mythological maiden Andromeda as depicted in Bayer's classic Uranometria *atlas. Here she is shown chained to seacoast rocks to be devoured by the sea monster Cetus. As so often happens in hero stories, her future husband Perseus rescued her just in time. Both Cetus and Perseus are separate constellations of their own. At lower left is that prominent stellar "W" formed by the main stars of Cassiopeia, who was Andromeda's mother. Unlike a number of his uranographical successors, Bayer usually showed the mythological outline picture surrounding only the stars of the particular constellation featured on a chart. U.S. Naval Observatory, Washington, D.C.*

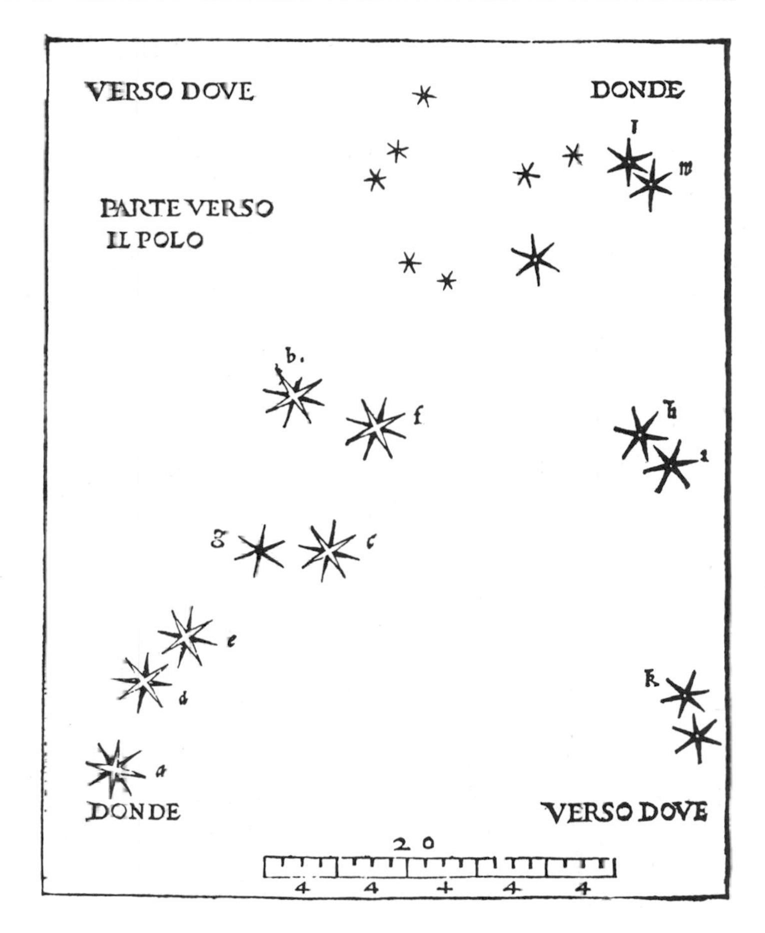

Fig. 6. Piccolomini's De le Stelle Fisse, *possibly the first printed atlas, introduced Roman-letter star labels. This chart shows Ursa Major, with the prominent Big Dipper asterism at lower left. Library of Congress, Washington, D.C.*

For the sake of historical completeness, it should be mentioned that Bayer was not the first to assign letter labels to stars—he was merely the first to assign Greek letters. He was preceded by Alessandro Piccolomini (1508–78), who in his atlas *De le Stelle Fisse* of 1540 introduced Roman-letter star labels. This work, which is said to be the first printed star atlas, used woodcuts instead of copper engravings—which Bayer's did—and was of considerably inferior accuracy and artistic quality when compared with the classic *Uranometria*.

Although we earlier said that the *Uranometria* represented in its era a new standard of star-chart accuracy, this did not apply to the plates showing constellations in the southern hemisphere of the sky invisible or poorly-seen from Europe such as Centaurus (the Centaur), and Argo Navis (the Ship) as well as a set of 12 entirely new groupings around the south celestial pole that were first charted in this work on plate 49. Here Bayer could not rely on Tycho Brahe's high quality positional measurements, but had to utilize considerably less accurate observations by explorers and mariners who wandered south of the Equator, which were largely those of the Dutchman Pieter Dirckszoon Keyzer (Petrus Theodori in Latin) who was responsible for those 12 new south polar

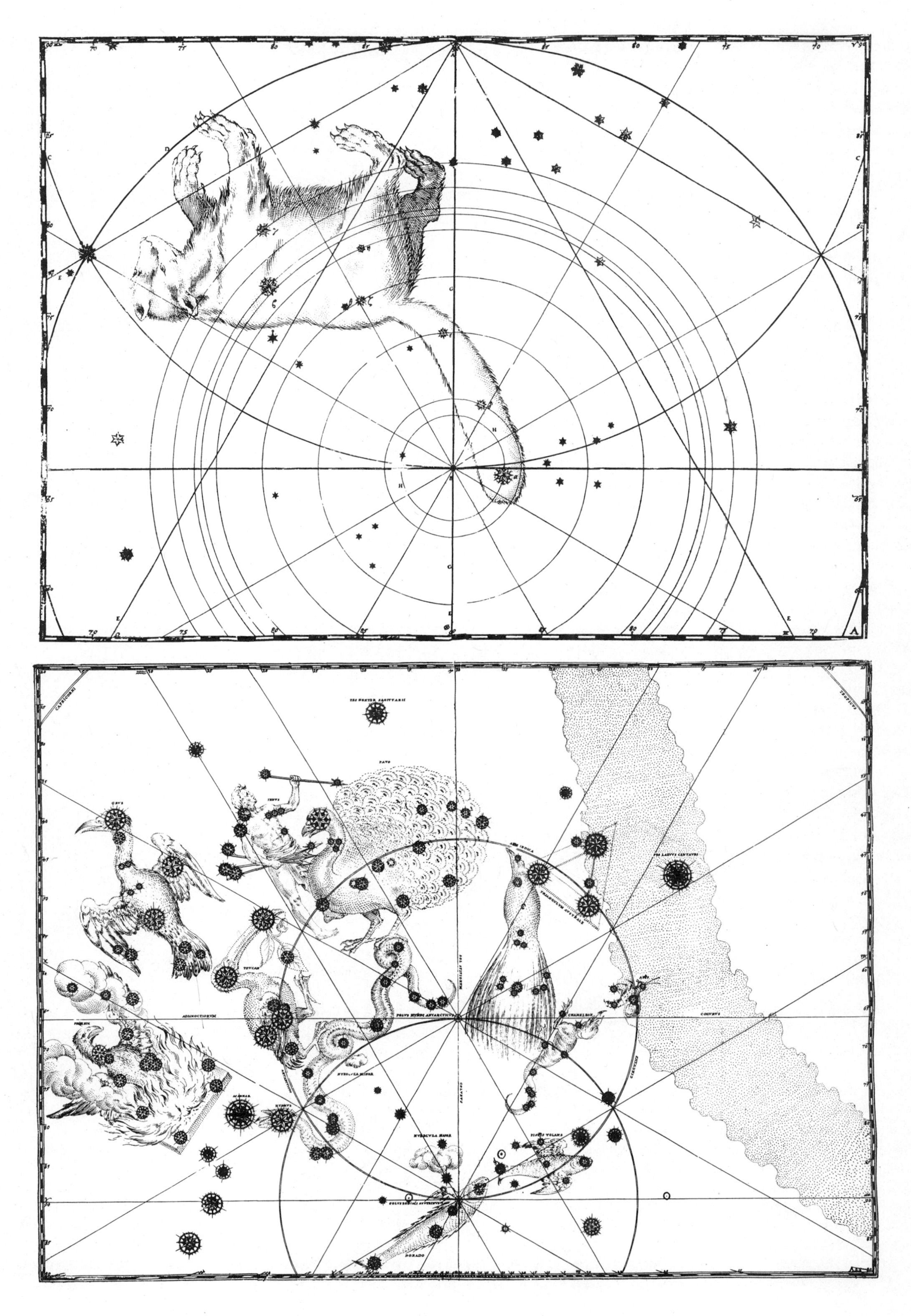

constellations in the *Uranometria*. The charts of Centaurus and Argo Navis, as well as the south polar one (plate 49), contain major positional and magnitude errors for the stars.

Bayer's chart 49, as well as Chart 1, also contains a mathematical and cartographic impossibility: double radial grids of both ecliptic and equatorial vertical coordinate circles 30° apart which radiate symmetrically from their respective poles. This can exist only on a globe (or the real sky), *never* on a flat map. This error points to the fact that certain aspects of cartography, particularly the principles of map projections and their proper use, were little known and understood in Bayer's day. (This has regrettably been a persistent problem with certain astronomers to the present time who create celestial charts. A leading example is the popular 20th century *Atlas Coeli* by Antonín Bečvář in which the intermediate declination charts have been plotted on improperly constructed conic projections which contain unnecessary distortion of star patterns; specifically, they are squeezed in an east-west direction. And quite a number of other star charts with unnecessary or excessive distortion have been issued over the years, which might have been avoided had their creators consulted any of a number of map-projection treatises or sought the assistance of a professional cartographer.)

However, in fairness to some of these old-time uranographers, it should be pointed out that many of today's numerous map-projection systems were unknown back then, largely because terrestrial cartography was a relatively infant discipline at the time. After all, it was still the Age of Exploration, prior to which there was little need for Earth maps covering large areas, while celestial cartography was still largely done on globes.

A well-known, often-commented-upon, and striking feature of the Bayer and just about all other atlases of the period (and for nearly three subsequent centuries) are the fanciful renditions of humans, animals, and objects surrounding the stars of the portrayed constellations. What was the reason for it—artistic, scientific, decorative? The answer is yes—to all of them. These figures were indeed intended to help visualize the arrangement, orientation, and extent of each constellation, although the star patterns in relatively few constellations come even close to suggesting a star picture. There are exceptions: the majestic main stars of Orion do suggest a standing human figure with that unmatched row of three almost equally bright and spaced stars in the middle of the pattern being his belt. And the stars of Ursa Major (the Great Bear) and Canis Major (the Great Dog) can without much difficulty be made to trace out stick-figure renditions of four-legged creatures.

Fig. 7. Shown on the opposite page are Charts 1 and 49 from Johann Bayer's Uranometria. *Chart 1 shows Ursa Minor, the Little Bear, which swings around the north pole by its un-bear-like tail, an anatomical anomaly of the two heavenly bears. Ursa Minor's principal stars also form the familiar Little Dipper asterism. The star at the end of the tail is Polaris, which in Bayer's day was some three degrees from the pole; since then precession has moved it to less than a degree away (see* URANOMETRIA 2000.0's *Charts 1-2 in the main atlas section to see its position for both the 1950 and 2000 epochs). The concentric circles on this Bayer plate are declination circles which do not represent particular degree intervals—the usual practice—but circles passing through the principal stars on the map. This chart shows both the north celestial and ecliptic poles, the latter being at the upper edge's center. Here, as well as with number 49, (which shows the south polar area) a major cartographic error is graphically depicted by showing equally spaced, 30° apart straight coordinate lines radiating from both poles. This is absolutely impossible on any kind of flat map—celestial or terrestrial; such symmetrical straight lines can radiate from only one pole on a particular plate. Note also on chart 49 the twelve additional constellations created by Bayer. U.S. Naval Observatory, Washington, D.C.*

But perhaps an even more important reason for these fanciful outlines was that during ancient and Medieval times astronomers regularly specified a star's position *by its location in the constellation figure*—even in star catalogues. Figure 9 is an example of this in the form of an excerpt from Ptolemy's star catalogue in his classic *Almagest* of the Second Century A.D. as printed in Volume 16 of the *Great Books of the Western World* set, published by Encyclopaedia Britannica; this

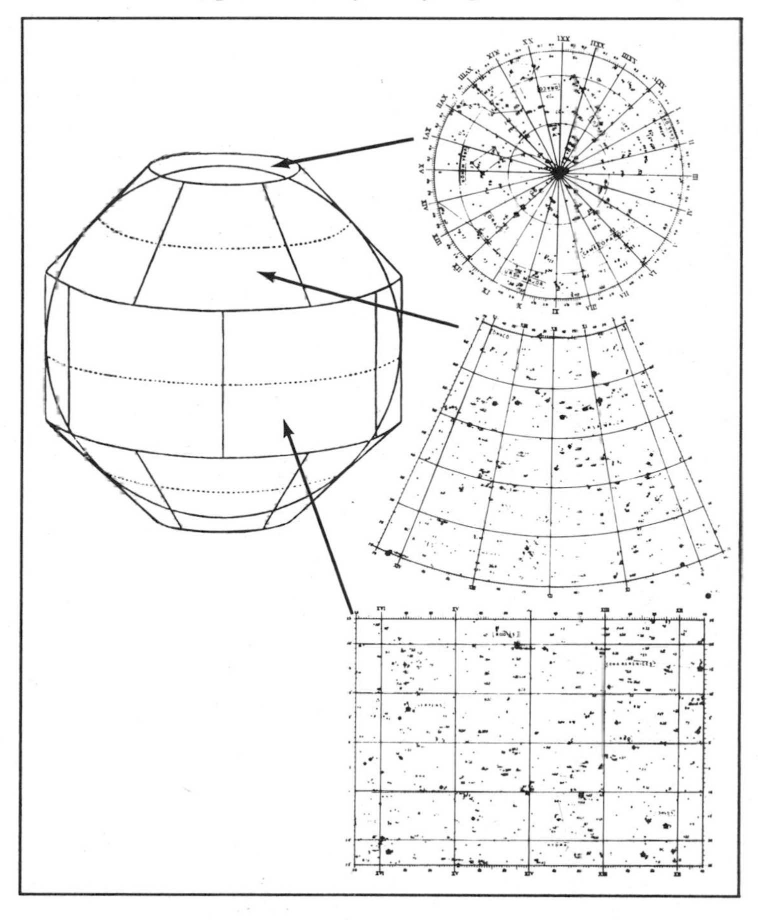

Fig. 8. How is the spherical sky flattened onto a flat sheet of paper to produce star charts and atlases? By techniques similar to terrestrial cartography, using any of a number of geometrical constructions called map projections. The diagram above, from Robert Ball's Atlas of Astronomy *(1892), explained those used in that atlas. There were three basic types: polar, conic, and cylindrical, each adapted for particular declination zones. This scheme has been utilized by a number of atlases, often with modifications such as additional conic zones (as is the case with* URANOMETRIA 2000.0*). The basic idea here is to use what cartographers call a developable surface, such as planes, cones, or cylinders, which can be flattened without stretching or distorting. These are but a few of the numerous map projections available to celestial or terrestrial cartographers, of which others are based on entirely different principles.*

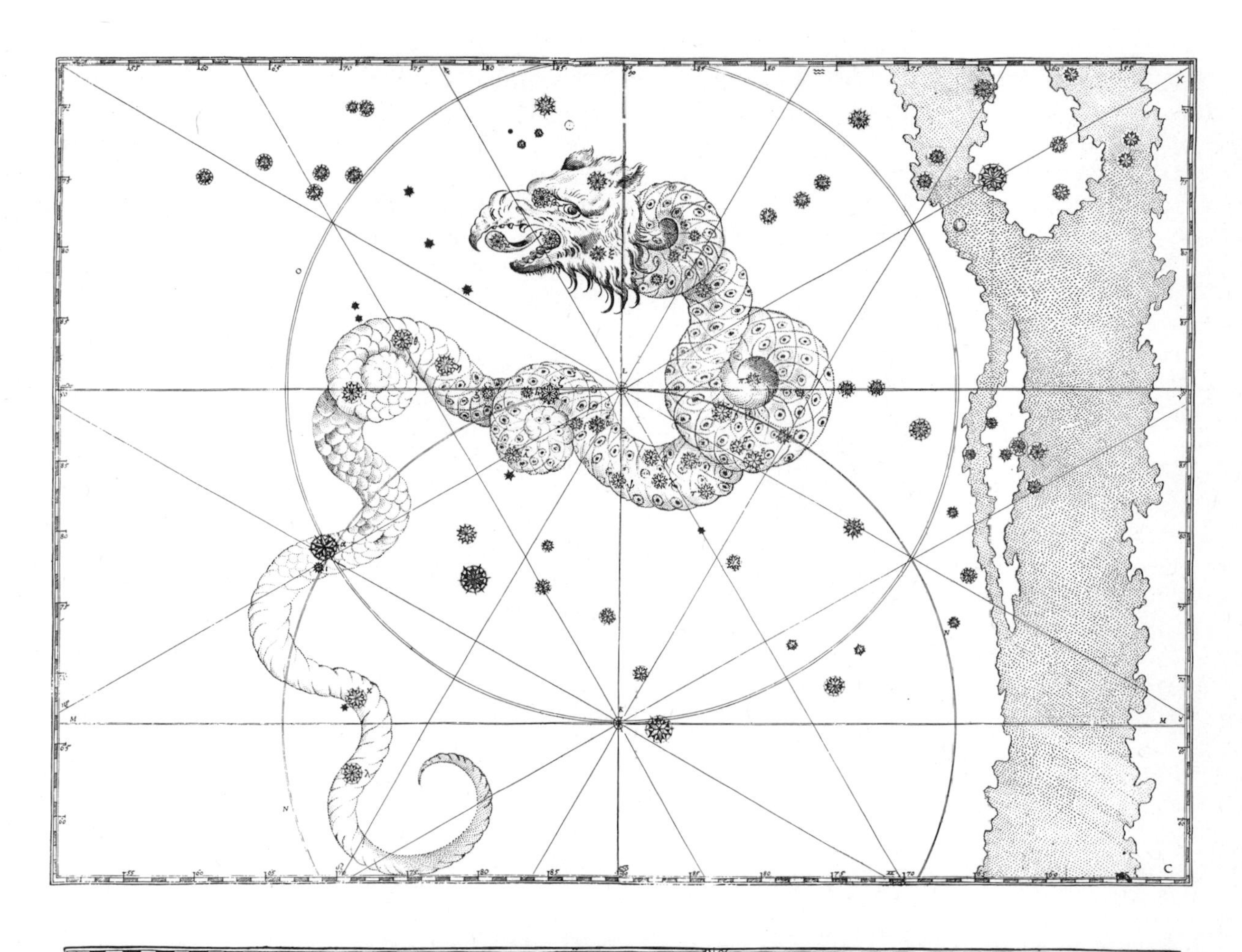

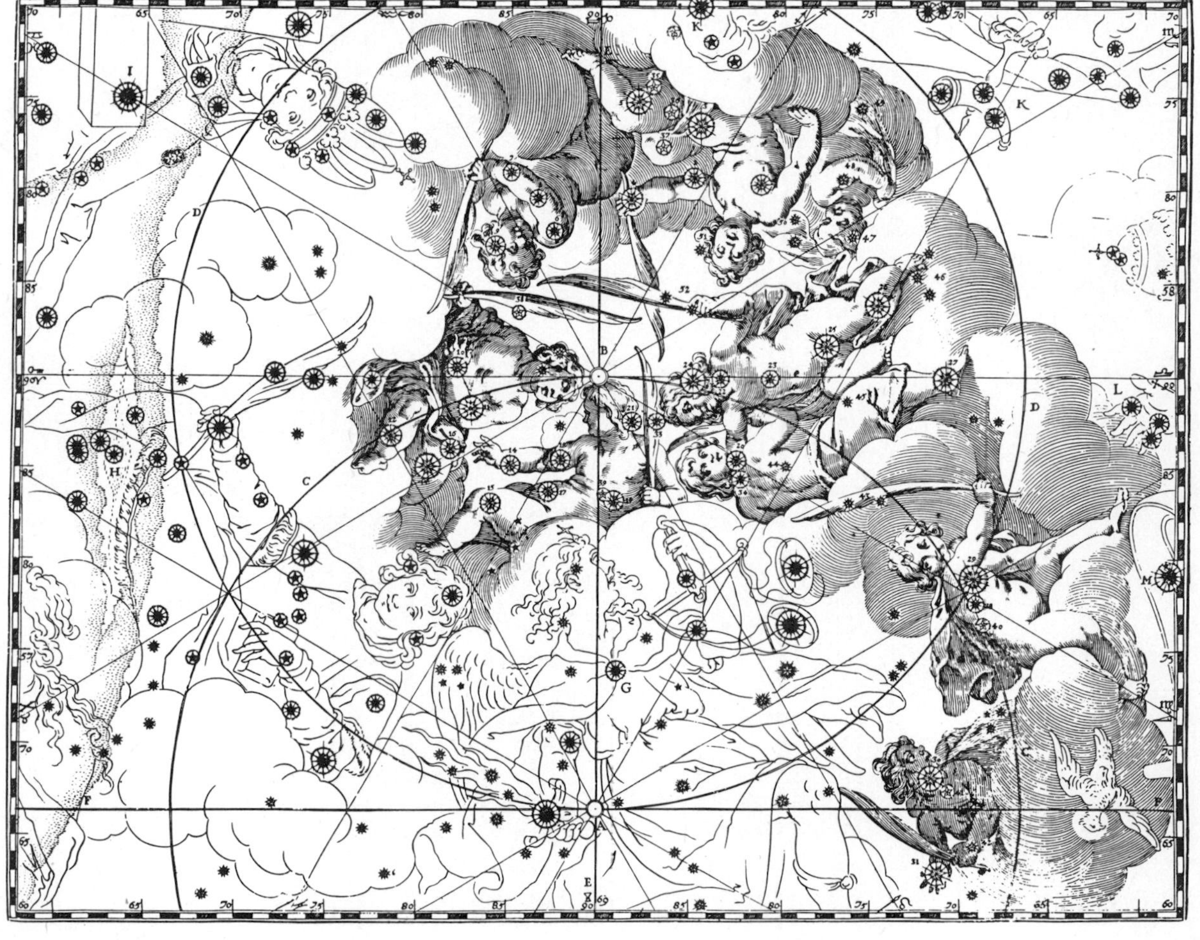

Configurations	*Longitude*	*Latitude*	*Magn.*
Constellation of the Little Bear			
The star on the tip of the tail	Twins 0°1/6′	N66°	3
The next one in the tail	Twins 2½°	N70°	4
The next one, before the beginning of the tail	Twins 16°	N74⅓°	4
The southern one on the western side of the rectangle	Twins 29⅔°	N75⅔°	4
The northern one on the same side	Crab 3⅔°	N77⅔°	4
The southern one of those on the eastern side	Crab 17½°	N72⅚°	2
The northern one on the same side	Crab 26⅙°	N74⅚°	2
In all, 7 stars of which 2 are of 2nd magnitude, 1 of 3rd, 4 of 4th.			
The unfigured star near it, that is the more southern star of first magnitude in a straight line with those on the eastern side	Crab 13°	N71⅙°	4
Constellation of the Big Bear			
The star at the tip of the muzzle	Twins 25⅓°	N39⅚°	4
The western star of those in the two eyes	Twins 25⅚°	N43°	5
The eastern one of these	Twins 26⅓°	N43°	5
The western star of the two in the forehead	Twins 26⅙°	N47⅙°	5
The eastern one of these	Twins 26⅔°	N47°	5
The star at the end of the western ear	Twins 28⅙°	N50½°	5
The western star of the 2 in the neck	Crab ½°	N43⅚°	4
The eastern one of these	Crab 2½°	N44⅓°	4
The northern star of the 2 in the breast	Crab 9°	N42°	4
The southern one of these	Crab 11°	N44°	4−
The star in the left knee	Crab 10⅔°	N35°	3
The northern star at the end of the left forefoot	Crab 5½°	N29⅓°	3
The southern one of these	Crab 6⅓°	N28⅓°	3
The star above the right knee	Crab 5⅔°	N36°	4
The star below the right knee	Crab 5⅚°	N33°	4
Of those in the quadrilateral, the star on the back	Crab 17⅔°	N49°	2
Of these, the star on the flank	Crab 22⅙°	N44½°	2
The star at the beginning of the tail	Lion 3⅙°	N51°	3
The remaining star in the left thigh	Lion 3°	N46½°	2
The western star of those at the end of the left hind-foot	Crab 22⅔°	N29⅓°	3
The star east of this one	Crab 24⅙°	N28¼°	3
The star in the left ham	Lion 1⅔°	N35¼°	4+
The northern star of those at the end of the right hindfoot	Lion 9⅚°	N25⅚°	3
The southern star of these	Lion 10⅓°	N25°	3
The first star of the 3 in the tail after the beginning	Lion 12⅙°	N53½°	2
The middle one of these	Lion 18°	N55⅔°	2
The third one at the end of the tail	Lion 29⅚°	N25°	3
In all, 27 stars of which 6 are of 2nd magnitude, 8 of 3rd, 8 of 4th, 5 of 5th.			

Fig. 9. Here is an excerpt from Ptolemy's star catalogue in the Almagest. *In addition to giving a star's position according to the ecliptic system of celestial longitude and latitude it gave the position in the constellation figure. Notice that some of the descriptive phrases are quite convoluted. For example, the 8th star in the Little Bear is described: "The unfigured star near it, that is the more southern star of first magnitude in a straight line with those of the eastern side." Reprinted by permission from* the Great Books of the Western World. *Copyright 1952 by Encyclopaedia Britannica, Inc.*

Fig. 10. On the opposite page is one of the most unusual star atlases to date, Coelum Stellatum Christianum, by Julius Schiller (d.1627), a contemporary of Johann Bayer who lived in the same city, Augsburg, and also like the latter earned his living from the law. Although Schiller's atlas was basically based on Bayer's, it represented a radical departure from the traditional classic constellations in that its purpose was to "de-paganize" the heavens by re-doing the constellations to represent Biblical figures. Here we have Schiller's S.S. Innocentibus, formed from the stars of Draco. One major difference from Bayer was the Schiller rendered the sky in the "mirror-reverse" backwards way they appear on celestial globes, which was still somewhat of a fashion in that era. Compare Schiller's S.S. Innocentibus with the accompanying plate of Bayer's Draco to see that practically the same stars were included in both renditions. Note also that Schiller shows surrounding constellations in outline form while Bayer draws only the principal character but does include surrounding stars. Library of Congress (Schiller) and U.S. Naval Observatory (Bayer), Washington, D.C.

volume also has the Copernicus star catalogue. Note those awkward primary descriptions of where in the outlined picture each star is located, so it was necessary to include these pictorial renditions on all star charts, even those used by professional astronomers. Indeed, the location of a celestial phenomenon, such as a new comet, might be announced as being presently "to the left of Andromeda's right knee," so every good astronomer had to know his celestial anatomy! And on the back of each chart folio in his atlas, even Bayer included such anatomical descriptions as his "gazetteer."

But Ptolemy also lists each star's position according to the old ecliptic system of celestial longitude and latitude. The first value is based on dividing the sky into twelve 30° segments extending to the north and south ecliptic poles, these segments being based on the signs of the zodiac, as we described earlier. So each star's longitude, no matter where in the sky even if far from the zodiac, is listed as being so many degrees of longitude into a segment named after a particular zodiacal sign. The vertical lines on the individual constellation charts in Bayer's atlas are these segment boundaries, so this quaint system was still in use then. However, Copernicus in his catalogue (also in the "Great Books" Vol. 16) ignores the sign segments and lists celestial longitude strictly on a continuous degree basis from 0° to 360°. It was not until the 18th century that our much-more-practical present-day equatorial system started being phased in, with star charts during the transitional period (lasting until the early-to-mid 19th century) featuring a double coordinate grid—both ecliptic and equatorial.

In the approximately two-and-a-half-century period following Bayer, quite a number of other atlases of the naked-eye stars was issued, virtually all containing those fanciful outline drawings of the constellation figures. Here we will cover only the leading and most significant works, yet for convenience of those who wish to pursue this subject further we have provided a carefully annotated bibliography at the end of this section.

The next major uranographical effort after Bayer occurred later in the same—17th—century, when another, and even more distinguished and urbane man-of-affairs, Johannes Hevelius of Danzig, that sometimes German-Polish-or-neither Baltic city now called Gdansk (and in Poland), produced his *Firmamentum Sobiescianum* atlas. This highly educated and cultured "Renaissance Man" was, in addition to being one of the leading astronomers of his day, among other things a brewer, municipal official, and skilled engraver (which he really used to advantage, as we shall see). The atlas was actually the second part of a catalogue of 1,564 naked-eye stars called *Prodromus Astronomiae*, issued in 1690 by Elizabeth Hevelius after her husband's death; this highly talented woman was his scientific collaborator, even in making celestial observations. Hevelius went Bayer one better in that he made his own positional observations of the stars, which were published in his catalogue and charted in his atlas, while Bayer had to rely on the work of others, largely Tycho Brahe.

There are some interesting similarities between Bayer's and Hevelius' efforts, both observationally and cartographically. The Danzig astronomer made his observations from his elaborate private observatory *Stellaburgum*, which in name and instrumentation was not too different from Tycho's *Uraniborg* of a century earlier. Both individuals used

Fig. 11. Commonly regarded as the next major star atlas after Bayer's Uranometria, *Johannes Hevelius' 1687* Firmamentum Sobiescianum sive Uranographia *(to use its full title) was a work of high accuracy and artistic merit with plates engraved by its author. Here we see Hevelius' rendition of Taurus, the Bull. The basic layout and arrangement of his charts share a number of similarities with Bayer; both used the simple—and cartographically naive—straight-line "trapezoidal" projection. Also, like Bayer, the atlas was plotted with ecliptic coordinates, since Hevelius (1611–1687) lived towards the end of the era when that system prevailed; the ecliptic itself runs across the center of this chart. A still further similarity with Bayer are those vertical longitude circles 30° apart representing the dividing lines between the classic signs (not constellations) of the zodiac. Around the time of the ancient Greeks, the zodiacal signs and constellations more-or-less coincided, yet in the intervening millenia the precessional movement of the heavens shifted the signs westward from the constellations of the same name. Here only the westernmost portion of the constellation of Taurus (at left) is still within the sign of Taurus; the Bull's head is already in the sign of Gemini, the Twins. Finally, the true major difference from Bayer is that Hevelius adopted the already archaic "mirror-reverse" mode of mapping the sky, or as it would appear on a globe. U.S. Naval Observatory, Washington D.C.*

measuring instruments with naked-eye sights, although by the time of Hevelius the telescope had been in existence for a half-century and more and was being used increasingly to aim positional instruments. But Hevelius totally rejected the telescopic sight, obstinately defending the naked-eye one. While it may seem odd to us today, such an attitude has not been that unusual in the history of astronomy or technology in general. Even as astrophotography became more widespread in the late 19th century, it was slow to gain universal acceptance among astronomers. However, in Hevelius' case he felt that instruments with naked-eye sights are more geometrically "pure" than one in which lenses or an attached instrument might introduce errors of their own. Whatever his reason, he was the last holdout of a then-dying tradition, and afterwards the use of telescopic instrumentation for astrometric (position measuring) activities became universal.

Yet Hevelius did remarkable work with his naked-eye instruments, obtaining a precision of better than one arc-minute for star positions, and the resulting accuracy of his atlas charts is in the same league as many present-day maps and atlases. (Even though star positions can today be observed to a small fraction of an *arc-second,* such hairsplitting precision cannot be utilized in plotting general star charts, for which 1 arc-minute accuracy—which Hevelius achieved—is quite ample.)

There are some major similarities between the atlases of

Fig. 12. Johannes Hevelius and his wife Elizabeth observing star positions with a highly precise six-foot radius brass sextant fitted with naked-eye sights. This well-to-do astronomer, merchant, municipal official, jack-of-many-trades built his own private observatory in the Baltic city of Danzig from which he made his own high-quality celestial observations which he utilized in his Firmamentum Sobiescianum *atlas. U.S. Naval Observatory, Washington, D.C.*

Bayer and Hevelius, and our Danzig dandy was equally ignorant of map-projection precepts. His main maps used the same straight-line "trapezoidal" projection as Bayer's. He even repeats Bayer's polar-chart blunder by including two symmetrical radial grids emanating from both the celestial and ecliptic poles.

Hevelius was a maverick in more ways than one, and not only by his stubborn adherence to an increasingly outmoded observing technique; he also adhered to an increasingly outmoded method of portraying the sky. This is in reference to his mirror-reverse star charts, which did not show the constellations as we see them when we look up, but as viewed from the fictitious "outside" of the celestial sphere—the way they are charted on celestial globes. While the geometry of a sky globe requires this, a star map *does not*. Indeed, a major advantage of flat charts, aside from their greater physical convenience, is that they can portray the starry firmament as we see it, while globes do not. Yet Hevelius *unnecessarily* depicted the sky in the awkward mirror-reverse manner of celestial globes. Why? Perhaps because he wished to maintain harmony with celestial globes, which were still widely used in his day.

Hevelius' atlas really stands out in terms of artistic quality, in which respect it equals or exceeds Bayer's, and Hevelius was able to engrave his own plates (an excellent zero-defect strategy!). Unlike Bayer, he showed the figure outlines of constellations adjoining the one featured on a particular plate by the use of a very pleasing subdued technique; see Fig. 11. In this work, Hevelius introduced some new constellations, including ones which are still recognized today such as Canes Venatici, Lacerta, Leo Minor, Lynx, Sextans, Scutum, and Vulpecula; these are, respectively, the Hunting Dogs, Lizard, Small Lion, Lynx, Sextant, Shield (of the Polish King Sobieski), and the Fox. Some of these originally had longer names: Vulpecula was first Vulpecula cum Anser (the Fox with Goose), and Scutum was Scutum Sobieski (Sobieski's Shield). A few other groups introduced by Hevelius have been dropped in modern times.

Less than a half century later, the next major uranographical milestone appeared: the *Atlas Coelestis* of John Flamsteed (1646–1719), England's first Astronomer Royal. Here celestial cartography took a quantum leap. Again the plotted precision of the stars is on a par with present-day general charts, and was based on observations made with telescopic instruments; these positions were listed in Flamsteed's *Britannic Catalogue*. The *Atlas Coelestis* has a finer and more detailed grid than previous works, with declination circles indicated for *each degree*; clearly, this was planned as an atlas of precision. Additionally, this is the first major atlas to adopt the more practical equatorial system, then starting to gain in prominence, as its primary grid, which corresponds to the apparent rotation of the sky (which, of course, is the reflex result of the Earth's rotation) and the movements of equatorially mounted telescopes that compensate for this motion. However, Flamsteed's maps retained the ecliptic coordinates as a secondary system, inaugurating the century-long era of double-gridded star charts to which we already referred.

Another major advance in this work is that Flamsteed used a mathematically "real" map projection to construct his charts: the sinusoidal, sometimes also called the "Sanson-Flamsteed" projection, since our English astronomer was long regarded as one of its creators, which we now know is not the case. This system correctly reproduces a main mathematical feature of the coordinate grid on the sphere (Earth and sky) whereby the vertical circles converge toward the poles according to the cosine of the declination or latitude. For example, on Earth the longitude circles are half as far apart at 60° latitude as they are at the Equator, since the cosine of 60° is 0.5. The sinusoidal projection (so named because the vertical circles are actually sine curves, which are identical in shape to cosine curves) preserves this property, but at the price of greater distortion with increasing distance from the central vertical circle.

It is rather interesting, however, that although the sinusoidal projection represents a closer simulation of the sphere than such simplified schemes as those earlier "trapezoidal" grids, it does not greatly reduce the distortion of star patterns (or continents) plotted upon it; this, despite the fact

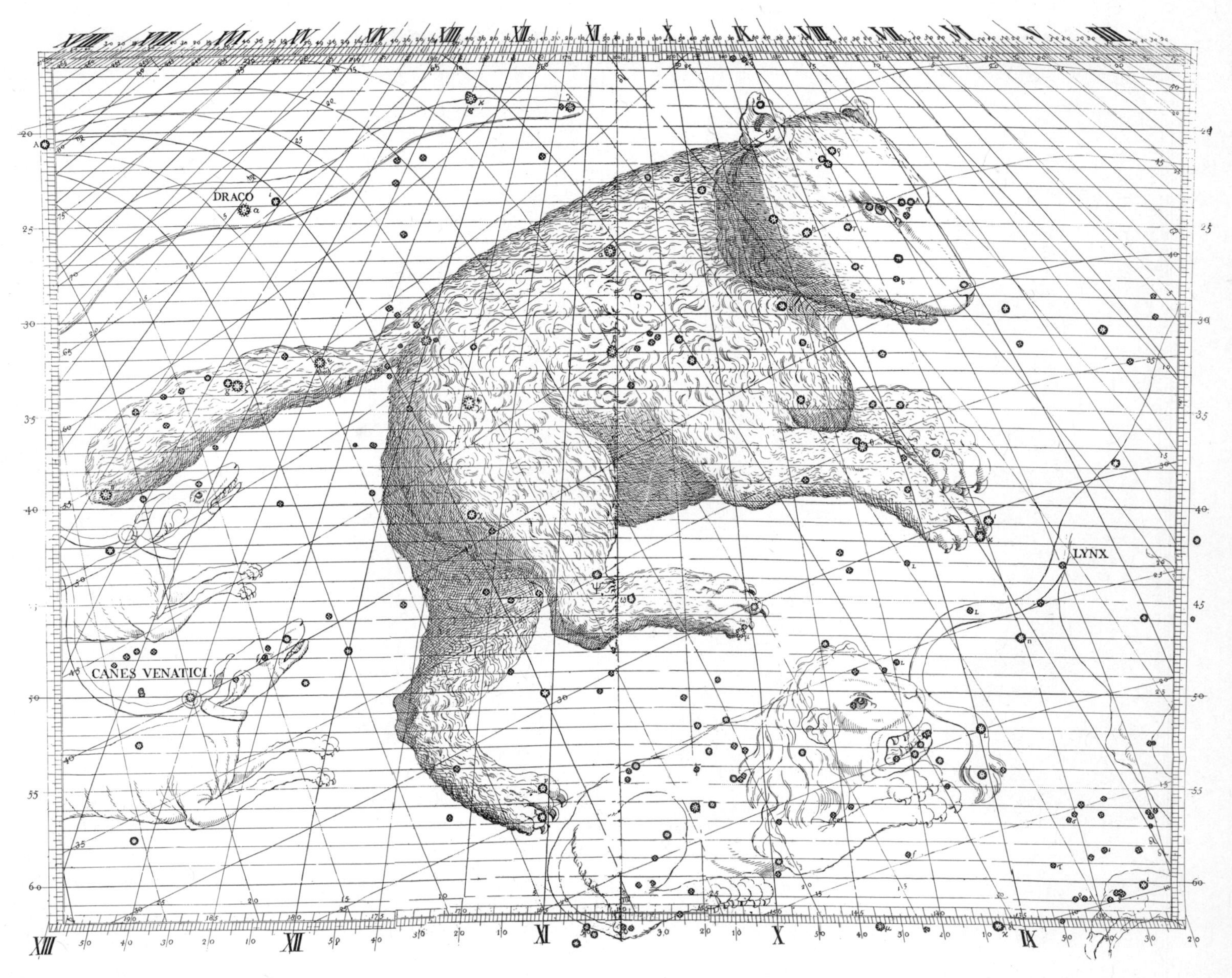

Fig. 13. John Flamsteed's Atlas Coelestis *was a landmark uranographical effort. Originally published in England in 1729, it became so popular that a number of editions, some foreign, were issued during the remainder of the 18th century. This is the Flamsteed portrayal of Ursa Major, the Great Bear. By the time this work appeared, the equatorial coordinate system started assuming ever-increasing importance, and it is here shown with declination circles for each degree. However, the older ecliptic grid is also included, running at a slant. Although* Atlas Coelestis *utilized the mathematically valid sinusoidal map projection for its primary (equatorial) grid, it often resulted in unnecessary distortion of star patterns, as is the case here. Those familiar with the constellations will notice that the well-known Big Dipper asterism (in the Bear's haunches and tail) looks stretched and warped, even though its stars are accurately plotted on the projection. An even more obvious indication of this grid's substantial shear distortion is the way the secondary celestial longitude circles make a sudden sharp turn above the Bear's tail. This would never occur on a low-distortion projection, such as one of the conic type, which Flamsteed would have been better off using, as did Bode in his* Uranographia. *However, one special property the sinusoidal projection does have is that it is equal-area; that is, if a coin were placed over any portion of the map it would cover the same number of square degrees. U.S. Naval Observatory, Washington, D.C.*

that Flamsteed himself stated that he chose this projection in order to portray the sky less distorted than by any projection he had hitherto seen. But it was a minimal improvement in this respect, as Fig. 13 which shows the prominent Big Dipper pattern, clearly indicates. This asterism looks so stretched and twisted as if it had passed through a wringer. And such distortion was so unnecessary! Even then other projections—particularly the conics—could have prevented this. However, the sinusoidal grid system was very much "the thing" in the 18th century, even for terrestrial maps.

Flamsteed's atlas was perhaps the best known and most popular set of celestial charts during that entire century, and enjoyed a number of editions, some simplified, which were not just English, but also French and German. And it was the basis for a number of imitations.

A common misconception about Flamsteed's atlas is that it introduced the "Flamsteed number" system of labeling the naked-eye stars of a particular constellation in the order of their right ascension; an inspection of his charts clearly shows it not to be so. What really happened is that the French astronomer Joseph Jerome de Lalande (1732–1807) assigned these numbers to the stars in a French edition of Flamsteed's

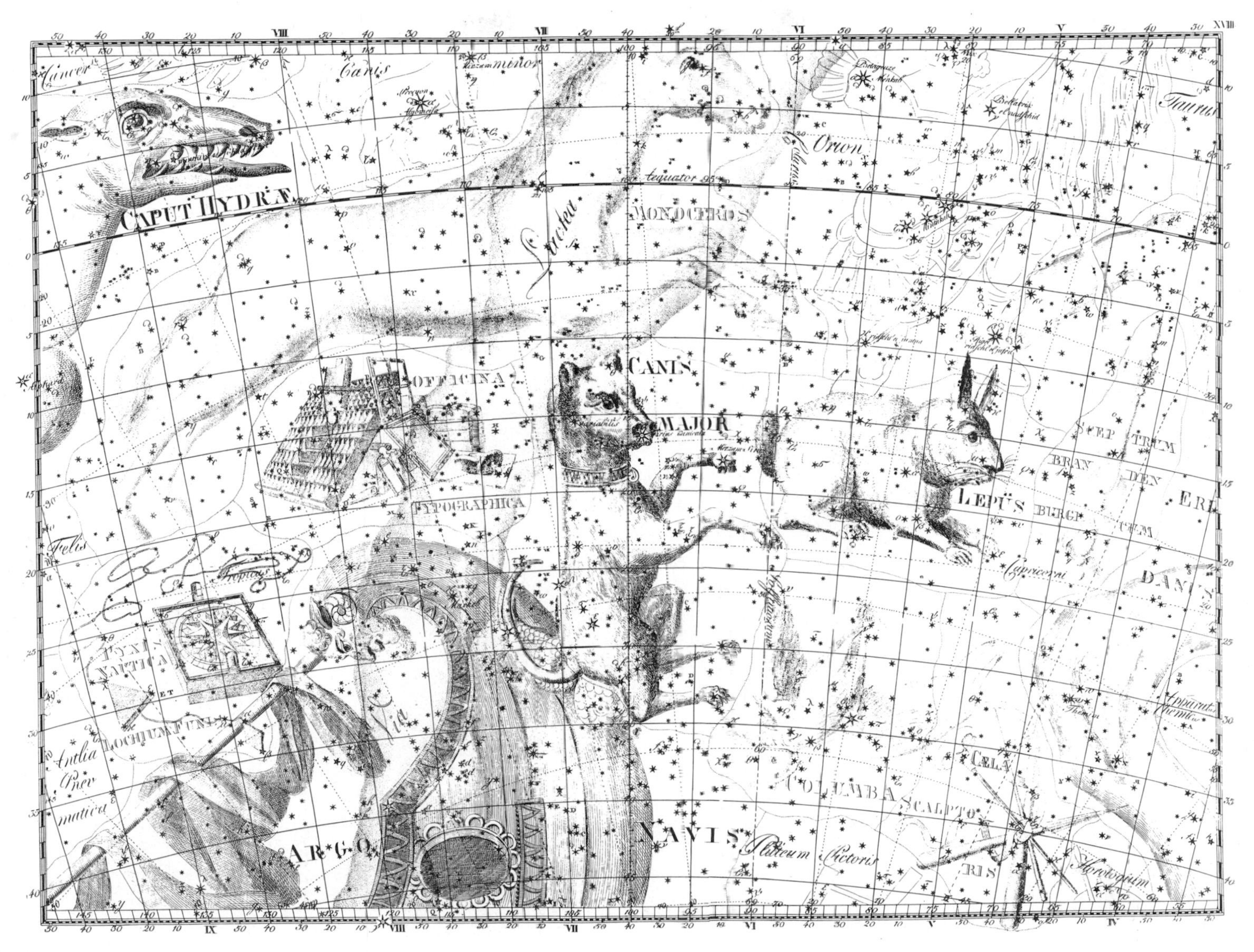

Fig. 14. The zenith of the oldtime pictorial star atlas was this elaborate, high-quality creation by the German astronomer Johann Elert Bode, his Uranographia *of 1801. Showing stars to quite a bit beyond naked-eye visibility (over 17,000 in all), it was a precisely-plotted work on excellent minimum-distortion conic projections. This plate shows the sky area in the vicinity of Canis Major, the Greater Dog, which also contains a few since-discarded groups such as Officina Typographica, the Printing Press; Lochium Funis, the Nautical Log Line; and Sceptrum Brandenburgicum, the Brandenburg Sceptre (in that era, astronomers, like composers, always sought ways to win royal favor and patronage). See also the front and back endpapers to this volume which show full size reverse printed portions from the equatorial zones of* Uranographia*'s two hemispheric maps. Trivia buffs might be interested in knowing that Bode (1747–1826), who was perhaps best known for his empirical "law" of planet distances, was a nearly exact contemporary of America's Renaissance Man President, Thomas Jefferson (1743–1826), who had a considerable knowledge of astronomy, among his numerous other interests, and who was inaugurated as President the same year Bode's* Uranographia *appeared. Did Jefferson see—or perhaps even own—this work? U.S. Naval Observatory, Washington, D.C.*

catalogue published in the 1780's. The stars in Flamsteed's atlas are labeled by the Bayer Greek letters.

The beginning of the 19th century—indeed, its very first year (1801)—saw the appearance of what could be regarded as the zenith of the old time atlases (those with pictured constellation outlines) when the noted German astronomer Johann Elert Bode (1747–1826) issued his splendid and monumental *Uranographia* atlas, a work of very high scientific and cartographic merit for its time. Its charts contained virtually all the naked-eye stars through the sixth magnitude and a number of fainter ones to the eighth.

Bode's atlas also represents the period when the sky was just about cluttered with the maximum number of constellations—just under 100—including a group introduced by Bode himself made up of rather faint naked-eye stars, since by then virtually all the brighter ones were pre-empted by other constellations. Among the Bode contributions are such quaint and now-defunct patterns as Officina Typographica (the Printing Press), Globus Aerostaticus (Montgolfier's Hot-Air Balloon), and Machina Electrica (Electrical Machine). Pity that he didn't live a bit later; he could also have given us something like "Ferroequus" (the Iron Horse, or Locomotive)! The name of one of the Bode groups, Quadrans Muralis (the Mural Quadrant) survives today in the form of the Quadrantid meteor shower that around January 4th radiates from the stars in this now-obsolete pattern in northern Bootes.

Bode was one of the first to draw constellation boundaries on his charts, a concept by which every star belongs to a definite group. Obvious as it may be to us today, prior to then—particularly in ancient times—there were gaps between the constellations, and stars in such areas were labeled as "unformed" or not part of any group. Similarly, back then it was an entirely accepted practice to overlap constellations, both their stars and outline pictures, which atlases such as Bayer's clearly shows as we compare its various charts. In fact, Bayer even assigned double Greek-letter labels to some stars, making them part of two adjacent constellations; examples are the star called Alpheratz which he labeled as both α Andromedae and δ Pegasi, and Elnath, originally β Tauri and γ Aurigae. Today these stars are strictly α Andromedae and β Tauri; no star is shared by two constellations, a concept Bode helped pioneer with his constellation boundaries. However, they were drawn as loose curvy lines and usually did not agree too closely from one atlas to the next in subsequent years. It wasn't until 1930 that the International Astronomical Union made the constellation boundaries—as well as the roster of recognized constellations—official. These IAU boundaries run strictly north-south and east-west (exactly so for epoch 1875.0), suggesting those of certain states in the western part of the continental United States.

And not the least of the virtues of Bode's *Uranographia* is that it was constructed on excellent, primarily conic, projections, thereby showing the constellation patterns with minimum distortion. A milestone has been reached! In fact, Bode abandoned the sinusoidal projection, which he had used in an earlier, simple atlas issued in 1782, at the urging of none other than Lalande.

THE TRANSITIONAL PHASE

The coming of the 19th century saw major revolutions in all areas of science, technology, and engineering: machines were increasingly performing the tasks that since the dawn of history depended on human or animal muscle. A major catalyst was the 18th-century discovery that steam is a powerful force and these developments helped spark the 19th-century Industrial Revolution. Moreover, the precision of fabricated articles increased substantially[1]; this was mandatory with the inception of mass production and the interchangeability of parts. These developments did not leave astronomy untouched. They gave rise to considerably more accurate position-measuring instrumentation. For example, in the 18th century the Englishman Jesse Ramsden developed a machine to calibrate astronomical instrument circles with considerably greater precision. All this resulted, by the early 19th century, in a significant increase in the accuracy with which star positions could be measured, and made it possible—by the 1830's—for the distances of a few nearby stars to be gauged by measuring their tiny apparent shifts against the sky, called parallax, resulting from the Earth's orbital motion. These parallactic shifts are minuscule; for most stars they are but a small fraction of an arc-second (1/3600th of a degree). By this point in time, stellar positional measurements began to approach today's standards of accuracy.

Of course, the greatest benefit of this was the ability to produce precision star catalogues. These listed stars to far greater positional accuracy than was needed for the construction of star charts, yet such exactness was crucial in the study of the sidereal universe, including, among numerous other things, the measurement of stellar motions, the structure and dynamics of our galaxy, as well as more"down to Earth" pursuits such as geodesy and navigation. Indeed, as we indicated, the positional accuracy observable by the early 18th century was quite adequate for constructing star charts with present-day standards of precision.

Other major changes were occurring in uranography during the first part of the 19th century. One of the most significant and obvious was the gradual phasing out of maps with fanciful constellation outlines which, as we also pointed out, were virtually universal on celestial charts of the 17th and 18th centuries.

It was increasingly felt that such artwork on sky maps is not only superfluous (and more expensive and time-consuming to produce), but undignified as well, making astronomy seem less of a science and more of a flight into fancy. Already at that time astronomers were at some pains to apprise the public that their discipline has nothing to do with astrology (celestial fortune telling), which pictorial star charts did tend to suggest to some. And, not least of all, these artistic augmentations to indicate a star's location were no longer needed, since celestial coordinates were far more precise, unambiguous. and scientific. It might be interesting to quote an excerpt where a leading 19th century astronomer, Sir John Herschel, discusses constellations in his popular, widely-read primer of the period, *Outlines of Astronomy*.

> Of course we do not here speak of those uncouth figures and outlines of men and monsters, which are usually scribbled over celestial globes and maps, and serve, in a rude and barbarous way, to enable us to talk of groups of stars, or districts in the heavens, by names which, though absurd or puerile in their origin, have obtained a currency from which it would be difficult to dislodge them.

Yet some star charts with "men and monsters" still managed to linger around for a while. A few post-Bode works of this type continued to appear, such as Alexander Jamieson's *A Celestial Atlas* of 1822, a British effort which within its covers included extensive textual material pertaining to the heavens; it was thus one of the earliest combined atlases and guidebooks, so popular today. Some printings of this work had charts with colored constellation figures, a major novelty at the time. Uranographically, Jamieson's atlas retained a decided 18th-century flavor, even to the point of utilizing that period's highly popular—and cartographically questionable—sinusoidal projection for its main maps. Jamieson was clearly influenced by Flamsteed of a century earlier, and his charts represent somewhat of a cartographic regression from Bode. Nevertheless, its charts appear less cluttered than Bode's and are thereby perhaps better suited for the novice, for whom they are adequately accurate and quite attractive.

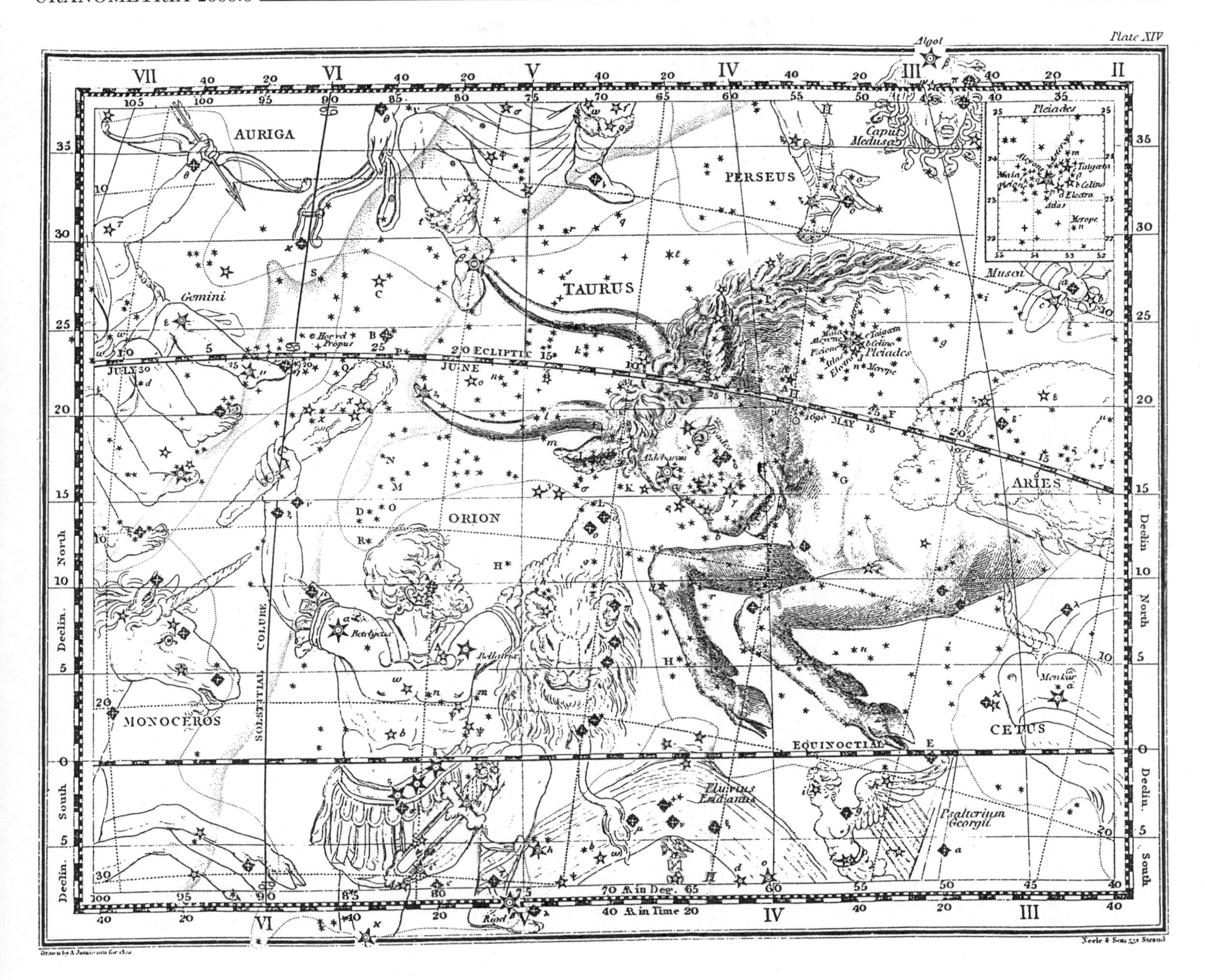

Fig. 15. Alexander Jamieson's A Celestial Atlas *appeared in England in 1822 as a popular guide to the heavens which included extensive textual material within its covers. Just over a decade later, a similar atlas-and-text celestial guide by Burritt appeared in the former colonies, which is on the next page. U.S. Naval Observatory, Washington, D.C.*

But to many the true swan song of the pictorial star atlas is the one which accompanied Elijah H. Burritt's highly popular guidebook to astronomy, *The Geography of the Heavens*, which first came out in 1835 and was the first "real" American uranographical effort. Elijah Burritt (1794–1838) was one of those classic old-time Yankee polymaths, dabblers, and students of a surprising variety of subjects. One of these was astronomy, which prompted him to produce this book-atlas celestial guide which became widely used among students of the sky in our then still-young Republic. It was reprinted several times, and this work is not that rare even today; in fact, its atlas has regularly been reproduced by facsimile.

However, despite the encomiums and general high regard it received, the Burritt *Geography of the Heavens* atlas is *not* a work of precision; its stellar positional accuracy is decidedly inferior to that with which the great majority of the stars were plotted in the Bayer *Uranometria* of more than two centuries previous, to say nothing of such later works as those of Hevelius and Flamsteed. It seems that whatever Burritt was, he was *not* a competent celestial cartographer and that his atlas closely resembles a work by Francis Wollaston.[2] The Burritt constellation outline figures were also heavily influenced by the already-mentioned Jamieson atlas, although Burritt's stellar positional accuracy was decidedly inferior to Jamieson.

Today we would regard Burritt as something of a plagiarist, but in earlier times such doings were regarded as less of a "crime"; indeed, on occasion it was considered an "honor" to be plagiarized!

So we must conclude that while Burritt hardly represented uranographical progress, nor did he intend to, he did represent progress in the popularization of astronomy because of the wide dissemination and use of his work. Countless individuals in the last century were introduced to the

Fig. 16. Elijah H. Burritt's Geography of the Heavens *atlas first appeared in 1835, along with an accompanying guidebook to the sky. It became a highly popular American "landmark" work, especially since in that era relatively few astronomical publications were available, particularly for the public. Burritt covers the entire celestial sphere on only six charts, two polar and four equatorial. Here is the equatorial plate covering the winter constellations, with Orion in the center. The top and bottom margins contain calendar scales which indicate, with the aid of the vertical right-ascension lines, which stars are on the meridian around 9 p.m. local time on the various dates. Notice the great similarity in the appearance of the constellation figures between Burritt and the Jamieson atlas on the preceding page. Burritt did indeed closely copy the Jamieson renditions except for star placement accuracy, where the latter's work is considerably superior.*

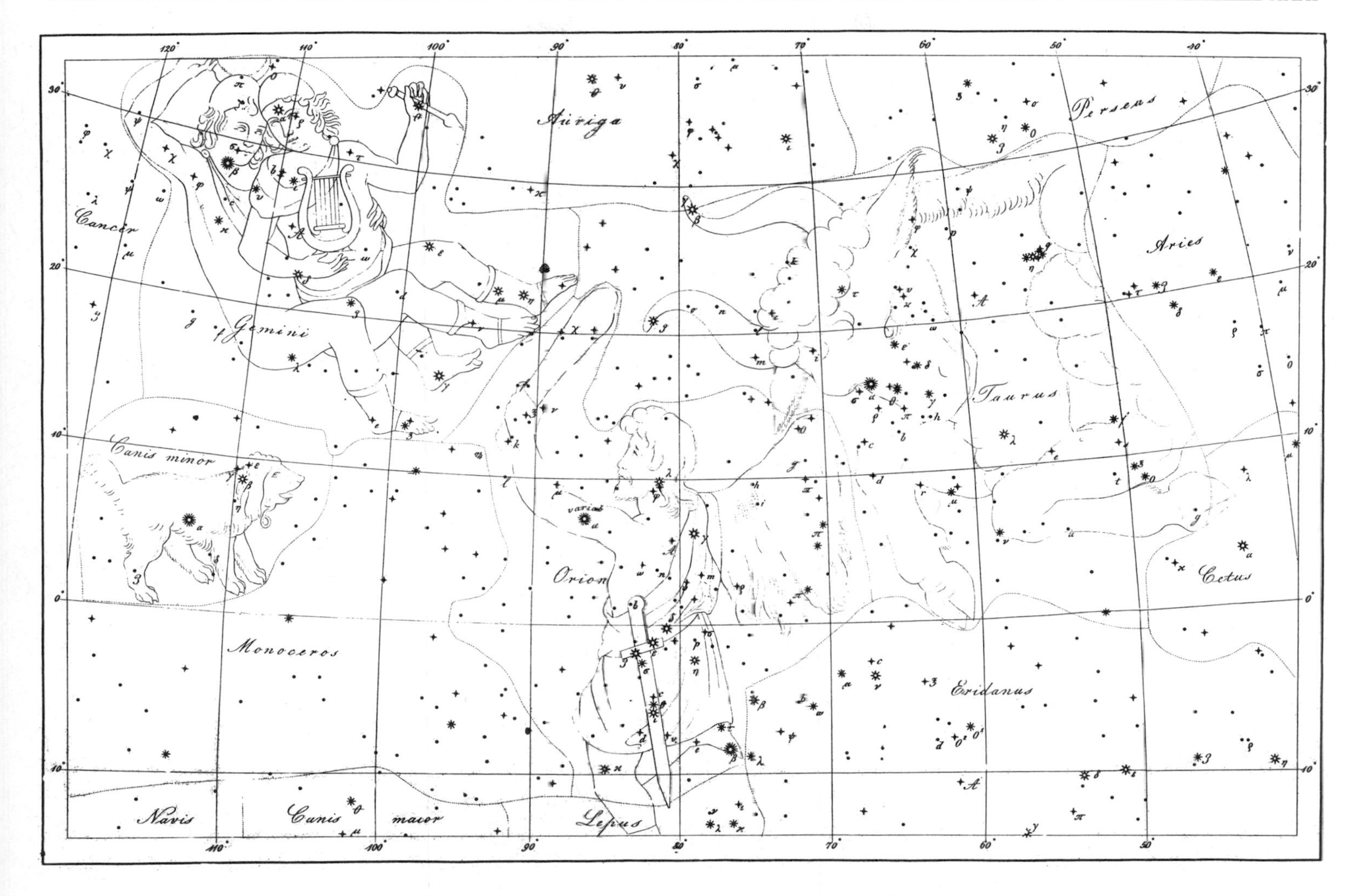

Fig. 17. What is often regarded as representing the transition from old to new is Friedrich Wilhelm Argelander's Uranometria Nova *of 1843. It is a very attractive and accurate rendition of the naked-eye stars and has constellation outline figures drawn in a relatively subdued manner (which seem to have been copied from Bayer) as if to herald the end of an era. Within two decades Argelander issued his monumental* Bonner Durchmusterung, *a far more detailed work which has the reputation of being the first totally modern celestial atlas. U.S. Naval Observatory, Washington, D.C.*

splendor of the "Bowl of Night" (to quote *The Rubaiyat of Omar Khayyam*) at a time when even city dwellers could experience a star-spangled firmament above their heads minus today's proliferating pollutants (including—and *especially*—light).

THE MODERN ERA

The present-day period of uranography is often regarded to have begun with the German astronomer Friedrich Wilhelm Argelander (1799–1875), a member of the 19th-century "German School" of astrometry, which included such luminaries as Friedrich Wilhelm Bessel (1784–1846) and Friedrich Georg Wilhelm Struve (1793–1864); the latter emigrated to Russia and helped found its famous Pulkovo Observatory, which became widely regarded as perhaps the 19th century's finest facility devoted to positional astronomy and still enjoys a high reputation today.

Perhaps the first modern star atlas could be regarded as Argelander's *Uranometria Nova* of 1843, a work containing 17 charts and accompanying catalogue of the naked-eye stars to the 6th magnitude. It can also be regarded as a segue between the old and new, because this Argelander work still had subdued constellation figures.

But Argelander's *Uranometria Nova* was but the prelude to his crowning work, the monumental *Bonner Durchmusterung*, the first comprehensive atlas and companion catalogue going substantially below the naked-eye limit; in this case it was to approximately 9th magnitude and beyond in some cases. This work has retained its value to the present day and is part of the source material used in compiling URANOMETRIA 2000.0. The *Bonner Durchmusterung* (commonly abbreviated as the BD) contains 37 charts and a catalogue with a total of approximately 325,000 stars from the north pole to −2° declination. It was complied by means of visual observations from 1852 to 1859 by Argelander and several assistants using a 78mm (3.1-inch) "comet seeker" type refracting telescope and was published in 1863. One of these assistants, Eduard Schönfeld (1828–91), extended Argelander's work after his mentor's death to −23° declination (which, interestingly, was what Argelander originally intended, but gave it up in order to get the BD out in "real" time as we would say today).

This BD extension was published in 1886, and became known as the *Südliche* [Southern] *Bonner Durchmusterung*

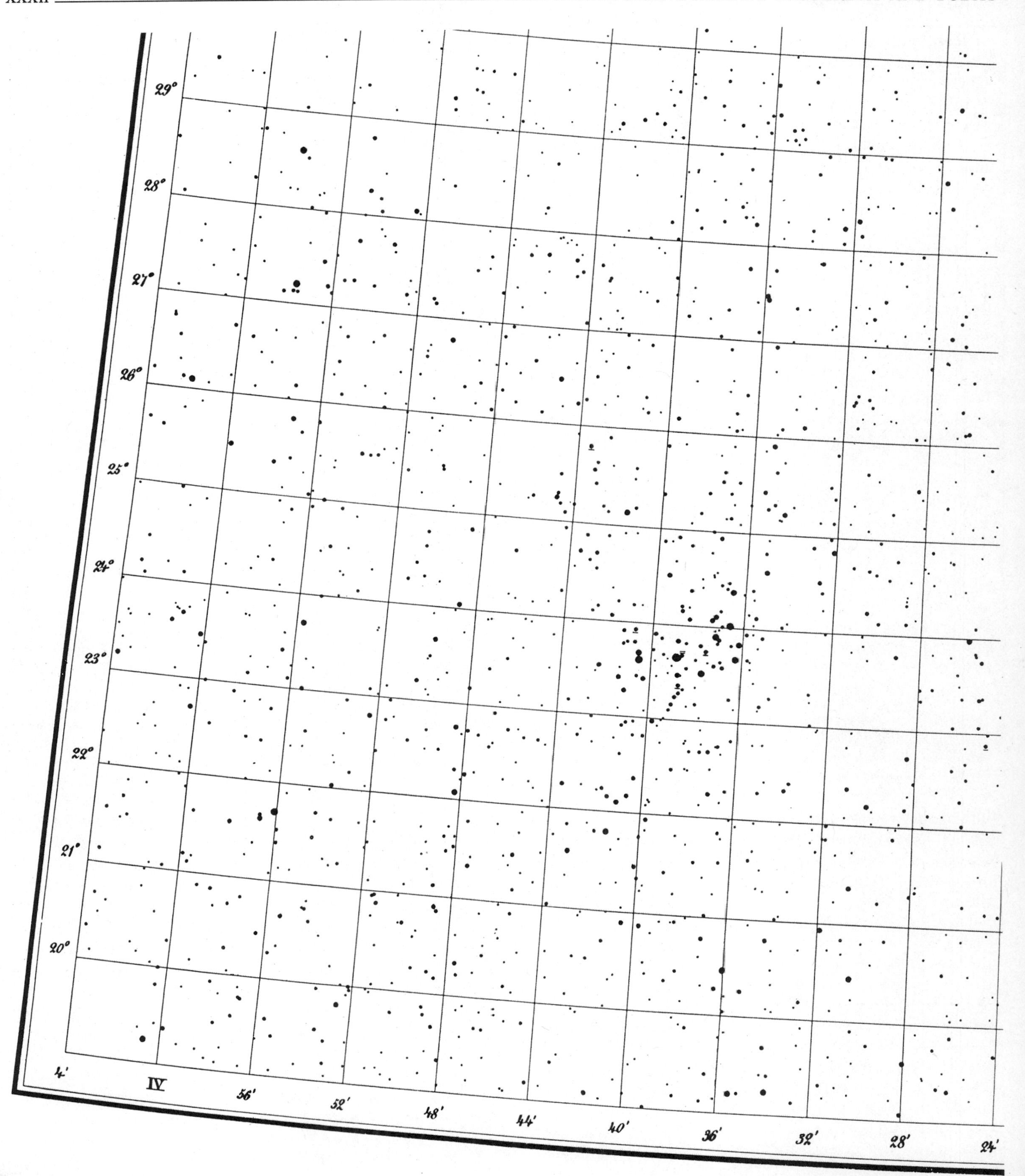

Fig. 18. Here is a corner of the chart showing the Pleiades, from Argelander's Bonner Durchmusterung, *which first appeared in 1863 (this plate is from the 1899 edition; this work was reprinted as late as after World War II). It represents a rather stark swing of the pendulum from the ornate artistic charts that still appeared only a relatively short time earlier. Perhaps the "swing" was too severe, because the charts are completely devoid of any form of star identification or labeling. However, the BD had an accompanying catalogue that listed each plotted star by one-degree declination zones. These BD listings remain to this day a primary way of designating telescopic stars of the 9th magnitude and brighter. Richmond Astronomical Society, Richmond, Virginia.*

or SBD and was compiled with a 159mm (6-inch) refractor when Schönfeld succeeded Argelander as director of Bonn Observatory. The SBD has been since regarded as being part of the BD and the two were sold together as a unit for many years; the SBD added about 135,000 stars to the total.

But a declination limit of −23° still leaves out a great deal of southern sky, that which is seen poorly or not at all from Germany. The next "Durchmusterung" was issued at Cordoba, Argentina, in 1908, going to −62°, and to the south pole by 1930; it became known as the *Cordoba Durchmusterung* or CoD, (The German word "Durchmusterung," roughly meaning in English "overall survey," was adopted in other languages as well for such charts.)

Towards the end of the 19th century a powerful new technology came to assist the astronomer and uranographer: photography. Its value for stellar mapping was soon realized and since then all atlases with the greatest number of stars have been photographic, either actual photographic prints or reproductions of the positives or negatives.

The first noteworthy effort in this direction was a well-known prematurely ambitious one planned at a special international conference held in Paris in 1887, the *Carte du Ciel* ("Map of the Sky"), which also was known as the *Astrographic Chart.* It turned into a stupendous undertaking involving eighteen observatories in both hemispheres, four of them French. They were each to photograph 2° by 2° squares of sky according to apportioned declination zones using a standard 13.5-inch photographic lens developed by Paul and Prosper Henry of the Paris Observatory. It was the unusually fine quality of this lens (by the standards of the day) that inspired Paris Observatory director Admiral Amédée Mouchez to propose the *Carte du Ciel* project.

So the work began ... and continued ... and dragged on ... for decades, tying up the energies and resources of those observatories well into the 20th century. The project called for more than 20,000 plates: two each for every 2° by 2° field, with one going down to about the 11th photographic magnitude from which precise star positions were to be measured and published in an accompanying *Astrographic Catalogue*; the second plate was simply a chart of the field to the 14th magnitude. These charts were printed not with the equatorial grid (right ascension and declination) of ordinary atlases, but with a rectangular astrometric net which is standard in such positional work. In order to express stellar locations in the conventional manner, these rectangular coordinates have to be converted to equatorial ones with the aid of formulae or tables, which were provided with the project. The *Carte du Ciel* was never really completed; one might say it sort of petered out eventually like "the One-Horse Shay."

Valuable as the results were to astronomy, the question remains: was the effort worth it? An apt summation of the situation was given by the late Joseph Ashbrook in his Astronomical Scrapbook[3] column in the June, 1958, *Sky and Telescope* entitled "The Brothers Henry" in which he discusses the *Carte du Ciel* project and concludes:[3]

> But when the work was started, astronomical photography was primitive. Wide-angle lenses for the observations, and computing machines for the reductions, were unknown when the rigid plans were adopted. In fact the venture was premature; the techniques of later years would have made more manageable the enormous labor of the enterprise. Some idea of the scale of the work is afforded by the experience at Oxford, one of the observatories that completed its share of the catalogue: 20 years and £34,000 were spent in observing, measuring, reducing, and publishing the places of almost 200,000 stars in its zone of declination, +25° to +31°. At Potsdam, Germany, if its original plan of publication had been carried to completion, that observatory's share of the catalogue would have consisted of 387 large volumes filling 45 feet of shelf space and weighing a ton!
>
> It is hardly to be wondered that the cooperating French observatories, Paris, Bordeaux, Algiers, and Toulouse, had their energies committed for decades, while the new science of astrophysics was rising in America, England, and Germany. The whole outlook of French astronomy was confined by the *Carte du Ciel* project. Perhaps astrophysical research in France would have begun to flower earlier if the brothers Henry had not made such good lenses!

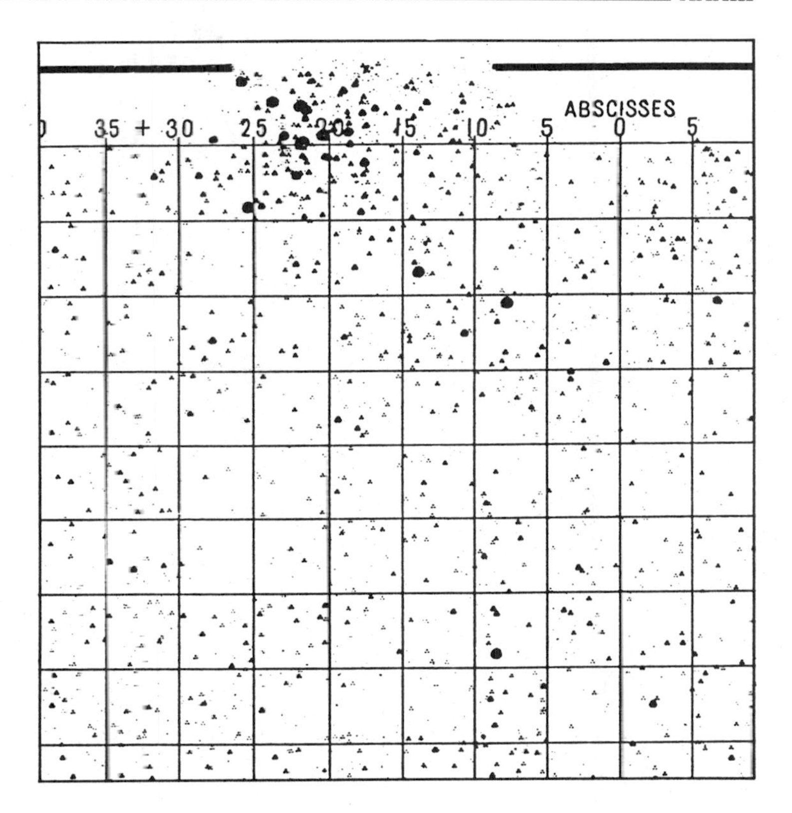

*Fig. 19. By far the most ambitious uranographical effort undertaken was the enormous–and never completed–*Carte du Ciel *set of 2° by 2° photographic charts, a project instigated by the French a century ago. Here is a portion of one of its plates from the constellation Auriga, of which each was exposed three times and shifted slightly so that each star formed a tiny triangle. This is one way of differentiating true stars from dust specks and other plate defects. This small section is from the top of the chart centered on R.A. 5^h28^m Dec. 33°, epoch 1900, taken at the Observatoire Royal de Belgique. U.S. Naval Observatory, Washington, D.C.*

When today's astronomer needs star positions, or other stellar data, to whatever magnitude, he or she generally pho-

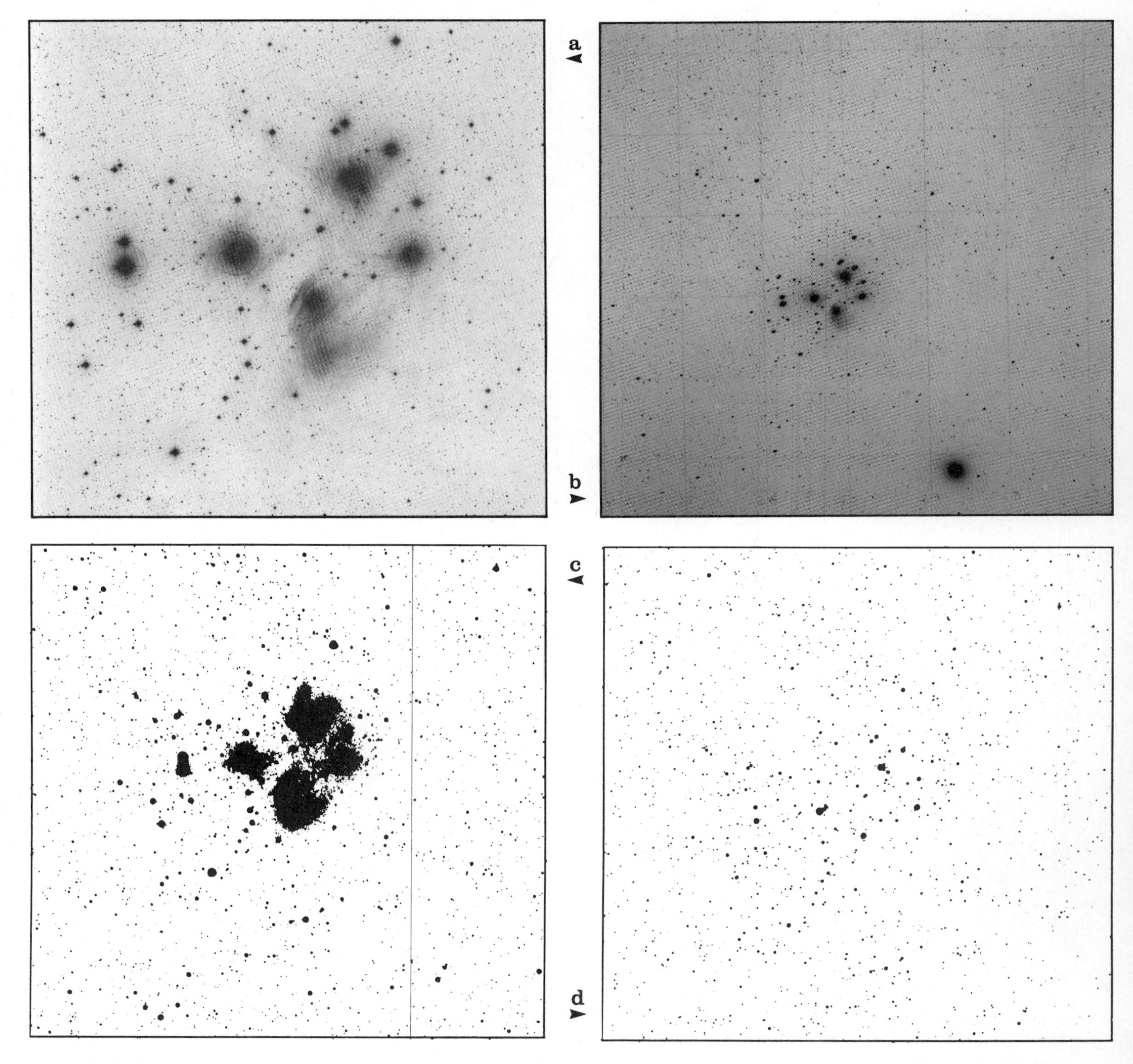

Fig. 20. Here are segments from four significant 20th century photographic atlases, three of which were done by amateurs! (a) The zenith of photographic star atlases is the classic National Geographical Society—Palomar Sky Survey, *which was largely funded by the former organization. When the Schmidt photographic telescope appeared in the 1930's, astronomers had an instrument which could cover a far larger sky area than previously without coma or other optical aberrations (notice the "pinpoint" star images over the entire plate area, whereas those taken with conventional telescopes often show "streaking" or other distortion near the plate edges). This is the "red light" plate of the Pleiades area; its blue-light companion shows the nebulosity surrounding this cluster so strongly as to practically conceal the stars. ©1960 National Geographic Society—Palomar Sky Survey. Reproduced by permission of the California Institute of Technology. (b) The Franklin-Adams Charts comprised 206 charts covering the entire sky. Like the* Palomar Sky Survey *these charts are photographic prints, each exposed from a negative—making for a very expensive atlas. (c) Hans Vehrenberg of West Germany is a prominent astrophotographer and astronomical publisher. He has produced a number of quality celestial photographs and atlases, of which the 450 chart* Atlas Stellarum *shown here is one of his latest examples. Asteroid observers, among others, like to plot the orbits of these bodies on Vehrenberg's photographic star atlases because their numerous faint stars help in visually locating these elusive objects. ©Treugesell-Verlag KG, Reproduced by permission of Dr. Hans Vehrenberg. (d) Christos Papadopoulos and Charles Scovil have compiled the 456 plates comprising the* True Visual Magnitude Photographic Star Atlas. *They used a special astrographic lens by Carl Zeiss, a VG10 (green) Schott filter and Kodak 103a-D spectroscopic plates to produce photographs with spectrum sensitivity very close to that of the eye. ©1979 Pergamon Books Ltd, reproduced by permission.*

tographs, analyzes, or measures the relevant stars or field *as needed*. (One also wonders if an ulterior motive of some of the participating *Carte du Ciel* astronomers might not have been to give themselves long-term job security through such a massive make-work effort.)

The 20th century saw the production of other, and mercifully much more manageably proportioned, photographic charts and atlases. One of the best-known early efforts was the *Franklin-Adams Charts* of 1914, a pole-to-pole representation of the sky by a skilled and enterprising British amateur astronomer and astrophotographer. It consists of 206 prints going down to the 15th magnitude and saw wide use for decades. After World War II, another skilled amateur, Hans Vehrenberg of West Germany, also produced well-known photographic atlases (in addition to quality individual pictures of deep-sky objects); these include his *Falkauer Atlas* of the early 1960's which goes down to approximately 13th magnitude, and his subsequent larger-scale *Atlas Stellarum 1950.0* which reaches the 14th magnitude.

The photographic atlases heretofore mentioned suffered basically from the problem that the stellar magnitudes are shown according to the photographic scale, making the orange and red stars quite a bit fainter, up to 2.5 magnitudes for the reddest stars, than they appear to the eye. This makes the identification of certain star fields, particularly in the densest portions of the Milky Way, difficult at times. By 1979 this problem was overcome with the issuance of C. Papadopoulos' and C. Scovill's *True Visual Magnitude Photographic Atlas*, a huge three-volume boxed set of charts showing the sky pretty much as the eye sees it.

But the "ultimate" photographic sky atlas, and highest quality one of them all from an imaging standpoint, is the *National Geographic Society-Palomar Sky Survey*, produced with the Mount Palomar Observatory's 48-inch Schmidt photographic telescope during the 1949–56 period. This covers the sky down to −33° declination with 935 pairs of plates (both blue-and red-sensitive emulsions) covering that many sky fields 6° square; the limiting stellar magnitude is about 21. This atlas is distributed as individually produced unbound photographic prints 14 by 17 inches in size. This work is typically used by professional astronomers and serious researchers (for which it was originally planned) so nothing other than actual high-quality prints which could be examined in detail could be considered; also, its substantial four-figure price made it pretty much available only to institutional purchasers. As this is written, a re-photographing of the Palomar Sky Survey is planned using the improved techniques and emulsions that have appeared in the intervening decades. Also, a similar-type photographic survey down to the south pole is being worked on with Australia's Siding Spring Observatory's 48-inch Schmidt instruments and the 40-inch Schmidt at Chile's European Southern Observatory.

Yet in a sense, another "ultimate" photographic atlas is Harvard College Observatory's enormous collection of photographic plates taken by patrol cameras (each field being repeatedly shot at regular intervals) from before the turn of the century until after World War II. These are the actual glass plates which were exposed; they are stored in metal cabinets on two floors of the Observatory's Cambridge, Massachusetts, headquarters, and have been of very great use to astronomers from all over the world in numerous areas of research.

Despite the major inroads photography has made during the past century in celestial mapping, it by no means spelled the end of the hand-plotted chart, and atlases constructed in this manner have continued to appear—getting better all the time—in this period and, in a sense, continuing the Argelander tradition. In fact, the immediate post-Argelander period saw the first major branching-off in uranography: hand-plotted and photographic charts.

The final decades of the 19th century saw significant improvements in celestial cartography spurred on not only by the ever-increasing amounts of amateur and professional astronomical activity, but also by major advances in all areas of the graphic arts, including printing, engraving (including photoengraving, by which artwork can be "shot" and made into printing plates), as well as lithography and everything else needed to produce quality charts. Much of this progress was a spin-off of the advances made in terrestrial cartography and the printing and reproduction of such maps.

During the decades following the American Civil War, which also saw relative peace and stability in Europe and increasing prosperity as the result of the Industrial Revolution, astronomy increasingly became one of those "gentleman's" pastimes, particularly in England. Members of the upper-class had the means and time to acquire some surprisingly sophisticated instrumentation, equipment, and even private observatories. This resulted in a relatively sudden profusion of astronomy books and observing aids, including star atlases and charts.

A catchy duo of British uranographers appeared at that time: Proctor and Peck (Richard A. Proctor and Sir William Peck). Between them they produced a number of atlases and handbooks of the sky; another such author was Sir Robert Ball.[4] This tradition of quality British atlases continued up to the well-known *Norton's Star Atlas*, originally by Arthur P. Norton, which first appeared in 1910 and whose most recent 17th edition came out in 1978. It is probably this century's best-known popular atlas, and it is also an observing handbook.

Notwithstanding this being so, Norton's atlas has to this day retained a decided 19th-century style and appearance, making it look like something from Proctor or Peck. For one thing, all deep-sky objects (clusters, nebulae, and galaxies) on its charts are shown with the same symbol, a cluster of dots. This reflects a common 19th-century attitude which made such objects but smudgy intruders onto the stellar scene; in fact, at the time all nebulae, whether gaseous or planetary, as well as galaxies were lumped into the catchall category of "nebulae" with galaxies being labeled "spiral nebulae." A modern atlas must differentiate the various deep-sky categories with different symbols for each, and the fact that Norton's does not do this makes it seriously deficient for the present-day deep-sky observer.

A word about these "smudgy intruders," or deep-sky objects as they are known today. Their systematic observation and cataloguing lagged considerably behind that of the stars, and, as a result, celestial atlases up to the latter part of the 19th century were largely lacking in any but the best-known such objects.

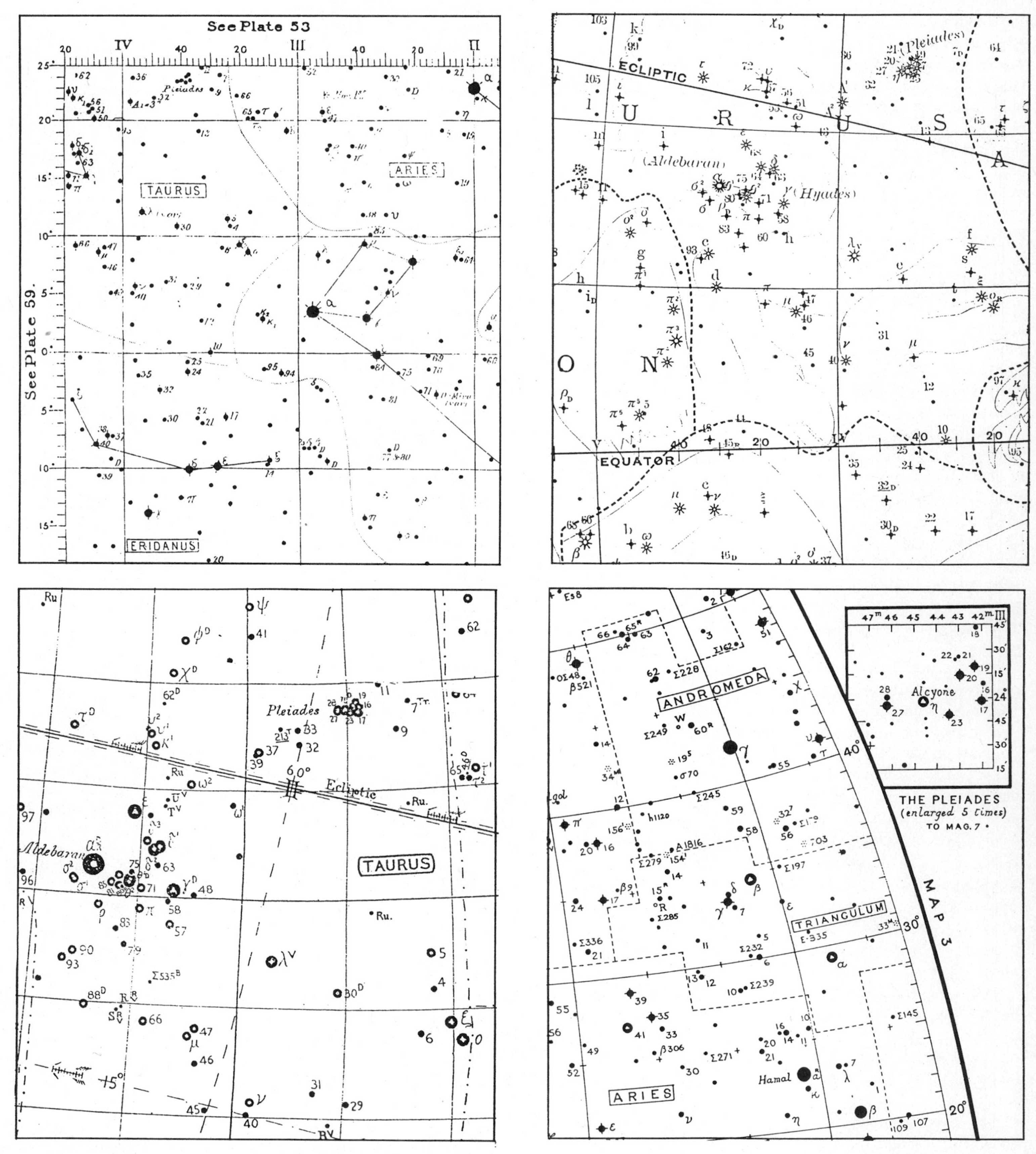

Fig. 21. Here are four typical popular star atlases dating from the late 19th to early 20th century. The first three are from the U.S. Naval Observatory collection: (a) Sir Robert Ball's 1892 Atlas of Astronomy *contained 72 plates. (b) Winslow Upton, an American, issued his Star Atlas in 1896 containing prominent curvy constellation boundaries. (c) The prolific Richard Proctor produced* A Star Atlas *in 1870. (d) Possibly the best-known 20th century book of sky charts is* Norton's Star Atlas *which has gone through 17 editions by 1978. It shows a remarkably large portion of the sky on each chart with little distortion on a grid system called the globular projection (except for the polar charts, where a conventional "polar" projection is used). Yet this work retains a resemblance to the other 19th century British atlases shown here, what with its undifferentiated symbolism for deep-sky objects and freehand lettering and labeling. Reproduced by permission of Longman Publishers, ©1978 R.M.G. Inglis.*

The first systematic effort was by Charles Messier (1730–1817) and Pierre Méchain (1744–1805). This was the famous *Messier Catalogue*, which lists some 110 deep-sky objects (there is some dispute as to the exact number). All of these objects are referred to even today primarily by their Messier number, such as M13, M31, and M42; these are, respectively the Hercules globular cluster, Andromeda galaxy, and Orion nebula. But, strangely enough, Messier's original intention was not to produce a deep-sky list. He was primarily a comet hunter and considered his compilation largely a "nuisance list" of objects that might be confused with comets!

Close on the tails of Messier and Méchain came William Herschel (1738–1822), the first of the "great" telescopic observers who discovered and recorded far more deep-sky objects than Messier, not as "nuisances," but deliberately. His son John (1792–1871), whom we earlier quoted with regard to constellation figures, extended his father's work, especially in the southern hemisphere where he observed from the Cape of Good Hope.

The Herschels' observations became the starting point of the classic *New General Catalogue of Nebulae and Clusters of Stars*, today simply called *New General Catalogue* (NGC), compiled by the Danish astronomer J.L.E. Dreyer (1852–1926) which was issued in 1888, supplemented by his *Index Catalogue* (IC) of 1895 and *Second Index Catalogue* (IC2) of 1908. Almost all these NGC objects (those that are missing could not be confirmed in the *Revised New General Catalog* which was one of the databases used in this atlas), plus a great many additional objects from other sources are plotted in URANOMETRIA 2000.0.

But England was not the only place quality atlases started appearing. In France, C. Dien in collaboration with the distinguished astronomer Camille Flammarion brought out a high quality work in the 1860's, the *Atlas Céleste* of 1865, and Flammarion was also noted for his excellent popular writings on astronomy. In Germany, the *Himmels Atlas* of R. Schurig, issued originally in 1886, became a standby, with its most recent version being the Schurig/Götz *Tabulae Caelestis* of 1960. Unfortunately, despite its high artistic quality and attractiveness, this latest edition contains so many errors as to be practically useless. In the 1920's two other significant German atlases appeared: two works with the same *Stern Atlas* title, one by M. Beyer and K. Graff and the other by P. Stuker. The Beyer-Graff atlas has sometimes been called "the Poor Man's *Bonner Durchmusterung*."

What about the United States? Strangely, relatively little significant uranography was done there during the decades discussed, some being of spotty quality. One such work, Upton's *Star Atlas* of the 1890's, became widely used, despite the fact that its plotting accuracy left something to be desired.

The next major uranographical milestone occurred after World War II, and came from a rather little-known observatory in Czechoslovakia at Skalnate Pleso (Rocky Lake). Despite some real postwar hardships and political upheavals in his country, Dr. Antonín Bečvář and his assistants produced the epochal *Atlas Coeli 1950.0*, which first appeared in 1948. Consisting of 16 charts from pole-to-pole to visual stellar magnitude 7.75, along with numerous deep-sky objects, it contained much more in the way of information and labeling than many previous works, including a number that went to fainter magnitude limits (many of which, including the famous *Bonner Durchmusterung*, were but unlabeled spatterings of dots). American rights to the Bečvář work were acquired by Sky Publishing Corporation, which issued it under the name *Atlas of the Heavens* (a translation of its Latin title) and by thus making it readily available at moderate cost made the Bečvář atlas the indispensable observer's standby for over three ensuing decades.

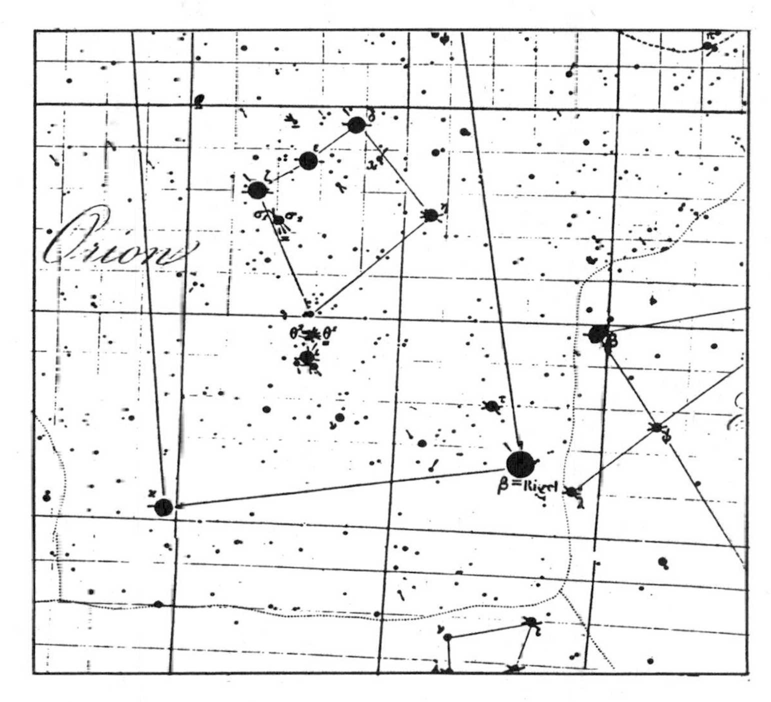

Fig. 22. Atlas Celéste of Dien and Flammarion was a comprehensive, precise, but today relatively little-known French effort of the 1860's. Note the unusually close coordinate net: a line for each degree of declination and four minutes of right ascension. However, this work contains a slight, but subtle "trap" for the unwary in that each chart shows stars down to the atlas's nominal magnitude limit (stated as 9th) for only the constellation or constellations "featured" on that particular plate. They are here delineated with those characteristic loose and curvy boundaries that were in existence throughout the 19th century and the first three decades of the 20th. These were never exactly the same from one atlas to the next, which resulted in considerable confusion, but at least were a start in establishing the limits and extents of each constellation. In 1930 the International Astronomical Union published official straight-line constellation boundaries which unambiguously established the limits of each group as well as those which are today officially recognized. Also, notice those straight "connecting lines" between a constellation's principal stars. These first appeared in the 19th century and remain to this day the standard way of tying together a constellation's main stars on general sky maps. U.S. Naval Observatory, Washington, D.C.

Despite its high quality, *Atlas Coeli* suffered from a significant cartographic flaw: improperly constructed conic projections for the charts running between 20° and 65° north and south declinations, causing the constellation patterns on those charts to be squeezed unnecessarily in an east-west direction, which we earlier indicated. This was unfortunate, because it marred the appearance of an otherwise very good work. In 1958, Bečvář issued a bound color version of *Atlas Coeli*, with deep-sky objects and the Milky Way being shown

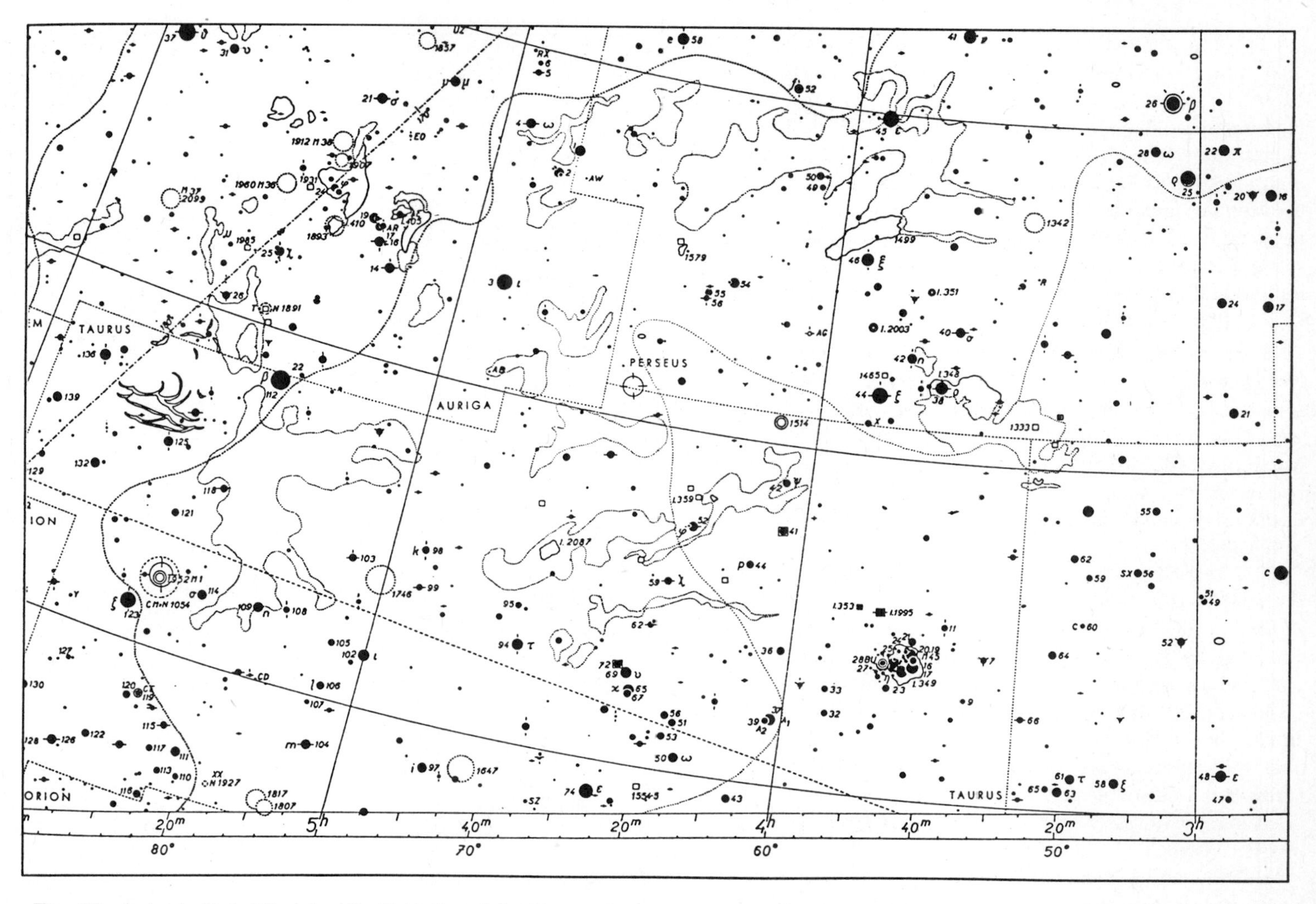

Fig. 23. Antonín Bečvář's Atlas Coeli (Atlas of the Heavens), *which appeared in 1948, was another of those truly "epochal" works. It offered a greater variety of information than perhaps any previous atlas (which mostly emphasized only stars) and became the indispensable observer's companion for more than three decades. Reproduced by permission of the Nakladatelství Československé akademie věd.*

in various, easily-distinguishable hues and thereby making it a very attractive set of star maps.

But Bečvář did not stop with *Atlas Coeli*. At about the same time in 1958 he issued his *Atlas Eclipticalis*, a large bound work covering the sky between +30° and −30°, while featuring something then unprecedented in star atlases: each star colored according to its spectral class. In 1962 Bečvář issued a companion *Atlas Borealis*, showing the sky from +30° to the north pole, and he completed his coverage of the sky from −30° to the south pole with his *Atlas Australis* of 1964. Both of these latter two atlases were done in the same style as *Eclipticalis*, with spectrally colored stars. One thing that should be noted about these three works is that they do not have an even approximate limiting magnitude; while they are more-or-less complete to 9th magnitude, they also show many stars fainter than that (in some cases to 12th and 13th magnitude); but below 9th the star field is increasingly incomplete, whereas in the actual sky the number of stars would increase explosively past that point. The reason is that these three atlases were plotted using various catalogues (Yale Zone, Boss General, and others) which have uneven cutoff points. Bečvář simply decided to include all stars "whose precise positions are known." There is also a magnitude inconsistency between the three works in that *Eclipticalis* and *Australis* use visual

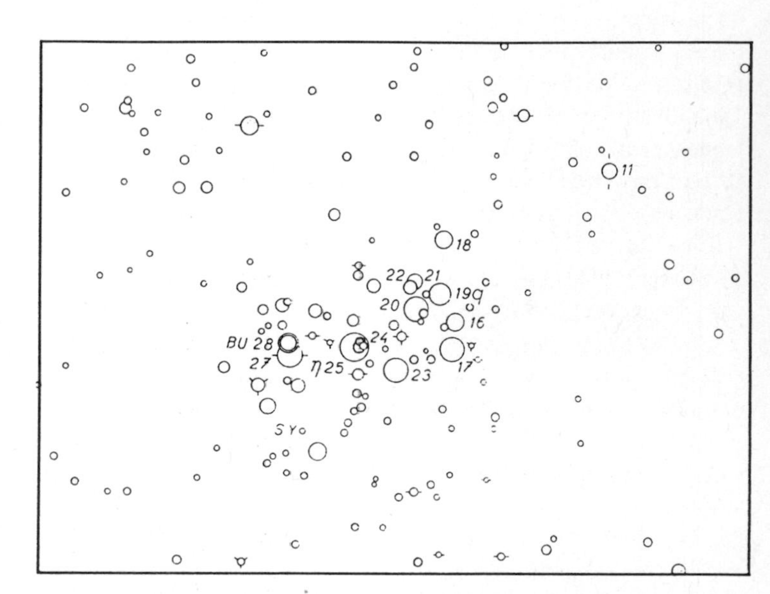

Fig. 24. This plot of the Pleiades is from Bečvář's Atlas Eclipticalis. *The stars are shown as open circles which were filled with a color to represent the star's spectral class. The primary requirement for inclusion of a star in this atlas was that the star's position had to be precisely known; therefore many faint stars below the "official" limit of 9th magnitude are included. Reproduced by permission of the Nakladatelství Československé akademie věd.*

magnitudes, whereas *Borealis* adopts photographic ones. In planning URANOMETRIA 2000.0, we sought to avoid such inconsistencies as much as possible, fully cognizant of the fact that data on the stars often leave much to be desired, especially with regard to the magnitudes of the fainter stars.

Also, the three abovementioned post-*Coeli* Bečvář atlases do not include any deep-sky objects, although, interestingly, a number of prominent galactic (open) clusters appear by reason of their brightest stars being plotted, but the clusters are not identified as such. Finally, the huge sizes of these works make them difficult to use at the telescope, although in their most recent editions of the 1970's, each sheet is folded, thus making the volumes considerably more manageable than the original unfolded bound printings.

The story now brings us to 1980. In late August of that year Sky Publishing received a brand new atlas by a then little known Dutch astronomy aficionado, commercial artist, and uranographer named Wil Tirion (who in the mid-1970's produced a naked-eye atlas for epoch 1950 now issued by the British Astronomical Association). Called *Sky Atlas 2000.0*, it was the first major work for that forthcoming coordinate epoch, which by 1980 was the more logical one to use instead of the then still-common 1950, which Norton's and the Bečvář atlases, among a number of others, are plotted for. *Sky Atlas 2000.0* was originally issued the following June in a pair of black-and-white editions: one with white background, the other with black (Desk and Field editions, respectively). Tirion then spent that entire summer personally producing color-separation films for a bound "Deluxe" color edition, which came out in the fall of 1981 as a most attractive work.

Tirion's *Sky Atlas 2000.0* enjoyed a huge sale upon its appearance, and promptly eclipsed Bečvář's *Atlas Coeli*, the observer's standby until then. The Tirion atlas went to magnitude 8.0 with some 43,000 stars, compared to *Coeli's* 7.75 limit and 32,500 stars. Tirion also showed some 2,500 deep-sky objects, all labeled, unlike Bečvář. And, not least of all, are *Sky Atlas 2000.0*'s considerably superior map projections. Worth noting also is that a dozen people were involved in producing *Atlas Coeli* with its 32,500 stars on 16 charts and it took them a year. Just one person did *Sky Atlas 2000.0* with its 43,000 stars on 26 charts and it required but 30 months. (Any time-motion or efficiency experts care to comment—or is there indeed something to the "too many cooks" syndrome?)

Sky Atlas 2000.0 made its author world famous as an exceptional celestial cartographer, and other publishers rapidly besieged him for other charts and atlases. Among these is a set of very attractive monthly maps in *Astronomy* magazine based on computer plots created by its editor, Richard Berry, another specialist in computerized uranography. Wil Tirion is now the world's leading uranographer and, in the writer's opinion, perhaps the greatest that ever lived considering the quality and quantity of his work. This is due in no small measure to his great skill and knowledge in *both* celestial cartography and the graphic arts, something virtually unprecedented.

The era of the hand-plotted atlas survived the introduction of astrophotography by a century, but, except for possibly simple general charts of naked-eye stars and constellations, very likely won't outlast the next technological quan-

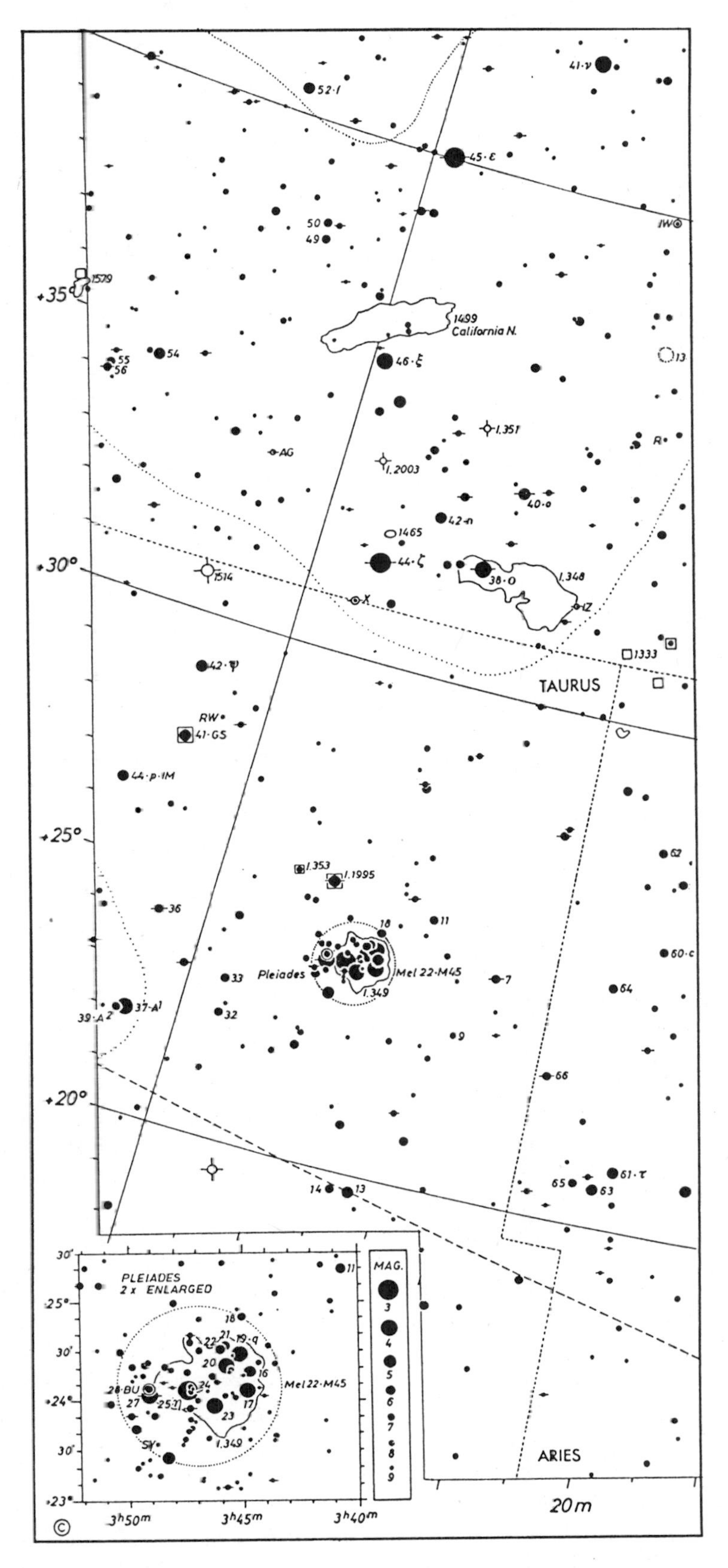

Fig. 25. The successor to Bečvář's Atlas Coeli *is* Sky Atlas 2000.0, *the first major effort by the present-day dean of uranographers, Wil Tirion, one of URANOMETRIA 2000.0's authors. This was the first popular atlas to be offered for epoch 2000.0. In addition to plotting 43,000 stars with a limiting magnitude of 8.0+ it displays about 2,500 deep-sky objects. It is available with white stars on a black background, black stars on a white background, and in full color. ©1981 Sky Publishing Corporation. Reproduced by permission.*

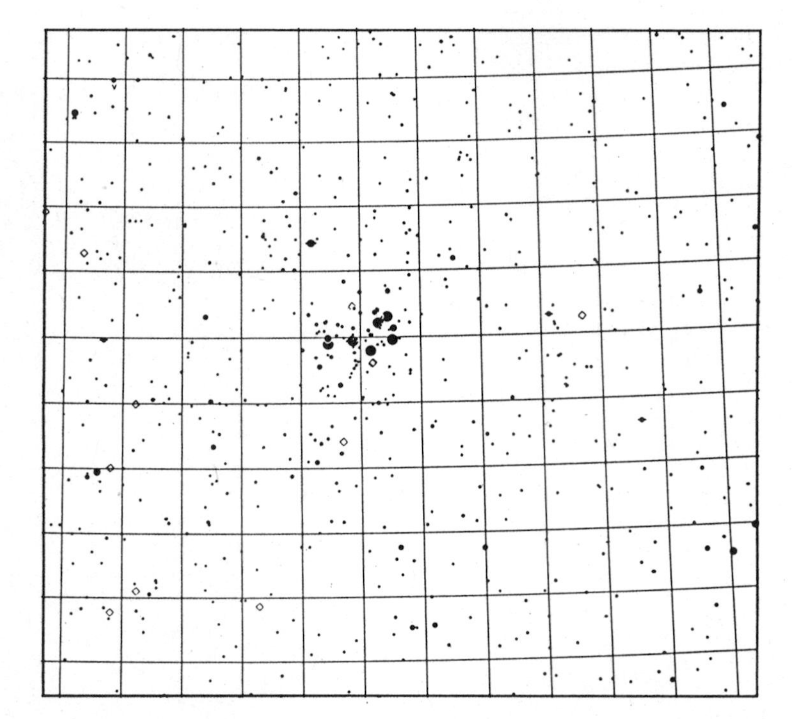

Fig. 26. The Smithsonian Astrophysical Observatory Star Atlas *(SAO) was the first major computer-plotted atlas. It appeared in the late 1960's and in a sense was something of a "20th century* Bonner Durchmusterung." *Its charts also have that severe, label-less style like the BD, and also came with an accompanying catalogue. However, it is primarily a special-purpose atlas to serve the needs of optical satellite and spacecraft tracking, rather than general observers and astronomers, who, nevertheless, still found it useful.* Smithsonian Astrophysical Observatory Star Atlas ©*1969. Reproduced by permission.*

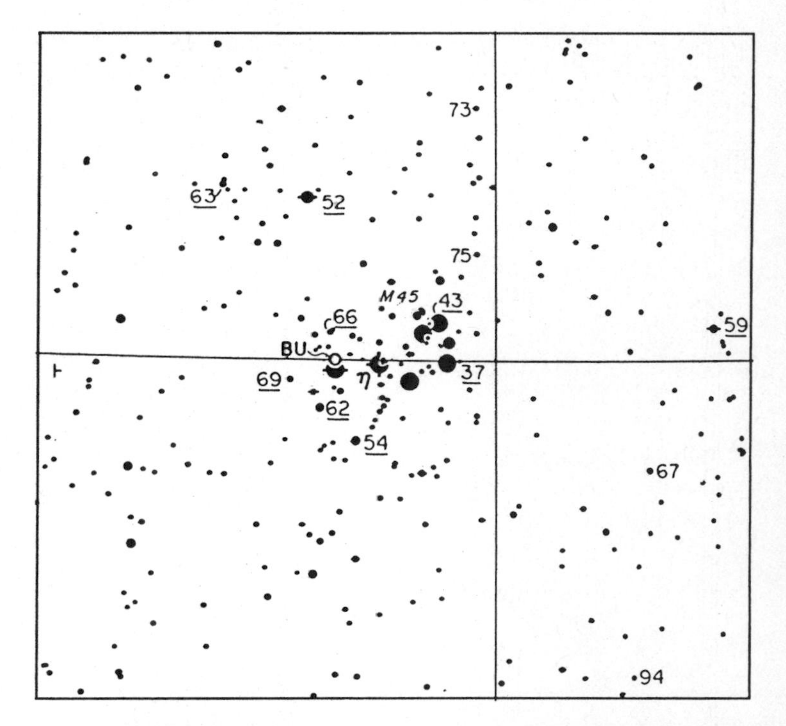

Fig. 27. Most of the atlases we have pictured here can be described as "general interest" as is the intent for URANOMETRIA 2000.0. *However, numerous examples exist of specialized works like the* SAO, *E.E. Barnard's* Photographic Atlas of Selected Regions of the Milky Way *and Charles Messier's earlier charts showing the paths of various comets. Above is a segment from the* The American Association of Variable Star Observers Star Atlas *by Charles Scovil. It is an adaptation of the* SAO Star Atlas *to meet the special needs of variable star observers.* ©*1980 AAVSO; reproduced by permission of AAVSO.*

tum jump in uranography: the computer-plotted atlas, of which the one you are now holding in your hands is the most comprehensive one yet produced. Even though the computer has been with us for some four decades, not many such efforts have appeared, the first one worth noting being the *Smithsonian Astrophysical Observatory Star Atlas* of the 1960's, which has a four-volume accompanying catalogue containing some 260,000 stars. This work goes basically to the 9th magnitude, but suffers from an even worse incompleteness problem than the detailed Bečvář's trio previously discussed. The Smithsonian (SAO) atlas and catalogue only included stars that had accurate proper-motion data and therefore positional information because the main purpose of this effort was not to serve general observers, but to assist satellite and space-vehicle trackers work up accurate orbits by providing highly precise stellar background reference points. The result is that a great many stars, including a number of naked-eye ones, are missing.

But the SAO atlas, being computer-plotted, achieved unprecedented accuracy for its star positions. A computer can produce far greater positioning precision than a human, unless, for example, the person is willing to work with some kind of microscope at an enormously slower pace than the machine is able to achieve. However, this is only one side of the story. A computer can process and plot data enormously faster and more accurately than any individual,but it is far from able to produce a finished star atlas, complete with attractive, and unambiguous labelling. Here the intervention of a skilled uranographer-artist such as Wil Tirion is vital, thus combining the accuracy and speed of the computer with the creativity of the artist.

URANOMETRIA 2000.0 is the second significant computer-plotted atlas to appear, and it has been produced with techniques two decades newer than that employed by the SAO atlas compilers. The massive assembly and plotting of its data has been brought about largely through the highly skilled efforts of Barry Rappaport who, among other things, is a specialist in computerized astronomical data and graphics. He has also produced *Cape Photographic Durchmusterung* and *Cordoba Durchmusterung* atlases with the aid of computers and modern-day plotters. Working with this same equipment he produced the base charts for URANOMETRIA 2000.0. Wil Tirion then took the computer plots, added labels, symbols, lines (such as constellation boundaries) and many additional deep-sky objects. His unequaled mental star atlas also proved indispensable in spotting errors, omissions, and "glitches" in the original machine plots.

URANOMETRIA 2000.0 is very much intended for the celestial observer, whether amateur or professional. It shows and labels far more deep-sky objects than any other previous atlas, computer-plotted or not. (The SAO work does indicate the presence of more deep-sky objects, but minus any labels. Also, its charts are a bit too small to be read easily.)And,among other things, URANOMETRIA 2000.0 shows full constellation boundaries, which to the best of our knowledge is another "first" for such a detailed atlas.

Further details concerning background information about URANOMETRIA 2000.0 and what it includes can be found in the Introduction.

Just as the original *Uranometria* by Bayer required good data and skill, both astronomical and graphical, so did our URANOMETRIA 2000.0. Back then Bayer had Tycho Brahe's unprecedentedly accurate observations to work from; we who worked on URANOMETRIA 2000.0 have had at our disposal far more extensive and accurate data than could ever be dreamed about in the early 17th century. Our information comes from sources dating back to the last century in some cases, as well as data compiled just "yesterday." We acknowledge with gratitude our debt to everyone from Tycho Brahe and Johann Bayer to those numerous present-day astronomers and their hard-working assistants, many of whom are still only students. Without all of them, this work would never have been possible.

We take great pleasure in presenting this new atlas to the astronomical community and hope it will become a constant companion to anyone who observes, studies, or merely looks at the Bowl of Night professionally or for pleasure.

GEORGE LOVI

FOOTNOTES

1. One way—not often realized—by which the precision of manufactured goods during different historical periods can be gauged pretty well is simply by looking at printed materials (books, newspapers, etc.). Prior to 1800 much of the type on a page was generally quite uneven and ragged both in alignment and shape; as the 19th century progressed, the quality of typography rapidly started approaching today's standards.
2. In the January 1985 issue of *Sky and Telescope,* Peggy Aldrich Kidwell of the National Museum of American History revealed, in "Elijah Burritt and the 'Geography of the Heavens' " that the Burritt atlas is closely copied from the 1811 British work *A Portraiture of the Heavens* by Francis Wollaston.
3. A collection of these Astronomical Scrapbooks has been reprinted in the book *Astronomical Scrapbook* by Joseph Ashbrook, Sky Publishing Corp, and Cambridge University Press, 1984.
4. Proctor, Peck, and Ball were all highly gifted British astronomy popularizers for the broadest lay public. The last two were professional astronomers, while Proctor was a highly prolific author as well as lecturer who died in the United States of yellow fever and was buried in Brooklyn, N.Y. Proctor lived from 1837 to 1888; Peck and Ball from 1862 to 1925 and 1840 to 1913, respectively.

A SELECTED ANNOTATED BIBLIOGRAPHY

The field of uranography is relatively specialized and arcane, combining as it does both astronomy and cartography, so the literature on it is not very extensive. The following is a specially selected, rather than exhaustive, compendium for the benefit of those who wish to pursue the subject further. A short description of each work is provided to help guide the reader and make this bibliography of maximum usefulness.

Brown, Basil. *Astronomical Atlases, Maps and Charts,* London, 1968: Dawsons of Pall Mall.

This is a reprint of a classic work, first issued in 1932, which has been a standby for uranography aficionados, connoisseurs, and collectors ever since. It covers not only star charts and atlases, but lunar and planetary cartography as well. The book is basically a running commentary of who did what and when, with not much in the way of detailed descriptions of individual works (as in Deborah Jean Warner's *The Sky Explored,* described further on). Also, the quality of the illustrations in this 1968 reprint is decidedly inferior to the original 1932 work.

Celestial Images. Boston, 1984: Boston University Art Gallery.

This elaborate and copiously illustrated booklet was put together in conjunction with an exhibition of old star charts at the Boston University Art Galley between January 24 and March 24, 1985 and later at the National Museum of American History, Smithsonian Institution, Washington, D.C. and Williams College Museum of Art, Williamstown, Massachusetts. It contains major articles as well as descriptions of individual works by leading specialists in this field.

Eichhorn, Heinrich, *Astronomy of Star Positions,* New York, 1974: Frederick Ungar Publishing Co.

One of the best general present-day treatises on astrometry (positional astronomy). In addition to discussing astrometric techniques and precepts, the book is especially good in its coverage of star catalogues, past and present. Included is an extensive commentary on the *Astrographic Catalogue (Carte du Ciel),* including the work done on the project by the various participating observatories.

Gingerich, Owen, "Astronomical Maps," *Encyclopaedia Britannica,* 15th edition, 1975: Vol. 2 (Macropaedia).

A very good general overview of celestial cartography, past and present, by an authority in both this area and astronomical history in general, which has appeared in this leading reference work since the above date. Includes a historical sketch of how some of the earliest ancients mapped the sky and also covers certain non-stellar celestial cartography such as galaxies and the Moon.

King, Henry C., with John R. Millburn, *Geared to the Stars* Toronto, 1978: University of Toronto Press.

This work is the *sine qua non* for those who interests include three-dimensional uranography—mechanical models. It's all here: the grand sweep of devices over the centuries and millenia from astrolabes, armillary spheres, orreries, astronomical clocks, all the way to today's projection planetarium and its development. The authors spared no effort to seek out as many illustrations as possible to support their extensive, authoritative text.

Rosenfeld, Rochelle Susan, *Celestial Maps and Globes and Star Catalogues of the Sixteenth and Early Seventeenth Centuries,* Ph.D Thesis, School of Education, Health, Nursing and Arts Professions, 1980: New York University.

Here is a uniquely valuable work that represents a real contribution to the history of uranography, and deserves wider dissemination. It is not just a historical sketch, but also a detailed technical and mathematical analysis of the cartography discussed, including their construction and the map projections employed—an aspect of celestial (and terrestrial) map-making relatively poorly understood.

Snyder, George Sergeant, *Maps of the Heavens,* New York, 1984: Abbeville Press.

This huge "art" book contains a number of attractive, well-reproduced plates which are representative rather than exhaustive, and includes accompanying textual descriptions. The author is a map specialist at Sotheby's in New York City and also wrote the accompanying text. However, certain factual errors managed to creep in and consequently it should not be considered definitive.

Warner, Deborah J., *The Sky Explored, Celestial Cartography 1500–1800,* New York, 1979: Alan R. Liss, Inc.

A truly indispensable compendium for the serious student of uranography. It covers just about all the published works at the height of the pictorial star chart era. There are detailed descriptions and listings of the contents and features of the numerous included charts and atlases, and at least one sample map in most cases. The author is a well-known and accomplished astronomical, science, and uranographical historian at the Smithsonian Institution's National Museum of History and Technology.

Werner, Helmut, and Felix Schmeidler, *Synopsis of the Nomenclature of the Fixed Stars,* Stuttgart, West Germany, 1986: Wissenschaftliche Verlagsgesellschaft mbH.

This valuable, much-needed compilation, constellation by constellation, lists the naked-eye stars in tables that permits their designations and labels in a number of leading star catalogues, ancient and modern, to be compared side-by-side. The first part of the book contains excellent discussions (in German and English) on various aspects of stellar labeling, including the leading catalogues, atlases, and their significance. The late Dr. Werner was a prominent specialist in star and constellation lore covering a wide variety of cultures.

URANOMETRIA 2000.0

Volume I - The Northern Hemisphere to -6°

Arrangement of Charts

URANOMETRIA 2000.0's charts are numbered and ordered in an orthodox star-atlas manner: by declination zones working downward from the north to the south pole. Within each zone the charts are numbered by increasing right ascension, or west to east.

These charts have generous overlap areas along their margins for maximum user convenience, as well as numbers in small ovals outside each margin which indicate the particular map that abuts there. Note: because of their overall arrangement, charts on facing pages do *not* necessarily abut along their nearest margins. Refer to the Index Chart at the back of the book to see the overall plan. Also, the listing on page X in the Introduction provides this information in tabular form.

Finally, for those who wish to read off directly the position of any plotted star or object in URANOMETRIA 2000.0, we provide in the inside back cover pocket two 8 by 10 inch clear acetate overlays with fine grid lines arranged by declination zones.

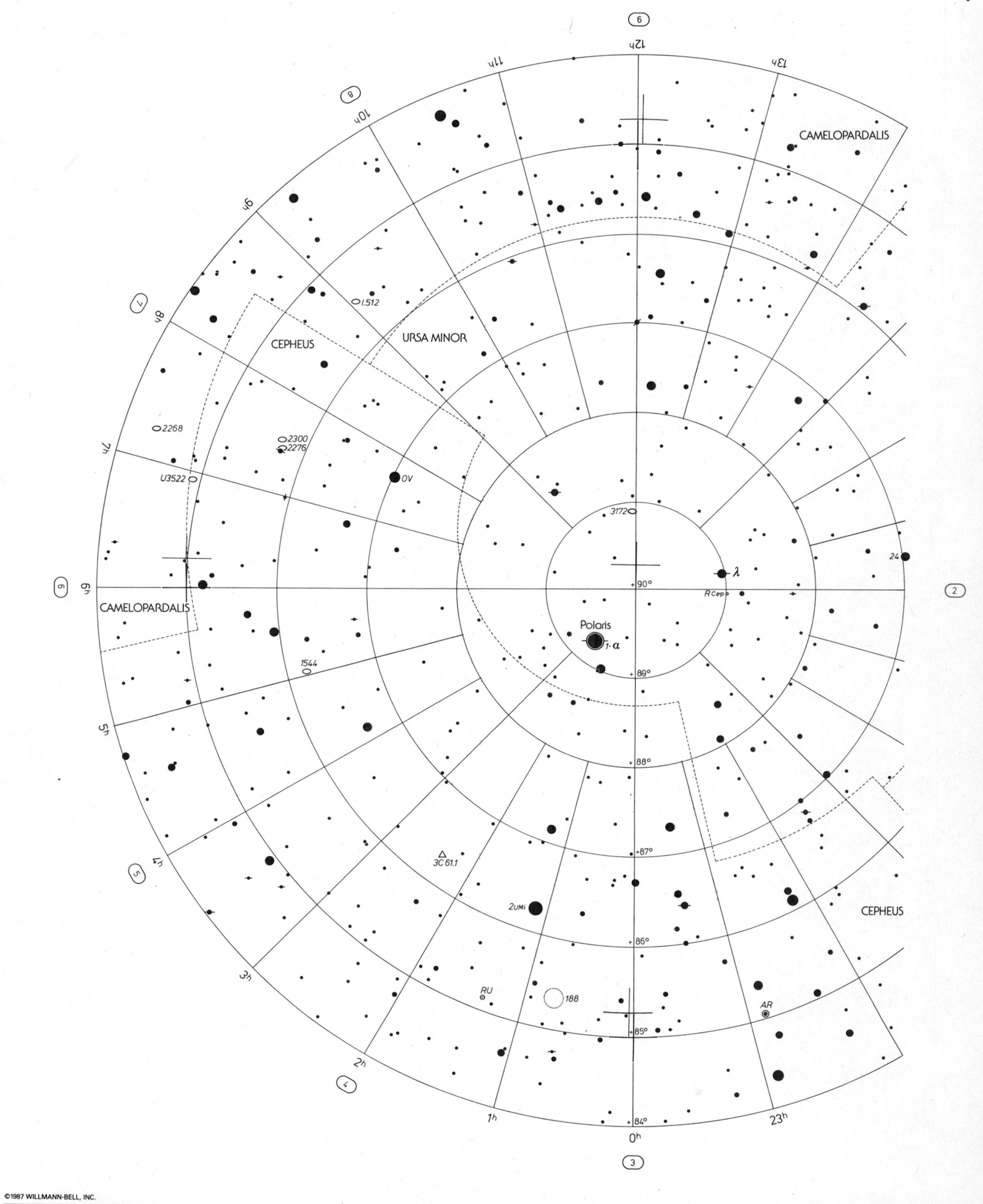

STELLAR MAGNITUDES	DOUBLE OR MULTIPLE STARS	OPEN STAR CLUSTERS	GLOBULAR STAR CLUSTERS	PLANETARY NEBULAE	BRIGHT NEBULAE	DARK NEBULAE	GALAXIES	QUASAR / RADIO SOURCE / X-RAY SOURCE
−1 0 1 2	VARIABLE STARS	to scale <5′	to scale <5′	>120″ 120″–60″ 60″–30″ <30″	to scale 10′–5′ <5′	to scale 10′–5′ <5′	to scale <5′	QUASAR; RADIO SOURCE; X-RAY SOURCE
3 4 5 6 7 8 9 >9.5								

Barry Rappaport & Wil Tirion

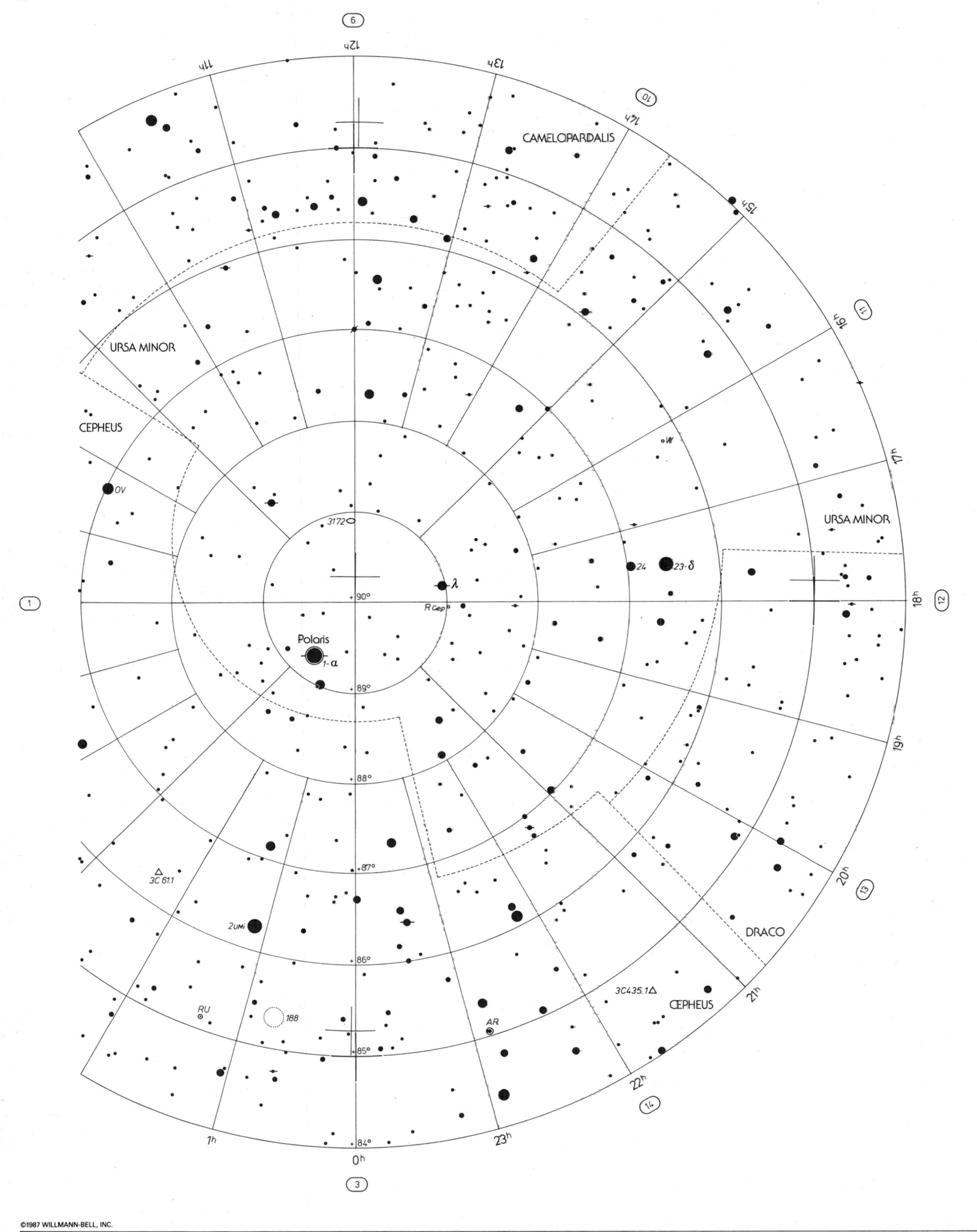

STELLAR MAGNITUDES: −1, 0, 1, 2, 3, 4, 5, 6, 7, 8, 9, >9.5

DOUBLE OR MULTIPLE STARS

VARIABLE STARS

OPEN STAR CLUSTERS: to scale, <5′

GLOBULAR STAR CLUSTERS: to scale, <5′

PLANETARY NEBULAE: >120″, 120″–60″, 60″–30″, <30″

BRIGHT NEBULAE: to scale, 10′–5′, <5′

DARK NEBULAE: to scale, 10′–5′, <5′

GALAXIES: to scale, <5′

QUASAR

RADIO SOURCE

X-RAY SOURCE

Barry Rappaport & Wil Tirion

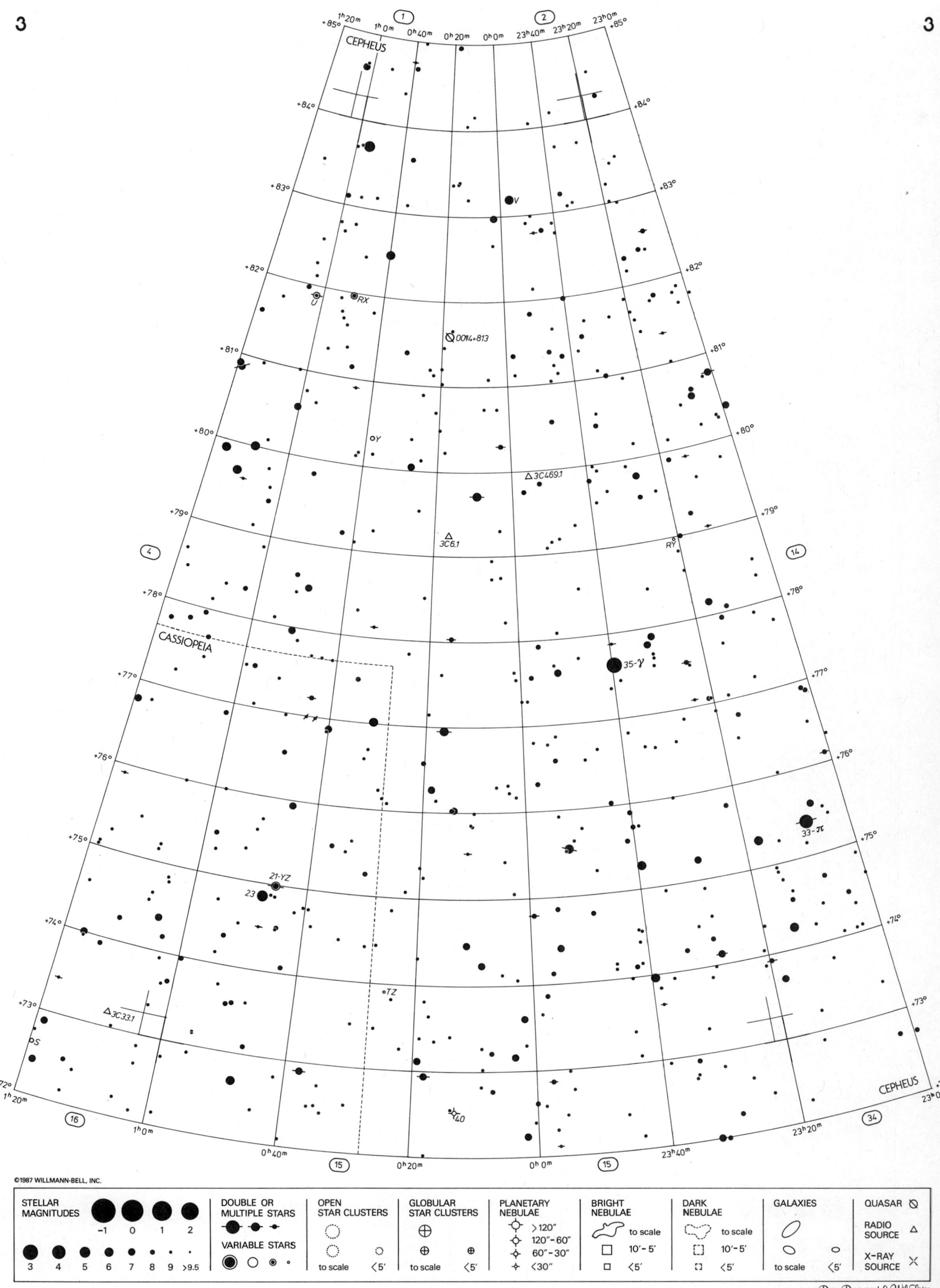

CEPHEUS
CASSIOPEIA
CEPHEUS
1h20m
1h0m
0h40m
0h20m
0h0m
23h40m
23h20m
23h0m
+85°
+84°
+83°
+82°
+81°
+80°
+79°
+78°
+77°
+76°
+75°
+74°
+73°
+72°
1
2
4
14
16
15
15
34
V
U
RX
0014+813
Y
3C469.1
3C6.1
RY
35-γ
33-π
21-YZ
23
TZ
3C33.1
S
40
©1987 WILLMANN-BELL, INC.
STELLAR MAGNITUDES
−1 0 1 2
3 4 5 6 7 8 9 >9.5
DOUBLE OR MULTIPLE STARS
VARIABLE STARS
OPEN STAR CLUSTERS
to scale <5′
GLOBULAR STAR CLUSTERS
to scale <5′
PLANETARY NEBULAE
>120″
120″–60″
60″–30″
<30″
BRIGHT NEBULAE
to scale
10′–5′
<5′
DARK NEBULAE
to scale
10′–5′
<5′
GALAXIES
to scale <5′
QUASAR
RADIO SOURCE
X-RAY SOURCE
Barry Rappaport & Wil Tirion

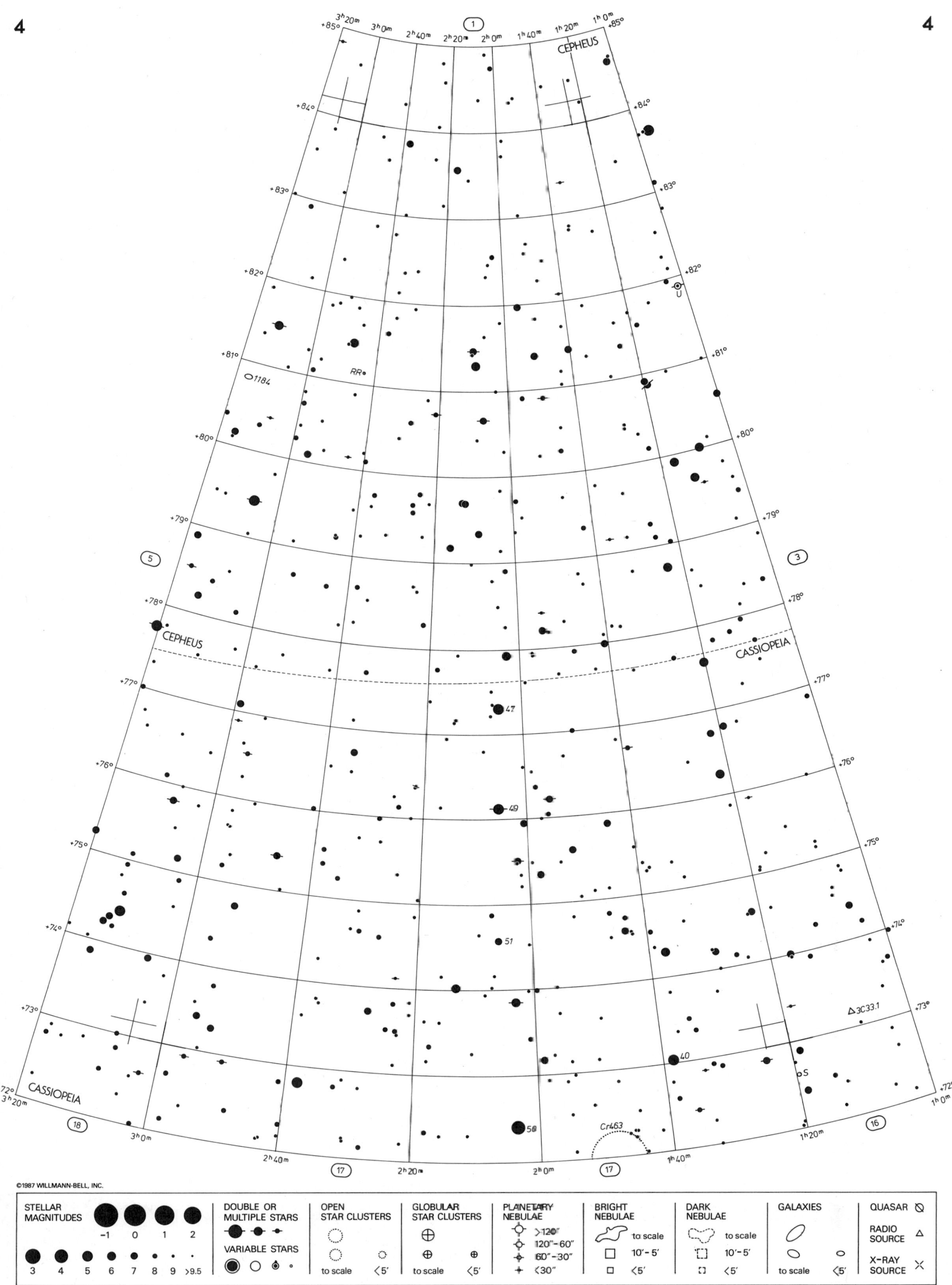
CEPHEUS
CASSIOPEIA
1184
RR
U
47
49
51
40
S
58
Cr463
3C33.1
©1987 WILLMANN-BELL, INC.
STELLAR MAGNITUDES
DOUBLE OR MULTIPLE STARS
VARIABLE STARS
OPEN STAR CLUSTERS
GLOBULAR STAR CLUSTERS
PLANETARY NEBULAE
BRIGHT NEBULAE
DARK NEBULAE
GALAXIES
QUASAR
RADIO SOURCE
X-RAY SOURCE
to scale
Barry Rappaport & Wil Tirion

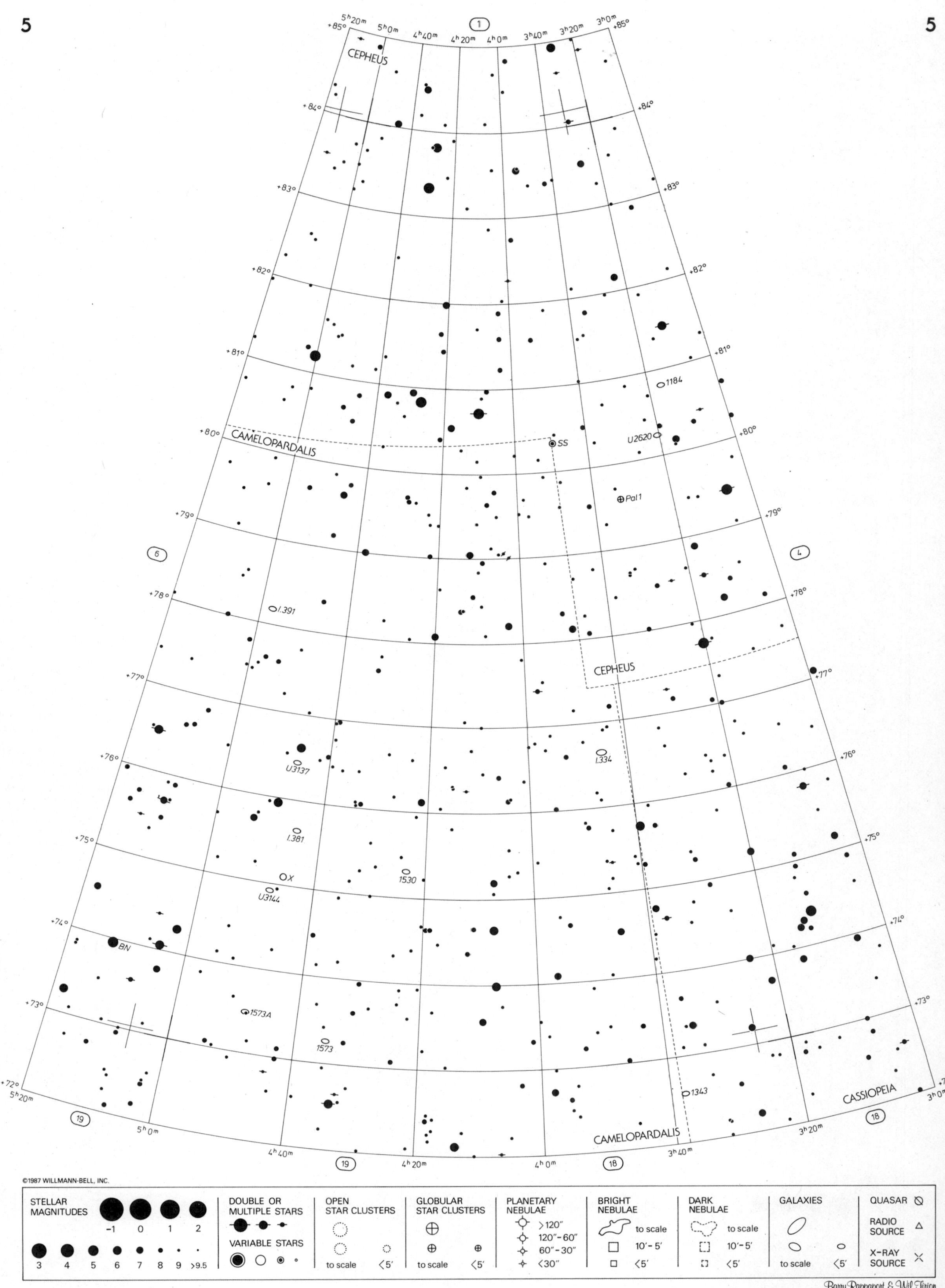
CEPHEUS
CAMELOPARDALIS
CASSIOPEIA
1184
U2620
SS
Pal 1
I.391
U3137
I.334
I.381
X
U3144
1530
BN
1573A
1573
1343
5h20m
5h0m
4h40m
4h20m
4h0m
3h40m
3h20m
3h0m
+85°
+84°
+83°
+82°
+81°
+80°
+79°
+78°
+77°
+76°
+75°
+74°
+73°
+72°
1
4
6
18
19
©1987 WILLMANN-BELL, INC.
STELLAR MAGNITUDES
-1 0 1 2 3 4 5 6 7 8 9 >9.5
DOUBLE OR MULTIPLE STARS
VARIABLE STARS
OPEN STAR CLUSTERS
to scale
<5′
GLOBULAR STAR CLUSTERS
to scale
<5′
PLANETARY NEBULAE
>120″
120″–60″
60″–30″
<30″
BRIGHT NEBULAE
to scale
10′–5′
<5′
DARK NEBULAE
to scale
10′–5′
<5′
GALAXIES
to scale
<5′
QUASAR
RADIO SOURCE
X-RAY SOURCE
Barry Rappaport & Wil Tirion

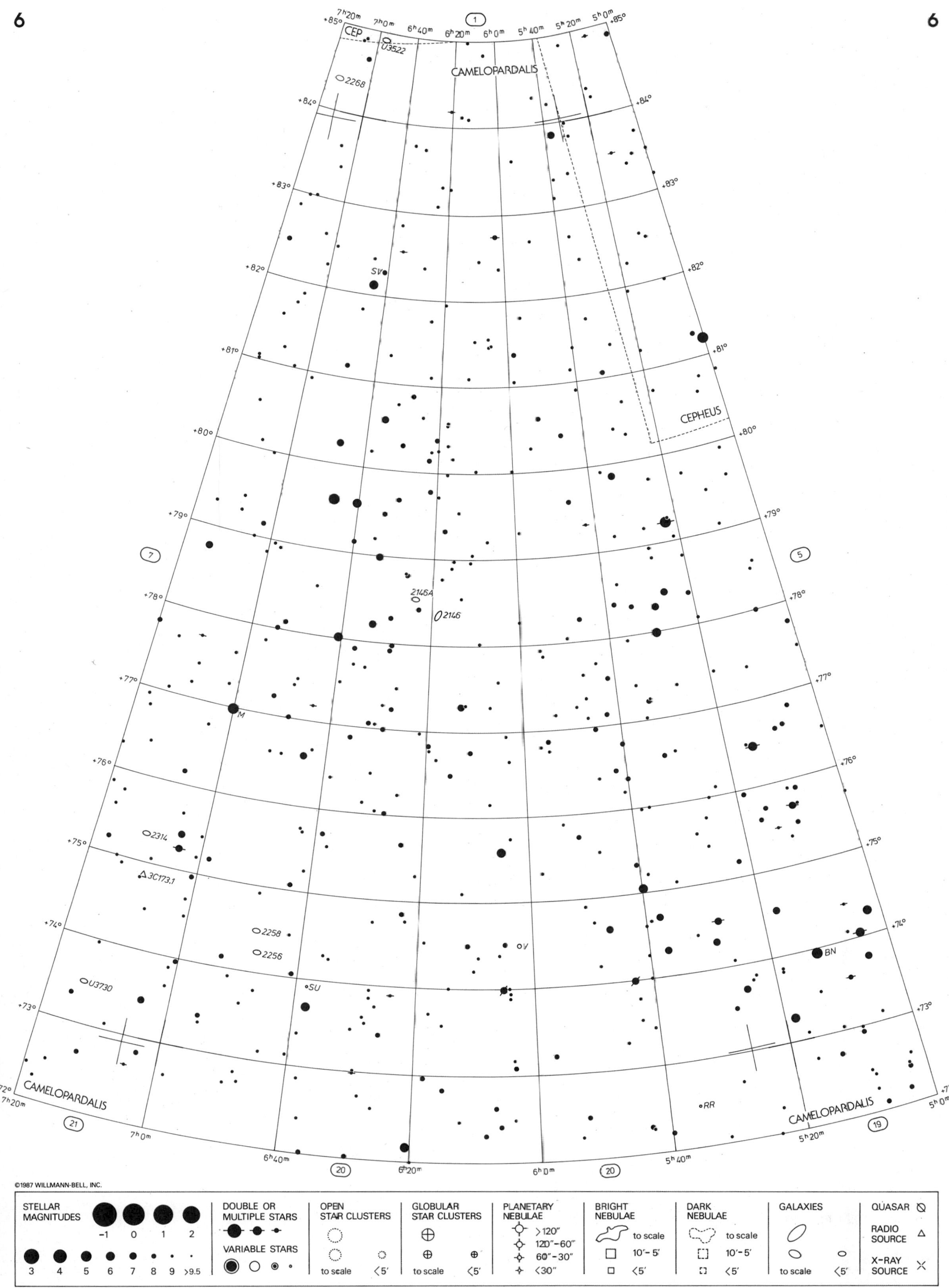

STELLAR MAGNITUDES -1 0 1 2 3 4 5 6 7 8 9 >9.5 | DOUBLE OR MULTIPLE STARS | VARIABLE STARS | OPEN STAR CLUSTERS to scale <5′ | GLOBULAR STAR CLUSTERS to scale <5′ | PLANETARY NEBULAE >120″ 120″–60″ 60″–30″ <30″ | BRIGHT NEBULAE to scale 10′–5′ <5′ | DARK NEBULAE to scale 10′–5′ <5′ | GALAXIES to scale <5′ | QUASAR | RADIO SOURCE | X-RAY SOURCE

Barry Rappaport & Wil Tirion

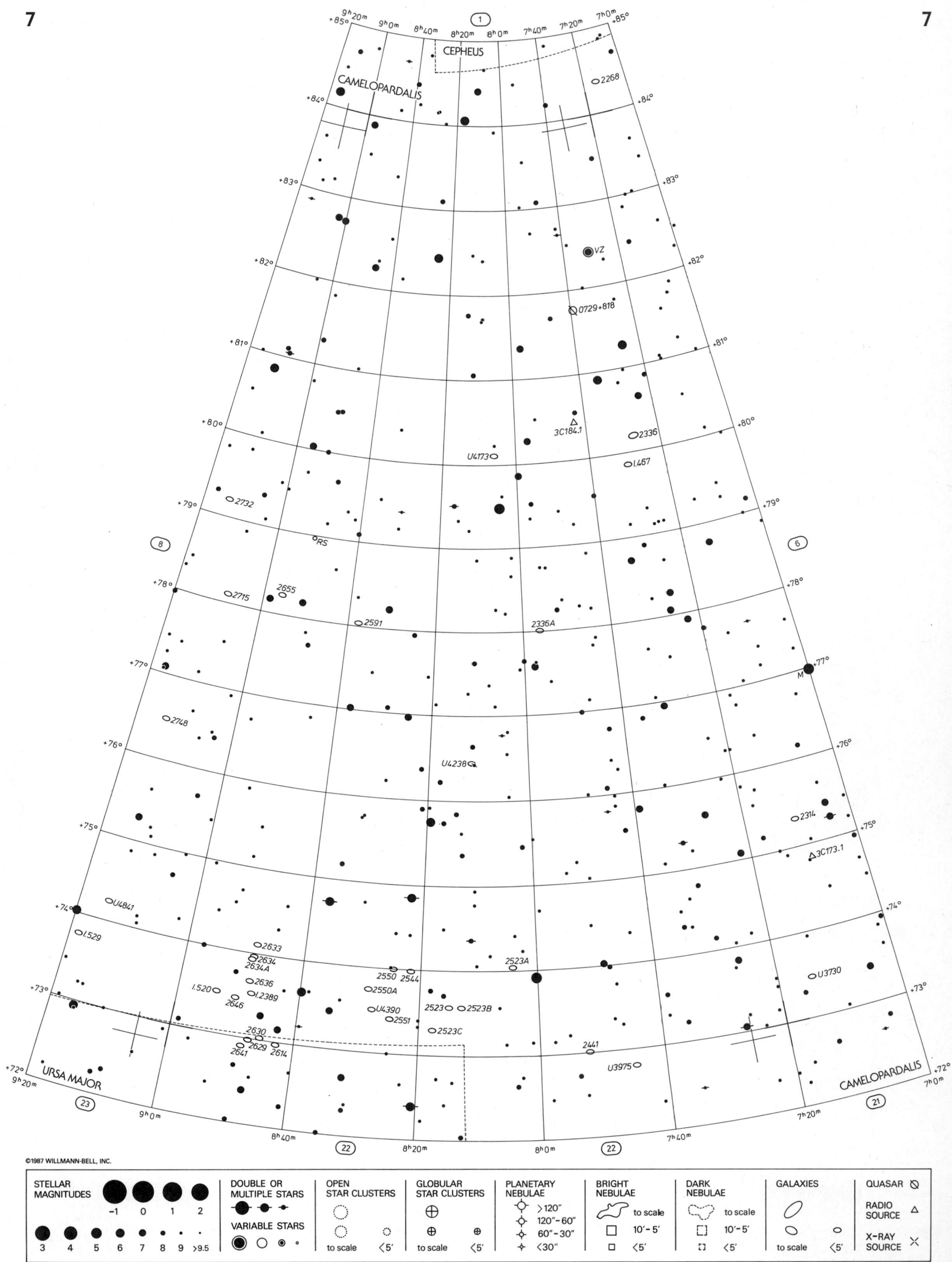
CEPHEUS
CAMELOPARDALIS
2268
VZ
0729+818
3C184.1
2336
U4173
I.467
2732
RS
2715
2655
2591
2336A
M
2748
U4238
2314
3C173.1
U4841
I.529
2633
2634
2634A
2636
I.520
I.2389
2646
2550
2544
2550A
2523A
U4390
2551
2523
2523B
2523C
2630
2629
2641
2614
2441
U3975
U3730
URSA MAJOR
CAMELOPARDALIS
©1987 WILLMANN-BELL, INC.
STELLAR MAGNITUDES
−1 0 1 2
3 4 5 6 7 8 9 >9.5
DOUBLE OR MULTIPLE STARS
VARIABLE STARS
OPEN STAR CLUSTERS
to scale <5′
GLOBULAR STAR CLUSTERS
to scale <5′
PLANETARY NEBULAE
>120″
120″–60″
60″–30″
<30″
BRIGHT NEBULAE
to scale
10′–5′
<5′
DARK NEBULAE
to scale
10′–5′
<5′
GALAXIES
to scale <5′
QUASAR
RADIO SOURCE
X-RAY SOURCE
Barry Rappaport & Wil Tirion

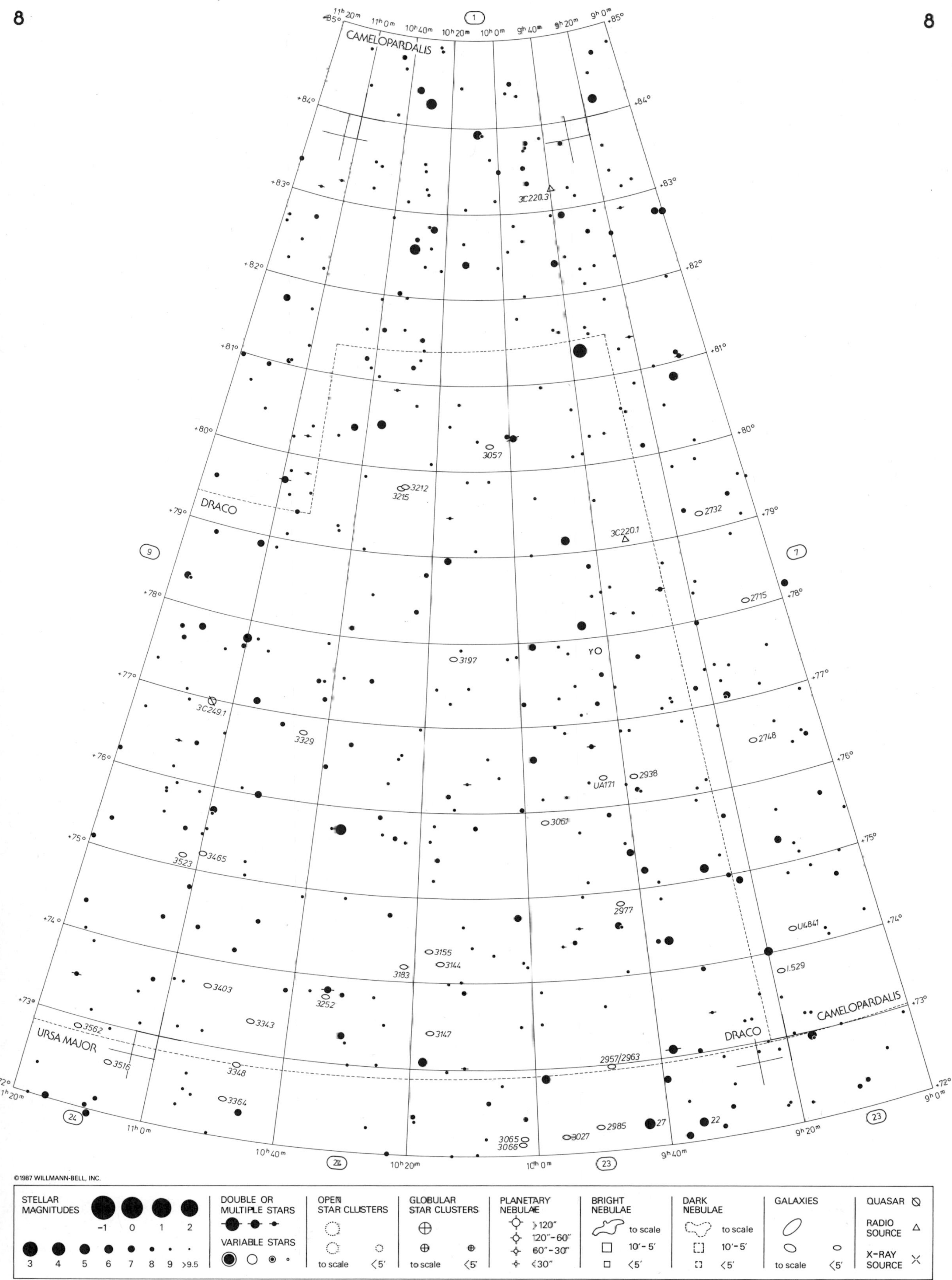

Barry Rappaport & Wil Tirion

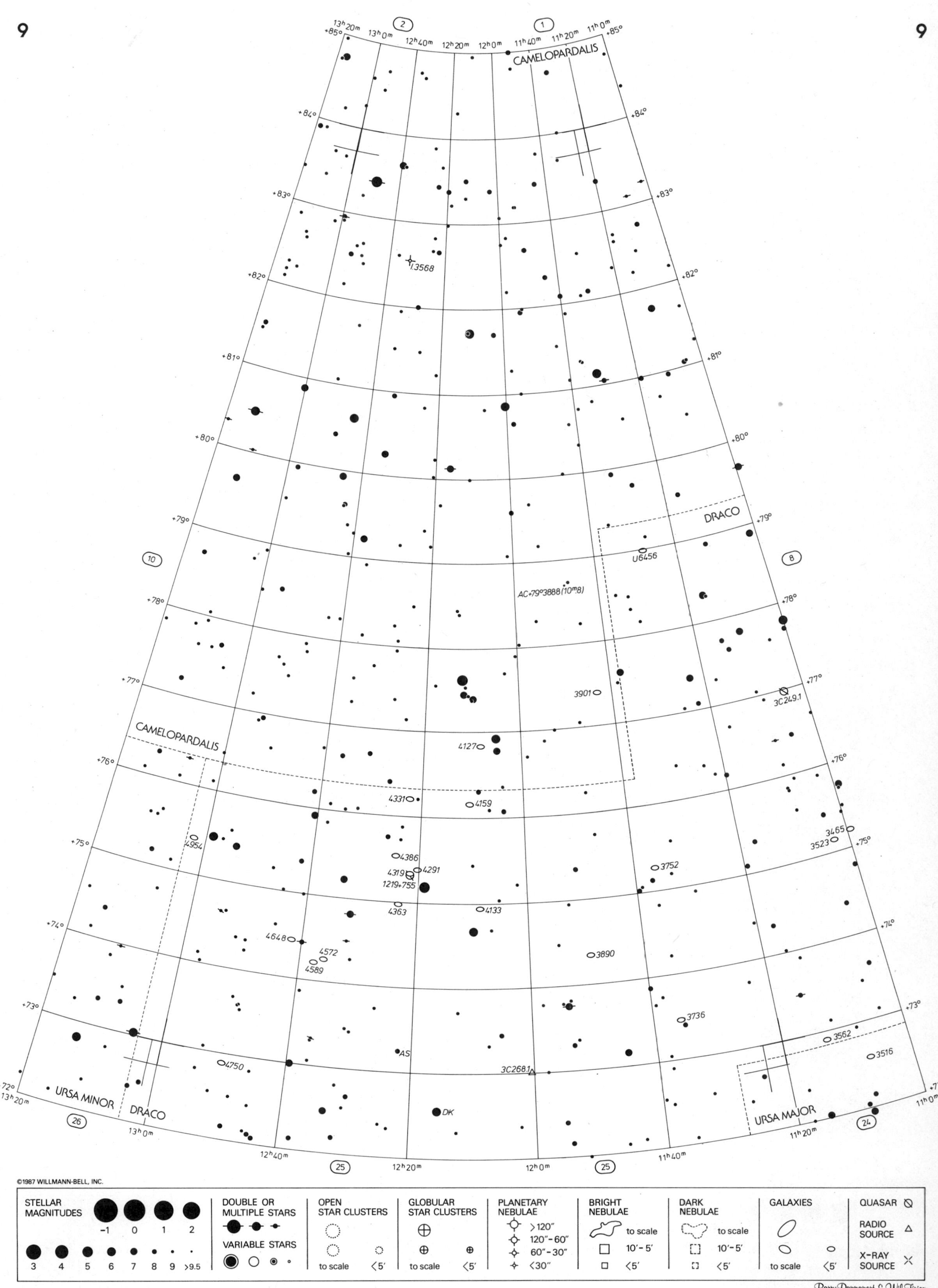
CAMELOPARDALIS
DRACO
URSA MINOR
URSA MAJOR
13h20m
13h0m
12h40m
12h20m
12h0m
11h40m
11h20m
11h0m
+85°
+84°
+83°
+82°
+81°
+80°
+79°
+78°
+77°
+76°
+75°
+74°
+73°
+72°
I.3568
U6456
AC+79°3888 (10m8)
3901
3C249.1
4127
4331
4159
3465
3523
4954
4386
4291
4319
1219+755
3752
4363
4133
4648
4572
4589
3890
3736
3562
3516
4750
AS
3C268.1
DK
©1987 WILLMANN-BELL, INC.
STELLAR MAGNITUDES
-1 0 1 2
3 4 5 6 7 8 9 >9.5
DOUBLE OR MULTIPLE STARS
VARIABLE STARS
OPEN STAR CLUSTERS
to scale
<5′
GLOBULAR STAR CLUSTERS
to scale
<5′
PLANETARY NEBULAE
>120″
120″-60″
60″-30″
<30″
BRIGHT NEBULAE
to scale
10′-5′
<5′
DARK NEBULAE
to scale
10′-5′
<5′
GALAXIES
to scale
<5′
QUASAR
RADIO SOURCE
X-RAY SOURCE
Barry Rappaport & Wil Tirion

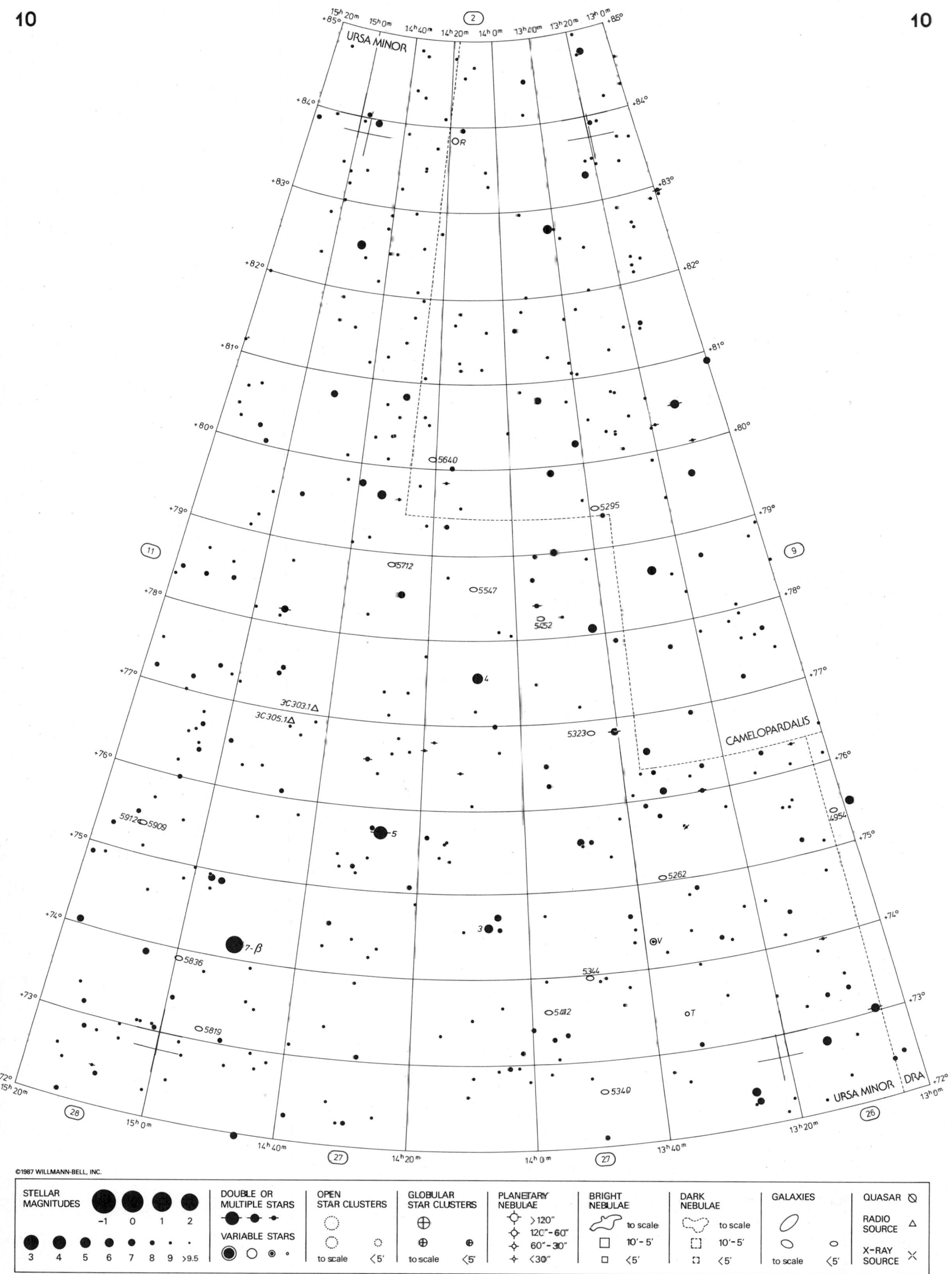

Barry Rappaport & Wil Tirion

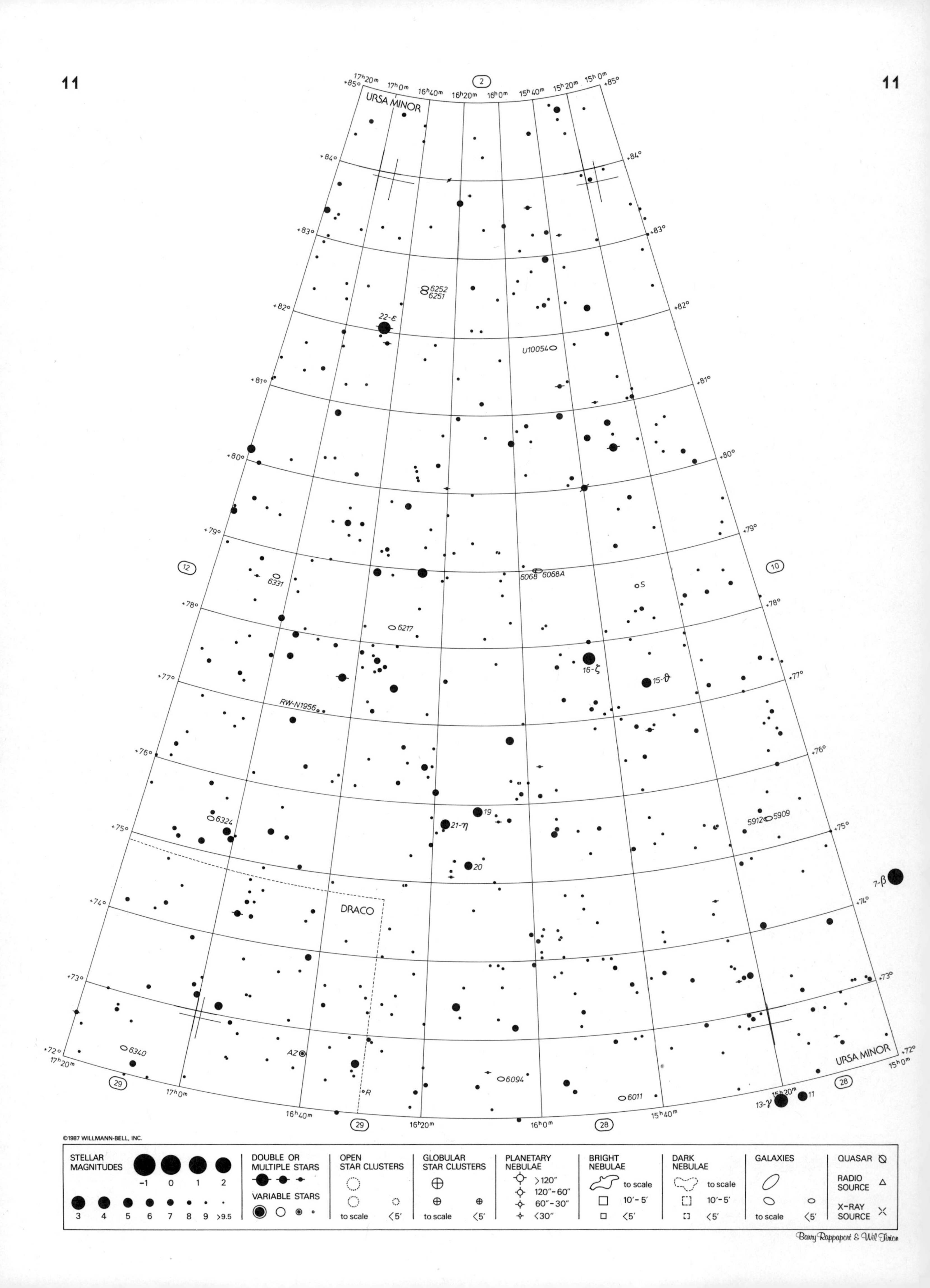

URSA MINOR
DRACO
URSA MINOR
22-ε
16-ζ
15-ϑ
21-η
19
20
7-β
13-γ
11
6252
6251
U10054
6068
6068A
6331
S
6217
RW-N1956
6324
5912
5909
6340
AZ
R
6094
6011
STELLAR MAGNITUDES
DOUBLE OR MULTIPLE STARS
VARIABLE STARS
OPEN STAR CLUSTERS
GLOBULAR STAR CLUSTERS
PLANETARY NEBULAE
BRIGHT NEBULAE
DARK NEBULAE
GALAXIES
QUASAR
RADIO SOURCE
X-RAY SOURCE
to scale
Barry Rappaport & Wil Tirion

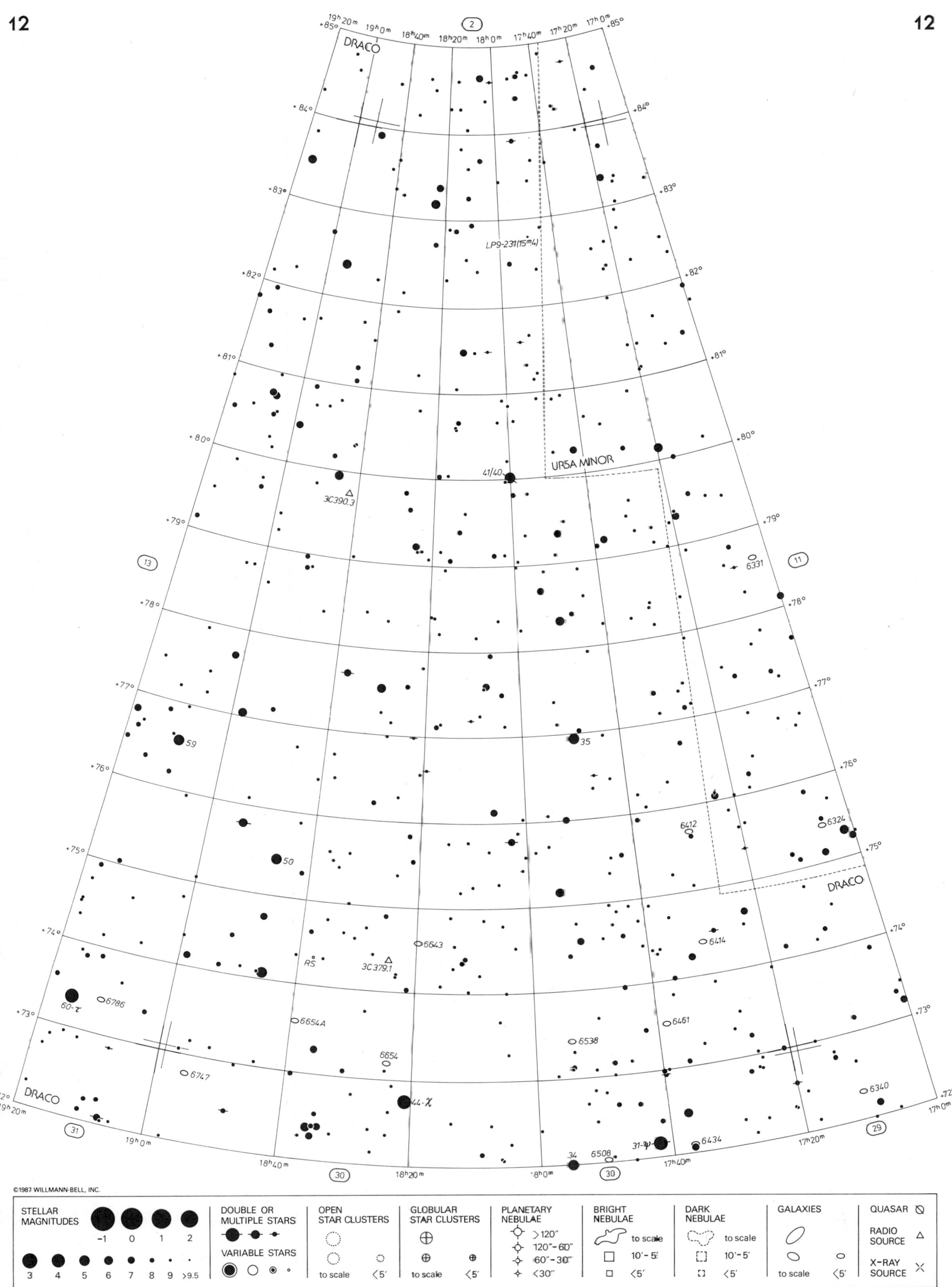

Barry Rappaport & Wil Tirion

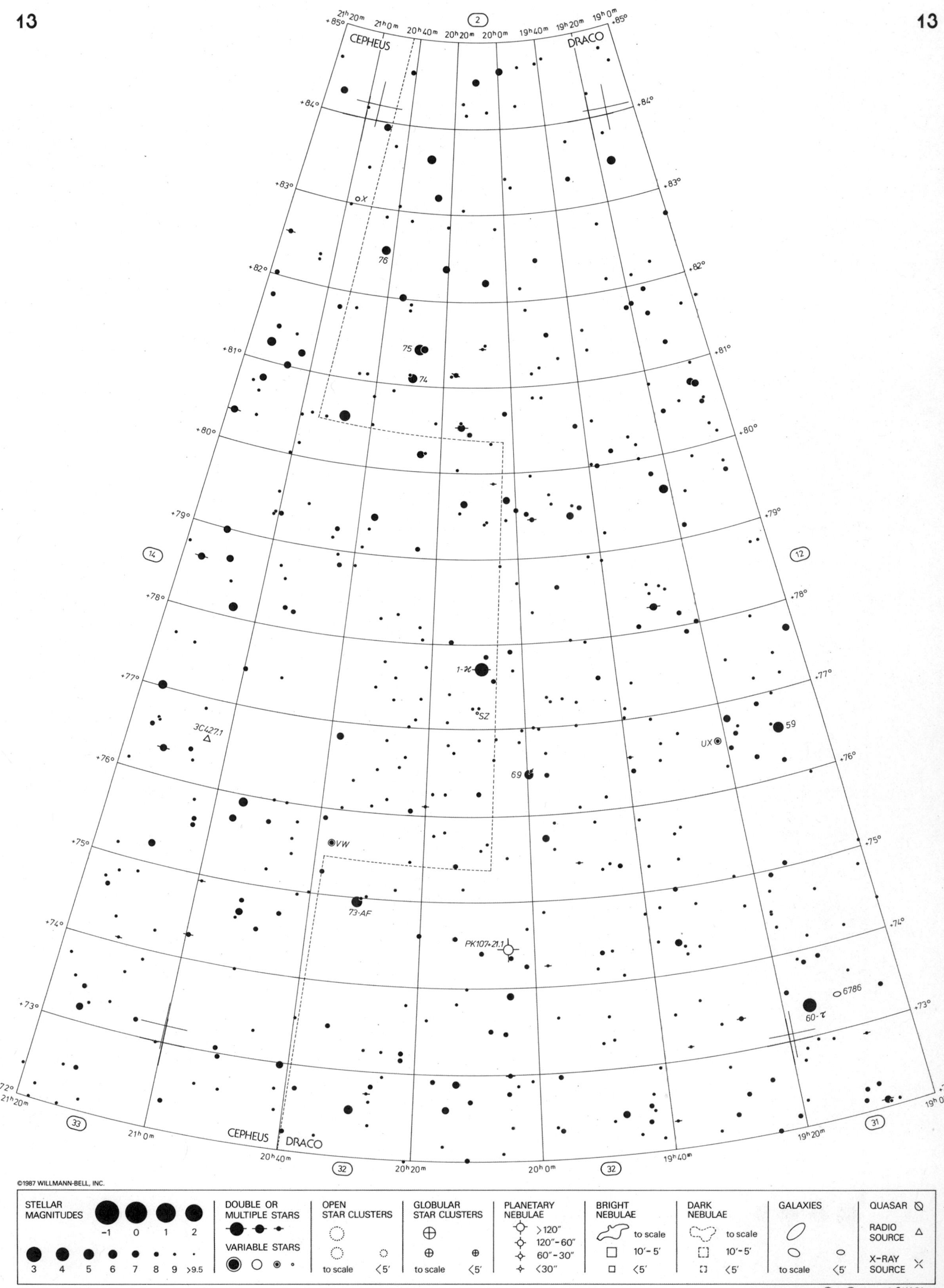
CEPHEUS
DRACO
76
75
74
1-κ
SZ
3C427.1
UX
59
69
VW
73-AF
PK107+21.1
6786
60-τ
STELLAR MAGNITUDES
DOUBLE OR MULTIPLE STARS
VARIABLE STARS
OPEN STAR CLUSTERS
GLOBULAR STAR CLUSTERS
PLANETARY NEBULAE
BRIGHT NEBULAE
DARK NEBULAE
GALAXIES
QUASAR
RADIO SOURCE
X-RAY SOURCE
to scale
Barry Rappaport & Wil Tirion

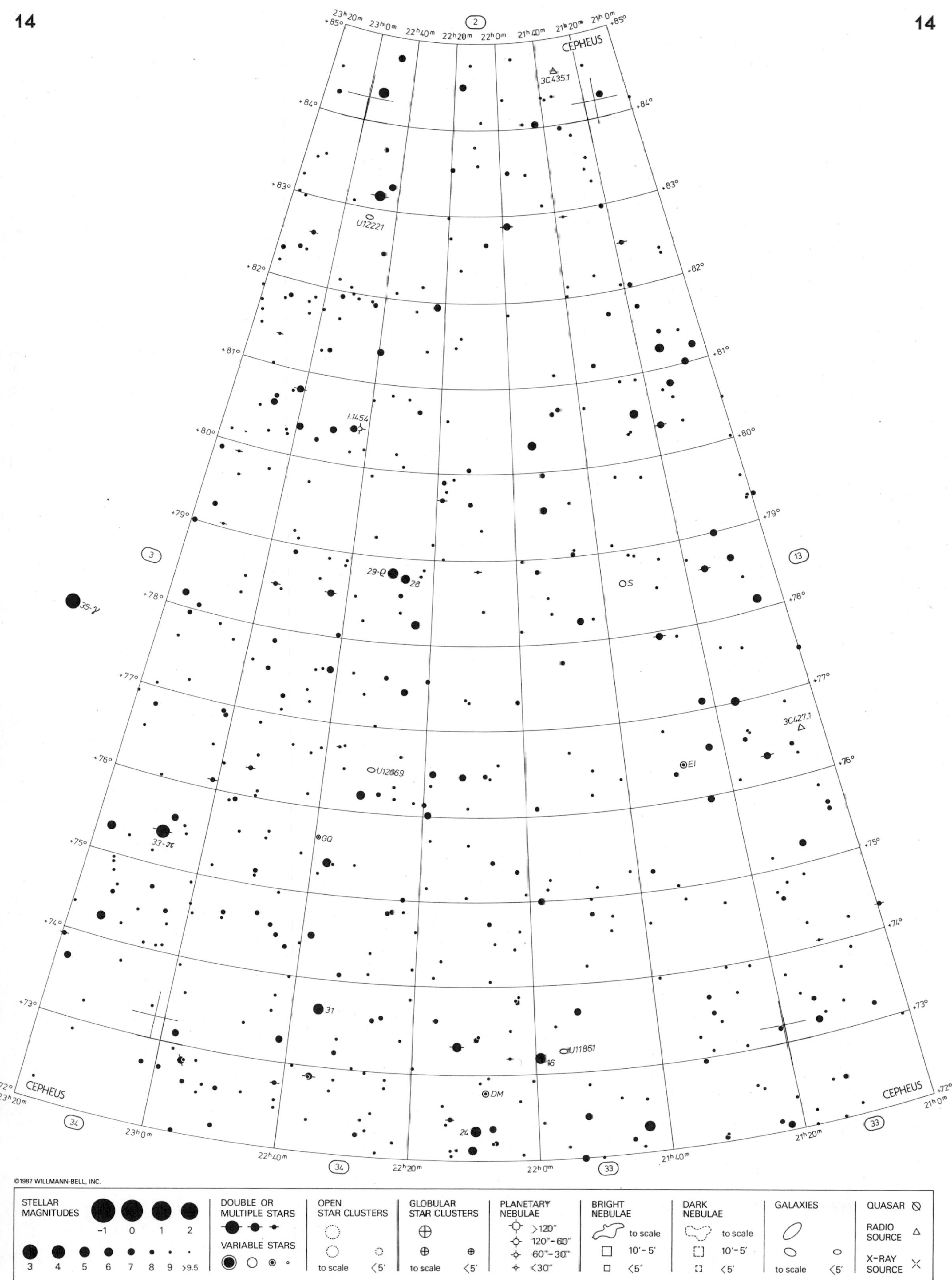

Barry Rappaport & Wil Tirion

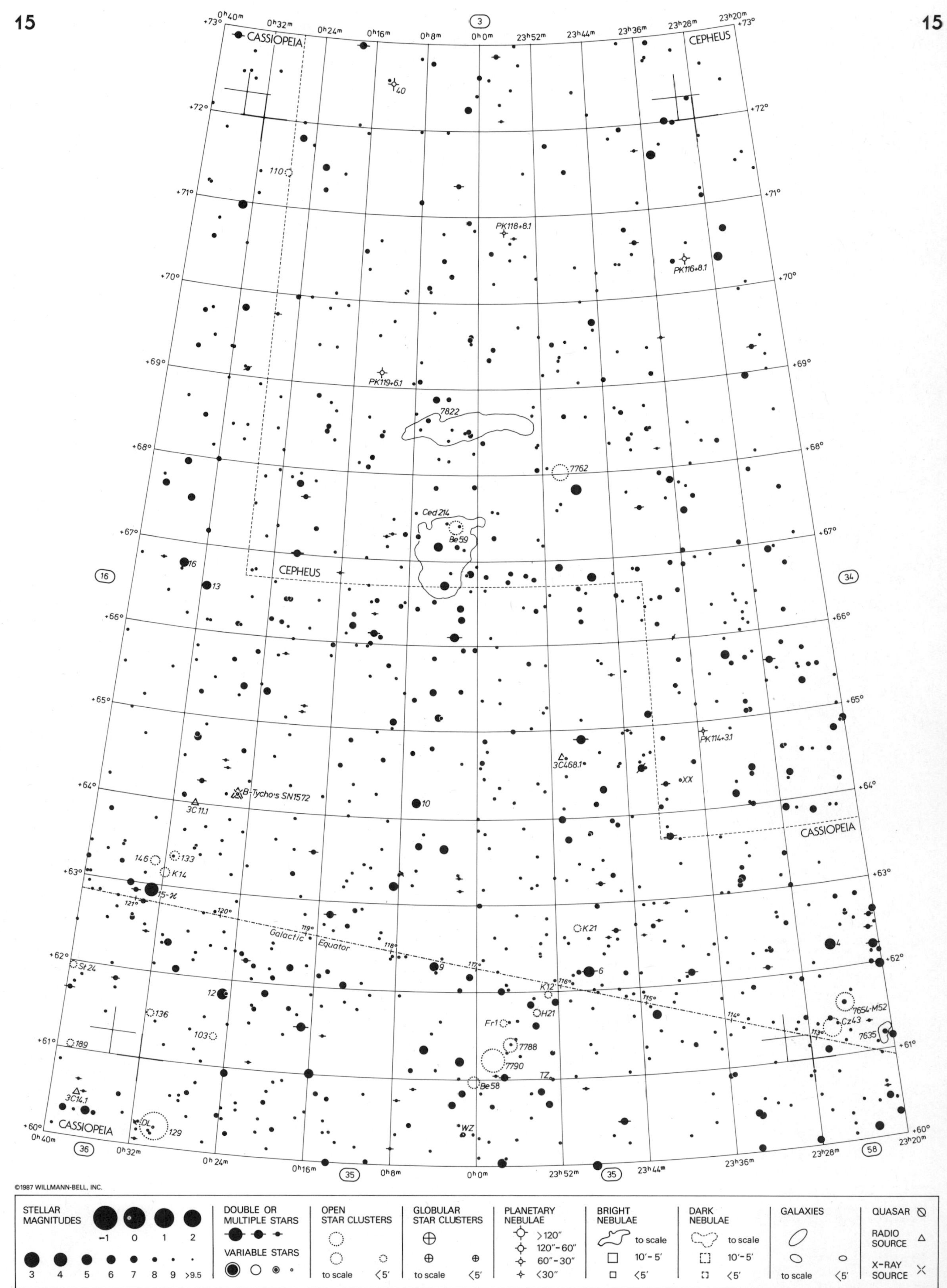

Barry Rappaport & Wil Tirion

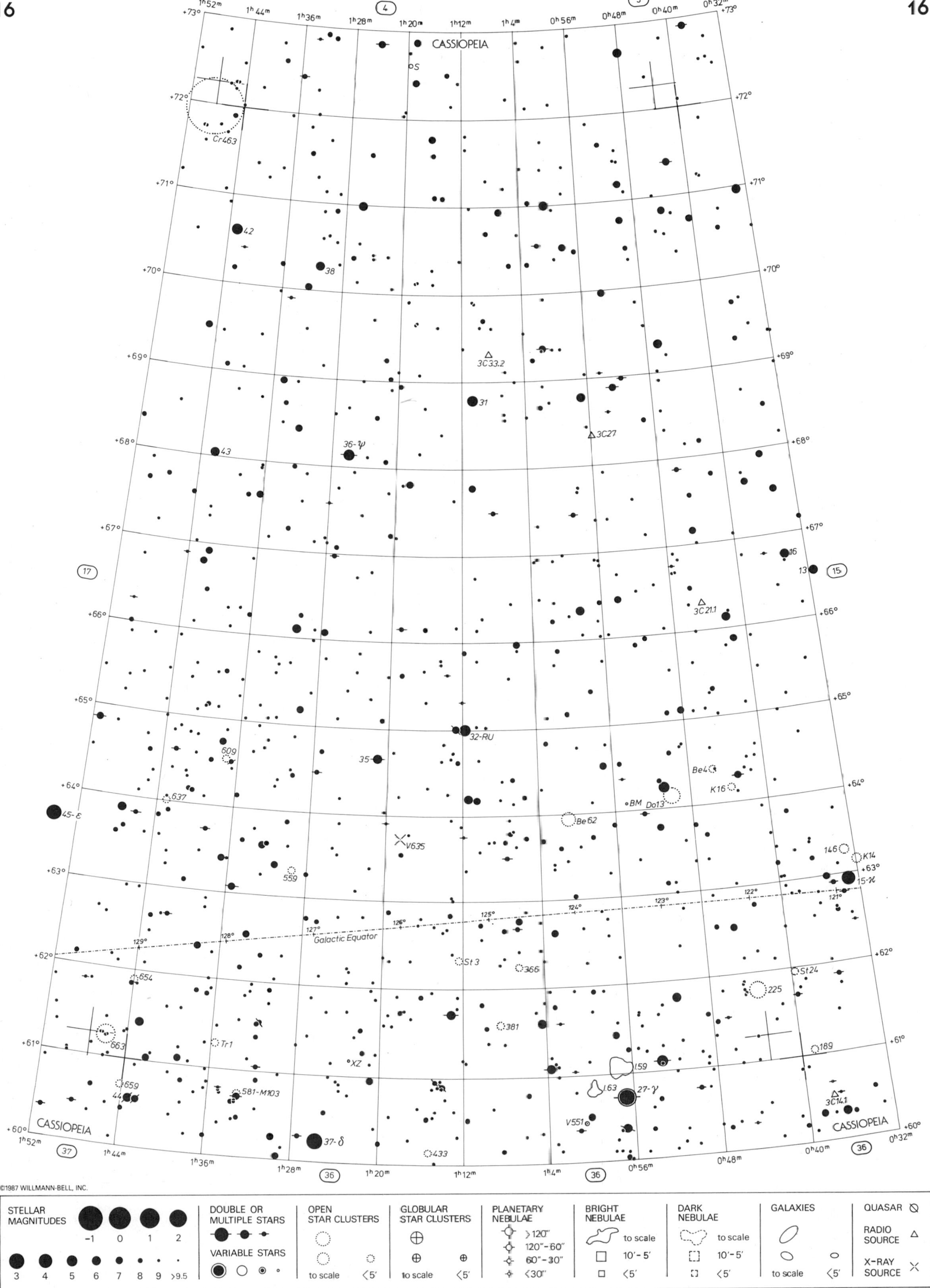

Barry Rappaport & Wil Tirion

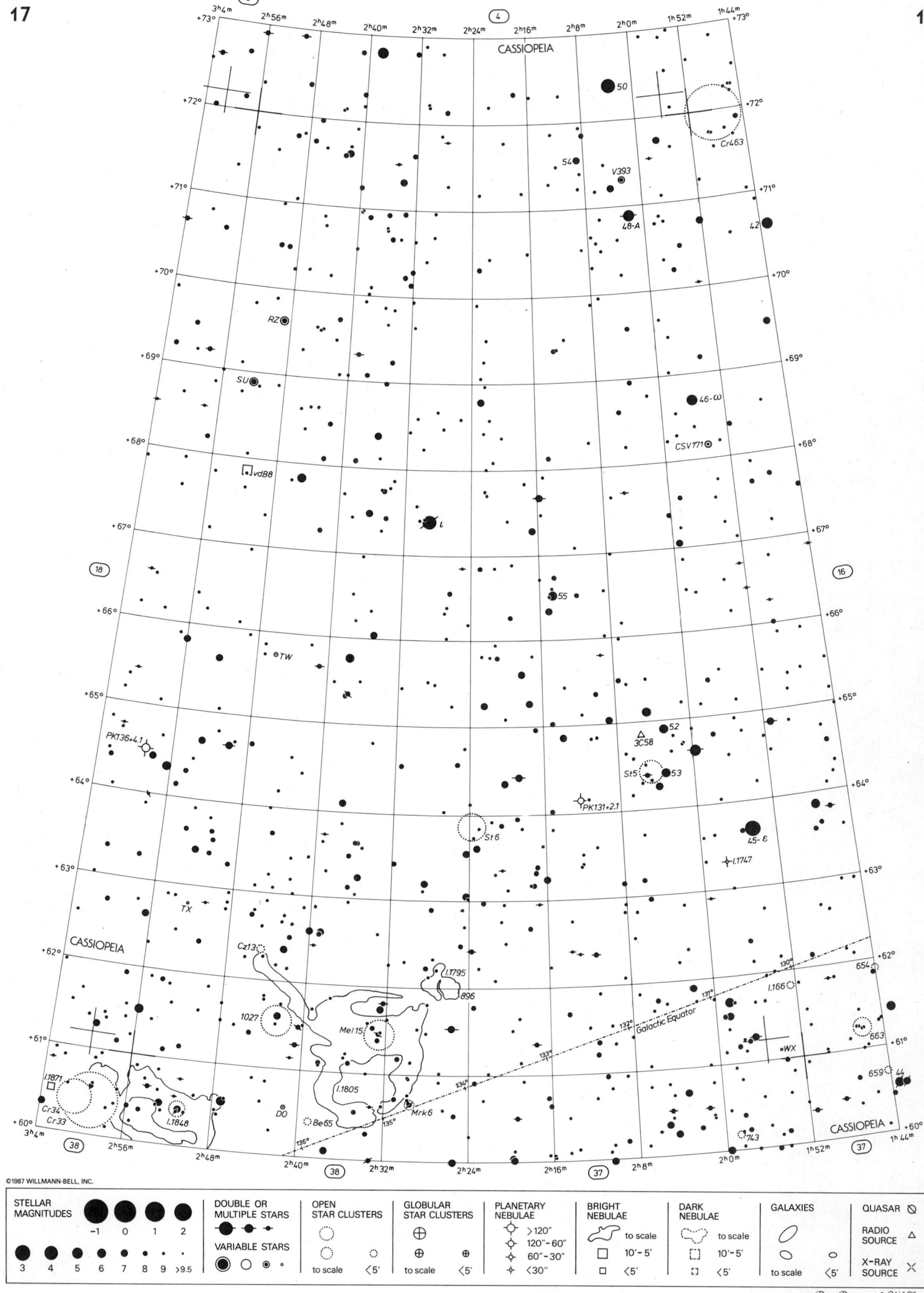

Barry Rappaport & Wil Tirion

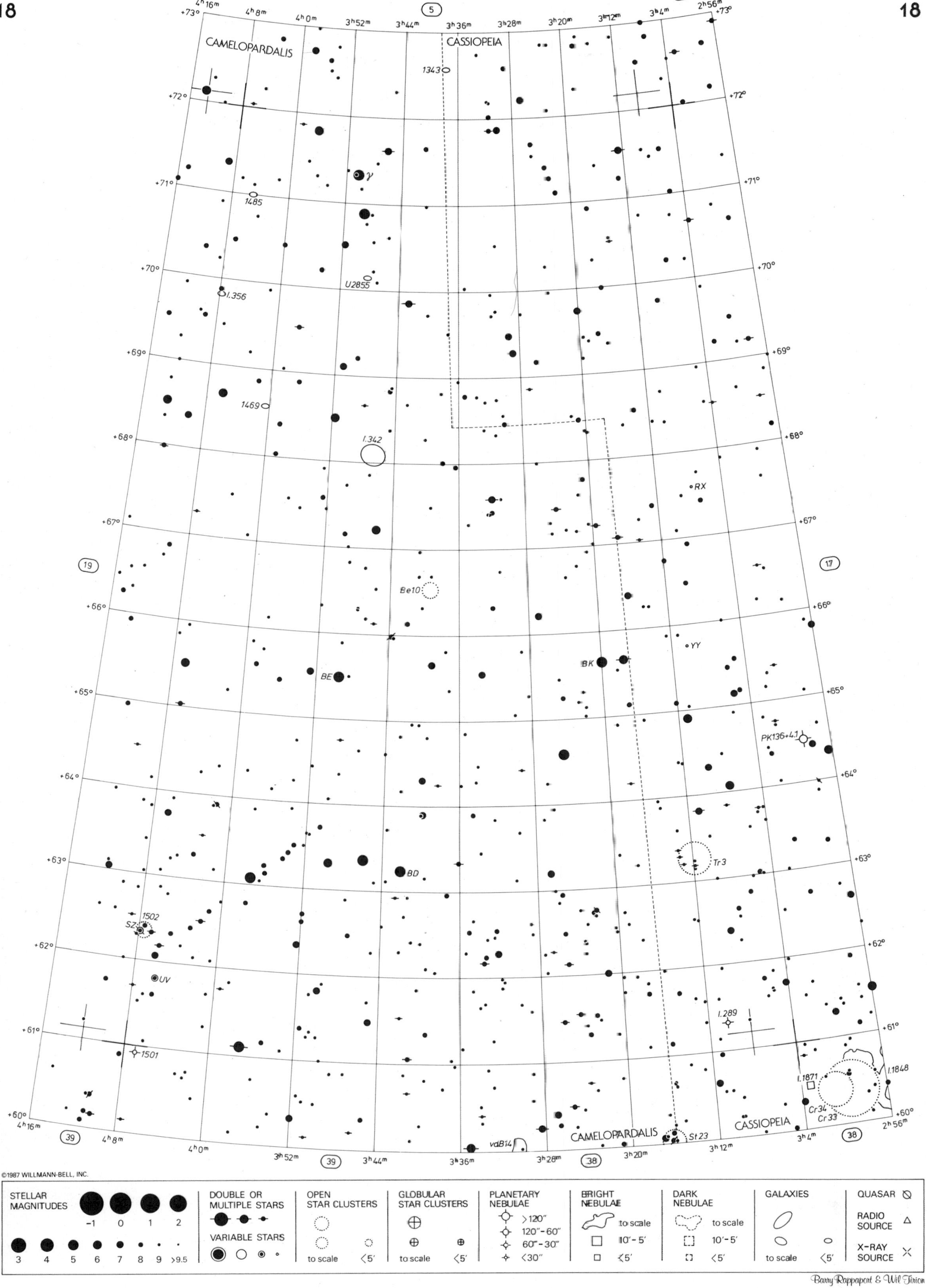
CAMELOPARDALIS
CASSIOPEIA
4h16m
4h8m
4h0m
3h52m
3h44m
3h36m
3h28m
3h20m
3h12m
3h4m
2h56m
+73°
+72°
+71°
+70°
+69°
+68°
+67°
+66°
+65°
+64°
+63°
+62°
+61°
+60°
5
4
19
17
39
38
1343
γ
1485
U2855
I.356
1469
I.342
RX
Be10
YY
BK
BE
PK136+4.1
Tr3
BD
1502
SZ
UV
I.289
1501
I.1871
I.1848
Cr34
Cr33
vdB14
St23
©1987 WILLMANN-BELL, INC.
STELLAR MAGNITUDES
-1 0 1 2
3 4 5 6 7 8 9 >9.5
DOUBLE OR MULTIPLE STARS
VARIABLE STARS
OPEN STAR CLUSTERS
to scale <5′
GLOBULAR STAR CLUSTERS
to scale <5′
PLANETARY NEBULAE
>120″
120″-60″
60″-30″
<30″
BRIGHT NEBULAE
to scale
10′-5′
<5′
DARK NEBULAE
to scale
10′-5′
<5′
GALAXIES
to scale <5′
QUASAR
RADIO SOURCE
X-RAY SOURCE
Barry Rappaport & Wil Tirion

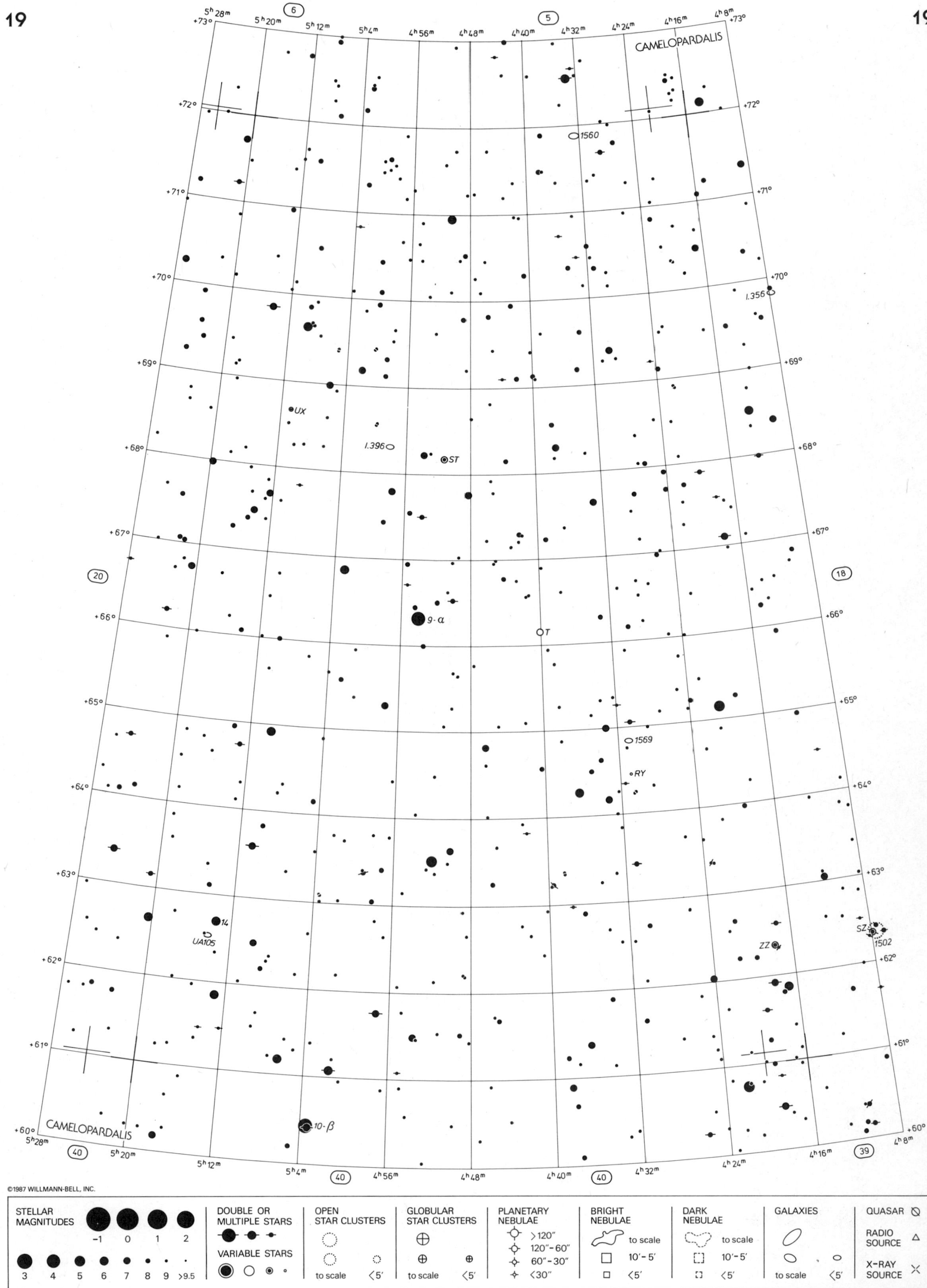

Barry Rappaport & Wil Tirion

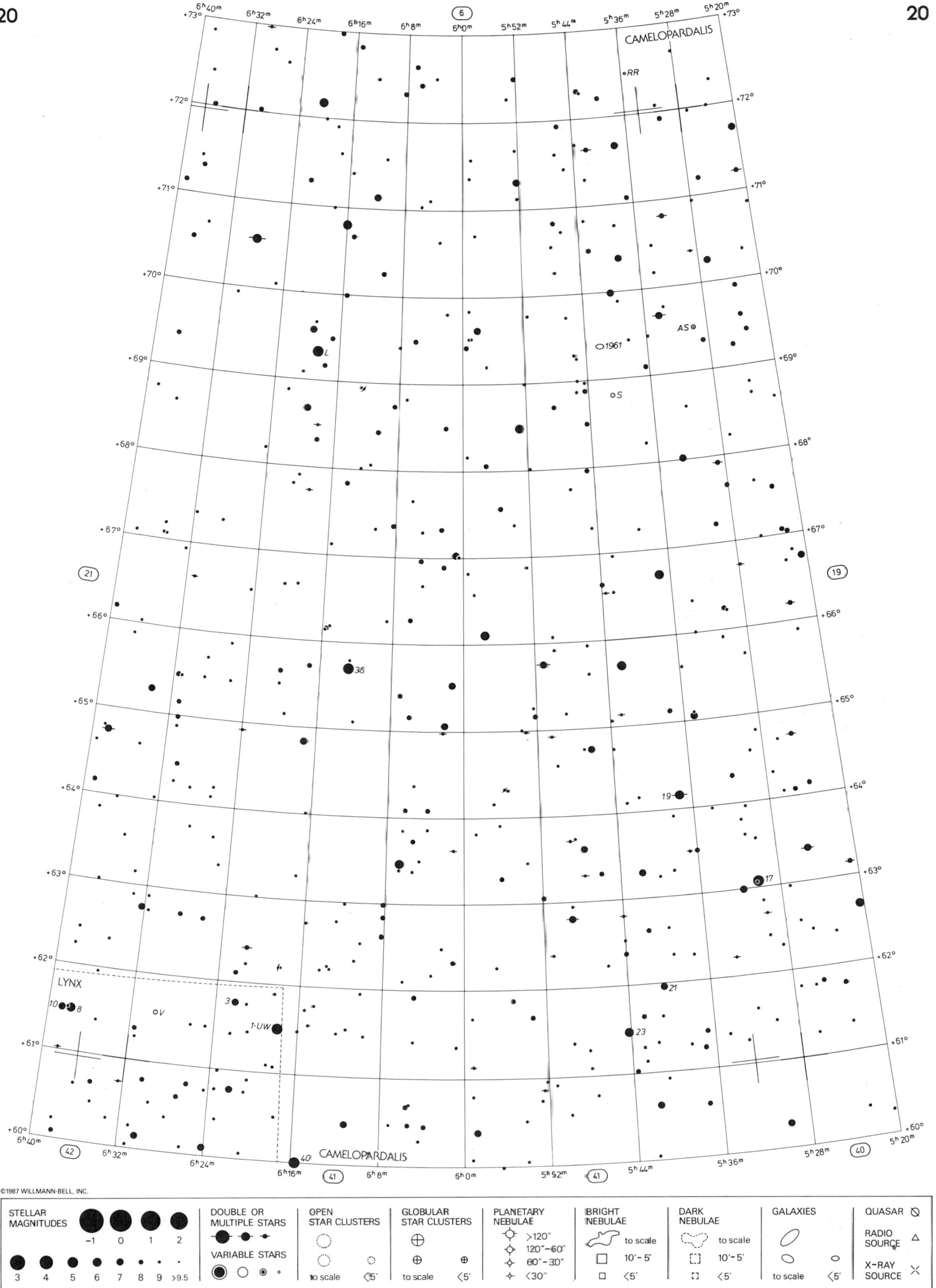
CAMELOPARDALIS
LYNX
CAMELOPARDALIS
6
21
19
42
41
41
40
©1987 WILLMANN-BELL, INC.
STELLAR MAGNITUDES
DOUBLE OR MULTIPLE STARS
VARIABLE STARS
OPEN STAR CLUSTERS
GLOBULAR STAR CLUSTERS
PLANETARY NEBULAE
BRIGHT NEBULAE
DARK NEBULAE
GALAXIES
QUASAR
RADIO SOURCE
X-RAY SOURCE
to scale
Barry Rappaport & Wil Tirion

21

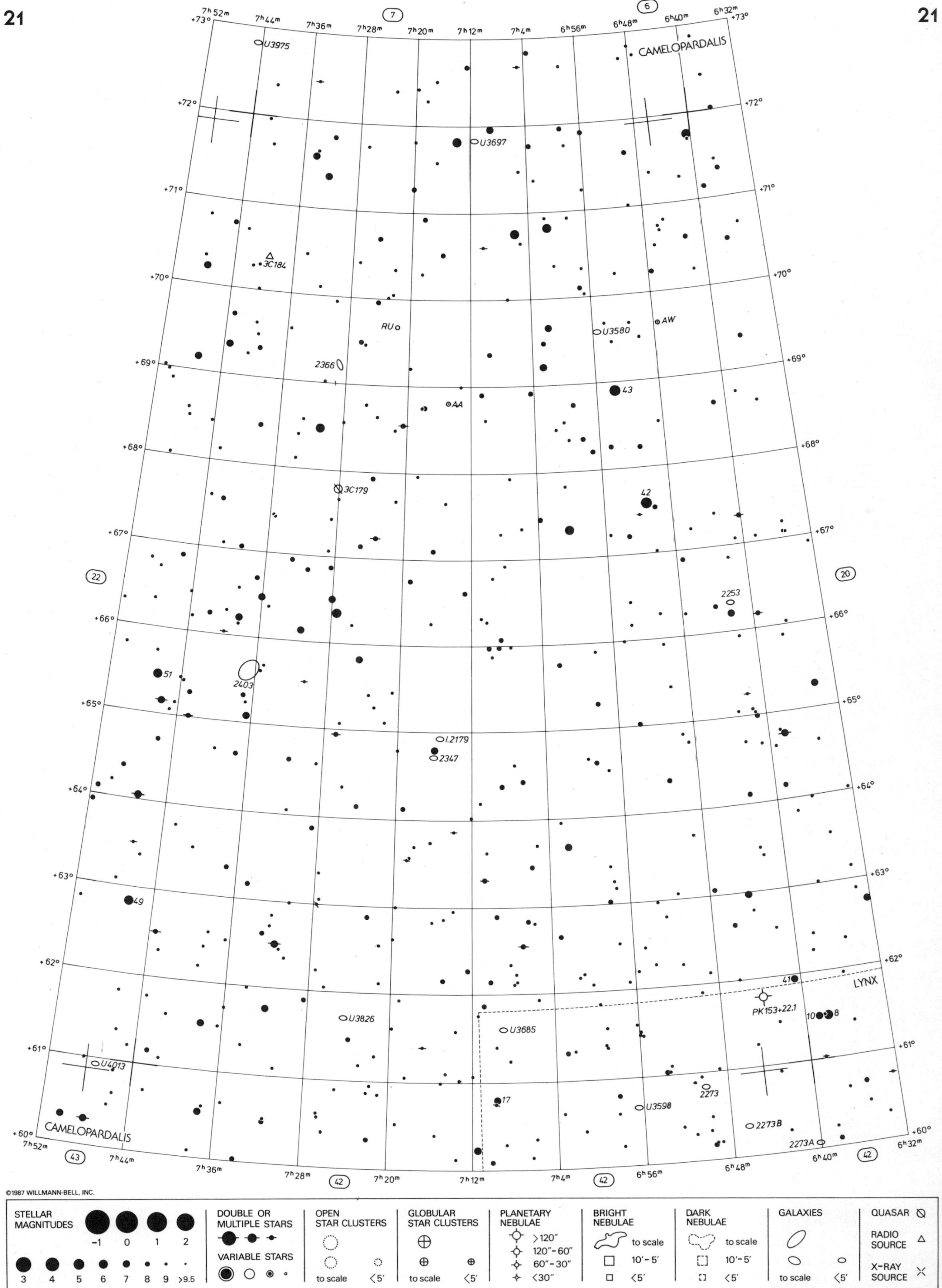

Barry Rappaport & Wil Tirion

22

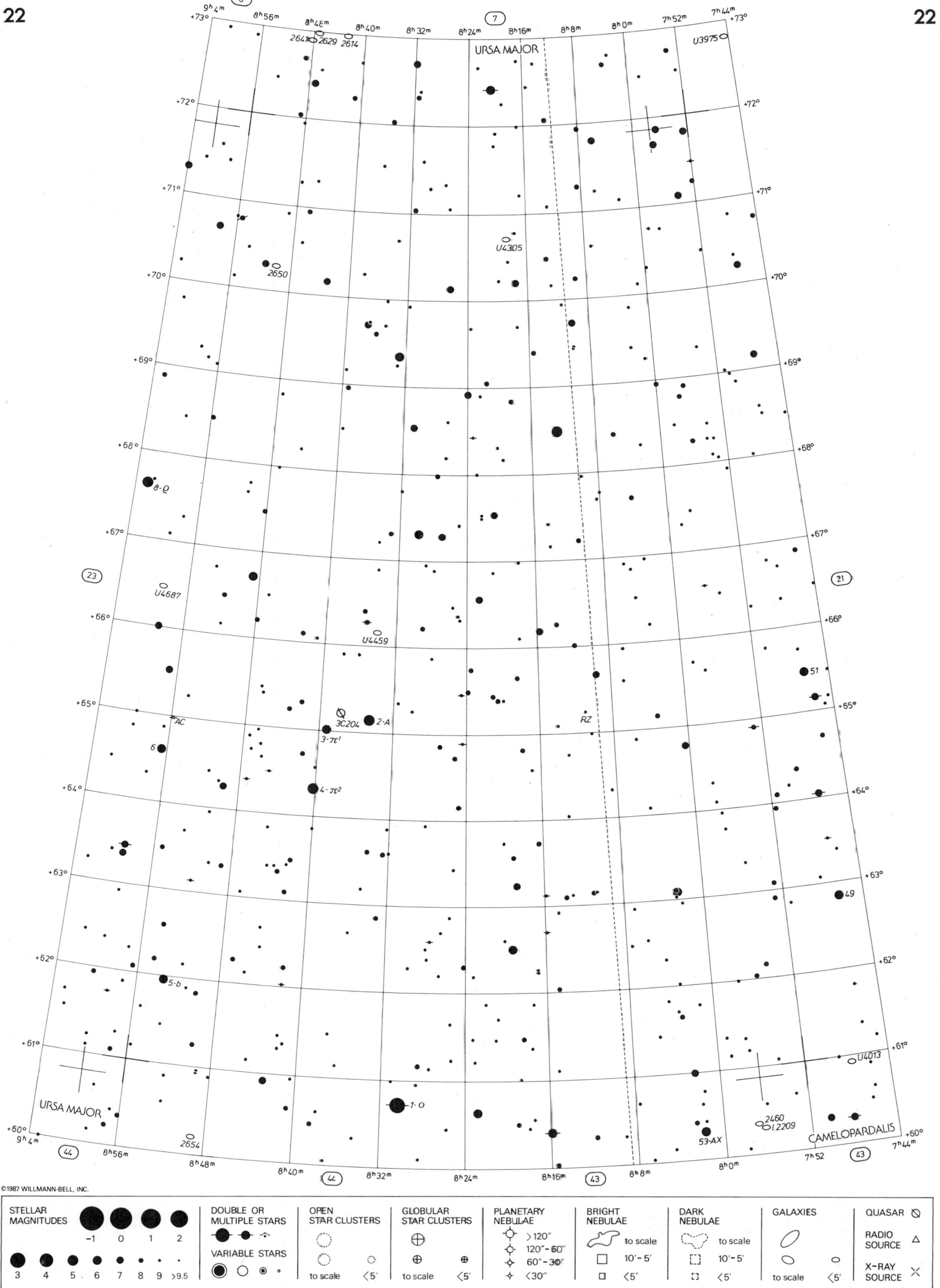

URSA MAJOR
CAMELOPARDALIS
U3975
2641
2629
2614
U4305
2650
8-ρ
U4687
U4459
3C204
2-A
3-π¹
4-π²
RZ
AC
6
51
49
5-b
1-o
2654
2460
I.2209
53-AX
U4013
STELLAR MAGNITUDES
DOUBLE OR MULTIPLE STARS
VARIABLE STARS
OPEN STAR CLUSTERS
GLOBULAR STAR CLUSTERS
PLANETARY NEBULAE
BRIGHT NEBULAE
DARK NEBULAE
GALAXIES
QUASAR
RADIO SOURCE
X-RAY SOURCE
to scale
Barry Rappaport & Wil Tirion

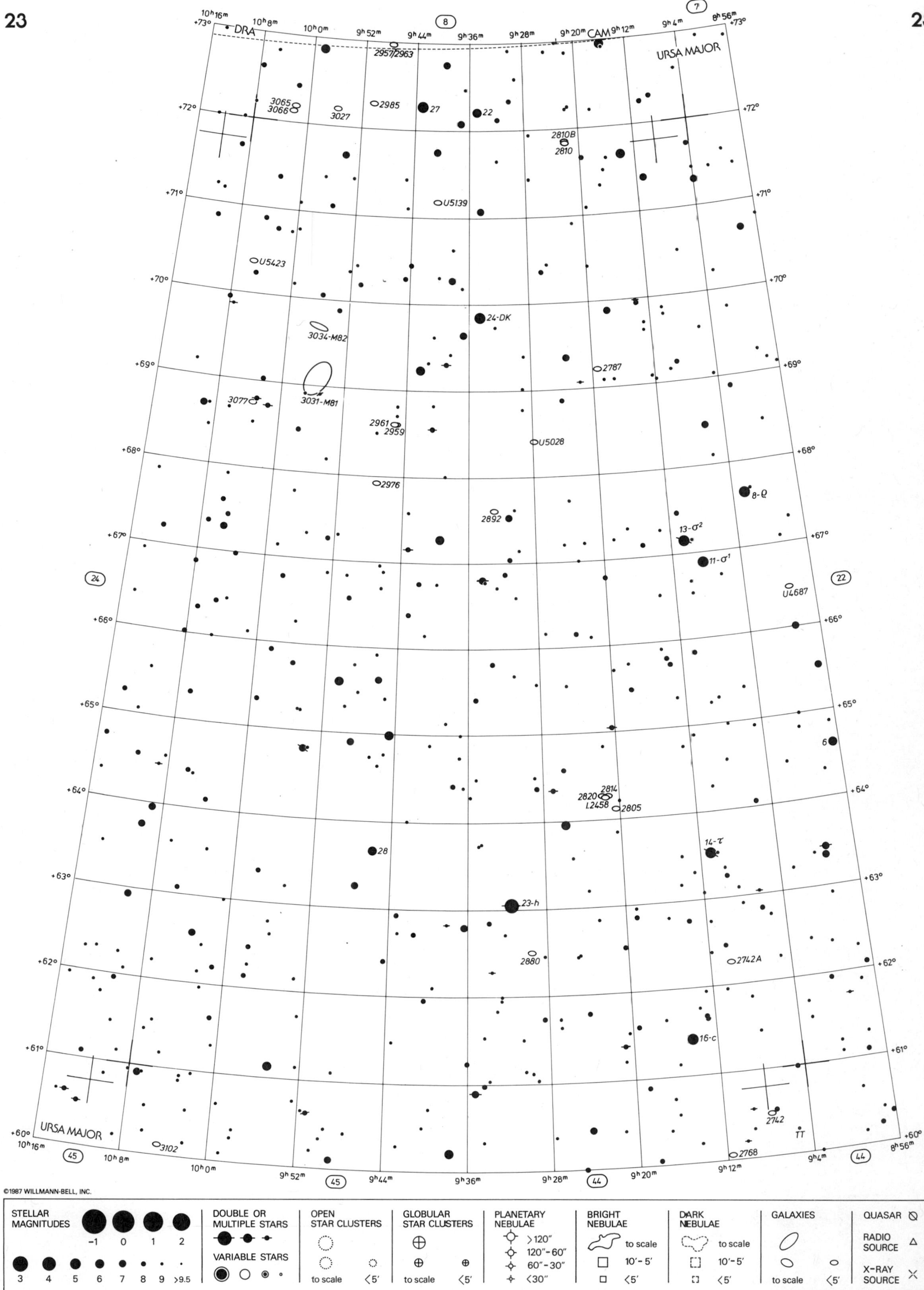

Barry Rappaport & Wil Tirion

24

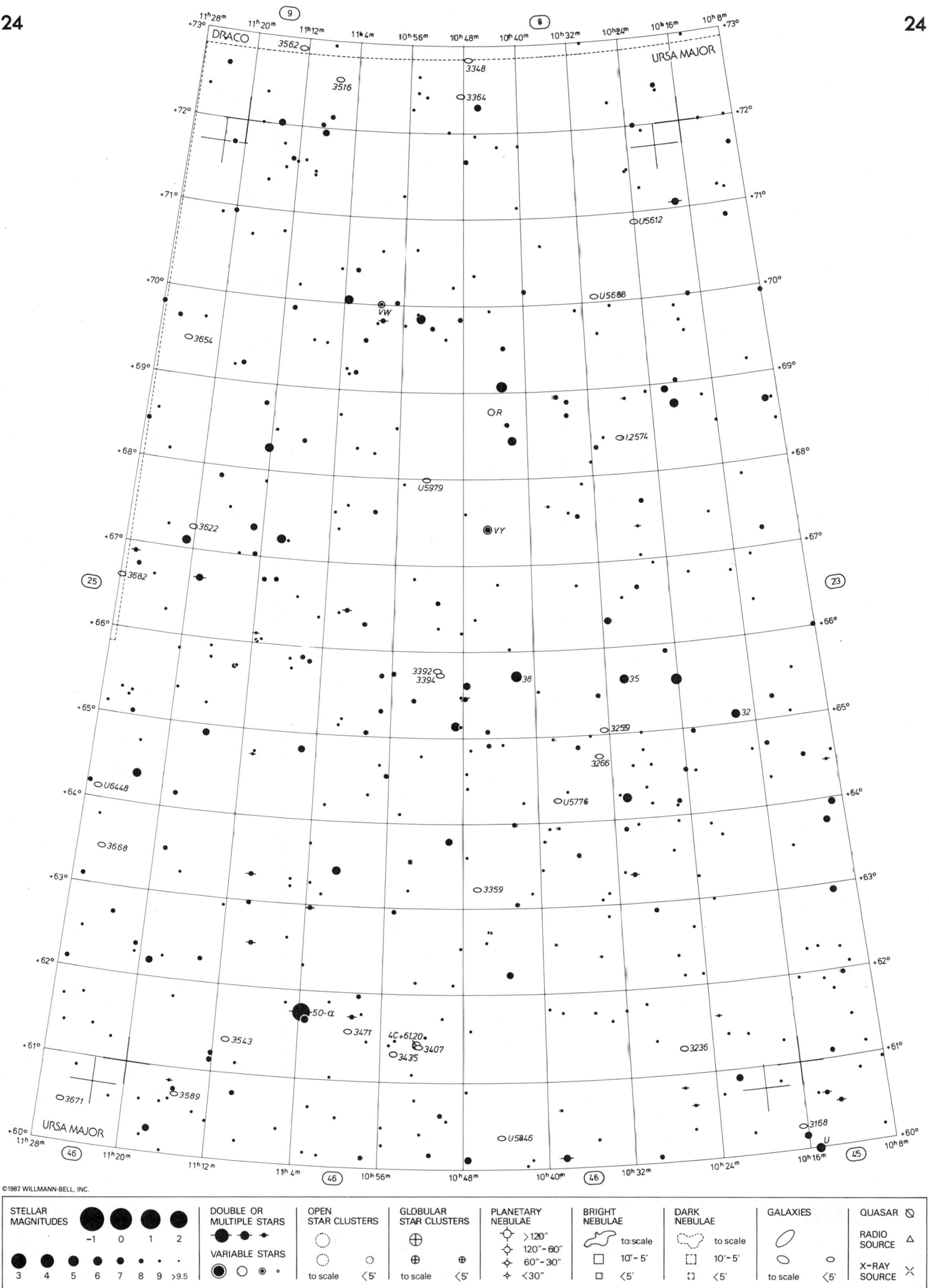

Barry Rappaport & Wil Tirion

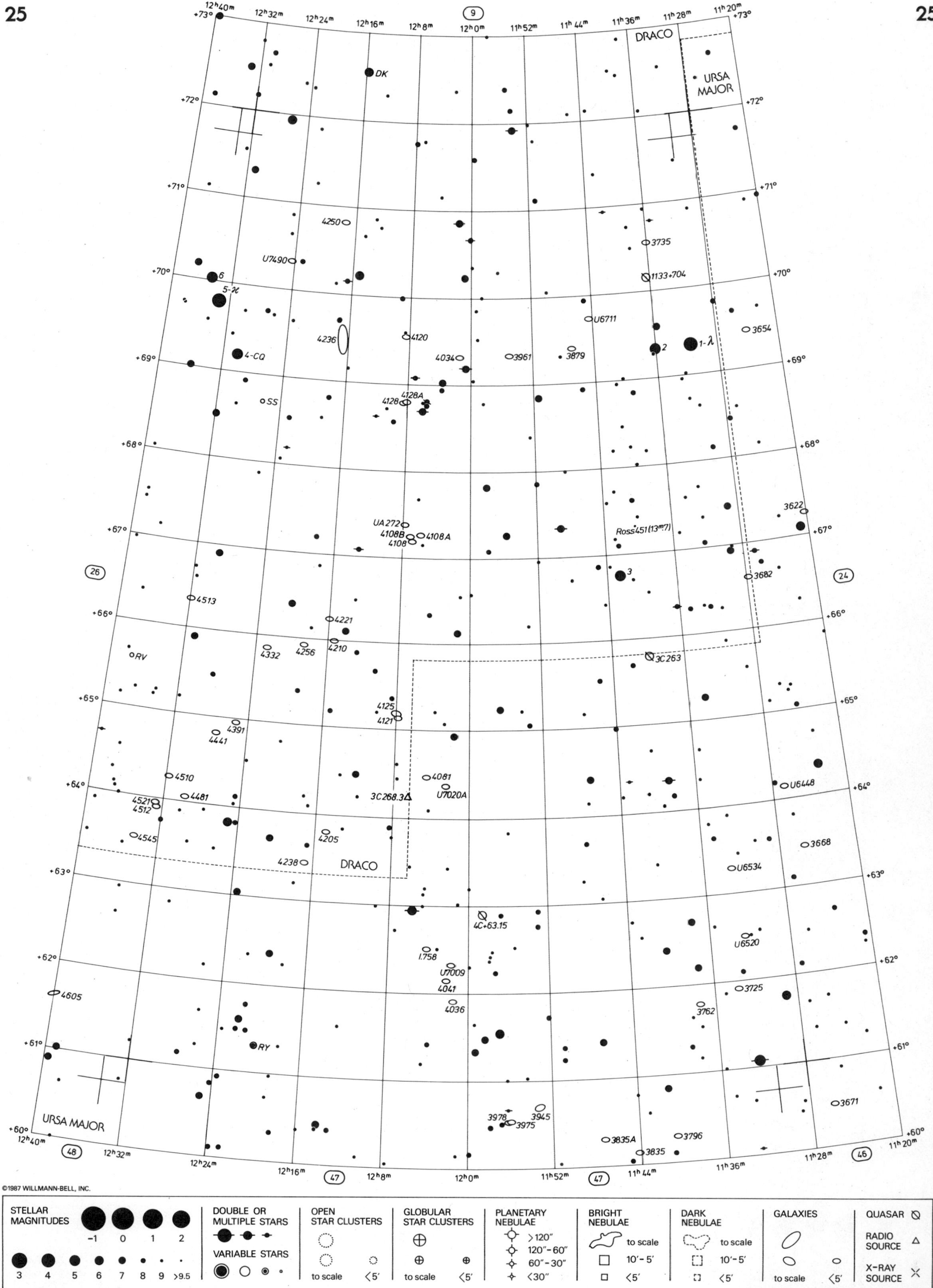

Barry Rappaport & Wil Tirion

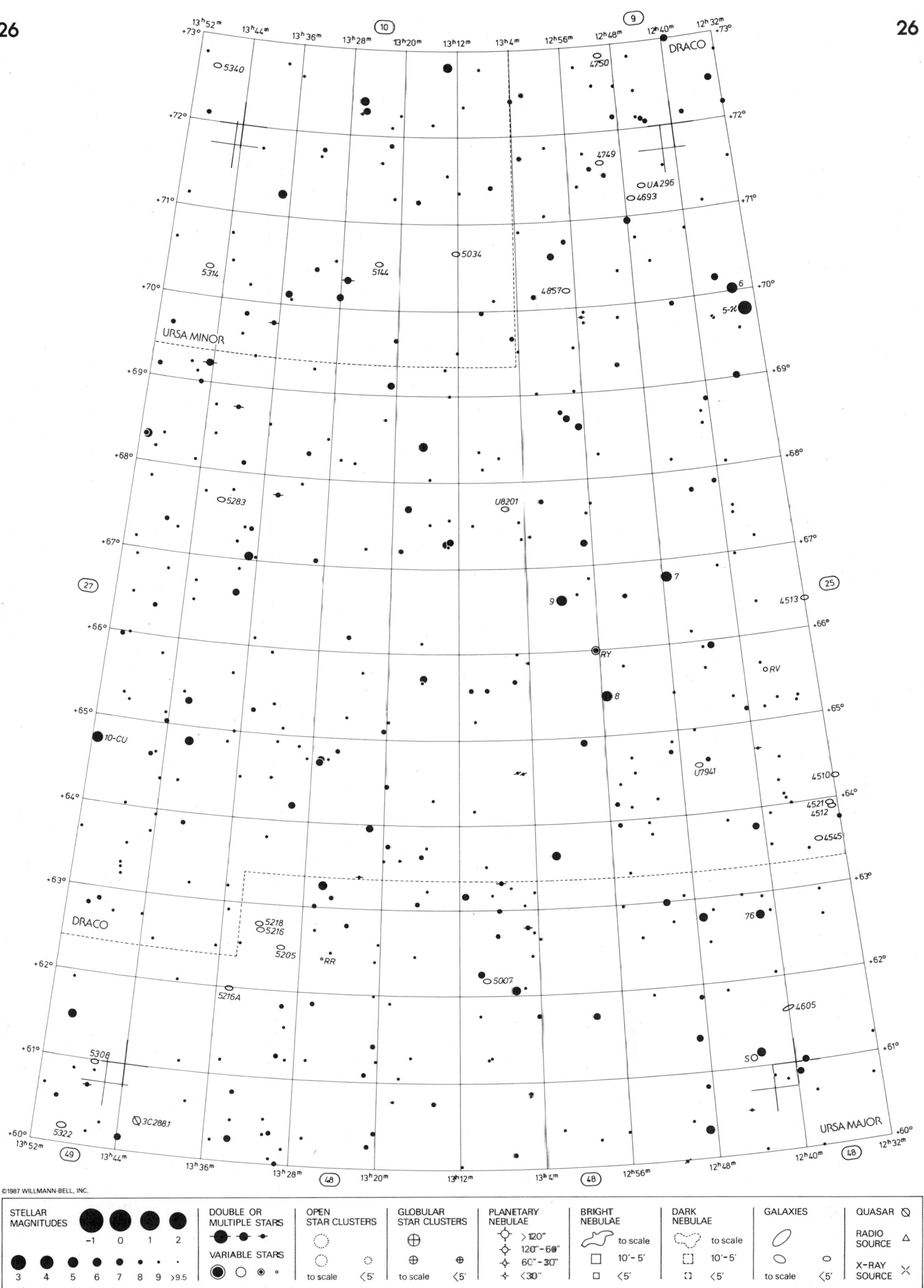

Barry Rappaport & Wil Tirion

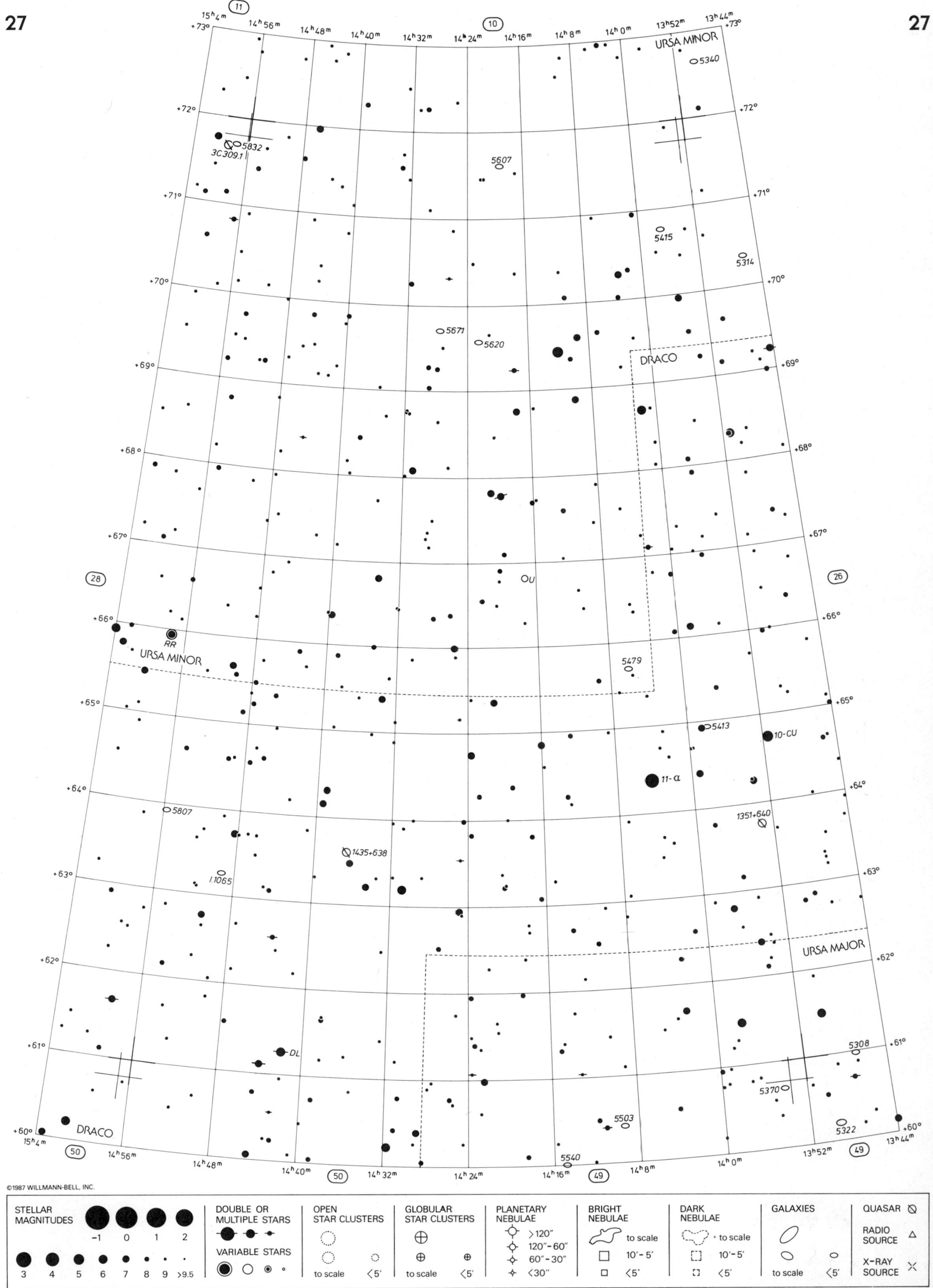

Barry Rappaport & Wil Tirion

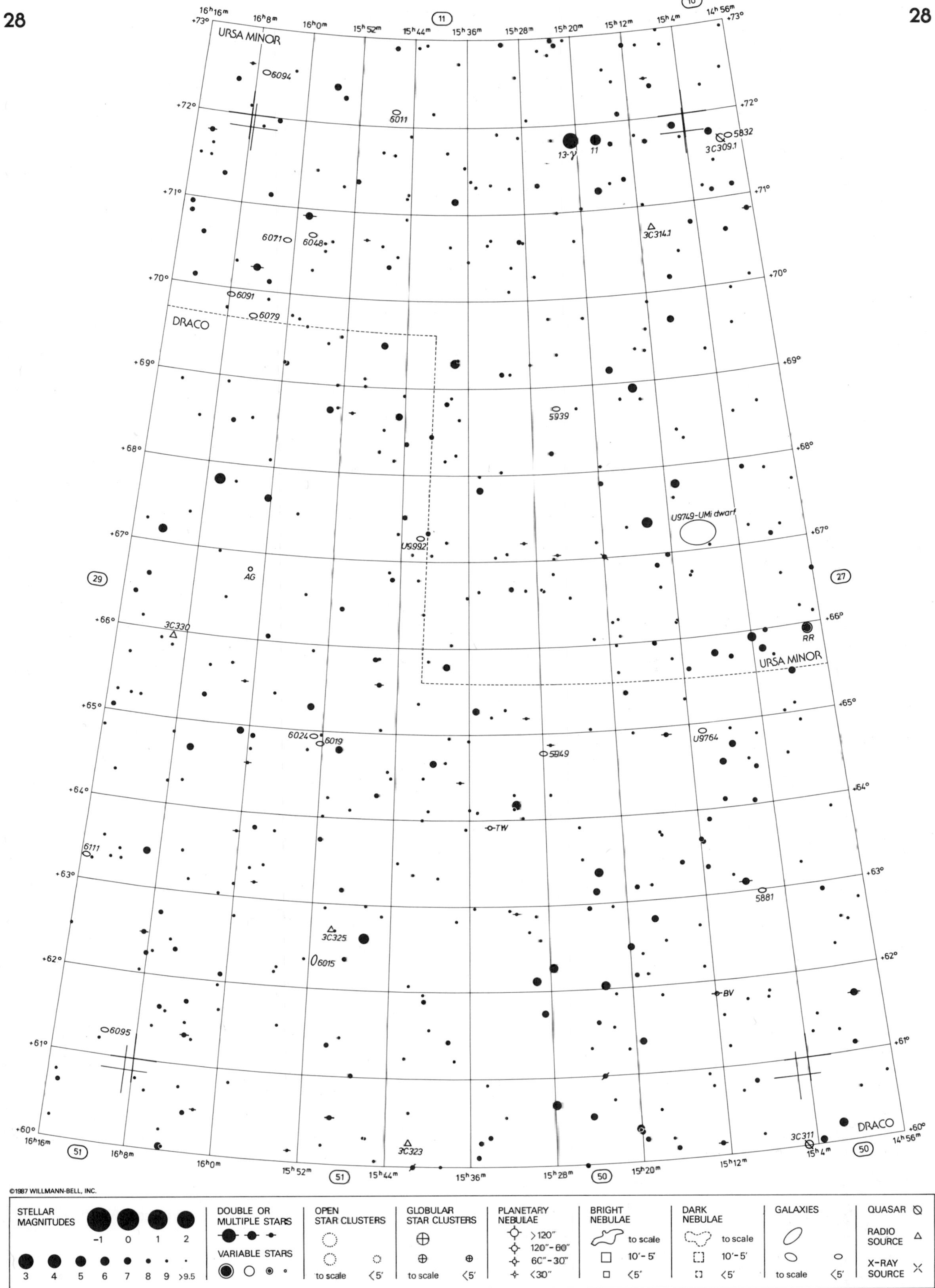

STELLAR MAGNITUDES: −1, 0, 1, 2, 3, 4, 5, 6, 7, 8, 9, >9.5
DOUBLE OR MULTIPLE STARS
VARIABLE STARS
OPEN STAR CLUSTERS: to scale, <5′
GLOBULAR STAR CLUSTERS: to scale, <5′
PLANETARY NEBULAE: >120″, 120″–60″, 60″–30″, <30″
BRIGHT NEBULAE: to scale, 10′–5′, <5′
DARK NEBULAE: to scale, 10′–5′, <5′
GALAXIES: to scale, <5′
QUASAR
RADIO SOURCE
X-RAY SOURCE

Barry Rappaport & Wil Tirion

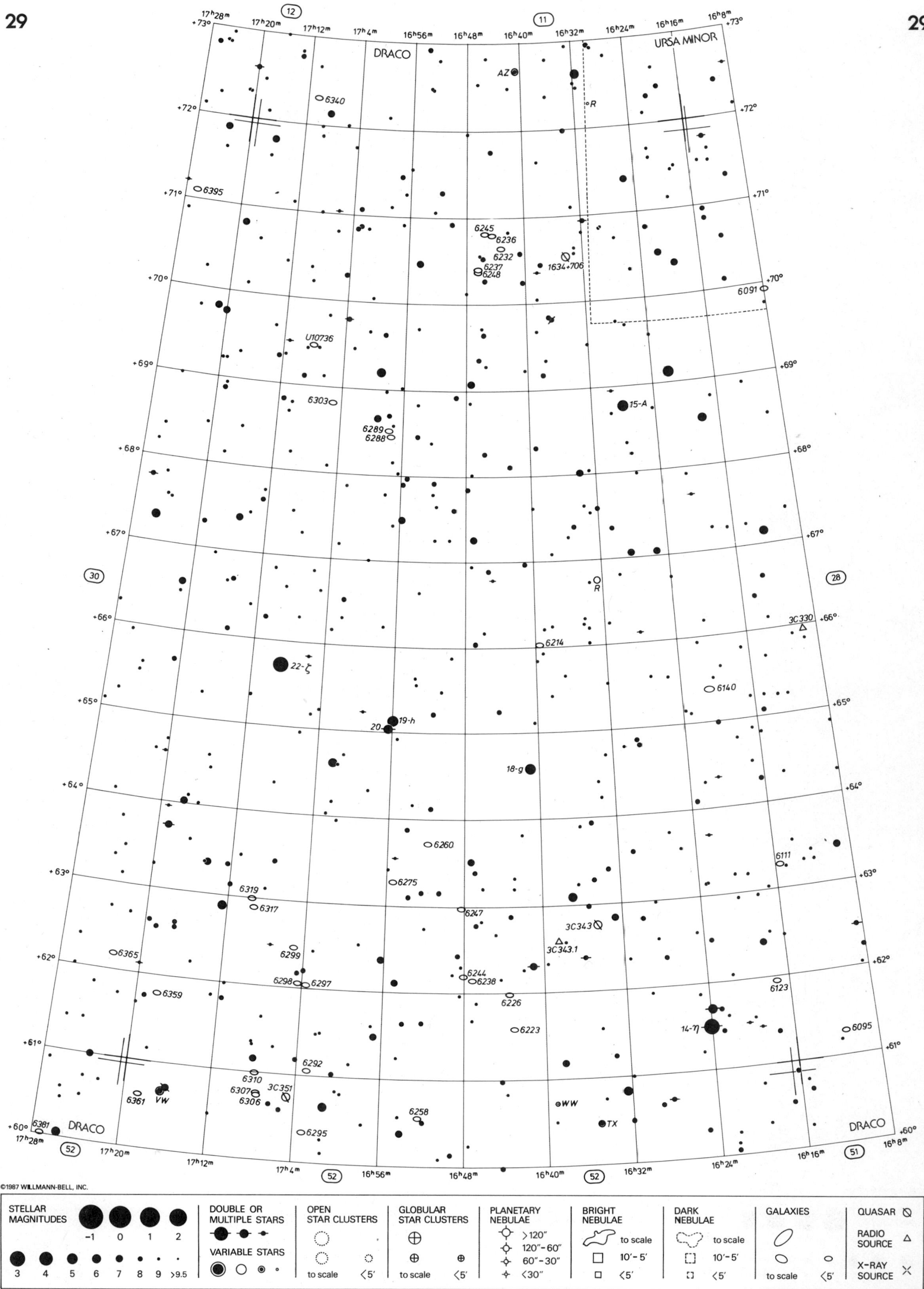

Barry Rappaport & Wil Tirion

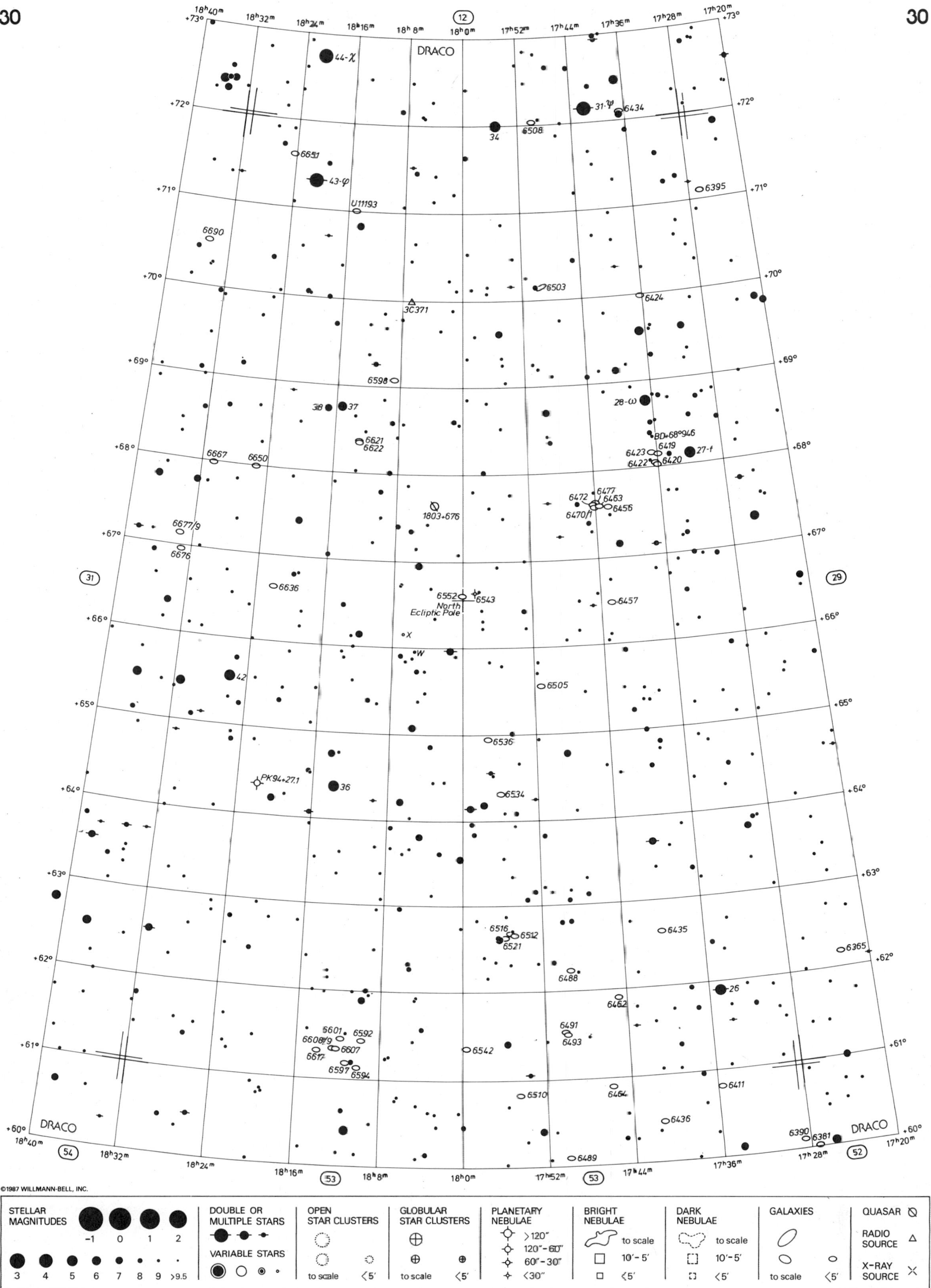
DRACO
44-χ
43-φ
31-ψ
34
28-ω
27-f
42
36
26
38
37
North Ecliptic Pole
3C371
1803+676
U11193
PK94+27.1
BD+68°946
X
W
6651
6690
6667
6650
6677/9
6676
6636
6598
6621
6622
6503
6508
6434
6395
6424
6419
6420
6423
6422
6472
6477
6463
6456
6470/1
6457
6552
6543
6505
6536
6534
6516
6512
6521
6435
6365
6488
6462
6491
6493
6542
6601
6592
6608/9
6607
6617
6597
6594
6510
6464
6411
6436
6390
6381
6489
18h40m
18h32m
18h24m
18h16m
18h8m
18h0m
17h52m
17h44m
17h36m
17h28m
17h20m
+73°
+72°
+71°
+70°
+69°
+68°
+67°
+66°
+65°
+64°
+63°
+62°
+61°
+60°
12
31
29
54
53
52
STELLAR MAGNITUDES
-1 0 1 2
3 4 5 6 7 8 9 >9.5
DOUBLE OR MULTIPLE STARS
VARIABLE STARS
OPEN STAR CLUSTERS
to scale <5'
GLOBULAR STAR CLUSTERS
to scale <5'
PLANETARY NEBULAE
>120"
120"-60"
60"-30"
<30"
BRIGHT NEBULAE
to scale
10'-5'
<5'
DARK NEBULAE
to scale
10'-5'
<5'
GALAXIES
to scale <5'
QUASAR
RADIO SOURCE
X-RAY SOURCE
Barry Rappaport & Wil Tirion

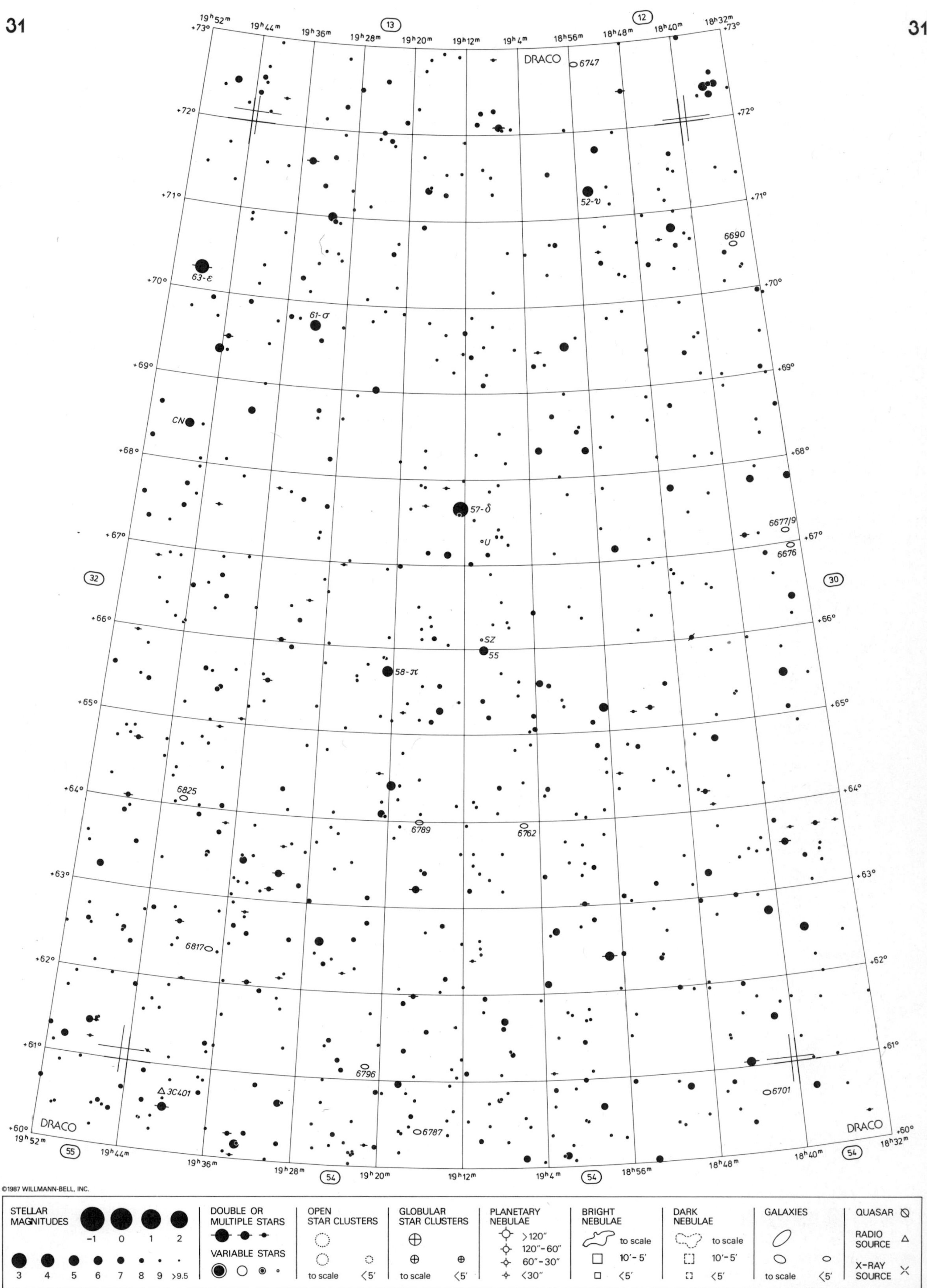

Barry Rappaport & Wil Tirion

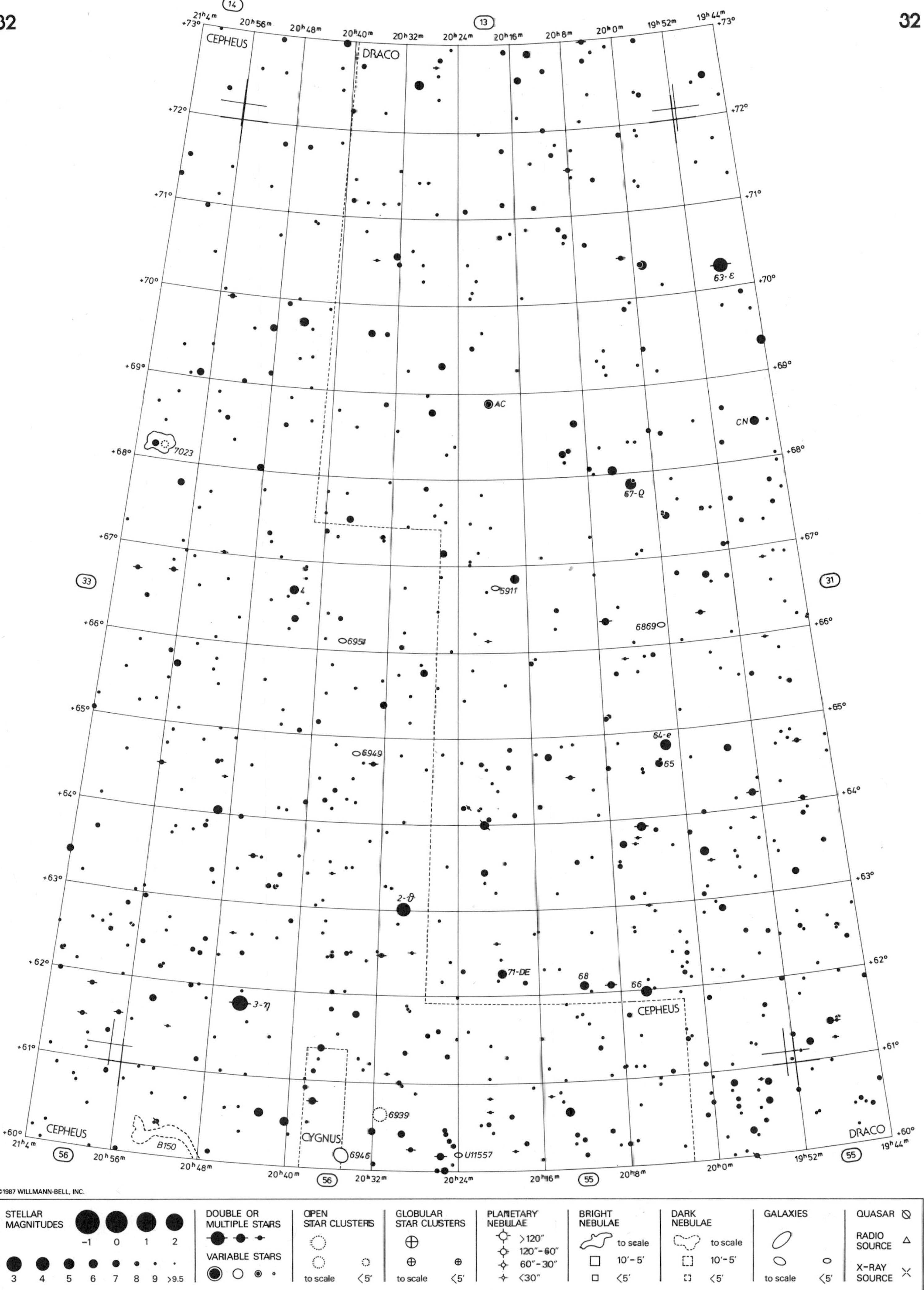

Barry Rappaport & Wil Tirion

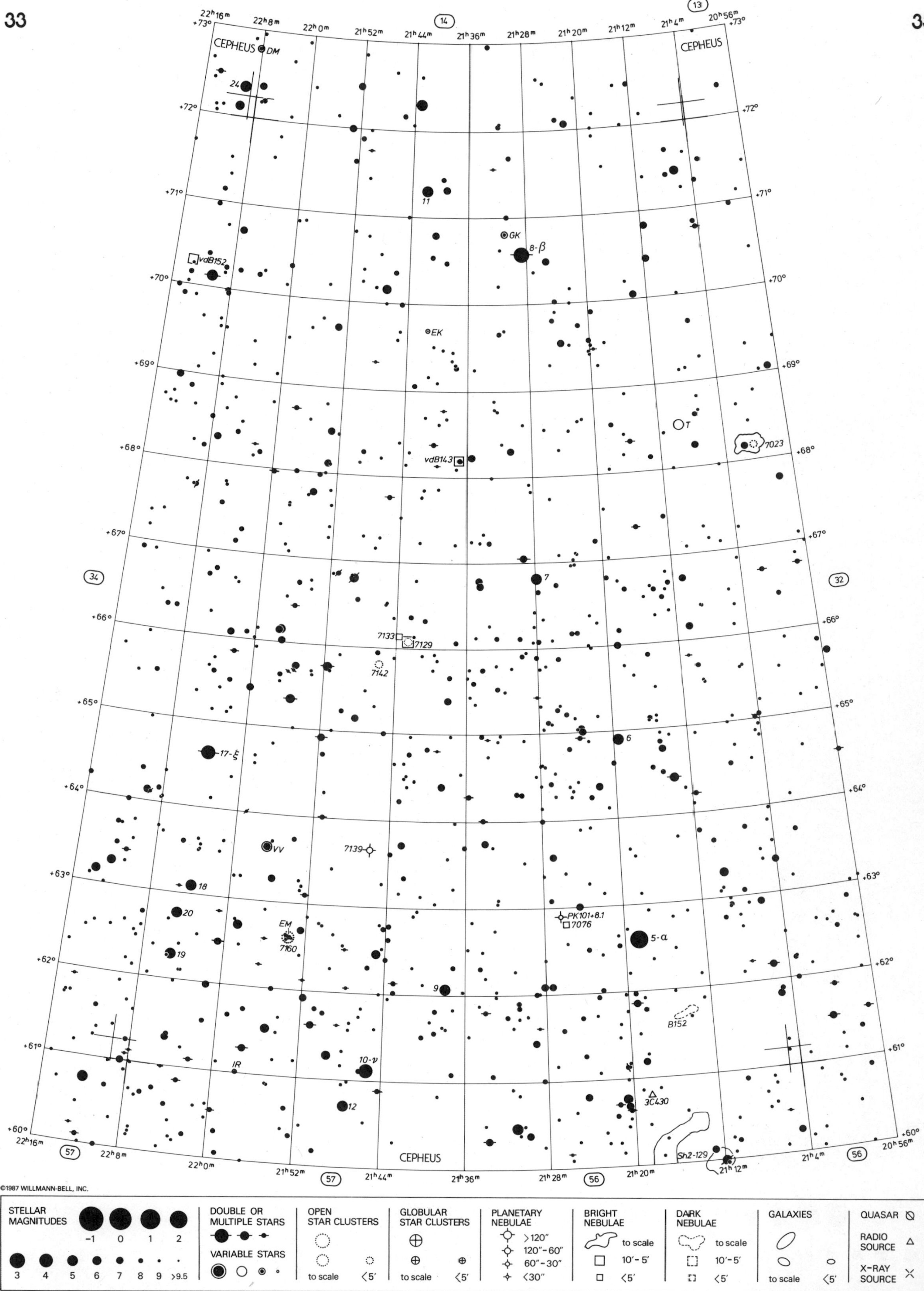
CEPHEUS
CEPHEUS
CEPHEUS
DM
24
11
GK
8-β
vdB152
EK
T
7023
vdB143
7
7133
7129
7142
6
17-ξ
VV
7139
18
20
EM
7160
19
PK101+8.1
7076
5-α
9
B152
IR
10-ν
12
3C430
Sh2-129
STELLAR MAGNITUDES
-1 0 1 2
3 4 5 6 7 8 9 >9.5
DOUBLE OR MULTIPLE STARS
VARIABLE STARS
OPEN STAR CLUSTERS
to scale <5′
GLOBULAR STAR CLUSTERS
to scale <5′
PLANETARY NEBULAE
>120″
120″-60″
60″-30″
<30″
BRIGHT NEBULAE
to scale
10′-5′
<5′
DARK NEBULAE
to scale
10′-5′
<5′
GALAXIES
to scale <5′
QUASAR
RADIO SOURCE
X-RAY SOURCE
Barry Rappaport & Wil Tirion

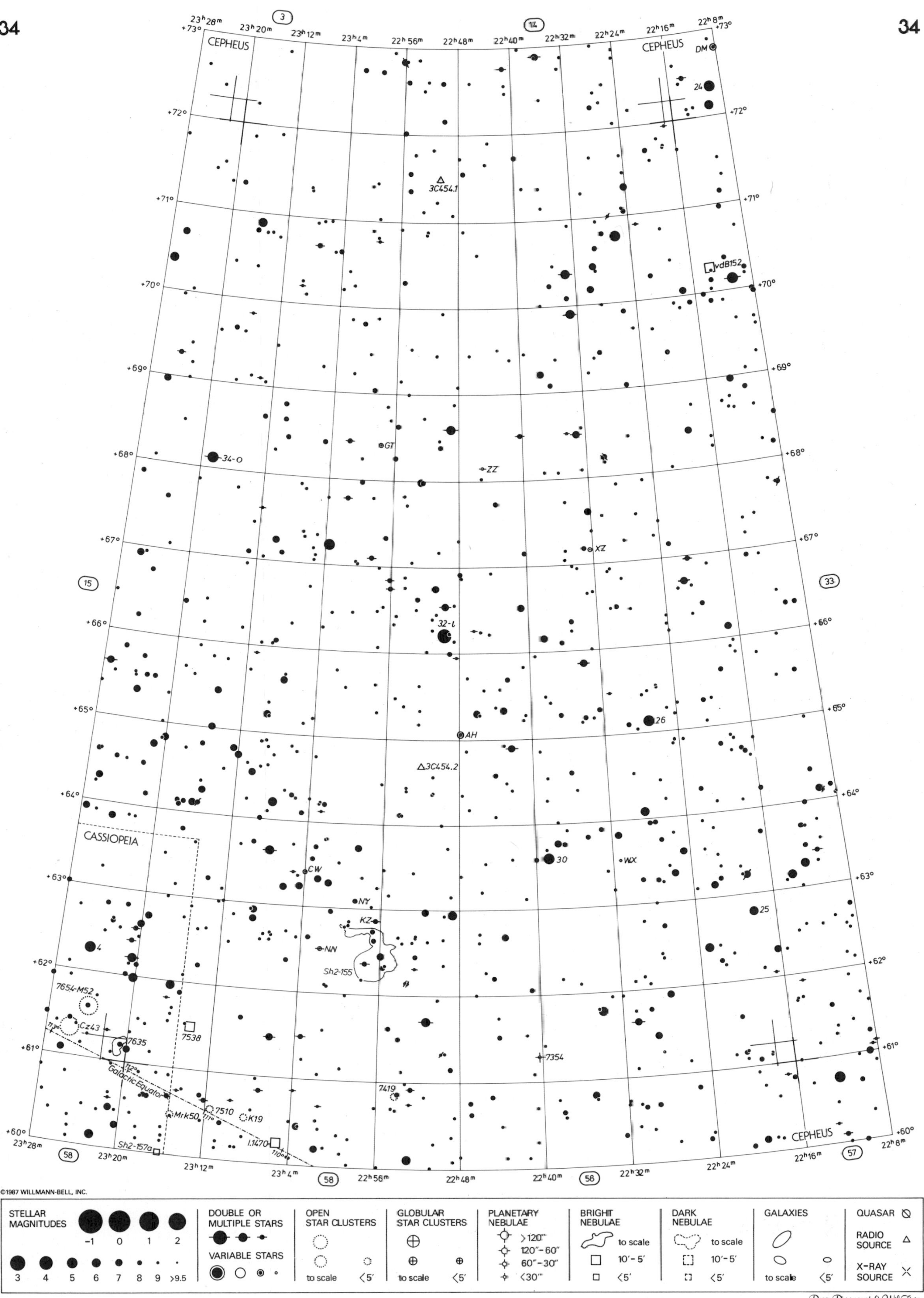
CEPHEUS
CASSIOPEIA
3C454.1
vdB152
34-O
GT
ZZ
XZ
32-ι
AH
3C454.2
26
30
WX
CW
NY
KZ
NN
Sh2-155
25
7654-M52
Cz43
7635
7538
7354
7419
7510
K19
Mrk50
Galactic Equator
I.1470
Sh2-157a
©1987 WILLMANN-BELL, INC.
STELLAR MAGNITUDES
DOUBLE OR MULTIPLE STARS
VARIABLE STARS
OPEN STAR CLUSTERS
GLOBULAR STAR CLUSTERS
PLANETARY NEBULAE
BRIGHT NEBULAE
DARK NEBULAE
GALAXIES
QUASAR
RADIO SOURCE
X-RAY SOURCE
to scale
Barry Rappaport & Wil Tirion

0h32m 0h24m 0h16m 0h8m 0h0m 23h52m 23h44m 23h36m 23h28m

(15)

+62° +61° +60° +59° +58° +57° +56° +55° +54° +53° +52° +51° +50° +49°

CASSIOPEIA 12 · 136 · 103 · Fr1 · K12 · H21 · 7788 · 7790 · Be58 · TZ · 115° · Galactic · 114° · Equator · CASSIOPEIA

DL · 129 · WZ · I.10 · TV · 11-β · vdB1 · RY · 5-τ · AR · V436 · 7-ρ · V373 · PK114-4.1 · 18-α · 7789 · Zo · FM · PK119-6.1 · T · St19 · Y · 8-σ · St11 · EQ · SX · 14-λ · PK113-6.1 · 17-ζ · RR · St12 · SV · TU · AO · SS · R · PK112-10.1 · 18 · 7686 · CASSIOPEIA · ANDROMEDA

(36) (58) (60) (59) (59) (88)

0h32m 0h24m 0h16m 0h8m 0h0m 23h52m 23h44m 23h36m 23h28m

STELLAR MAGNITUDES	DOUBLE OR MULTIPLE STARS	OPEN STAR CLUSTERS	GLOBULAR STAR CLUSTERS	PLANETARY NEBULAE	BRIGHT NEBULAE	DARK NEBULAE	GALAXIES	QUASAR
−1 0 1 2	VARIABLE STARS	to scale <5′	to scale <5′	>120″ 120″–60″ 60″–30″ <30″	to scale 10′–5′ <5′	to scale 10′–5′ <5′	to scale <5′	RADIO SOURCE
3 4 5 6 7 8 9 >9.5								X-RAY SOURCE

Barry Rappaport & Wil Tirion

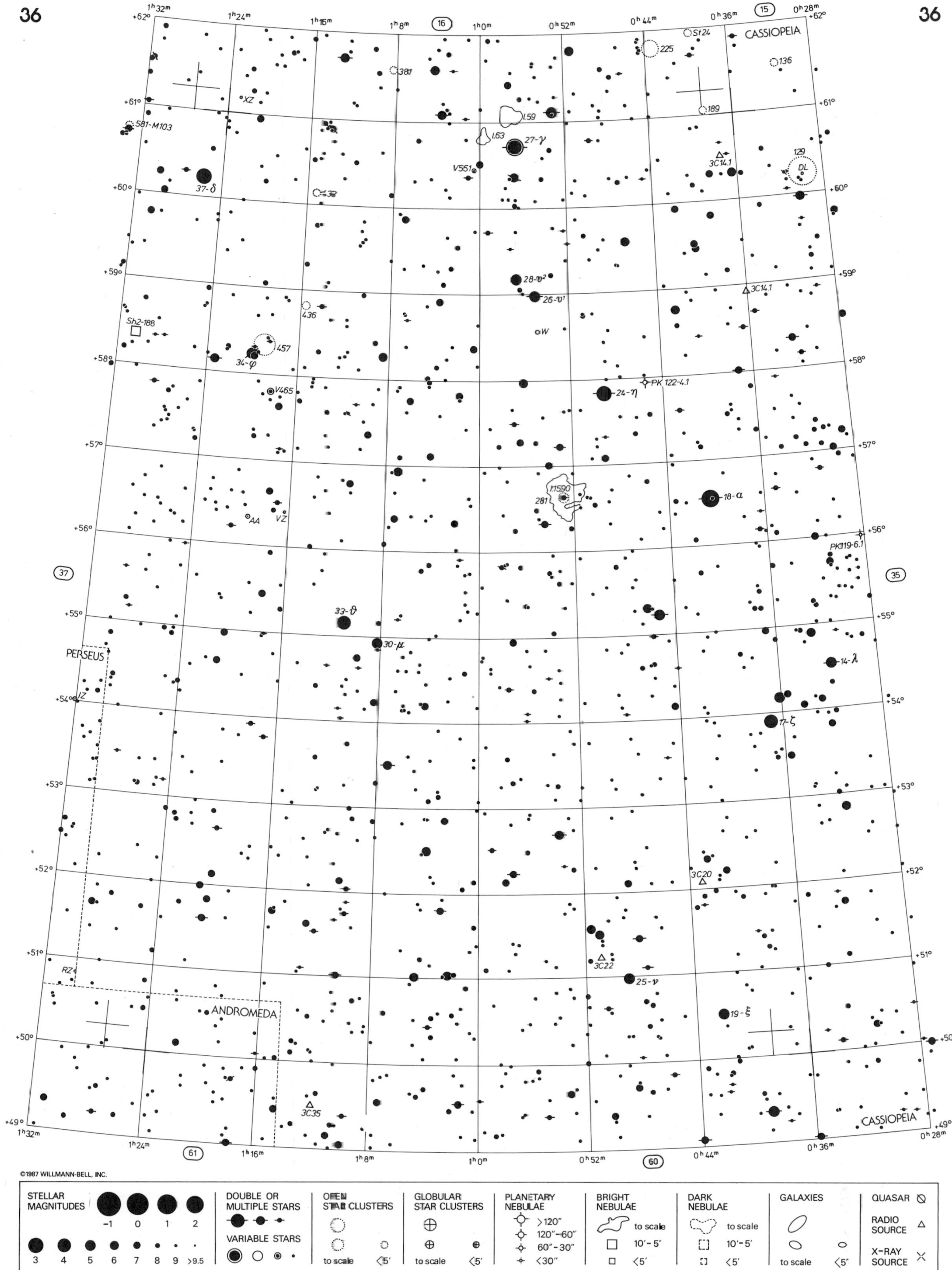

Barry Rappaport & Wil Tirion

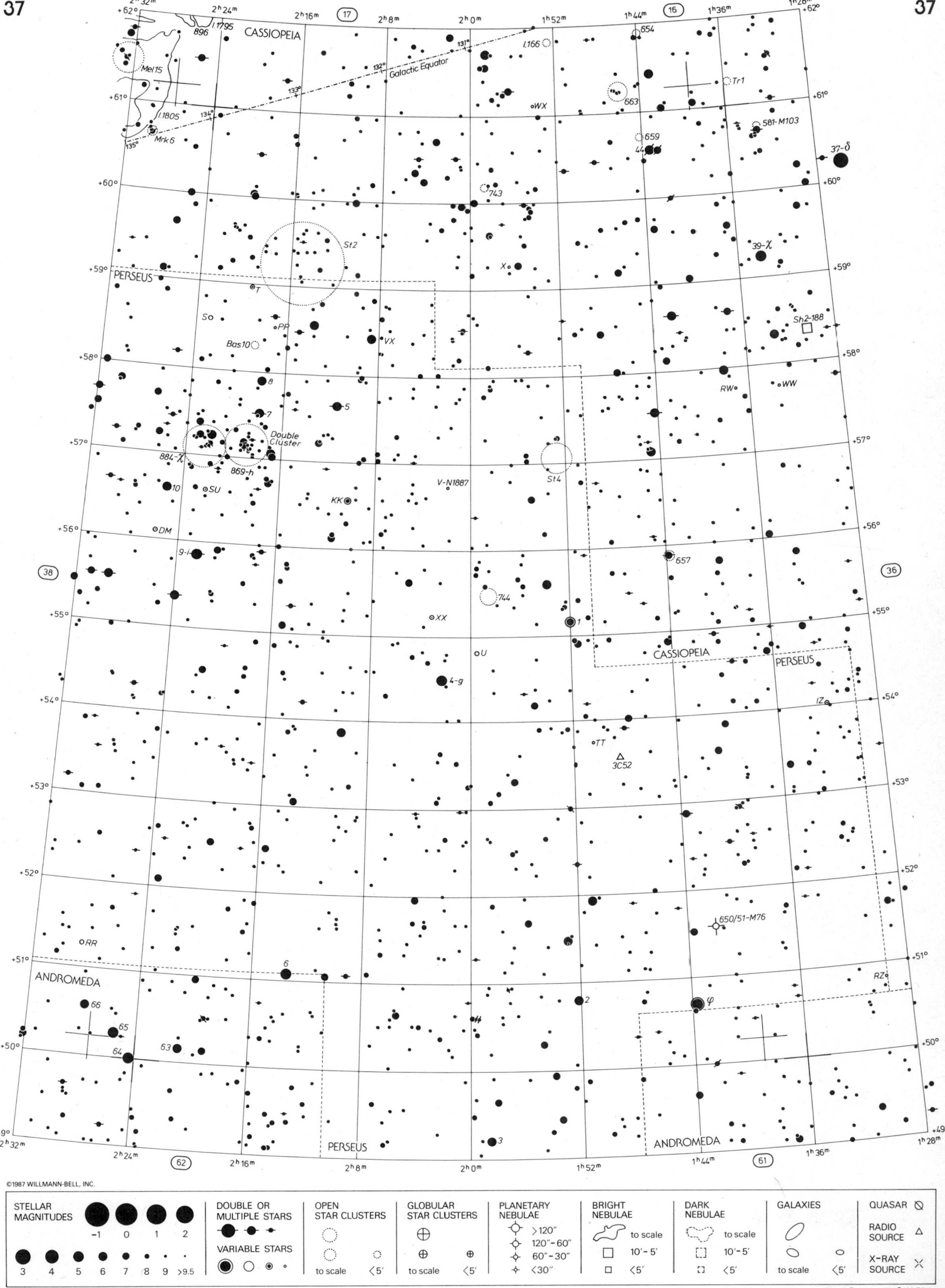

Barry Rappaport & Wil Tirion

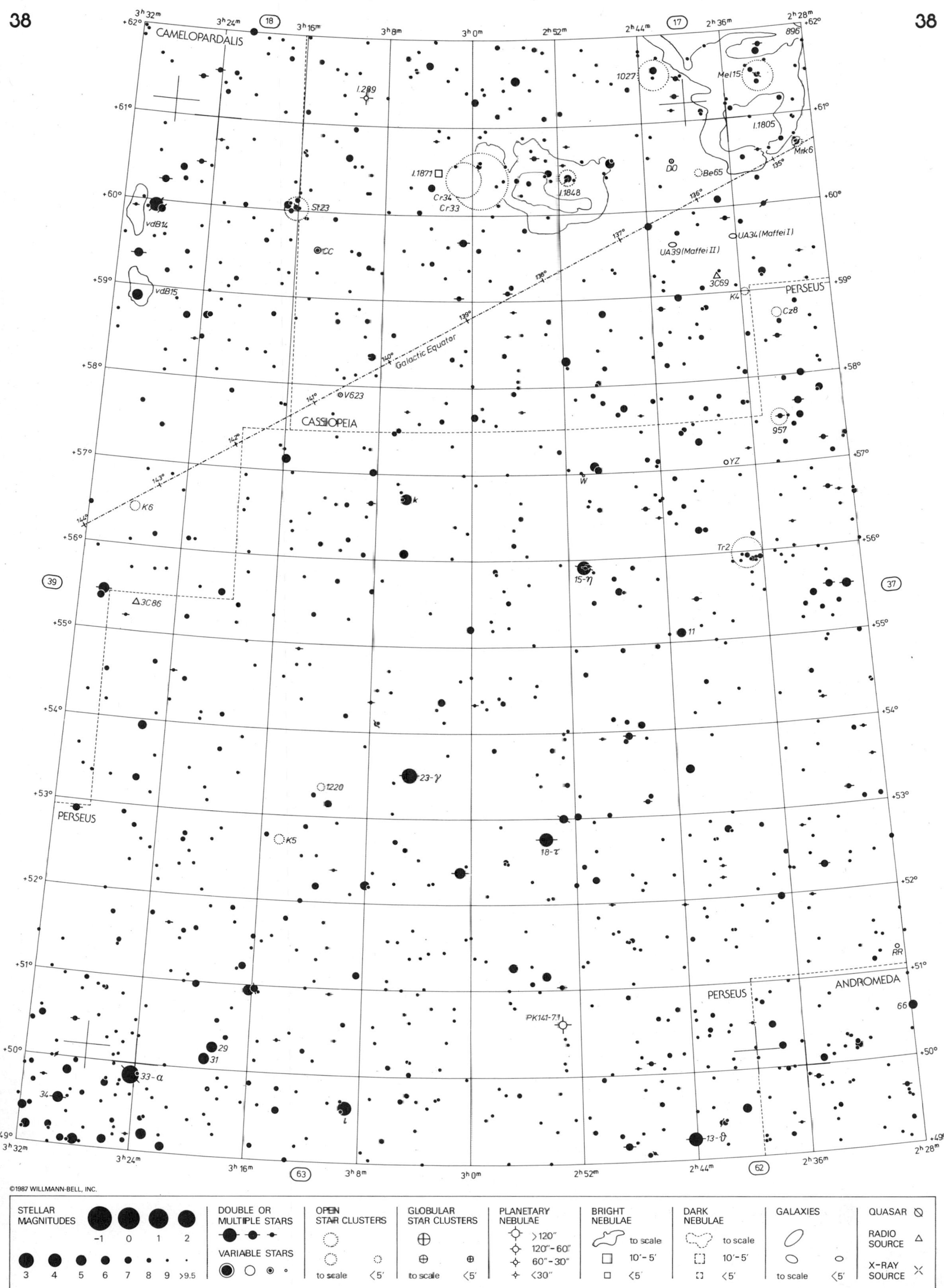

Barry Rappaport & Wil Tirion

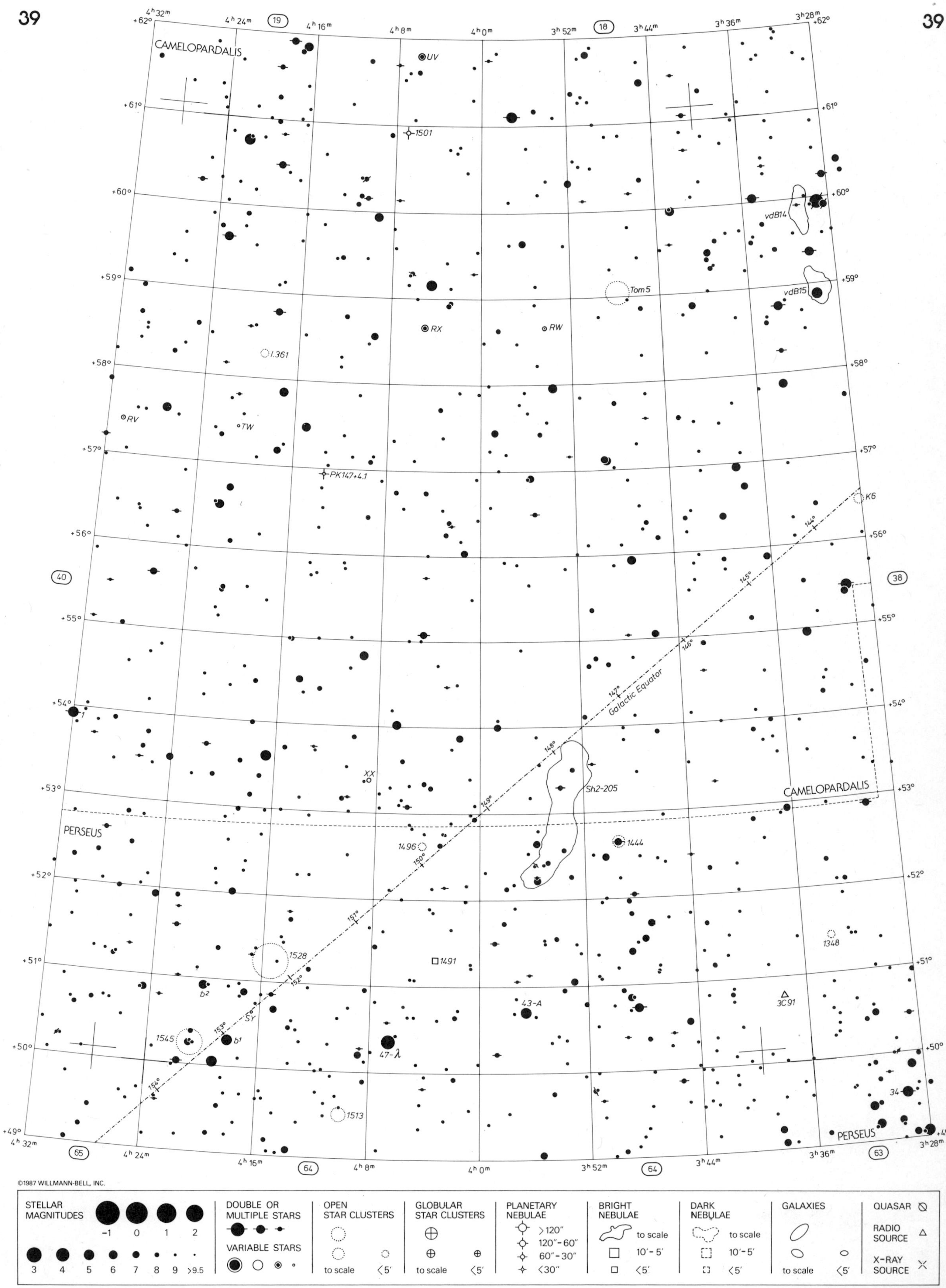

Barry Rappaport & Wil Tirion

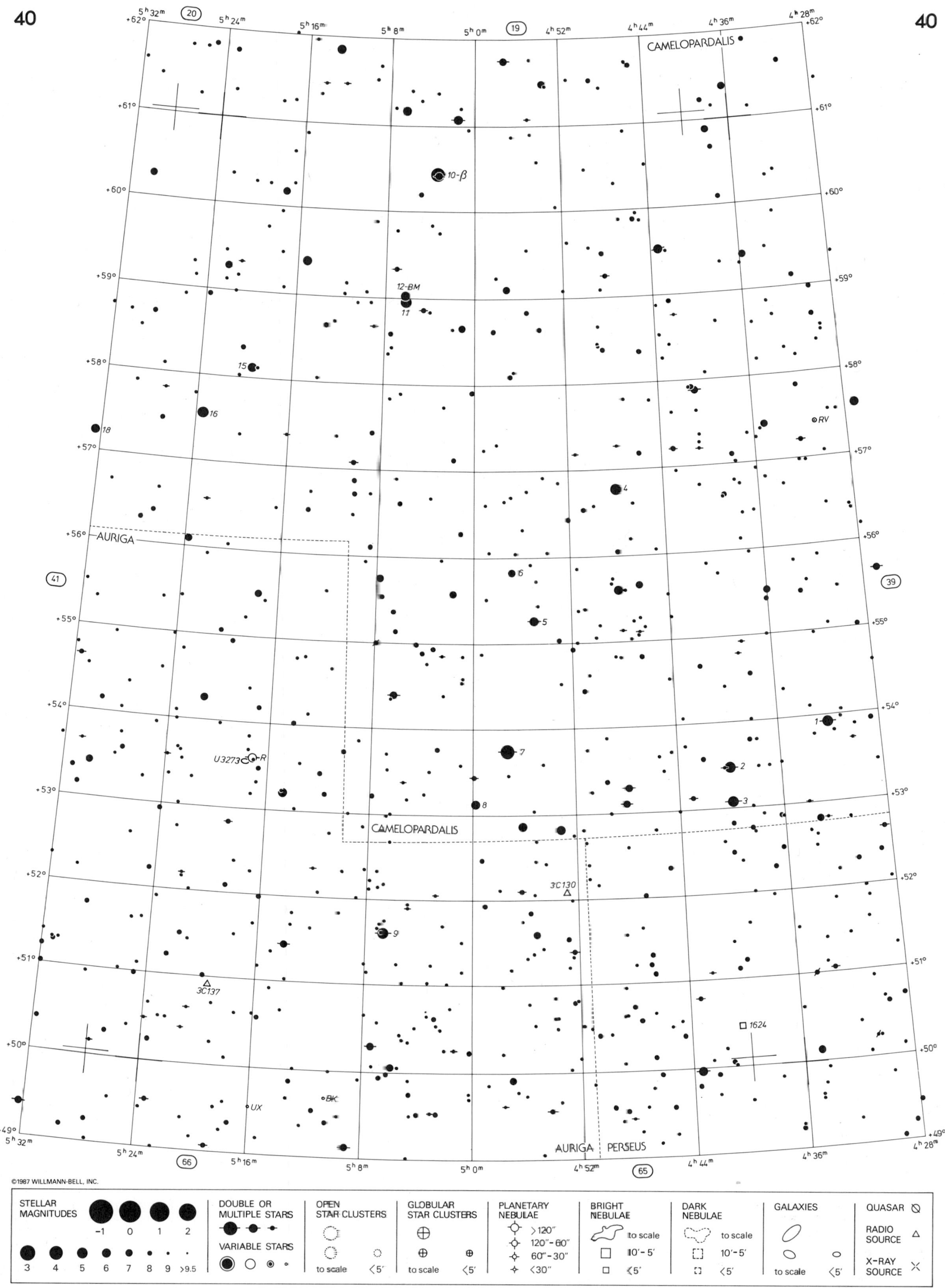

Barry Rappaport & Wil Tirion

41

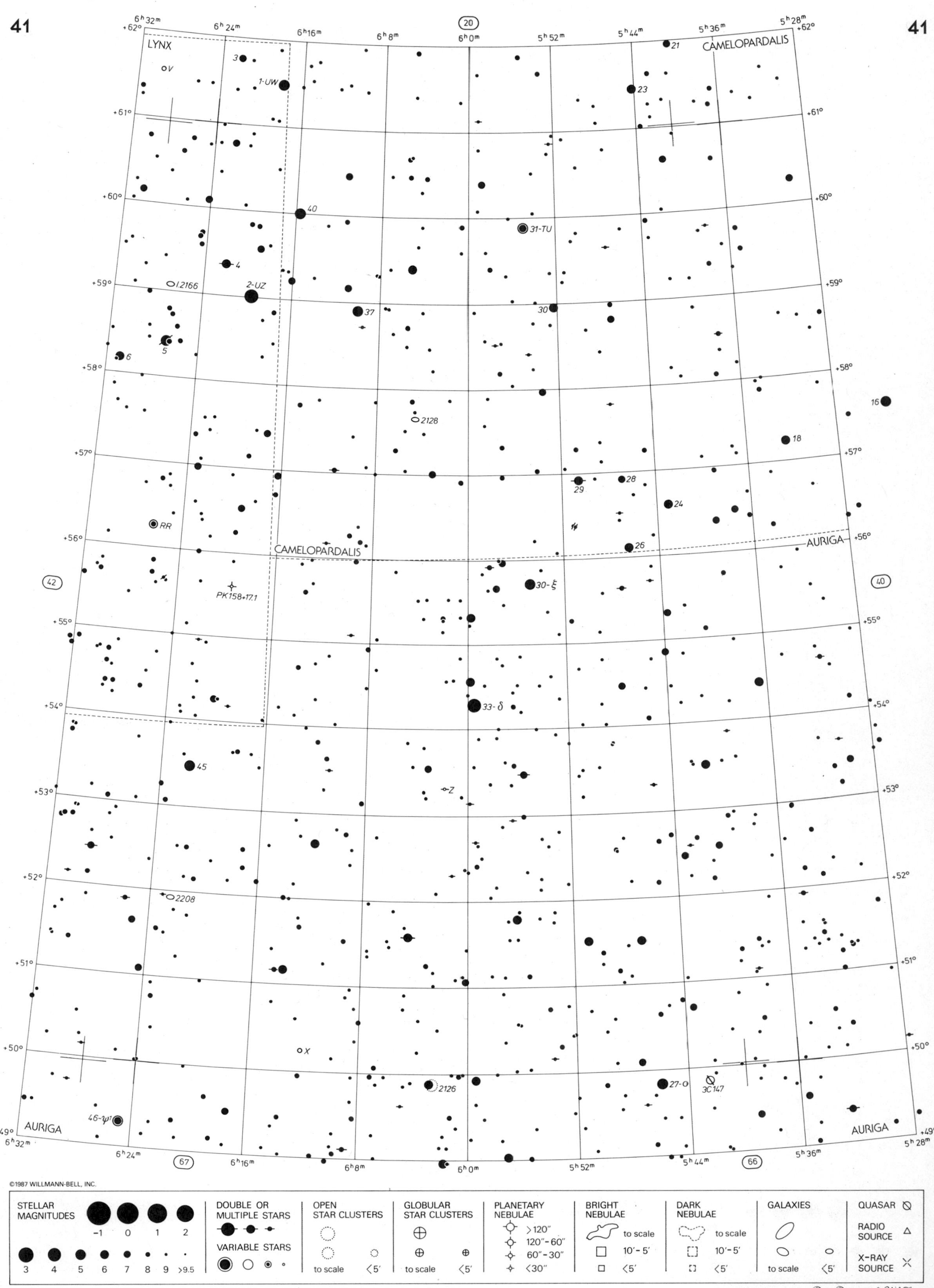

Barry Rappaport & Wil Tirion

7h32m +62° 7h24m 7h16m 7h8m (21) 7h0m 6h52m 6h44m 6h36m (20) 6h28m +62°

LYNX

CAMELOPARDALIS

LYNX

U3826 U3685 41 PK153+22.1 V 10 8 17 2273 U3598 2273B 2273A U 47 18 14 12 5 15 6 S U3574 13 11 U3647 9 7 19 R 3C171 UY 2326 2326A 2321 2320 2322 2315 2330 2340 2332 20 22 21

+61° +60° +59° +58° +57° +56° +55° +54° +53° +52° +51° +50° +49°

(43) (41)

AURIGA

LYNX

7h32m 7h24m 7h16m (68) 7h8m 7h0m 6h52m 6h44m (67) 6h36m 6h28m

STELLAR MAGNITUDES -1 0 1 2 3 4 5 6 7 8 9 >9.5

DOUBLE OR MULTIPLE STARS

VARIABLE STARS

OPEN STAR CLUSTERS to scale <5'

GLOBULAR STAR CLUSTERS to scale <5'

PLANETARY NEBULAE >120" 120"-60" 60"-30" <30"

BRIGHT NEBULAE to scale 10'-5' <5'

DARK NEBULAE to scale 10'-5' <5'

GALAXIES to scale <5'

QUASAR

RADIO SOURCE

X-RAY SOURCE

Barry Rappaport & Wil Tirion

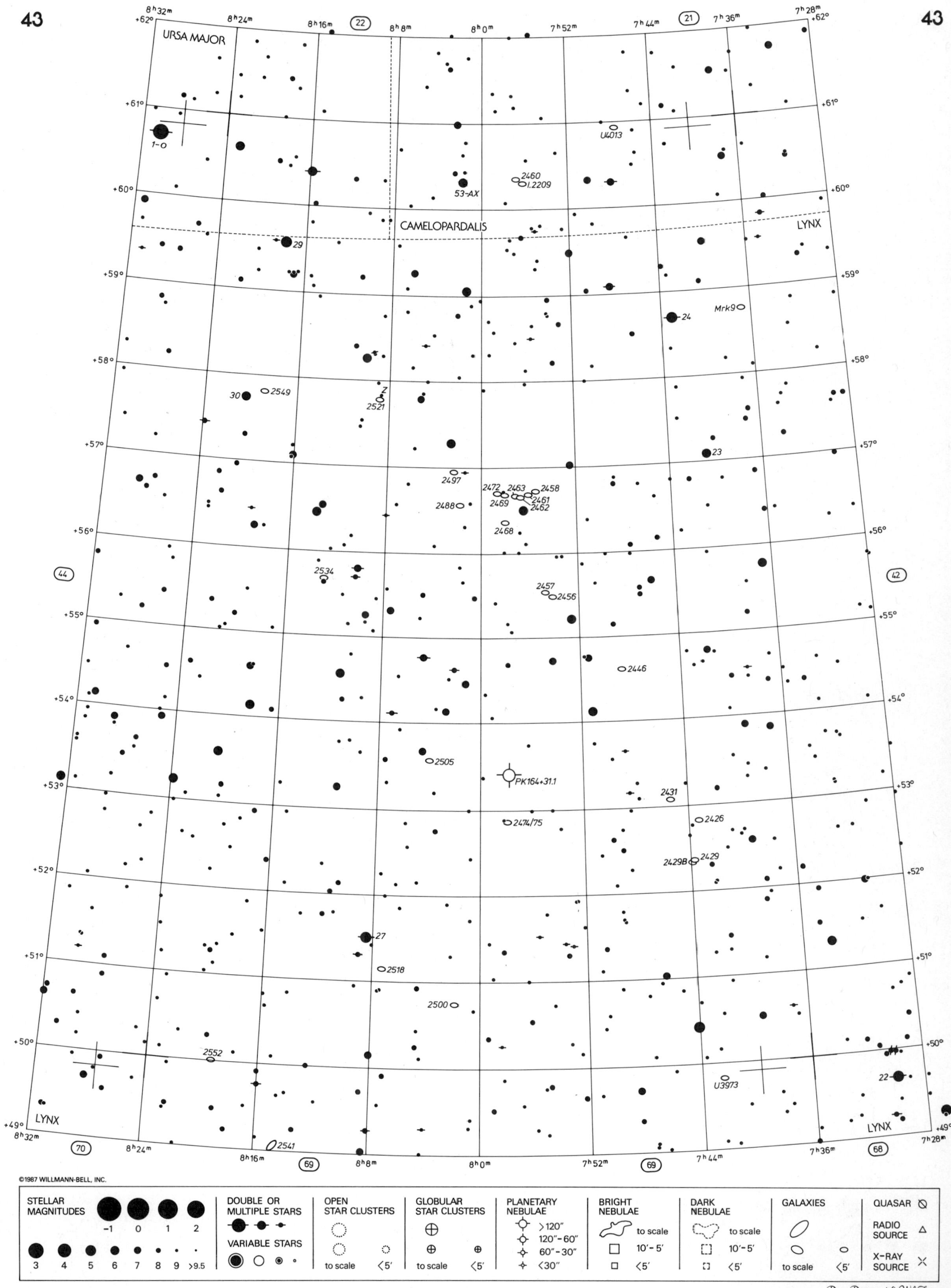
URSA MAJOR
CAMELOPARDALIS
LYNX
LYNX
LYNX
LYNX
8h32m
8h24m
8h16m
8h8m
8h0m
7h52m
7h44m
7h36m
7h28m
+62°
+61°
+60°
+59°
+58°
+57°
+56°
+55°
+54°
+53°
+52°
+51°
+50°
+49°
22
21
44
42
70
69
69
68
1-O
53-AX
2460
I.2209
U4013
29
24
Mrk9
30
2549
Z
2521
23
2497
2472
2463
2458
2469
2461
2462
2488
2468
2534
2457
2456
2446
2505
PK164+31.1
2431
2474/75
2426
2429B
2429
27
2518
2500
2552
U3973
2541
STELLAR MAGNITUDES
DOUBLE OR MULTIPLE STARS
VARIABLE STARS
OPEN STAR CLUSTERS
GLOBULAR STAR CLUSTERS
PLANETARY NEBULAE
BRIGHT NEBULAE
DARK NEBULAE
GALAXIES
QUASAR
RADIO SOURCE
X-RAY SOURCE
to scale
<5'
>120"
120"-60"
60"-30"
<30"
10'-5'
Barry Rappaport & Wil Tirion

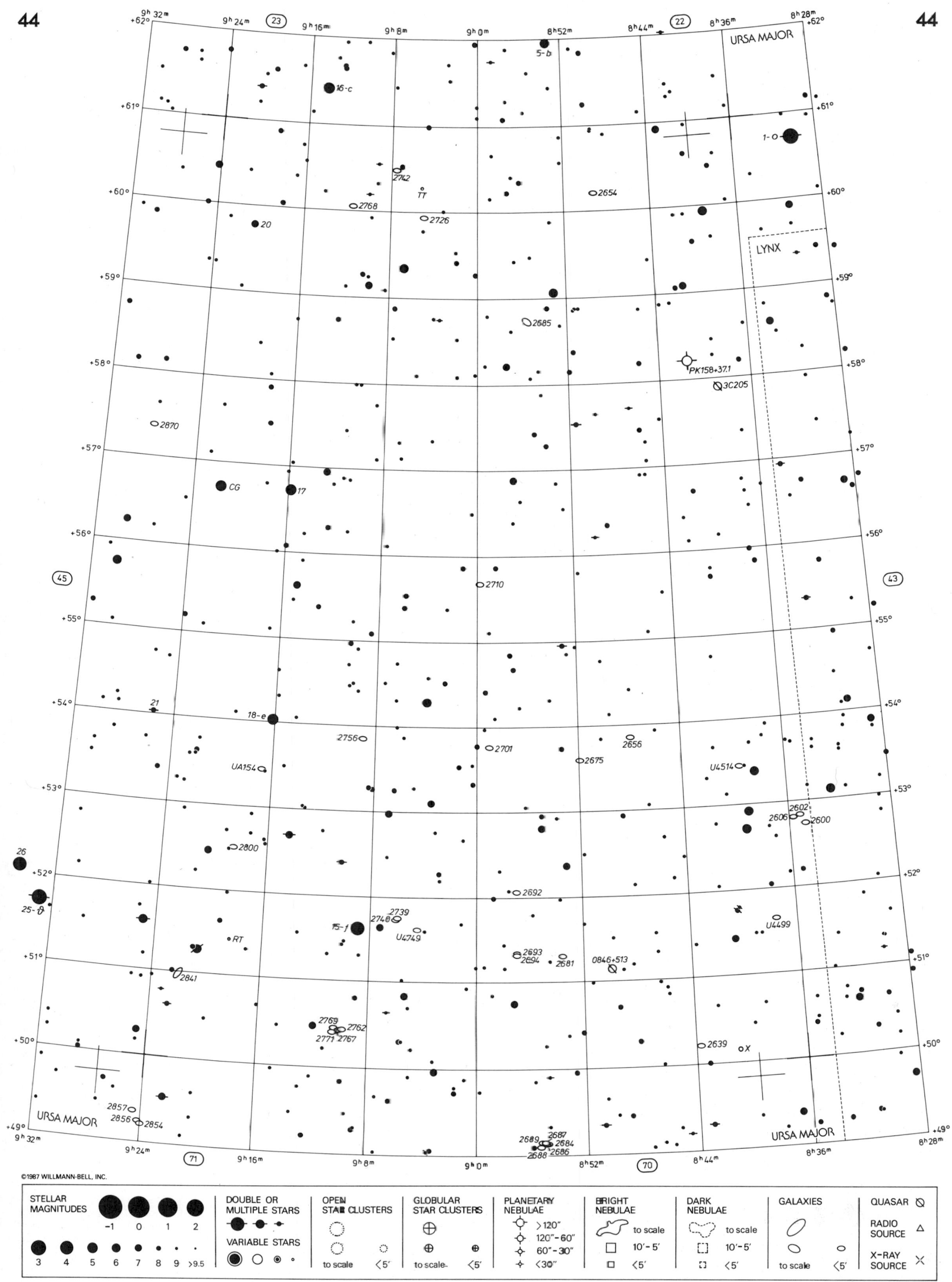

Barry Rappaport & Wil Tirion

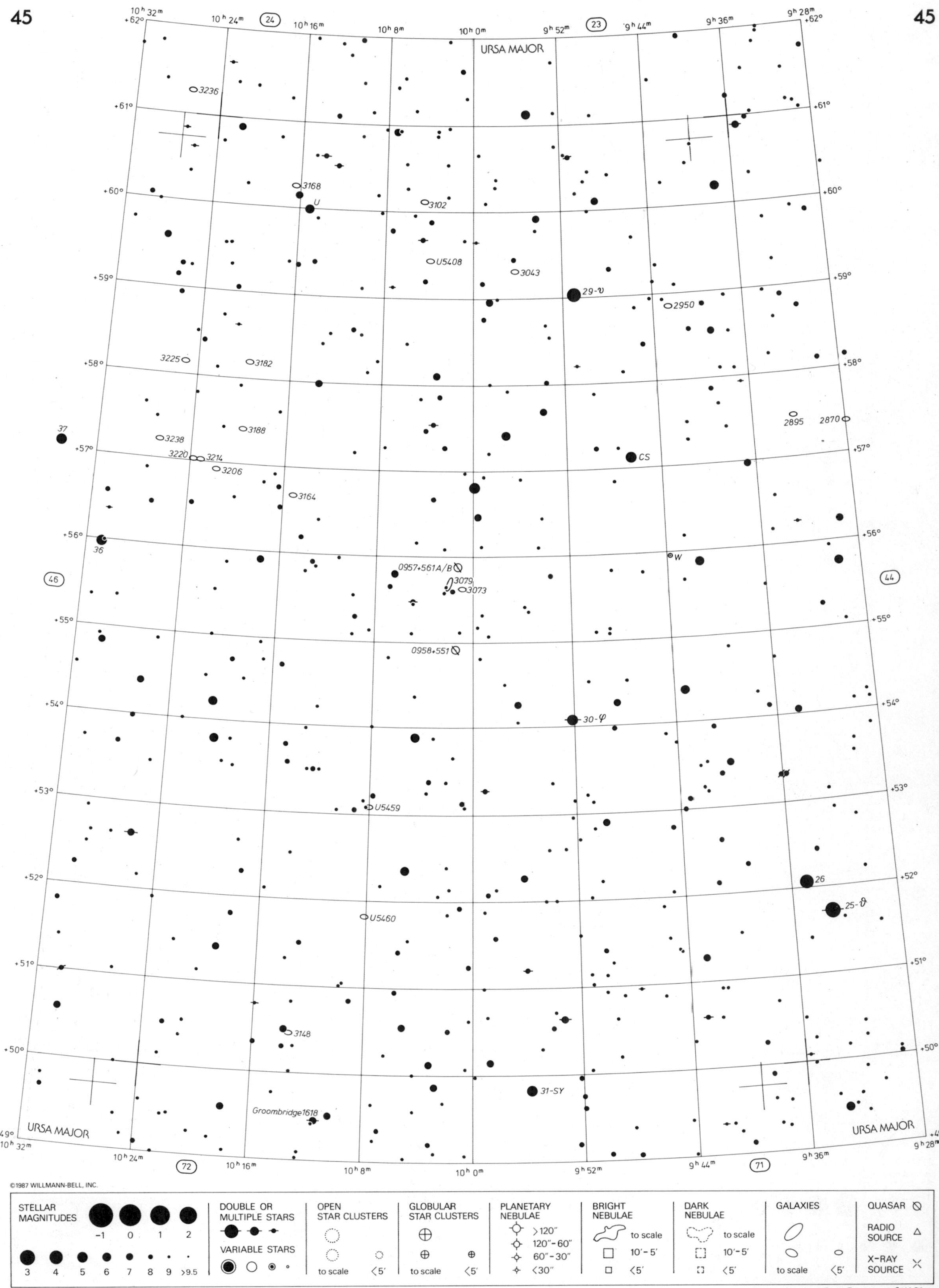

Barry Rappaport & Wil Tirion

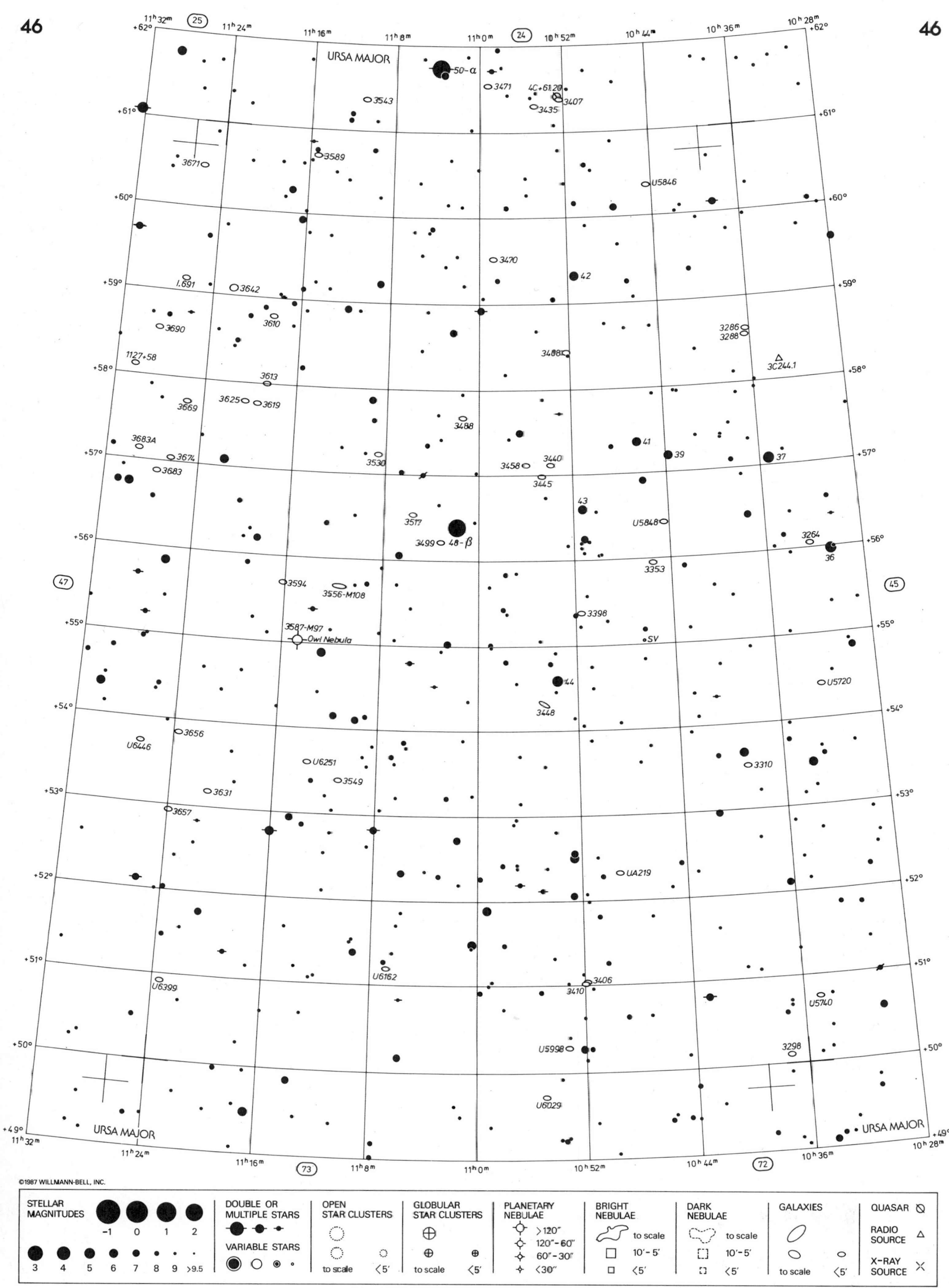

Barry Rappaport & Wil Tirion

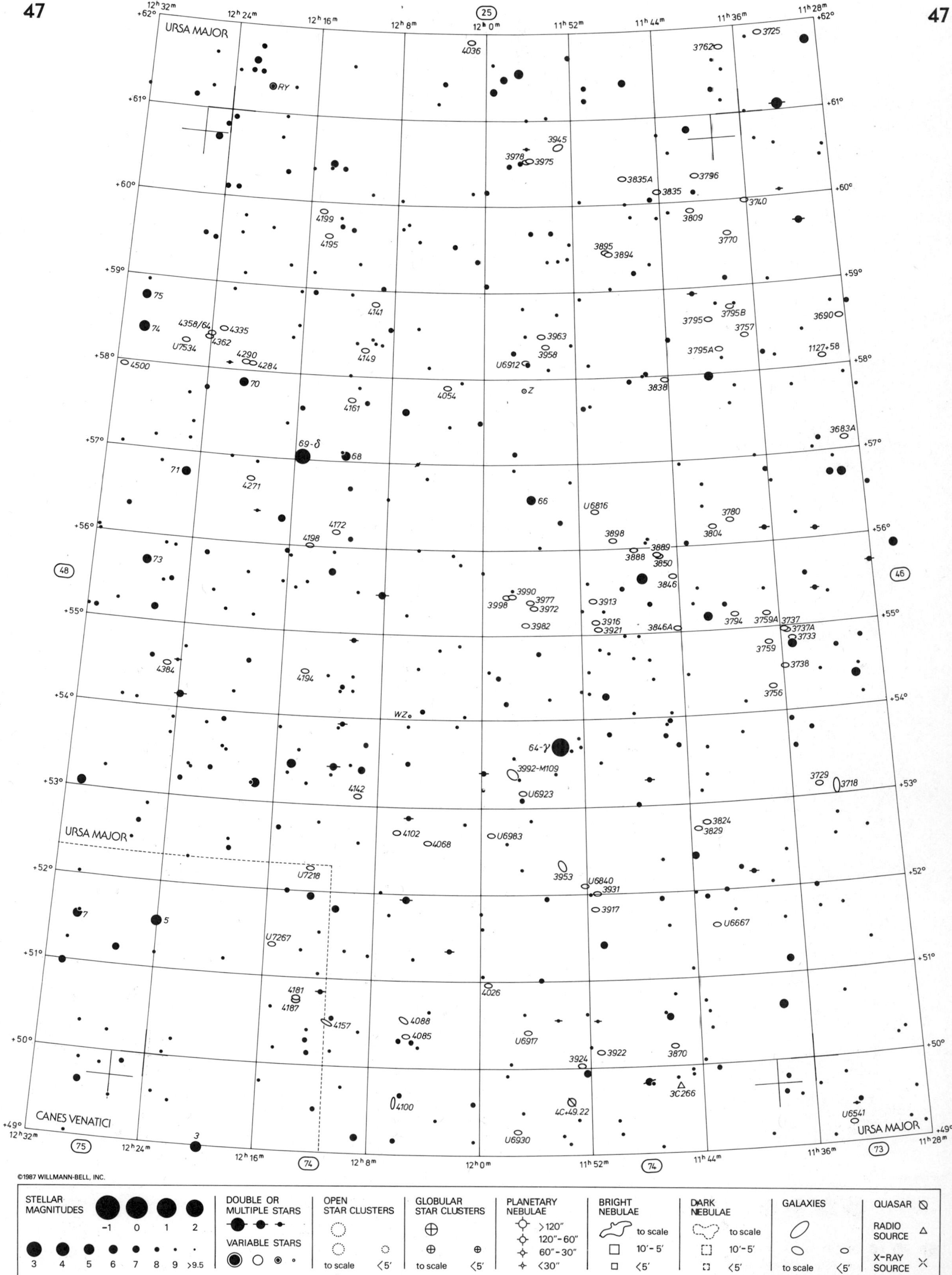

Barry Rappaport & Wil Tirion

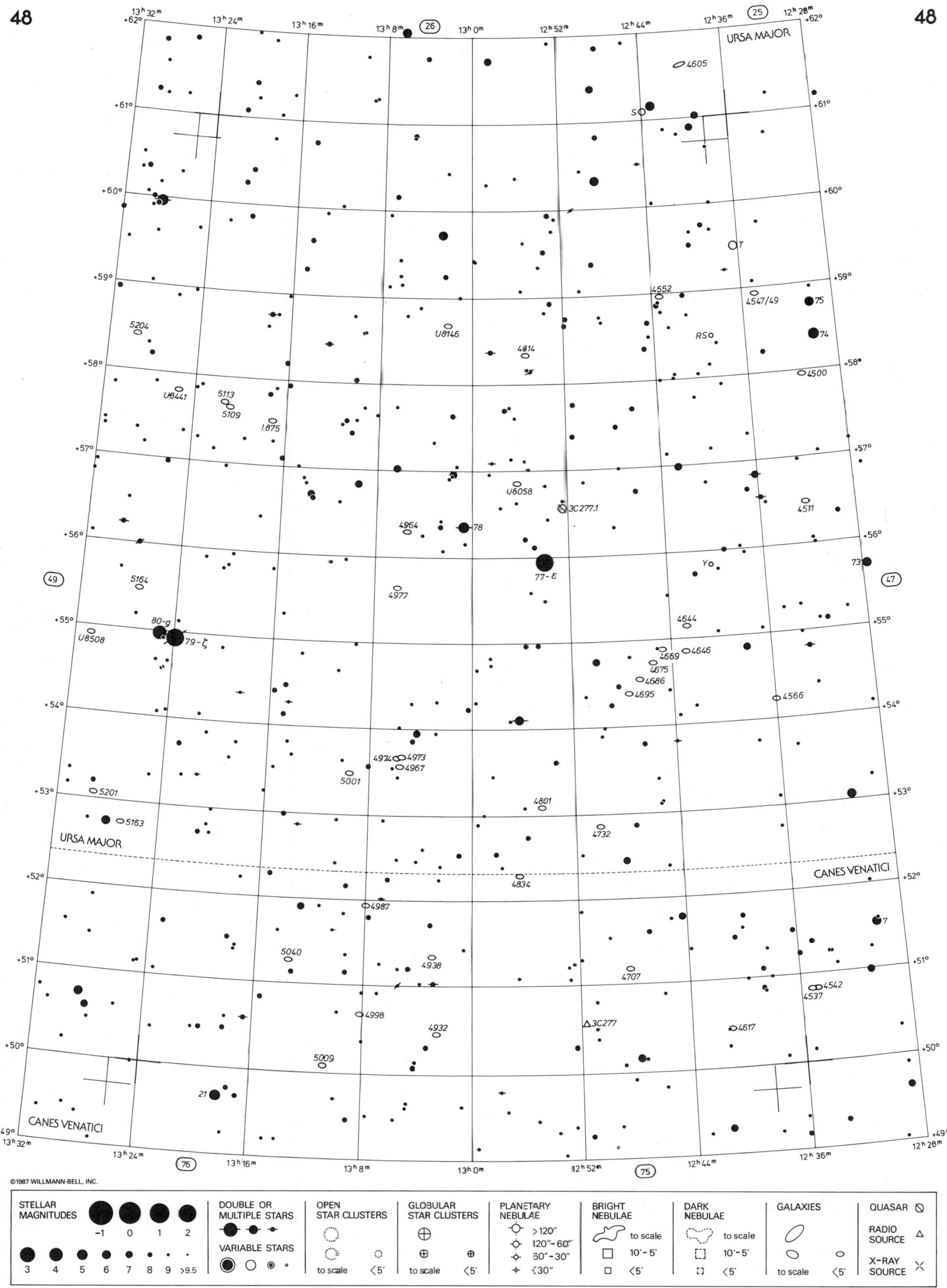

Barry Rappaport & Wil Tirion

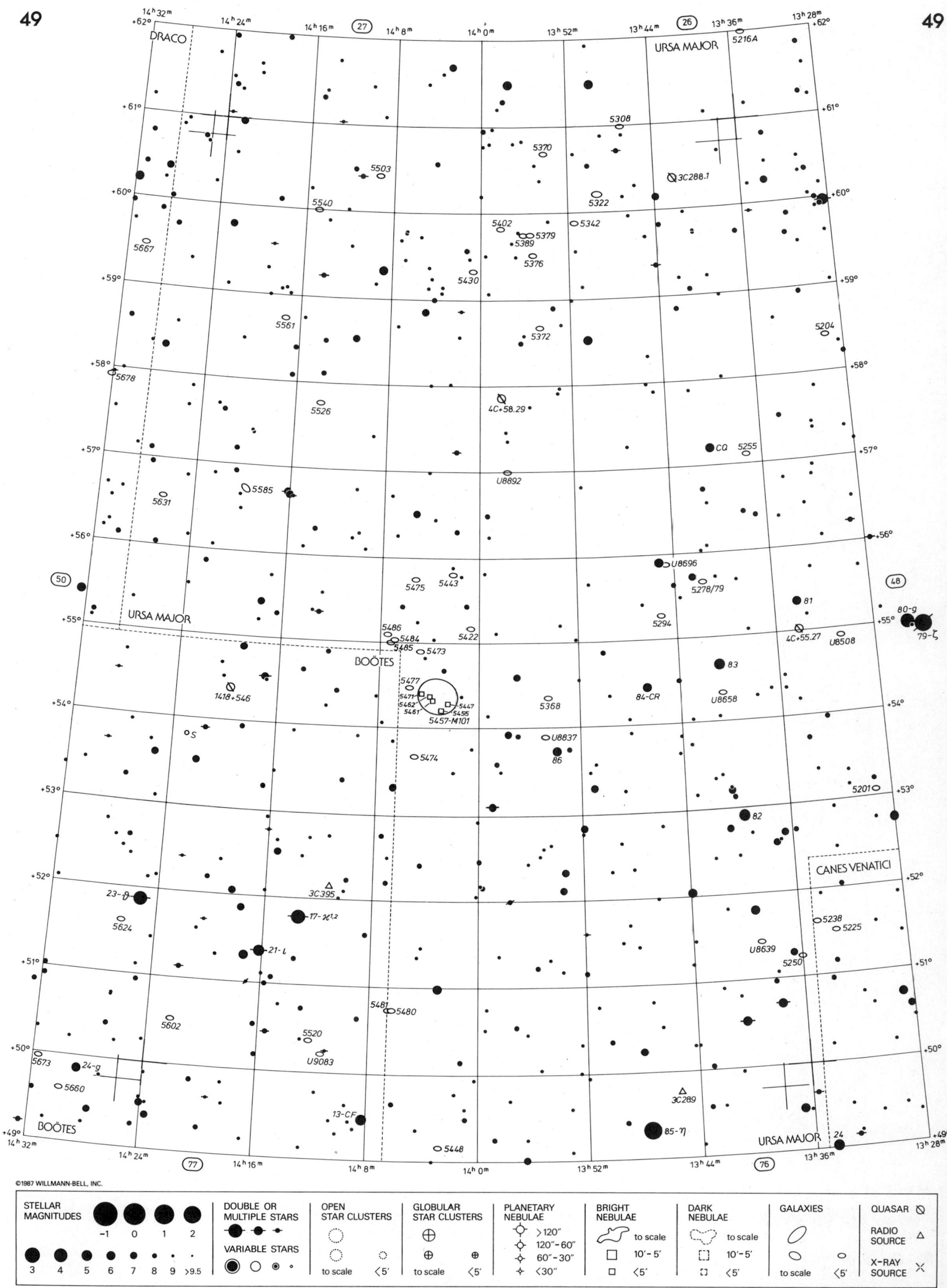

14h 32m
14h 24m
14h 16m
14h 8m
14h 0m
13h 52m
13h 44m
13h 36m
13h 28m
27
26
50
48
77
76
+62°
+61°
+60°
+59°
+58°
+57°
+56°
+55°
+54°
+53°
+52°
+51°
+50°
+49°
DRACO
URSA MAJOR
BOÖTES
CANES VENATICI
5216A
5308
5370
3C288.1
5503
5322
5540
5402
5342
5379
5389
5667
5376
5430
5561
5372
5204
5678
4C+58.29
5526
CQ
5255
U8892
5585
5631
U8696
5475
5443
5278/79
81
80-g
79-ζ
5294
4C+55.27
U8508
5486
5484
5485
5422
5473
83
5477
1418+546
84-CR
U8658
5368
5471
5462
5461
5447
5455
5457-M101
S
U8837
5474
86
5201
82
3C395
23-θ
17-κ1,2
5238
5225
5624
21-ι
U8639
5250
5481
5480
5602
5520
U9083
5673
24-g
5660
3C289
13-CF
85-η
BOÖTES
URSA MAJOR
24
5448
STELLAR MAGNITUDES
-1
0
1
2
3
4
5
6
7
8
9
>9.5
DOUBLE OR MULTIPLE STARS
VARIABLE STARS
OPEN STAR CLUSTERS
to scale
<5′
GLOBULAR STAR CLUSTERS
to scale
<5′
PLANETARY NEBULAE
>120″
120″-60″
60″-30″
<30″
BRIGHT NEBULAE
to scale
10′-5′
<5′
DARK NEBULAE
to scale
10′-5′
<5′
GALAXIES
to scale
<5′
QUASAR
RADIO SOURCE
X-RAY SOURCE
Barry Rappaport & Wil Tirion

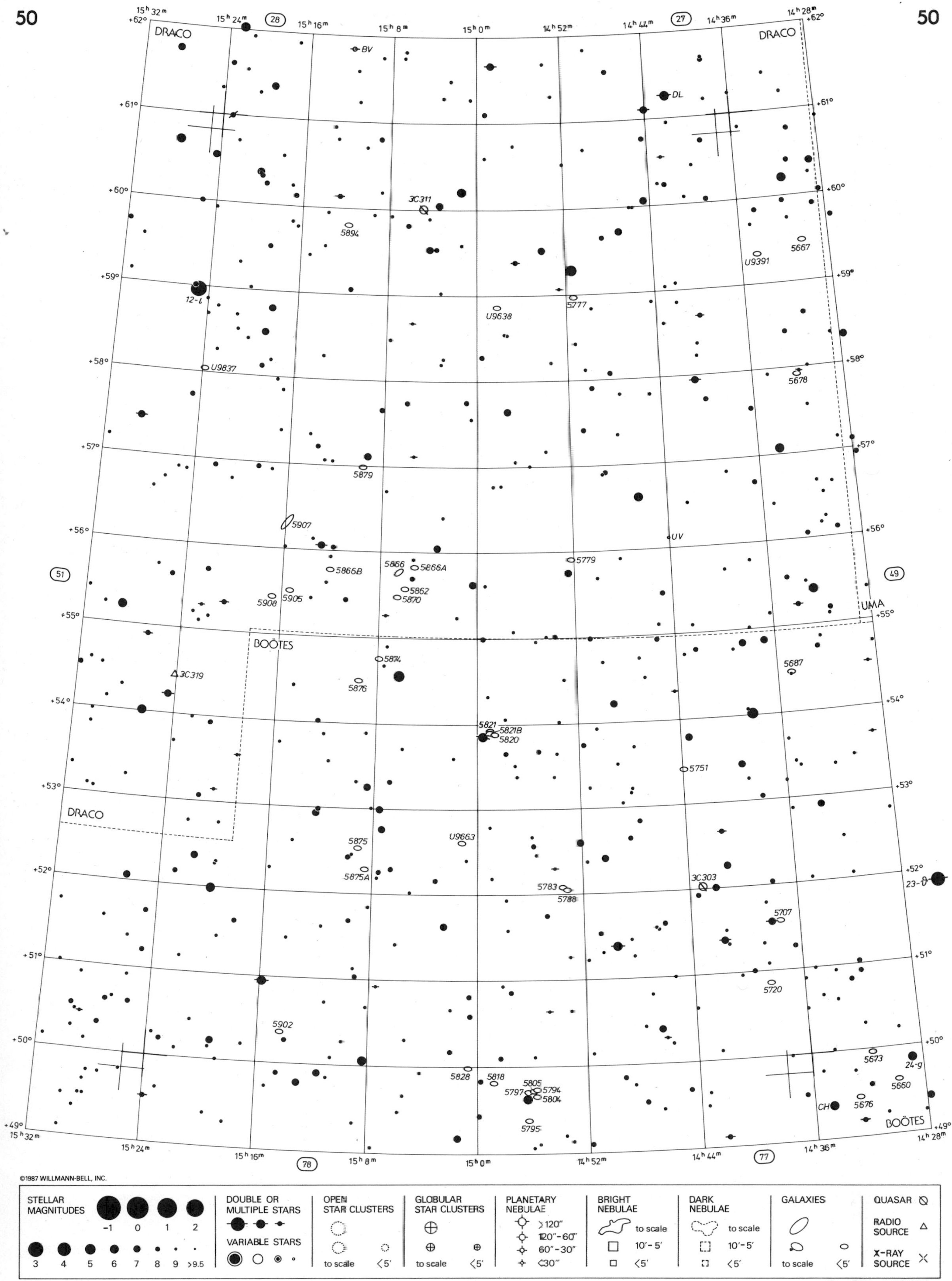

DRACO
BOÖTES
UMA
3C311
3C319
3C303
5894
5667
U9391
U9638
5777
U9837
5678
5879
5907
UV
5779
5866
5866A
5866B
5862
5870
5905
5908
5874
5687
5876
5821
5821B
5820
5751
5875
5875A
U9663
5783
5788
5707
5720
5902
5673
5660
5828
5818
5805
5794
5797
5804
5676
5795
STELLAR MAGNITUDES
DOUBLE OR MULTIPLE STARS
VARIABLE STARS
OPEN STAR CLUSTERS
GLOBULAR STAR CLUSTERS
PLANETARY NEBULAE
BRIGHT NEBULAE
DARK NEBULAE
GALAXIES
QUASAR
RADIO SOURCE
X-RAY SOURCE
to scale
Barry Rappaport & Wil Tirion

16h32m +62° 16h24m (29) 16h16m 16h8m 16h0m 15h52m (28) 15h44m 15h36m 15h28m +62°

DRACO

14-η

6123

6095

+61°

3C323

+60°

6189

6176

AT

5989

5976A

5981

5976

5985

5982

+59°

12-ι

6190

13-θ

+58°

6127

5987

6187

6130

6088

+57°

5965

5963

5971

5969

+56°

6136

(52)

6182

6157

3C322

(50)

6143

UA410

+55°

CL

+54°

+53°

BOÖTES

6090

BP

+52°

+51°

DRACO

RR

+50°

6154

+49° 16h32m

34 HERCULES

(80) 16h24m 16h16m (79) 16h8m 16h0m 15h52m (79) 15h44m 15h36m (78) 15h28m +49°

STELLAR MAGNITUDES	DOUBLE OR MULTIPLE STARS	OPEN STAR CLUSTERS	GLOBULAR STAR CLUSTERS	PLANETARY NEBULAE	BRIGHT NEBULAE	DARK NEBULAE	GALAXIES	QUASAR
−1 0 1 2				>120″	to scale	to scale		RADIO SOURCE
3 4 5 6 7 8 9 >9.5	VARIABLE STARS	to scale <5′	to scale <5′	120″–60″ 60″–30″ <30″	10′–5′ <5′	10′–5′ <5′	to scale <5′	X-RAY SOURCE

Barry Rappaport & Wil Tirion

52

17h32m (30) 17h24m 17h16m 17h8m 17h0m (29) 16h52m 16h44m 16h36m 16h28m

+62° +62°

DRACO DRACO

6359 6226 6223 14-η

+61° +61°

6310 6292 3C351 6307 6306 6361 VW 6258 WW TX

6390 6381 6295

+60° +60°

6399 6393 6394 6189

+59° +59°

6373 6376/77 6391 6290 6291 6285 6286 6206 6190

+58° +58°

U10822-Dra dwarf 6387 6385 6213 6211 AH 6187 6198 6338 6345 6346

+57° +57°

6382 6370

+56° +56°

(53) (51)

6182 24-ν1 25-ν2 6246 6246A

+55° +55°

21-μ

+54° +54°

1715-533

+53° +53°

6386 6358 AI 17 16 23-β WZ

+52° +52°

1700-518

+51° +51°

3C356 DRACO 1727+502

+50° +50°

U10806 6283

HERCULES 34 42

+49° +49°

17h32m 17h24m (81) 17h16m 17h8m 17h0m 16h52m (80) 16h44m 16h36m 16h28m

STELLAR MAGNITUDES	DOUBLE OR MULTIPLE STARS	OPEN STAR CLUSTERS	GLOBULAR STAR CLUSTERS	PLANETARY NEBULAE	BRIGHT NEBULAE	DARK NEBULAE	GALAXIES	QUASAR
-1 0 1 2	VARIABLE STARS	to scale <5'	to scale <5'	>120" 120"-60" 60"-30" <30"	to scale 10'-5' <5'	to scale 10'-5' <5'	to scale <5'	RADIO SOURCE
3 4 5 6 7 8 9 >9.5								X-RAY SOURCE

Barry Rappaport & Wil Tirion

53

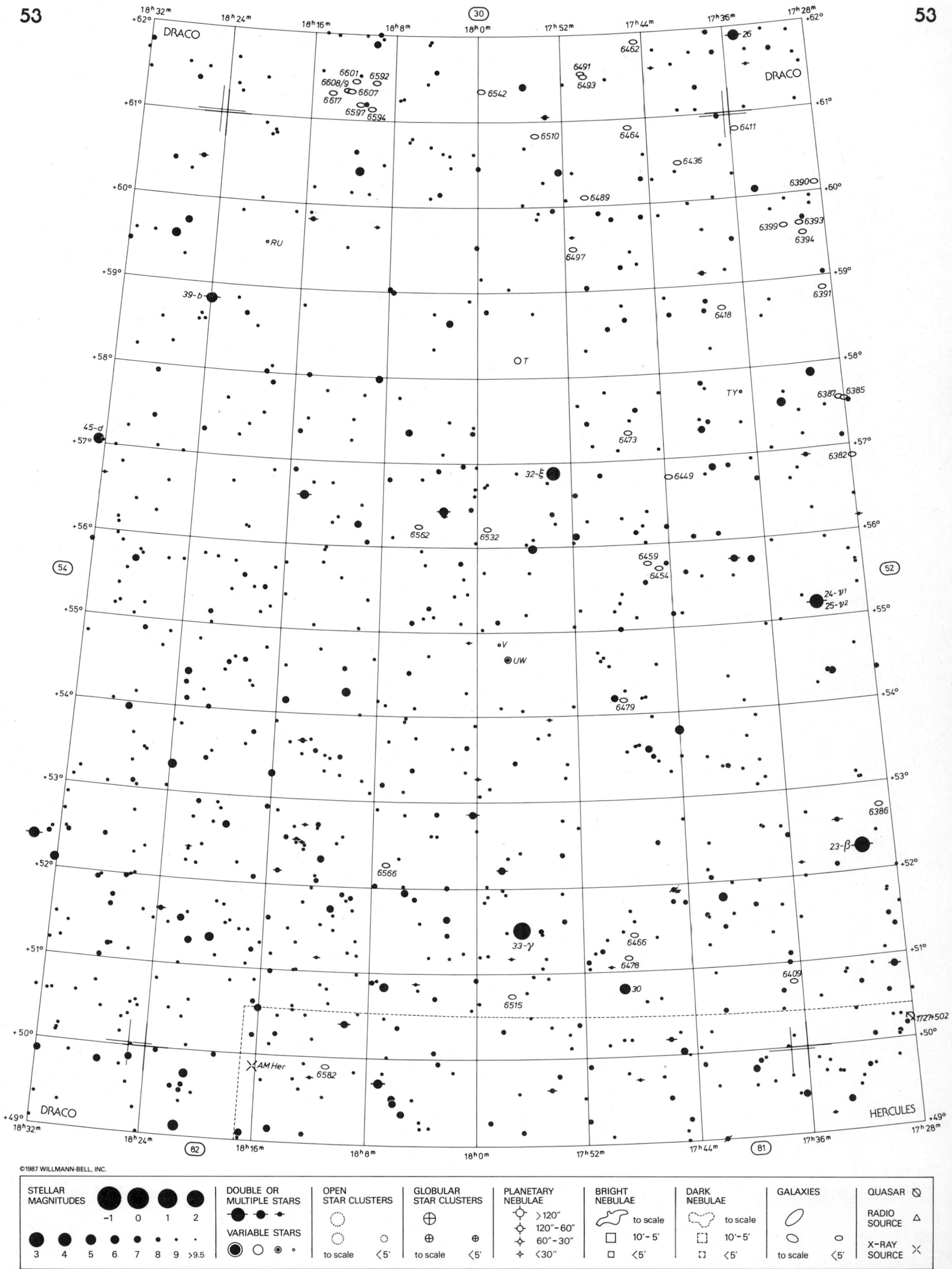

Barry Rappaport & Wil Tirion

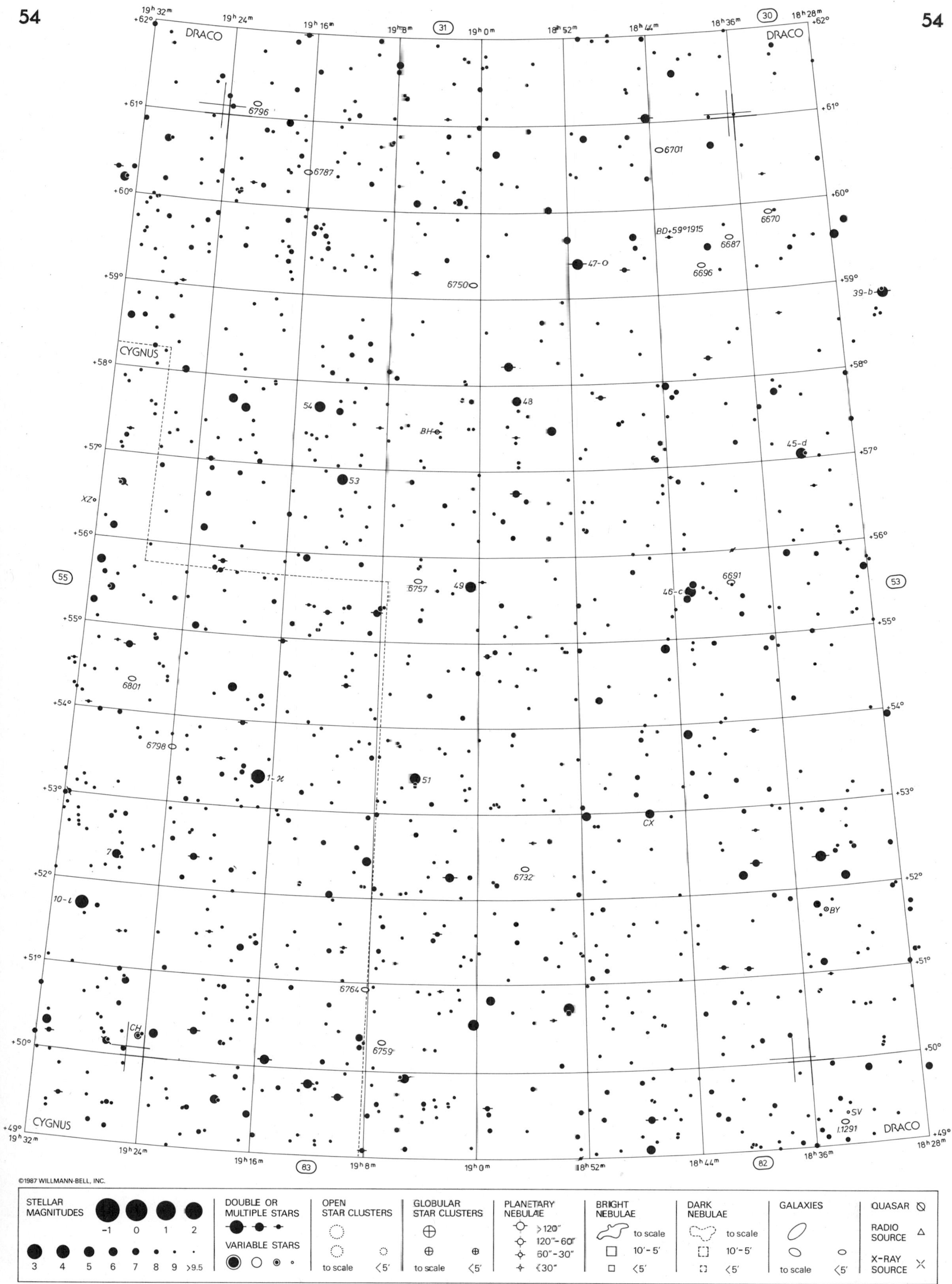
DRACO
CYGNUS
6796
6787
6701
6670
BD+59°1915
6687
6696
6750
47-o
39-b
54
48
BH
45-d
53
XZ
6691
6757
49
46-c
6801
6798
1-κ
51
CX
7
6732
10-ι
BY
6764
CH
6759
SV
I.1291
STELLAR MAGNITUDES
DOUBLE OR MULTIPLE STARS
VARIABLE STARS
OPEN STAR CLUSTERS
GLOBULAR STAR CLUSTERS
PLANETARY NEBULAE
BRIGHT NEBULAE
DARK NEBULAE
GALAXIES
QUASAR
RADIO SOURCE
X-RAY SOURCE
to scale
Barry Rappaport & Wil Tirion

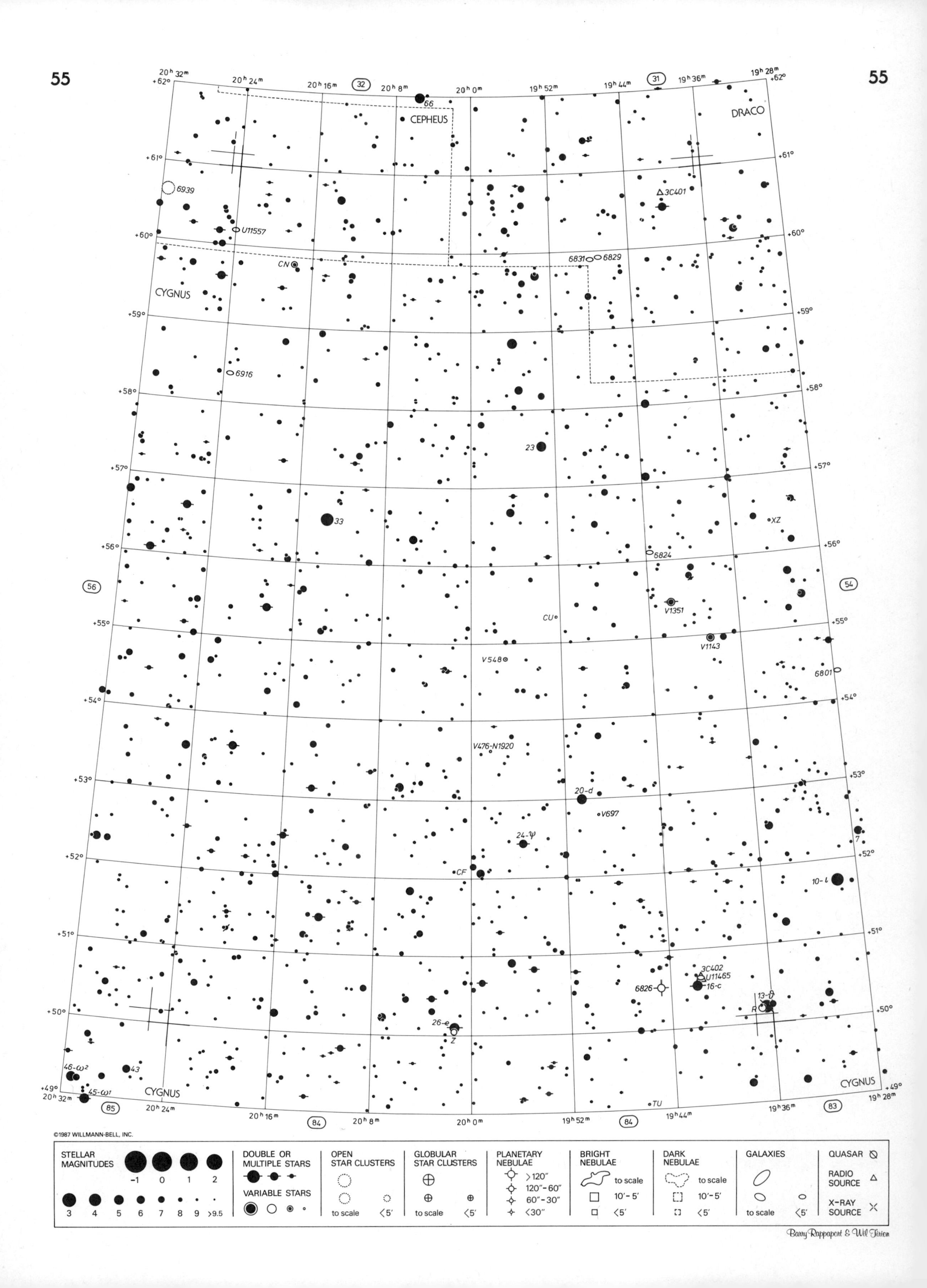
55
55
20h 32m
+62°
20h 24m
20h 16m
32
20h 8m
20h 0m
19h 52m
19h 44m
31
19h 36m
19h 28m
+62°
66
CEPHEUS
DRACO
+61°
+61°
6939
3C401
U11557
+60°
+60°
6831
6829
CN
CYGNUS
+59°
+59°
6916
+58°
+58°
23
+57°
+57°
33
XZ
+56°
+56°
6824
56
54
V1351
CU
+55°
+55°
V1143
V548
6801
+54°
+54°
V476-N1920
+53°
+53°
20-d
V697
24-ψ
7
+52°
+52°
CF
10-ι
+51°
+51°
3C402
U11465
16-c
6826
13-θ
R
+50°
+50°
26-e
Z
46-ω2
43
45-ω1
+49°
20h 32m
CYGNUS
CYGNUS
+49°
19h 28m
85
20h 24m
20h 16m
84
20h 8m
20h 0m
19h 52m
84
TU
19h 44m
19h 36m
83
©1987 WILLMANN-BELL, INC.
STELLAR MAGNITUDES
-1
0
1
2
3
4
5
6
7
8
9
>9.5
DOUBLE OR MULTIPLE STARS
VARIABLE STARS
OPEN STAR CLUSTERS
to scale
<5′
GLOBULAR STAR CLUSTERS
to scale
<5′
PLANETARY NEBULAE
>120″
120″-60″
60″-30″
<30″
BRIGHT NEBULAE
to scale
10′-5′
<5′
DARK NEBULAE
to scale
10′-5′
<5′
GALAXIES
to scale
<5′
QUASAR
RADIO SOURCE
X-RAY SOURCE
Barry Rappaport & Wil Tirion

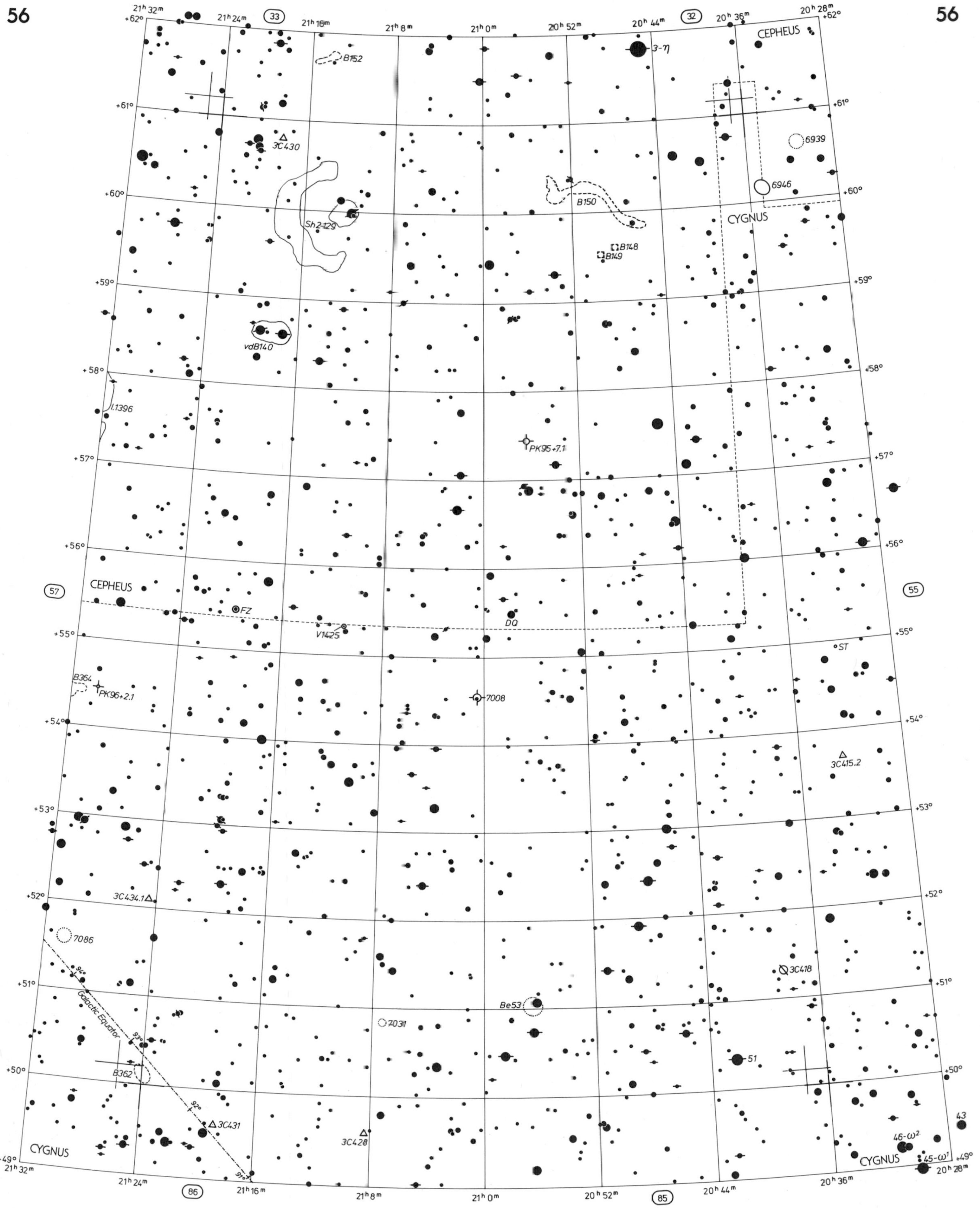

21h 32m
21h 24m
21h 16m
21h 8m
21h 0m
20h 52m
20h 44m
20h 36m
20h 28m
33
32
57
55
86
85
+62°
+61°
+60°
+59°
+58°
+57°
+56°
+55°
+54°
+53°
+52°
+51°
+50°
+49°
CEPHEUS
CYGNUS
3-η
B152
3C430
Sh2-129
B150
B148
B149
6939
6946
vdB140
I.1396
PK95+7.1
FZ
V1425
DQ
ST
B364
PK96+2.1
7008
3C415.2
3C434.1
7086
3C418
Galactic Equator
94°
93°
92°
91°
Be53
7031
B362
51
3C431
3C428
43
46-ω2
45-ω1

STELLAR MAGNITUDES
-1
0
1
2
3
4
5
6
7
8
9
>9.5
DOUBLE OR MULTIPLE STARS
VARIABLE STARS
OPEN STAR CLUSTERS
to scale
<5′
GLOBULAR STAR CLUSTERS
to scale
<5′
PLANETARY NEBULAE
>120″
120″-60″
60″-30″
<30″
BRIGHT NEBULAE
to scale
10′-5′
<5′
DARK NEBULAE
to scale
10′-5′
<5′
GALAXIES
to scale
<5′
QUASAR
RADIO SOURCE
X-RAY SOURCE

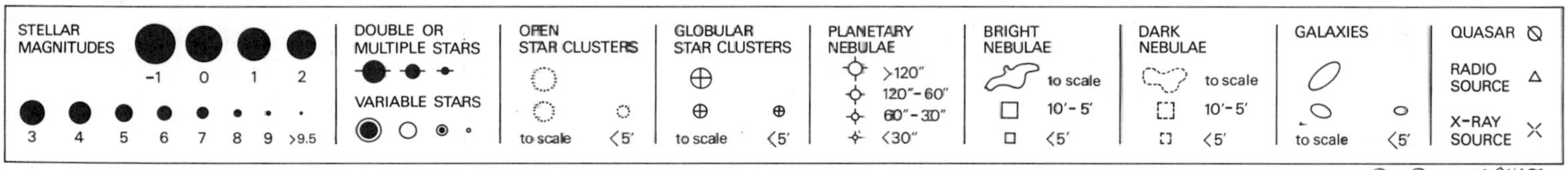

Barry Rappaport & Wil Tirion

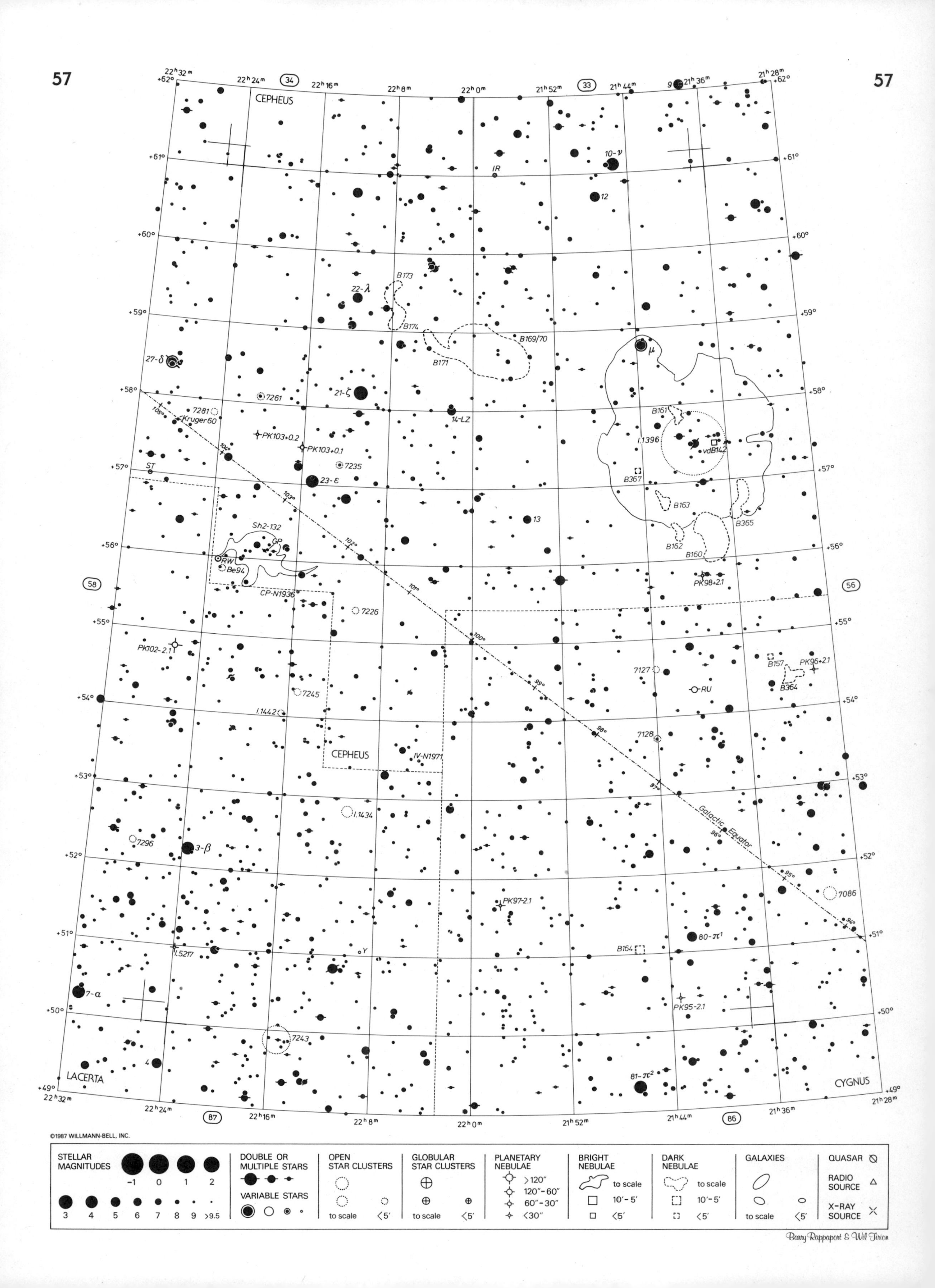
CEPHEUS
10-ν
IR
12
B173
22-λ
B174
B169/70
B171
μ
27-δ
21-ζ
7261
7281
Kruger 60
14-LZ
B161
I.1396
vdB142
PK103+0.2
PK103+0.1
7235
ST
23-ε
B367
13
B163
B365
Sh2-132
GP
RW
Be94
CP-N1936
B162
B160
PK98+2.1
7226
PK102-2.1
7127
B157
PK96+2.1
B364
RU
7245
I.1442
7128
CEPHEUS
IV-N1971
I.1434
Galactic Equator
7296
3-β
7086
PK97-2.1
80-π¹
I.5217
Y
B164
7-α
PK95-2.1
7243
4
81-π²
LACERTA
CYGNUS
STELLAR MAGNITUDES
DOUBLE OR MULTIPLE STARS
VARIABLE STARS
OPEN STAR CLUSTERS
GLOBULAR STAR CLUSTERS
PLANETARY NEBULAE
BRIGHT NEBULAE
DARK NEBULAE
GALAXIES
QUASAR
RADIO SOURCE
X-RAY SOURCE
to scale
Barry Rappaport & Wil Tirion

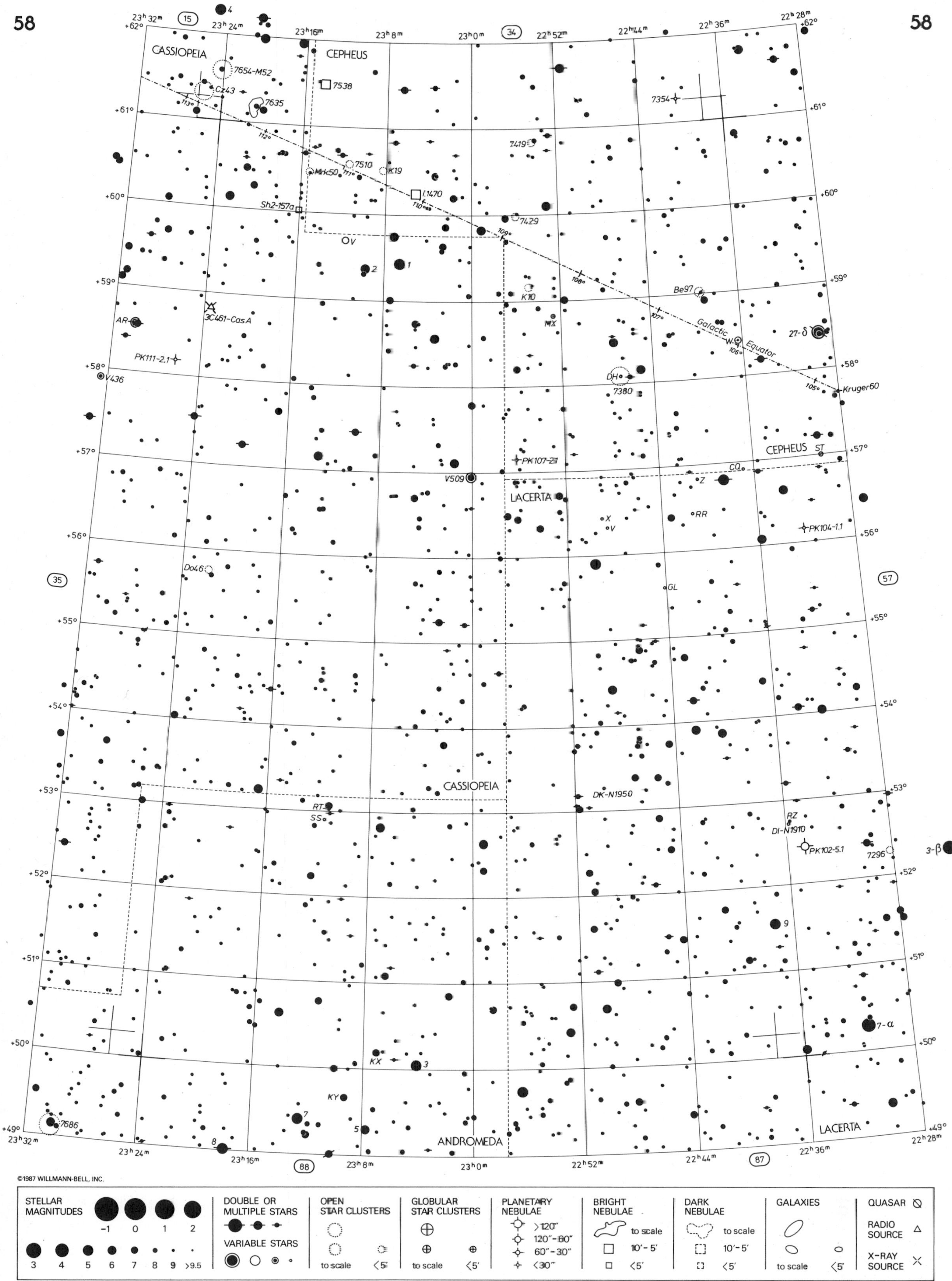

Barry Rappaport & Wil Tirion

59

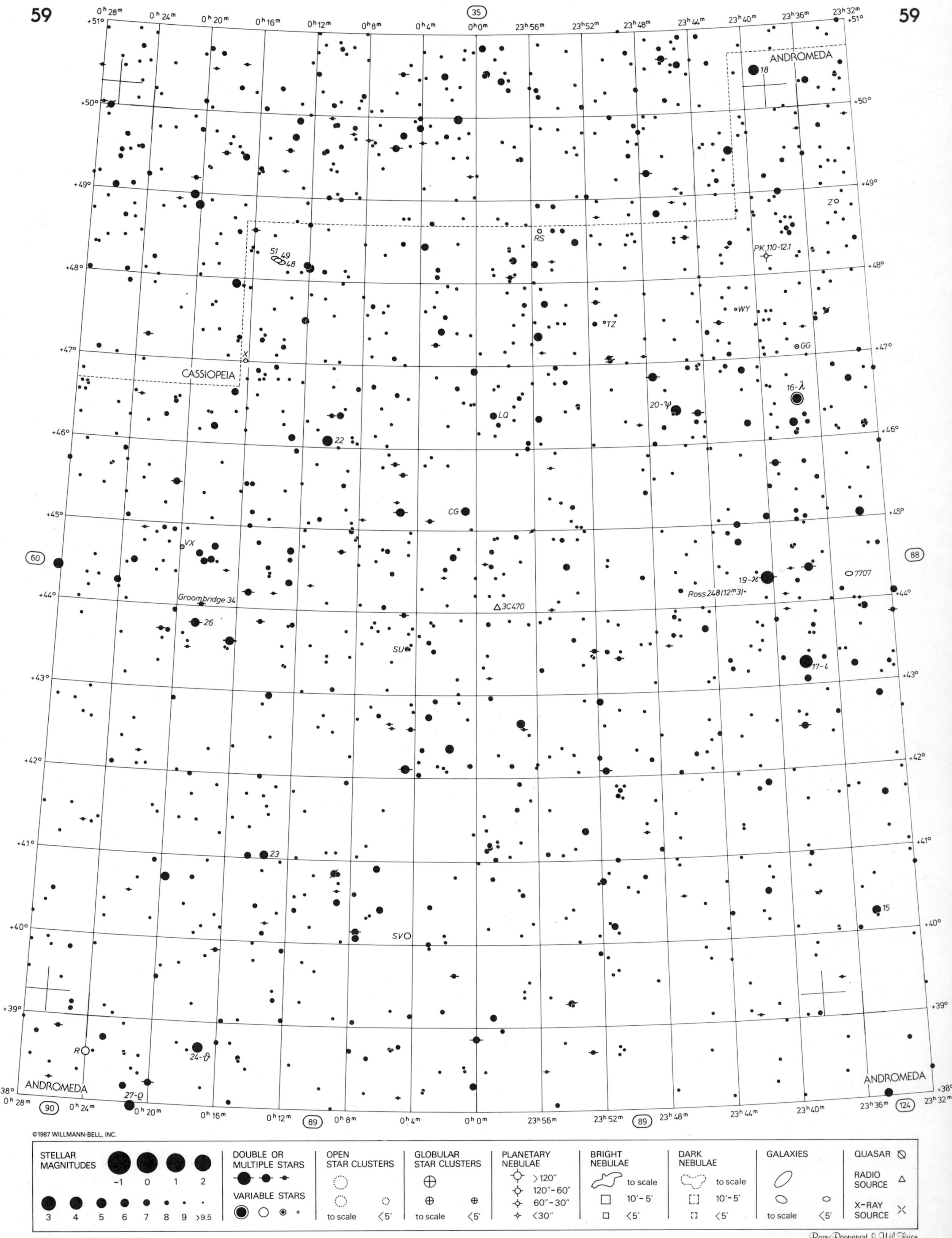
ANDROMEDA
CASSIOPEIA
ANDROMEDA
ANDROMEDA
18
Z
RS
51 49 48
PK 110-12.1
WY
TZ
X
GG
16-λ
20-ψ
LQ
22
CG
VX
60
88
7707
19-κ
Ross 248 (12.m3)
Groombridge 34
3C470
26
SU
17-ι
23
15
SV
R
24-θ
27-ρ
35
90
89
89
124
©1987 WILLMANN-BELL, INC.
STELLAR MAGNITUDES
-1 0 1 2
3 4 5 6 7 8 9 >9.5
DOUBLE OR MULTIPLE STARS
VARIABLE STARS
OPEN STAR CLUSTERS
to scale <5'
GLOBULAR STAR CLUSTERS
to scale <5'
PLANETARY NEBULAE
>120"
120"-60"
60"-30"
<30"
BRIGHT NEBULAE
to scale
10'-5'
<5'
DARK NEBULAE
to scale
10'-5'
<5'
GALAXIES
to scale <5'
QUASAR
RADIO SOURCE
X-RAY SOURCE
Barry Rappaport & Wil Tirion

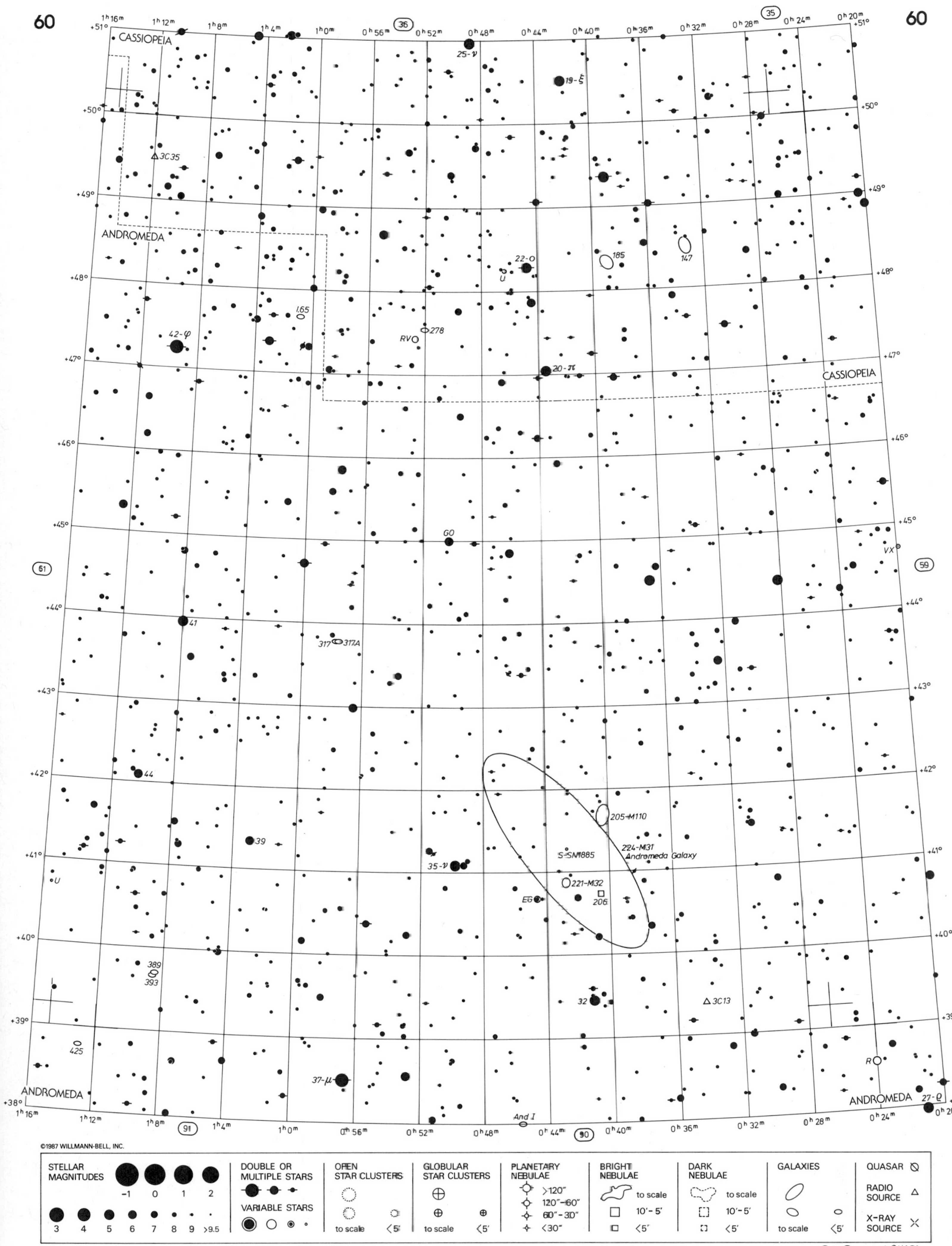

CASSIOPEIA
ANDROMEDA
25-ν
19-ξ
3C35
22-ο
185
147
U
1.65
278
RV
42-φ
20-π
GO
VX
41
317
317A
44
205-M110
224-M31
Andromeda Galaxy
S-SN1885
39
35-ν
221-M32
EG
206
389
393
32
3C13
425
R
37-μ
And I
27-ρ
©1987 WILLMANN-BELL, INC.
STELLAR MAGNITUDES
DOUBLE OR MULTIPLE STARS
VARIABLE STARS
OPEN STAR CLUSTERS
GLOBULAR STAR CLUSTERS
PLANETARY NEBULAE
BRIGHT NEBULAE
DARK NEBULAE
GALAXIES
QUASAR
RADIO SOURCE
X-RAY SOURCE
to scale
Barry Rappaport & Wil Tirion

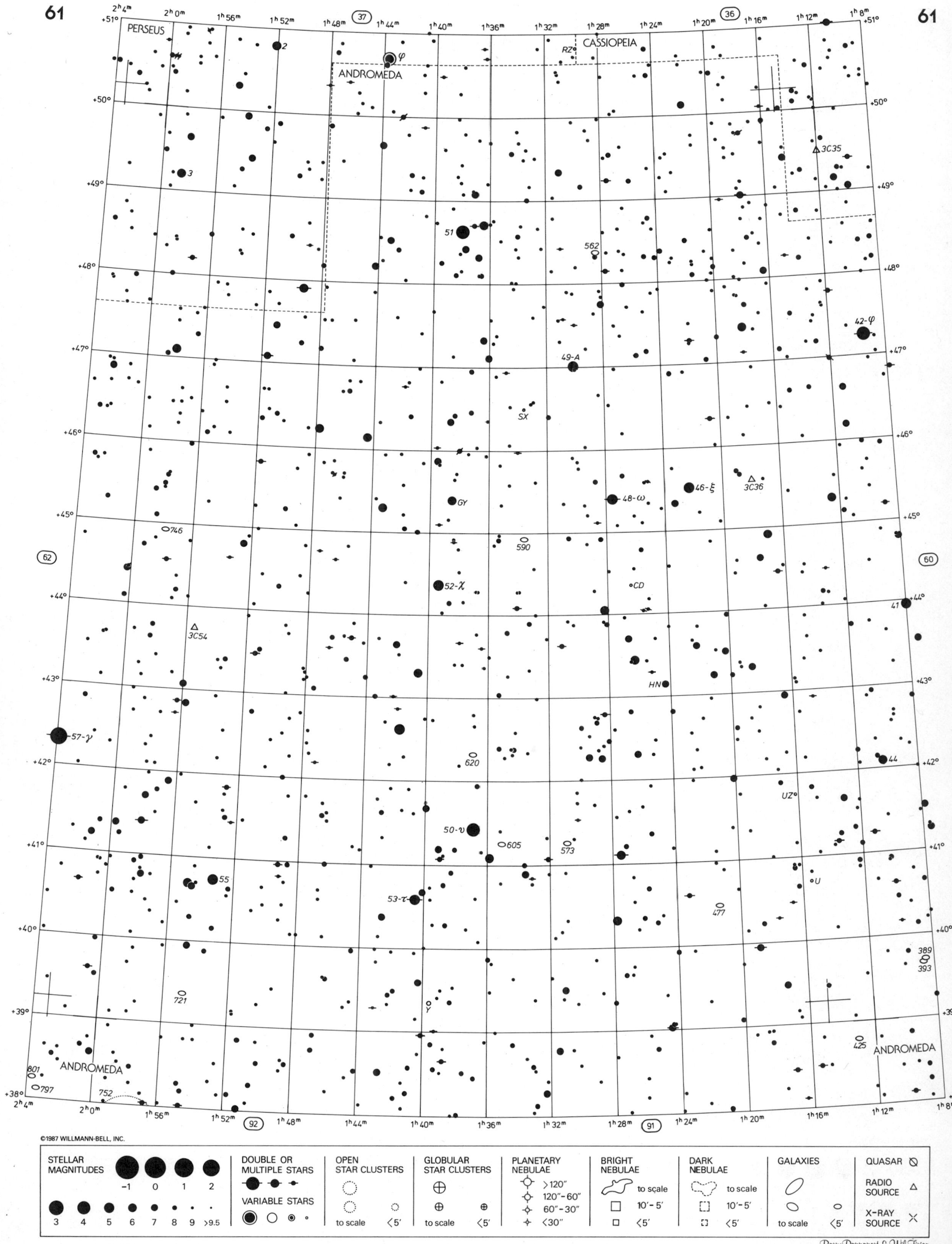
PERSEUS
CASSIOPEIA
ANDROMEDA
3C35
3C36
3C54
42-φ
49-A
51
48-ω
46-ξ
52-χ
57-γ
50-υ
53-τ
55
41
44
GY
HN
CD
SX
RZ
UZ
U
Y
562
590
746
620
605
573
477
721
389
393
425
801
797
752
STELLAR MAGNITUDES
DOUBLE OR MULTIPLE STARS
VARIABLE STARS
OPEN STAR CLUSTERS
GLOBULAR STAR CLUSTERS
PLANETARY NEBULAE
BRIGHT NEBULAE
DARK NEBULAE
GALAXIES
QUASAR
RADIO SOURCE
X-RAY SOURCE
to scale
Barry Rappaport & Wil Tirion

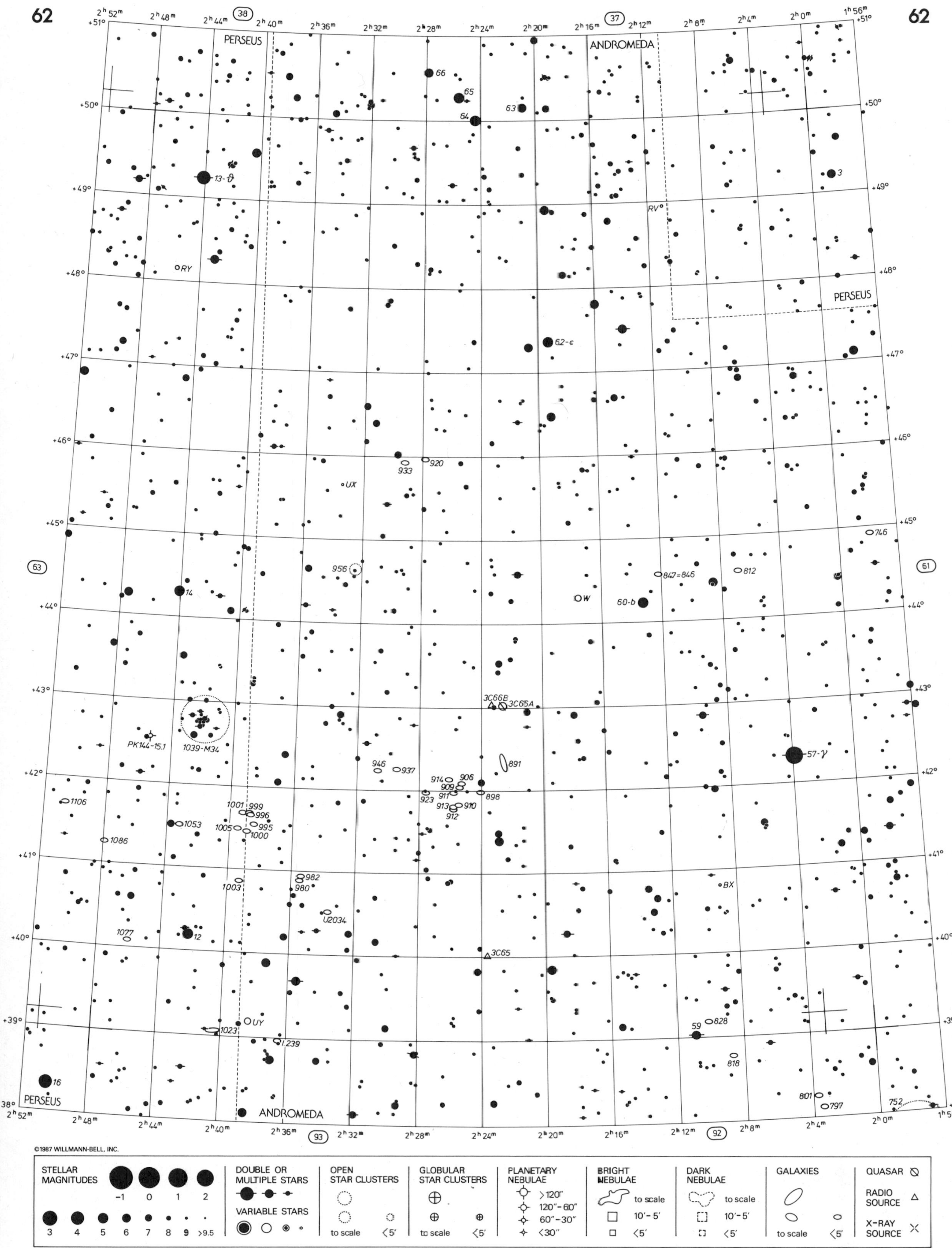
62
62
PERSEUS
ANDROMEDA
PERSEUS
PERSEUS
ANDROMEDA
38
37
63
61
93
92
+51°
+50°
+49°
+48°
+47°
+46°
+45°
+44°
+43°
+42°
+41°
+40°
+39°
+38°
2h 52m
2h 48m
2h 44m
2h 40m
2h 36m
2h 32m
2h 28m
2h 24m
2h 20m
2h 16m
2h 12m
2h 8m
2h 4m
2h 0m
1h 56m
66
65
64
63
3
13-θ
RY
RV
62-c
920
933
UX
746
956
847=846
812
14
W
60-b
3C66B
3C65A
PK144-15.1
1039-M34
57-γ
891
946
937
914
906
909
911
923
898
913
910
912
1106
1001
999
996
1053
1005
995
1000
1086
982
980
1003
BX
U2034
1077
12
3C65
UY
1023
1.239
59
828
818
16
801
797
752
©1987 WILLMANN-BELL, INC.
STELLAR MAGNITUDES
−1 0 1 2
3 4 5 6 7 8 9 >9.5
DOUBLE OR MULTIPLE STARS
VARIABLE STARS
OPEN STAR CLUSTERS
to scale
<5′
GLOBULAR STAR CLUSTERS
to scale
<5′
PLANETARY NEBULAE
>120″
120″-60″
60″-30″
<30″
BRIGHT NEBULAE
to scale
10′-5′
<5′
DARK NEBULAE
to scale
10′-5′
<5′
GALAXIES
to scale
<5′
QUASAR
RADIO SOURCE
X-RAY SOURCE
Barry Rappaport & Wil Tirion

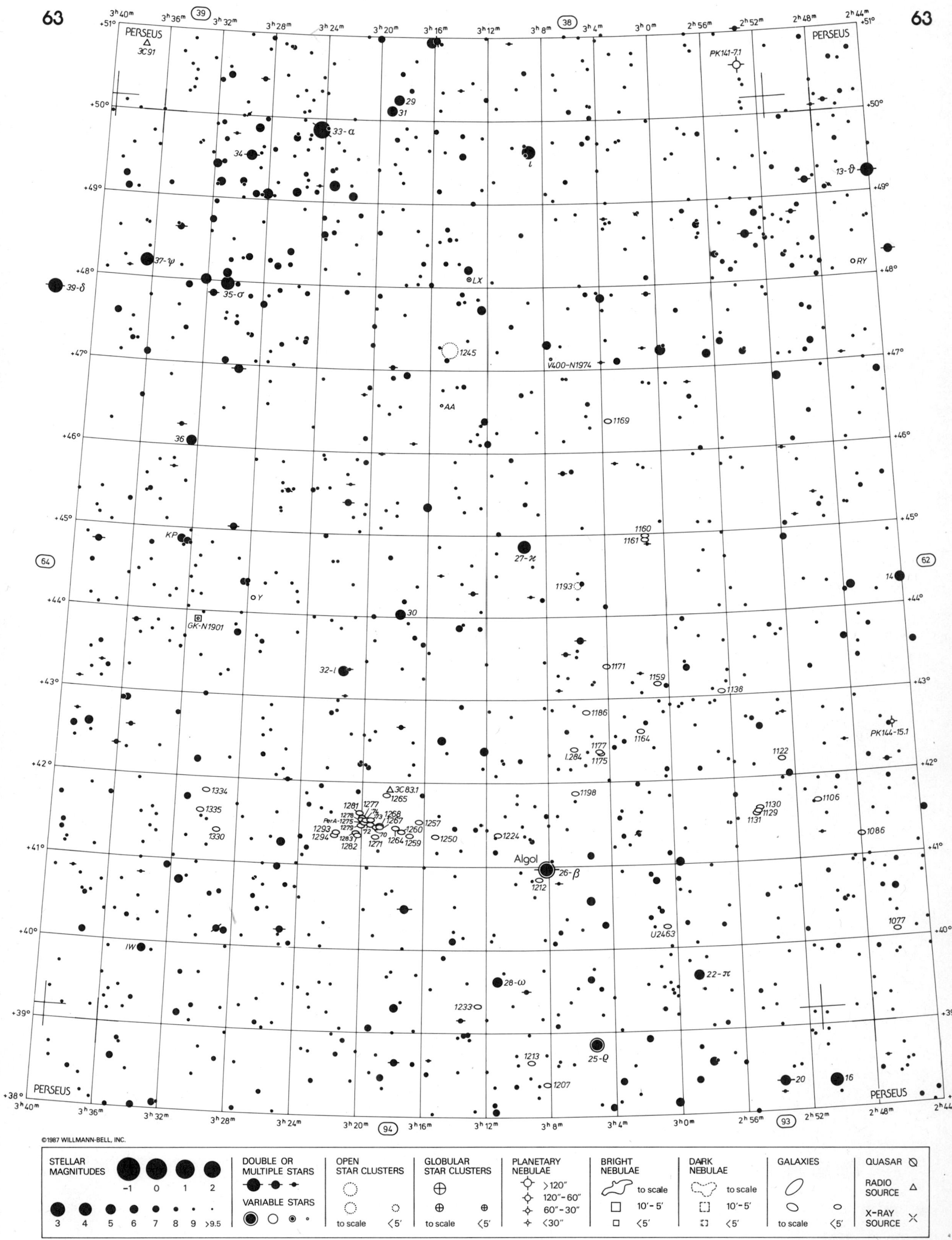

Barry Rappaport & Wil Tirion

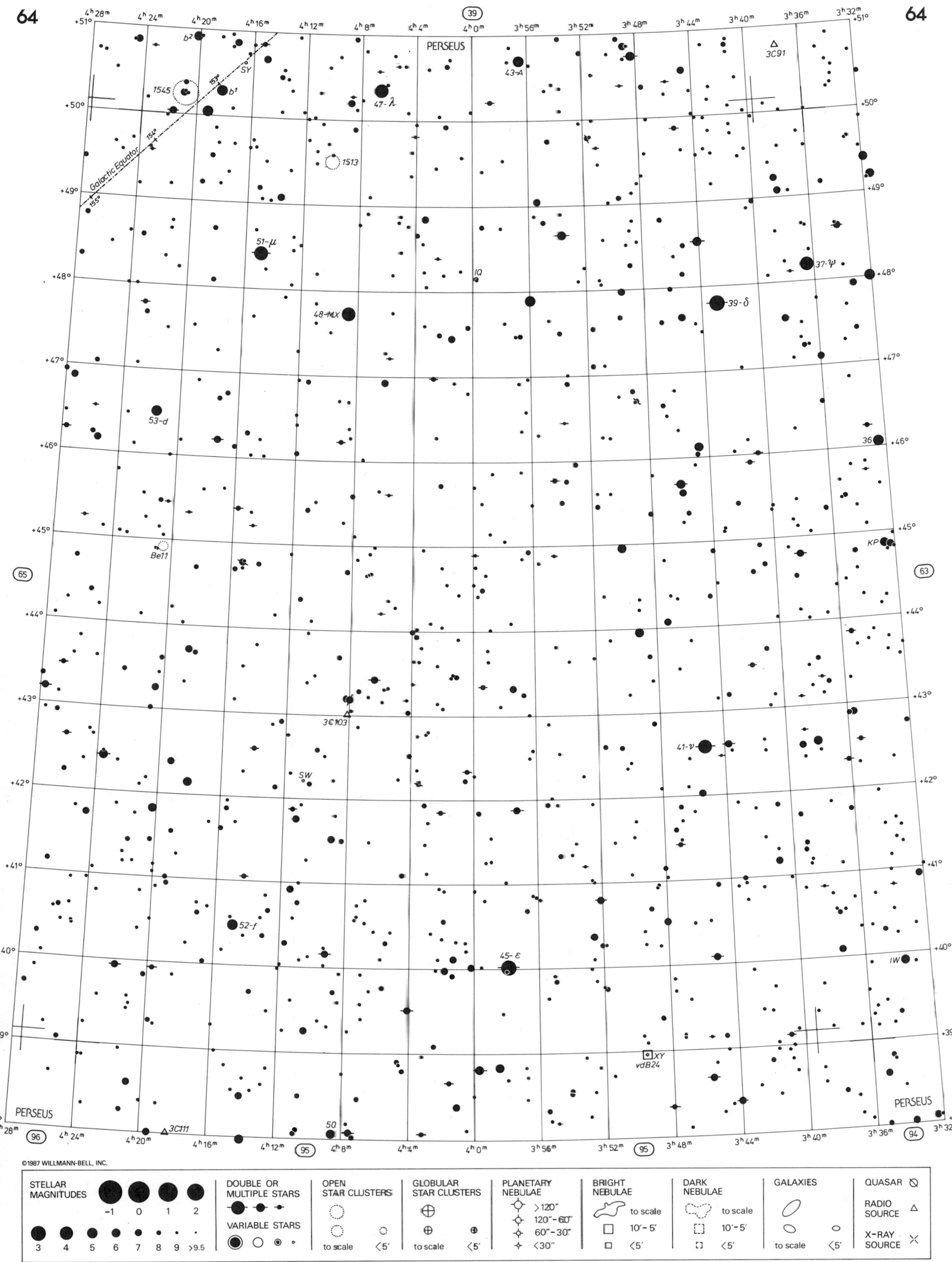

STELLAR MAGNITUDES	DOUBLE OR MULTIPLE STARS	OPEN STAR CLUSTERS	GLOBULAR STAR CLUSTERS	PLANETARY NEBULAE	BRIGHT NEBULAE	DARK NEBULAE	GALAXIES	QUASAR
−1 0 1 2	VARIABLE STARS	to scale <5′	to scale <5′	>120″ 120″–60″ 60″–30″ <30″	to scale 10′–5′ <5′	to scale 10′–5′ <5′	to scale <5′	RADIO SOURCE
3 4 5 6 7 8 9 >9.5								X-RAY SOURCE

Barry Rappaport & Wil Tirion

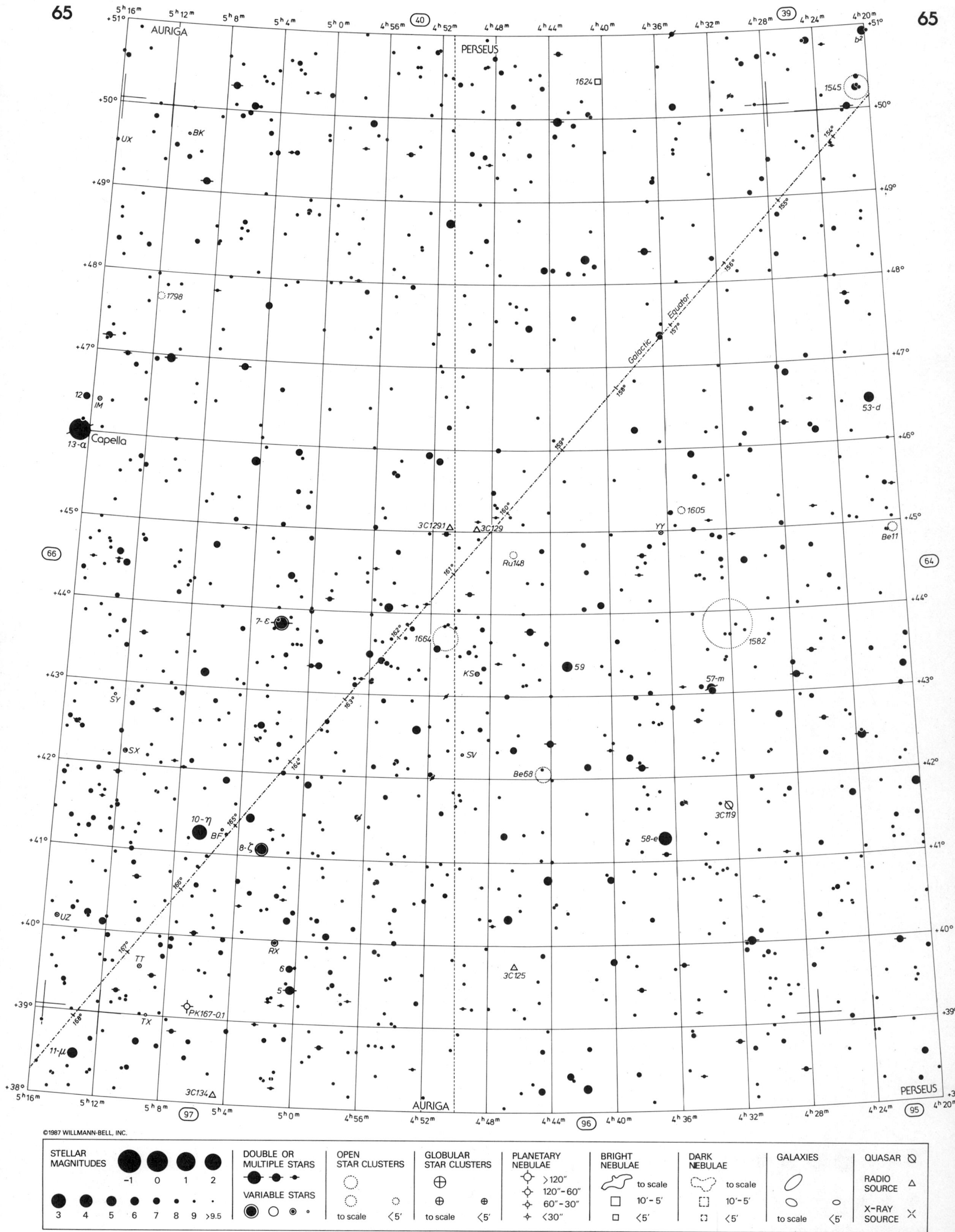

AURIGA
PERSEUS
1624
1545
b^2
BK
UX
1798
Galactic Equator
12
IM
53-d
Capella
13-α
1605
YY
3C129.1
3C129
Ru148
Be11
7-ε
1664
1582
59
KS
57-m
SY
SX
SV
Be68
3C119
10-η
BF
8-ζ
58-e
UZ
RX
TT
3C125
PK167-0.1
TX
11-μ
3C134
66
64
40
39
97
96
95
STELLAR MAGNITUDES
−1 0 1 2
3 4 5 6 7 8 9 >9.5
DOUBLE OR MULTIPLE STARS
VARIABLE STARS
OPEN STAR CLUSTERS
to scale <5′
GLOBULAR STAR CLUSTERS
to scale <5′
PLANETARY NEBULAE
>120″
120″–60″
60″–30″
<30″
BRIGHT NEBULAE
to scale
10′–5′
<5′
DARK NEBULAE
to scale
10′–5′
<5′
GALAXIES
to scale <5′
QUASAR
RADIO SOURCE
X-RAY SOURCE
Barry Rappaport & Wil Tirion

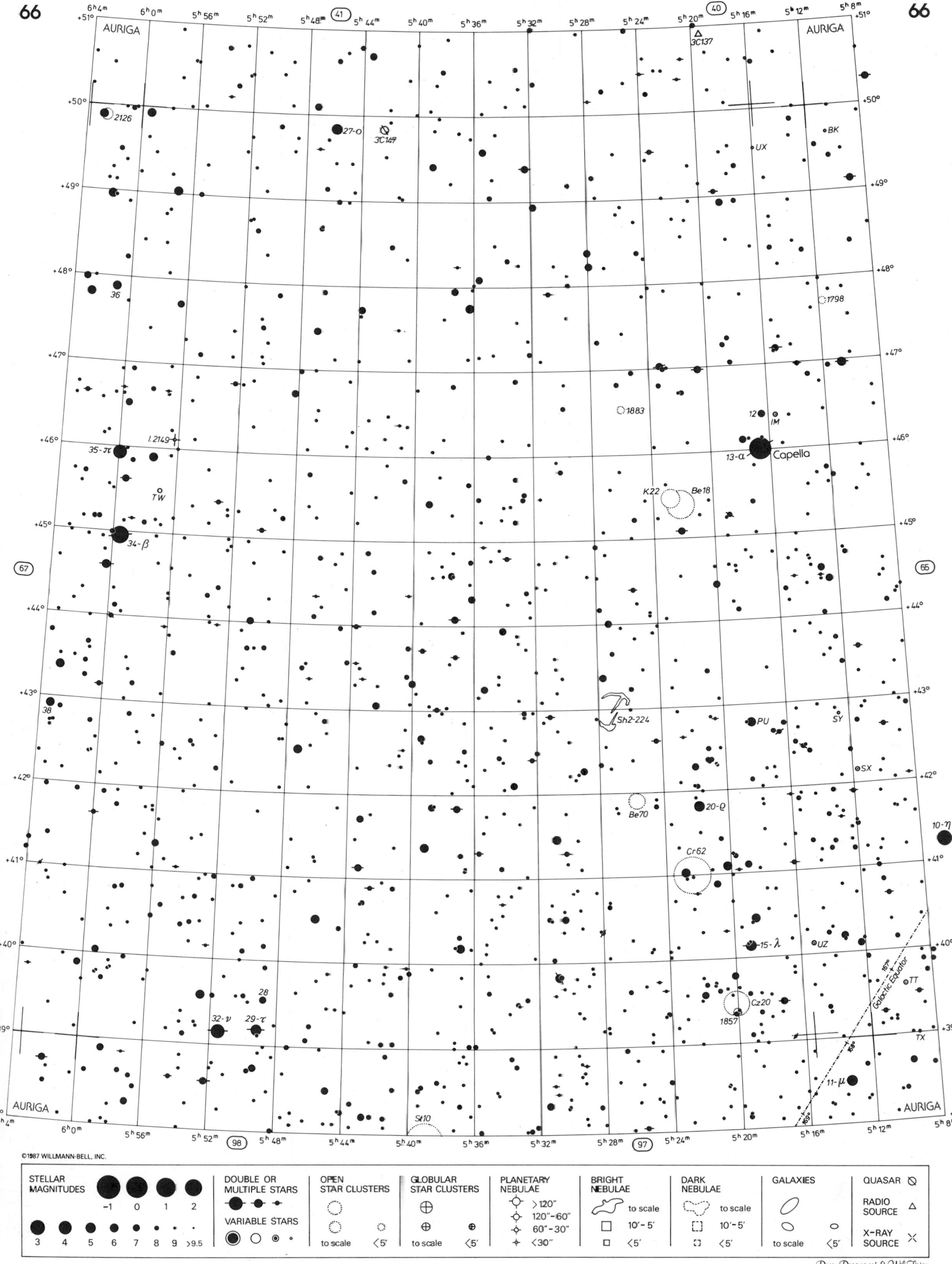
AURIGA
3C137
2126
27-o
3C147
UX
BK
36
1798
1883
12
IM
1.2149
35-π
13-α
Capella
TW
K22
Be18
34-β
38
Sh2-224
PU
SY
SX
Be70
20-ϱ
10-η
Cr62
15-λ
UZ
TT
Galactic Equator
28
32-ν
29-τ
Cz20
1857
TX
11-μ
St10
STELLAR MAGNITUDES
DOUBLE OR MULTIPLE STARS
VARIABLE STARS
OPEN STAR CLUSTERS
GLOBULAR STAR CLUSTERS
PLANETARY NEBULAE
BRIGHT NEBULAE
DARK NEBULAE
GALAXIES
QUASAR
RADIO SOURCE
X-RAY SOURCE
to scale
Barry Rappaport & Wil Tirion

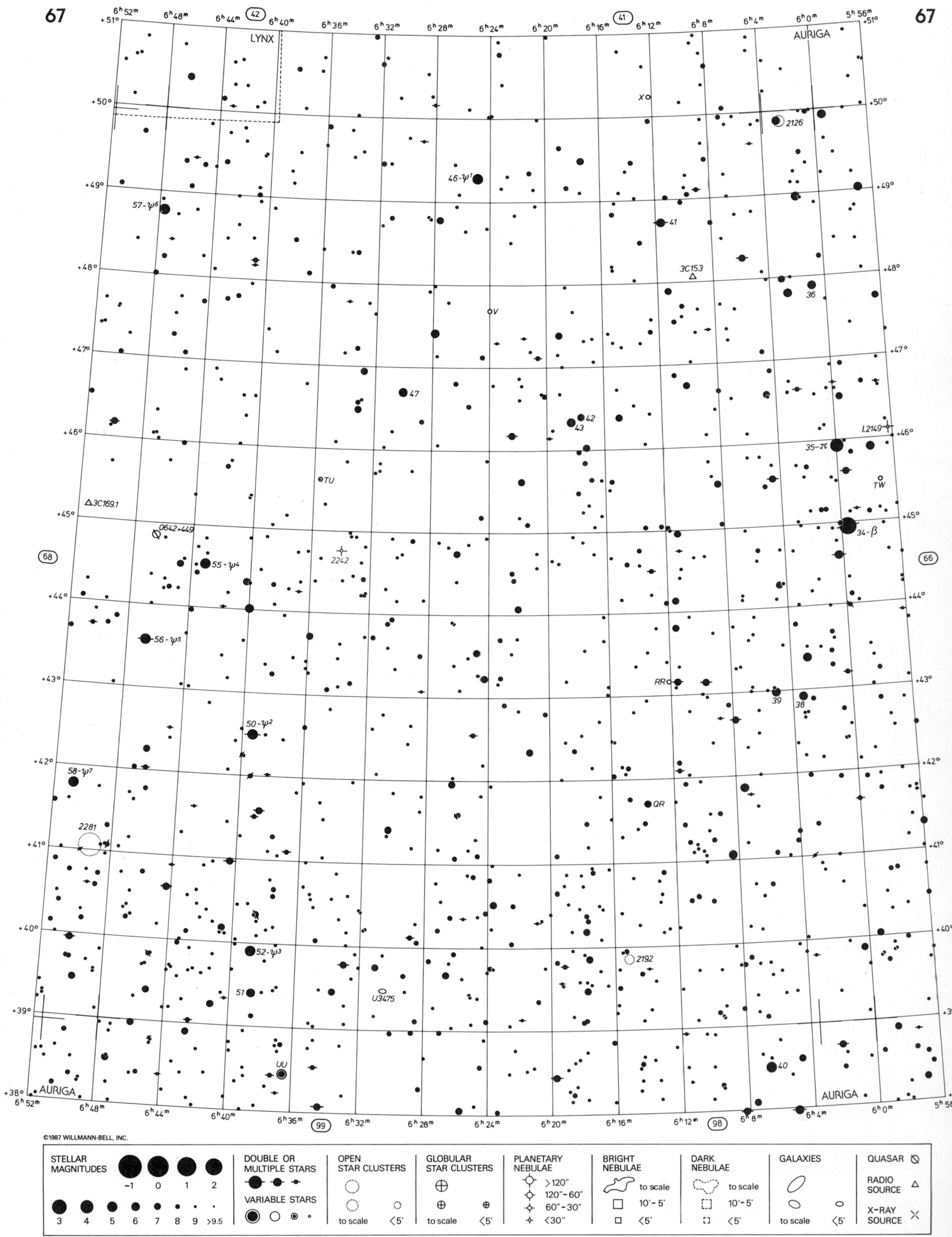

Barry Rappaport & Wil Tirion

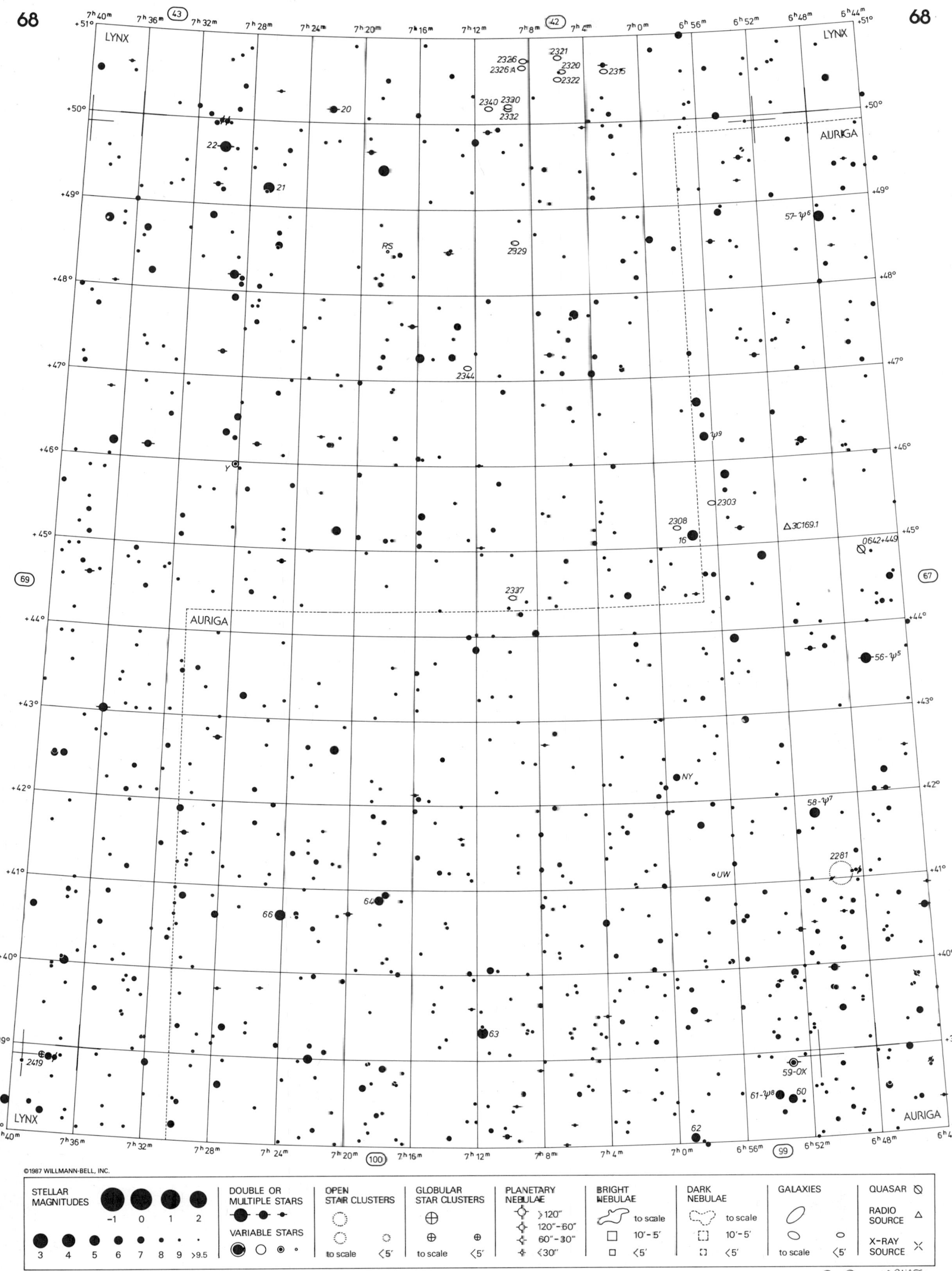

STELLAR MAGNITUDES	DOUBLE OR MULTIPLE STARS	OPEN STAR CLUSTERS	GLOBULAR STAR CLUSTERS	PLANETARY NEBULAE	BRIGHT NEBULAE	DARK NEBULAE	GALAXIES	QUASAR
-1 0 1 2	VARIABLE STARS	to scale <5'	to scale <5'	>120" 120"-60" 60"-30" <30"	to scale 10'-5' <5'	to scale 10'-5' <5'	to scale <5'	RADIO SOURCE
3 4 5 6 7 8 9 >9.5								X-RAY SOURCE

Barry Rappaport & Wil Tirion

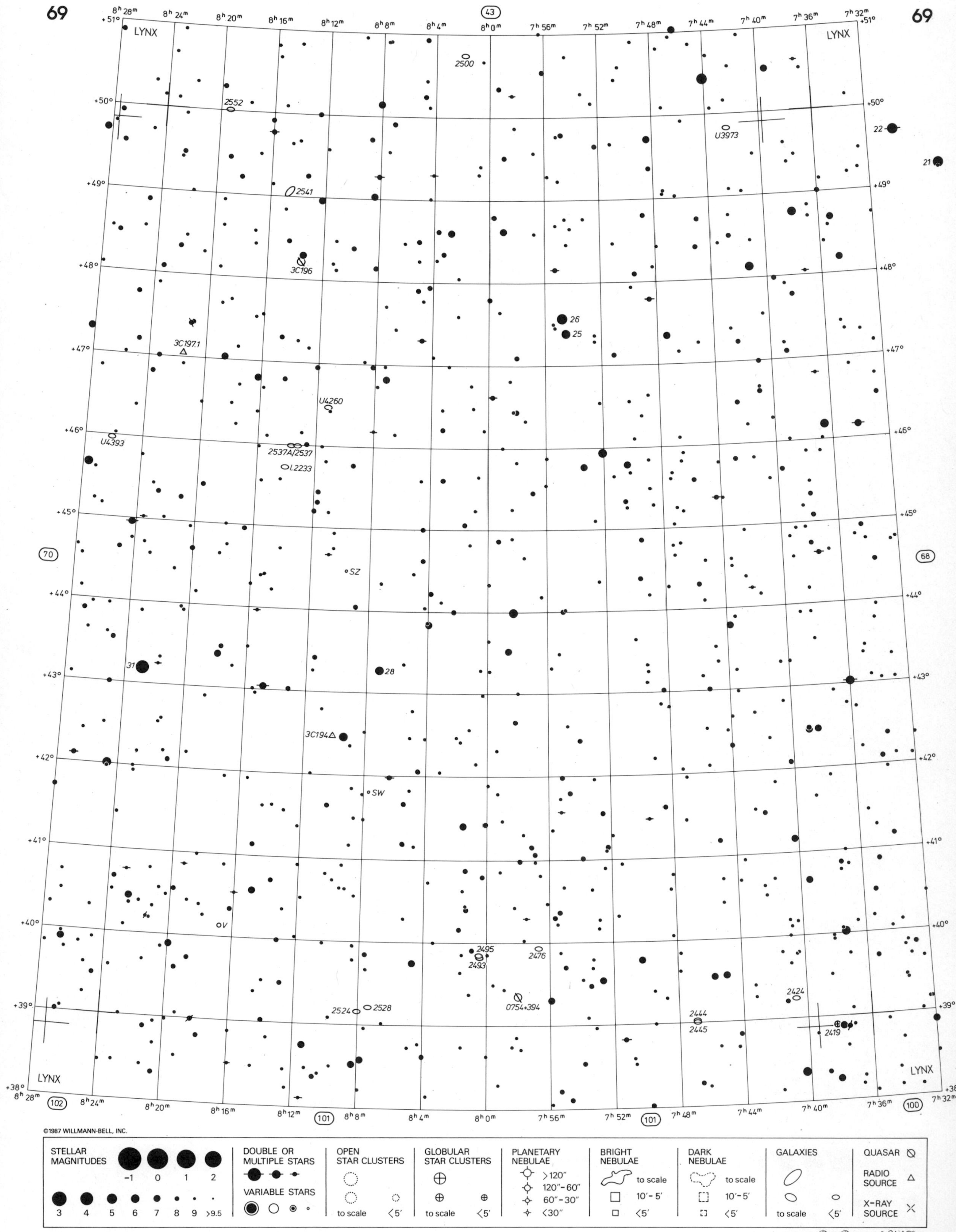
69
LYNX
8h 28m
8h 24m
8h 20m
8h 16m
8h 12m
8h 8m
8h 4m
8h 0m
7h 56m
7h 52m
7h 48m
7h 44m
7h 40m
7h 36m
7h 32m
+51°
+50°
+49°
+48°
+47°
+46°
+45°
+44°
+43°
+42°
+41°
+40°
+39°
+38°
43
70
68
102
101
100
2500
2552
U3973
22
21
2541
3C196
26
25
3C197.1
U4260
U4393
2537A/2537
I.2233
SZ
31
28
3C194
SW
V
2495
2493
2476
0754+394
2524
2528
2444
2445
2424
2419
©1987 WILLMANN-BELL, INC.
STELLAR MAGNITUDES
-1
0
1
2
3
4
5
6
7
8
9
>9.5
DOUBLE OR MULTIPLE STARS
VARIABLE STARS
OPEN STAR CLUSTERS
to scale
<5′
GLOBULAR STAR CLUSTERS
to scale
<5′
PLANETARY NEBULAE
>120″
120″-60″
60″-30″
<30″
BRIGHT NEBULAE
to scale
10′-5′
<5′
DARK NEBULAE
to scale
10′-5′
<5′
GALAXIES
to scale
<5′
QUASAR
RADIO SOURCE
X-RAY SOURCE
Barry Rappaport & Wil Tirion

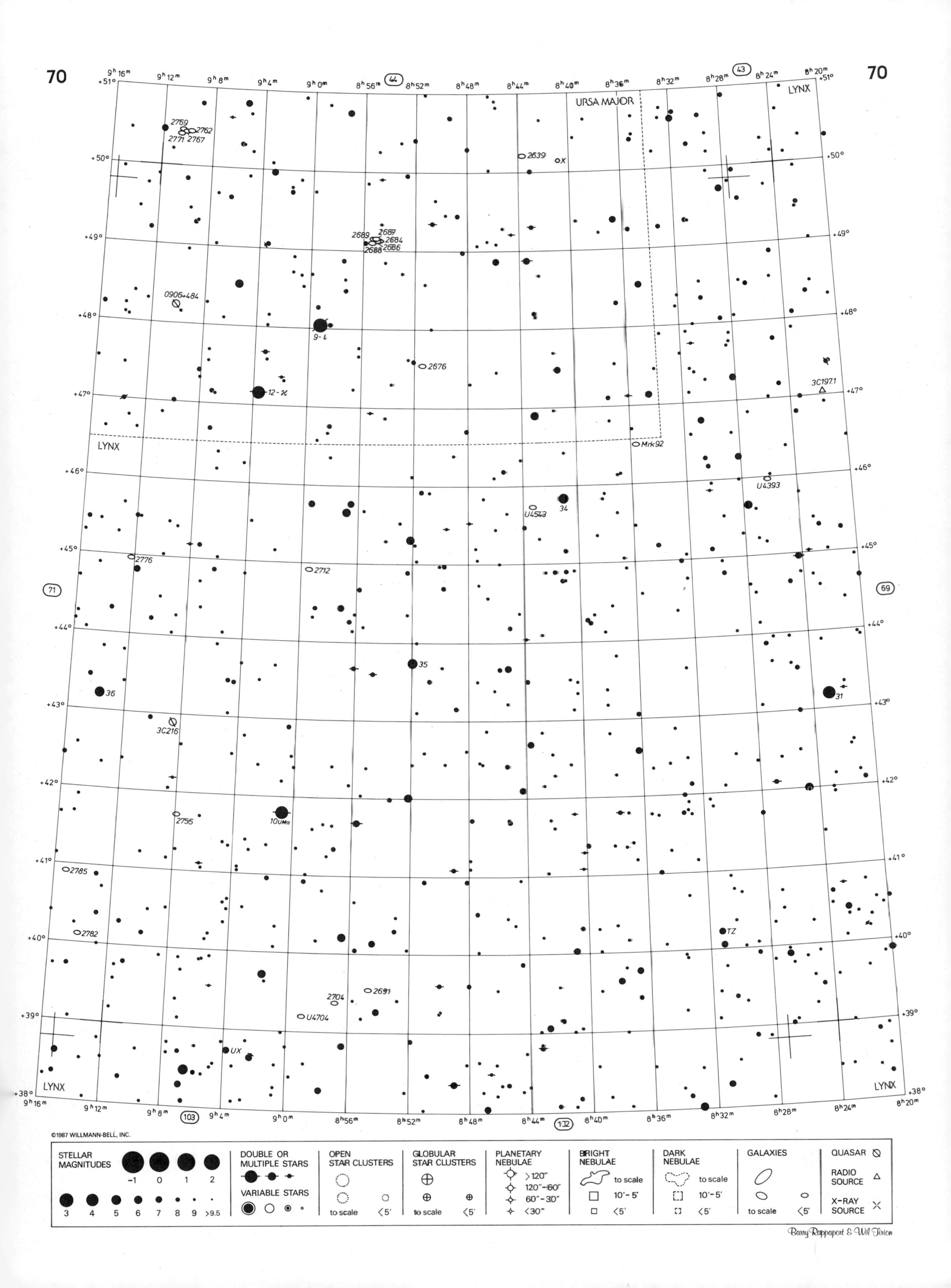
URSA MAJOR
LYNX
2769
2762
2771 2767
2639
X
2689
2687
2684
2686
2688
0906+484
9-ι
2676
3C197.1
12-κ
Mrk92
U4393
U4543
34
2776
2712
71
69
35
36
31
3C216
2755
10UMa
2785
2782
TZ
2691
2704
U4704
UX
103
102
STELLAR MAGNITUDES
DOUBLE OR MULTIPLE STARS
VARIABLE STARS
OPEN STAR CLUSTERS
GLOBULAR STAR CLUSTERS
PLANETARY NEBULAE
BRIGHT NEBULAE
DARK NEBULAE
GALAXIES
QUASAR
RADIO SOURCE
X-RAY SOURCE
to scale
Barry Rappaport & Wil Tirion

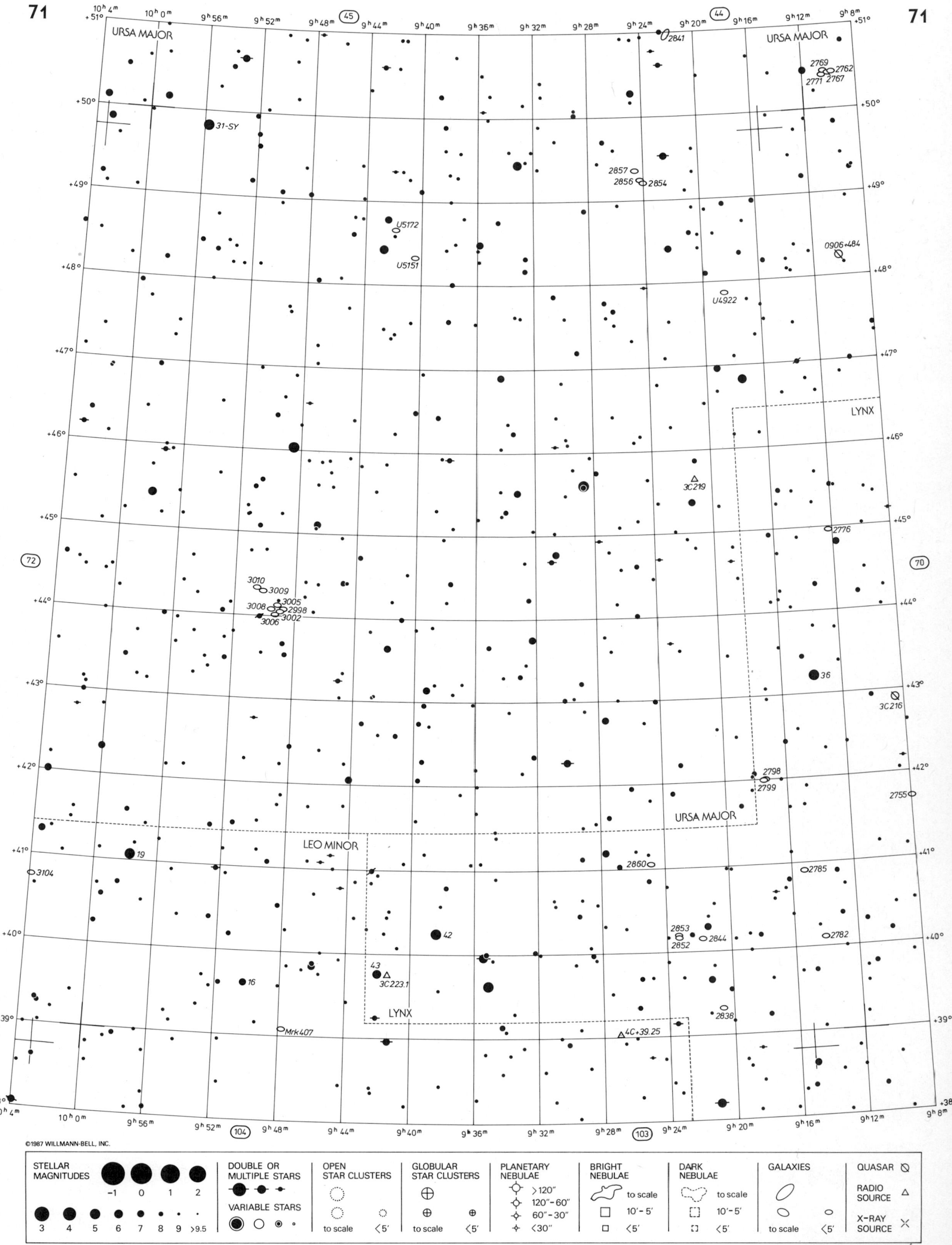

71
71
URSA MAJOR
URSA MAJOR
LYNX
LEO MINOR
URSA MAJOR
LYNX
31-SY
2841
2769
2762
2771
2767
2857
2856
2854
U5172
U5151
0906+484
U4922
3C219
2776
3010
3009
3005
3008
2998
3002
3006
36
3C216
2798
2799
2755
19
3104
2860
2785
2853
2852
2844
2782
42
43
3C223.1
16
2838
Mrk407
4C+39.25
©1987 WILLMANN-BELL, INC.
STELLAR MAGNITUDES
DOUBLE OR MULTIPLE STARS
VARIABLE STARS
OPEN STAR CLUSTERS
GLOBULAR STAR CLUSTERS
PLANETARY NEBULAE
BRIGHT NEBULAE
DARK NEBULAE
GALAXIES
QUASAR
RADIO SOURCE
X-RAY SOURCE
to scale
Barry Rappaport & Wil Tirion

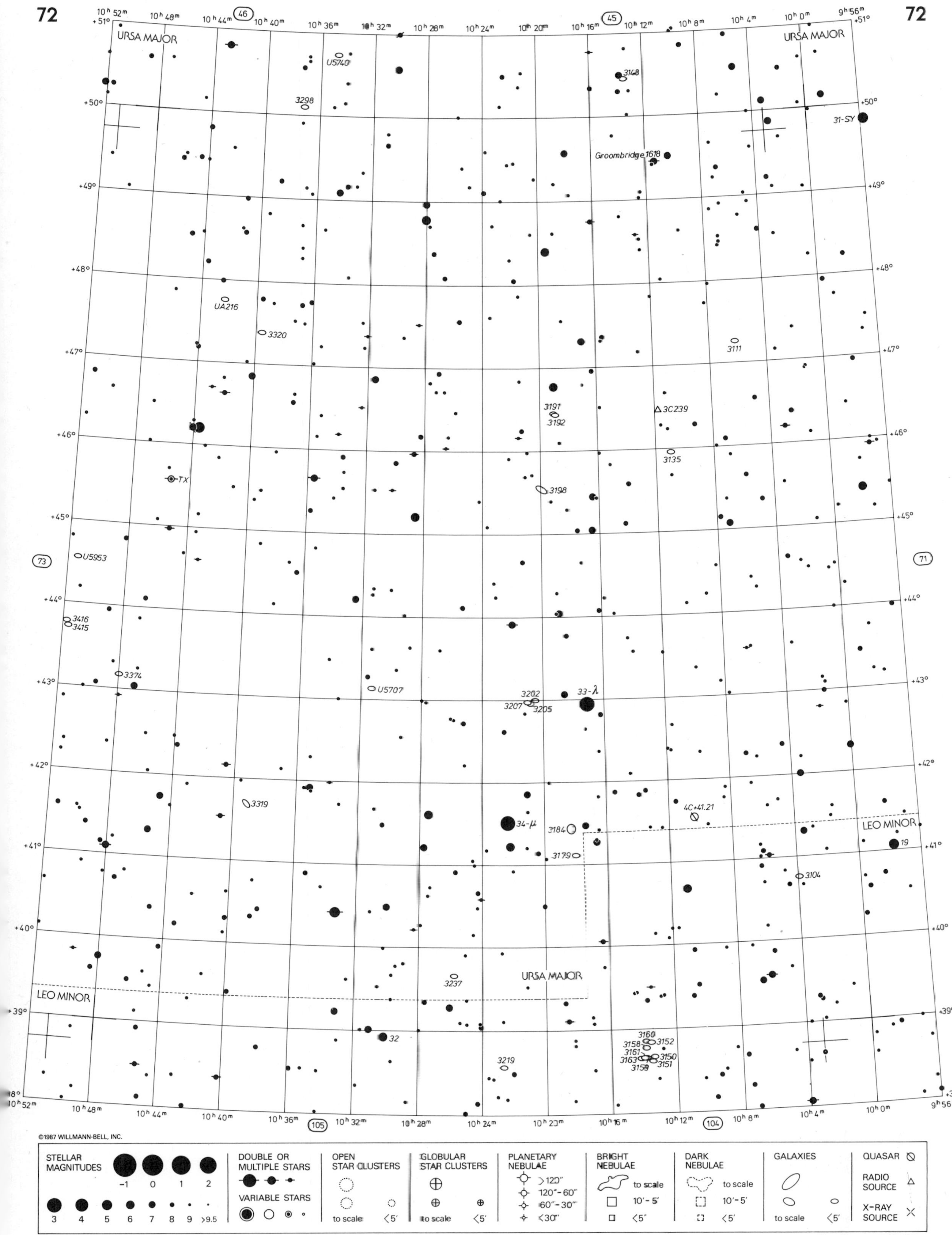

Barry Rappaport & Wil Tirion

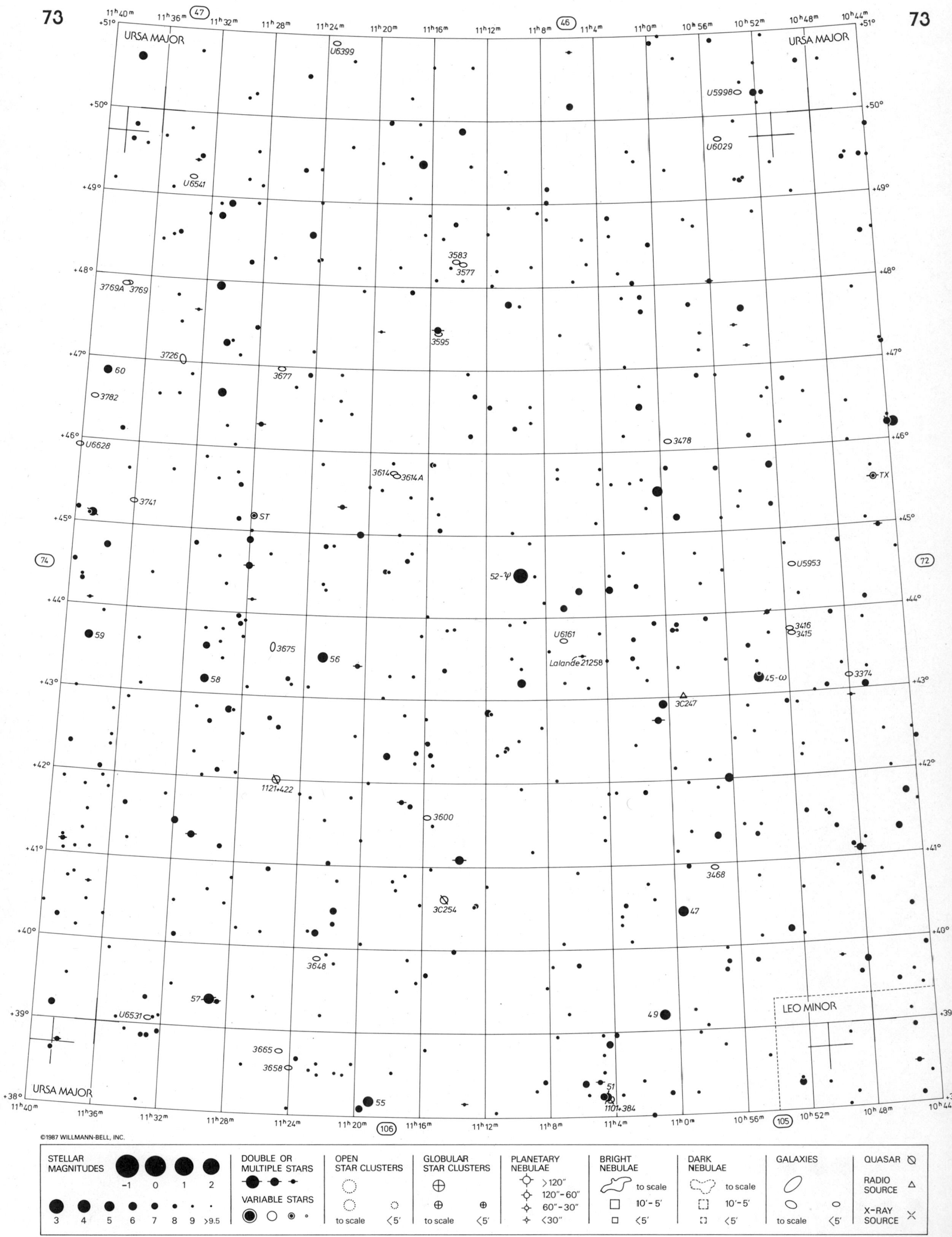

Barry Rappaport & Wil Tirion

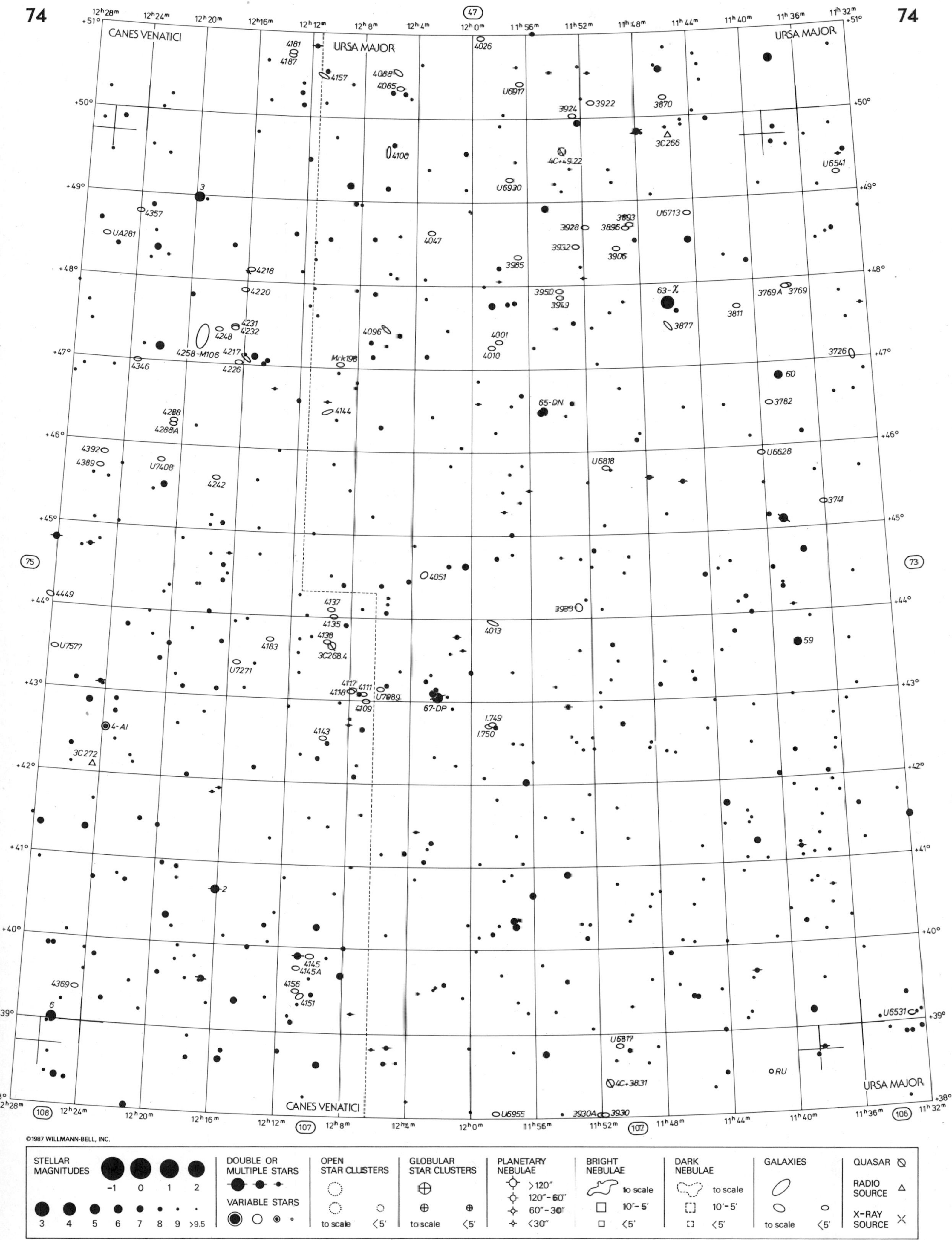

Barry Rappaport & Wil Tirion

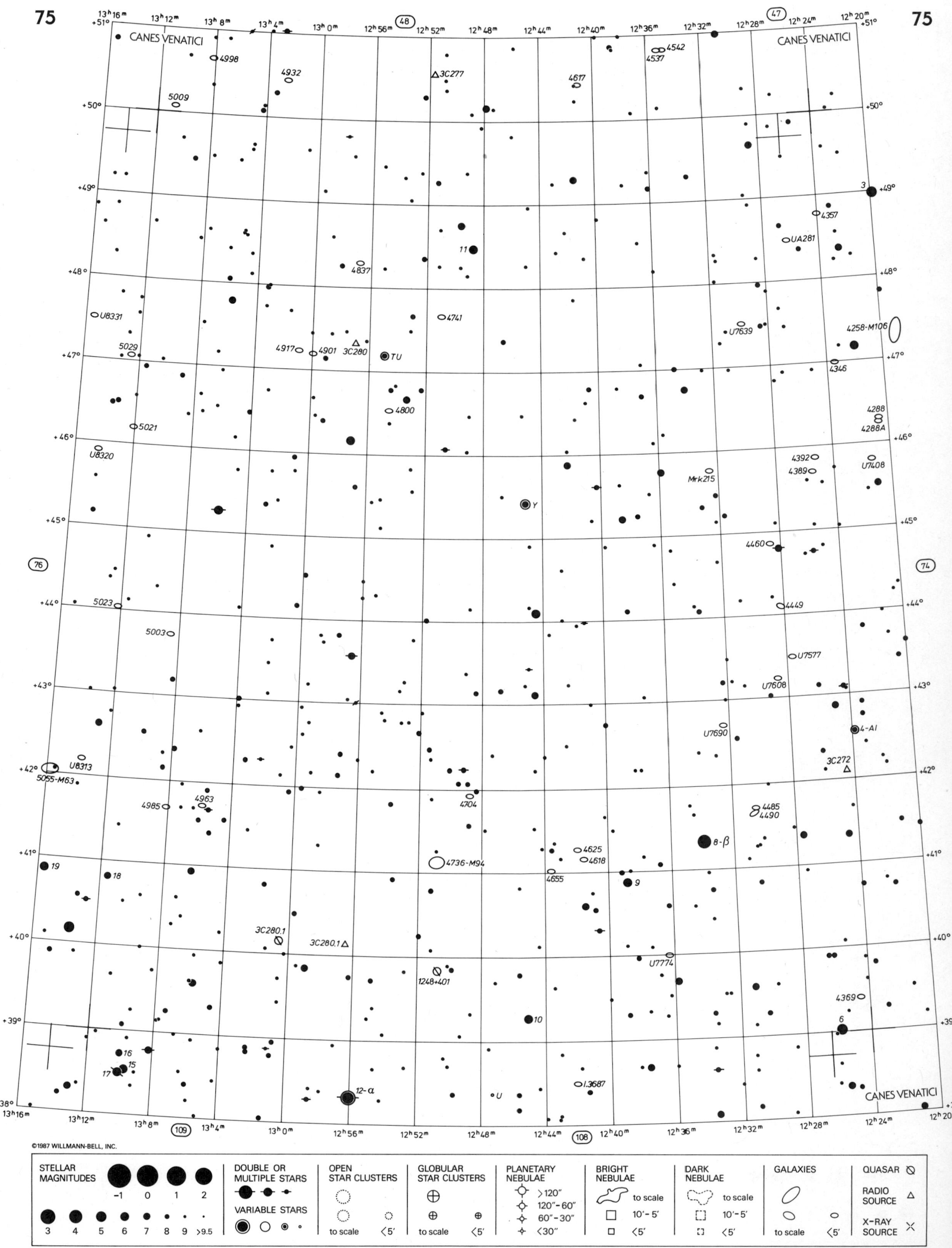

CANES VENATICI
13h16m 13h12m 13h8m 13h4m 13h0m 12h56m 12h52m 12h48m 12h44m 12h40m 12h36m 12h32m 12h28m 12h24m 12h20m
+51° +50° +49° +48° +47° +46° +45° +44° +43° +42° +41° +40° +39° +38°
48
47
76
74
109
108
4998
4932
5009
3C277
4617
4537
4542
4357
UA281
11
4837
U8331
4741
U7639
4258-M106
5029
4917
4901
3C280
TU
4346
4800
5021
4288
4288A
U8320
4392
4389
U7408
Mrk215
Y
4460
5023
4449
5003
U7577
U7608
U7690
4-AI
U8313
3C272
5055-M63
4985
4963
4704
4485
4490
8-β
4625
4618
4736-M94
4655
19
18
9
3C280.1
U7774
1248+401
4369
10
6
16
15
17
I.3687
12-α
U
3
©1987 WILLMANN-BELL, INC.
STELLAR MAGNITUDES
-1 0 1 2
3 4 5 6 7 8 9 >9.5
DOUBLE OR MULTIPLE STARS
VARIABLE STARS
OPEN STAR CLUSTERS
to scale
<5′
GLOBULAR STAR CLUSTERS
to scale
<5′
PLANETARY NEBULAE
>120″
120″-60″
60″-30″
<30″
BRIGHT NEBULAE
to scale
10′-5′
<5′
DARK NEBULAE
to scale
10′-5′
<5′
GALAXIES
to scale
<5′
QUASAR
RADIO SOURCE
X-RAY SOURCE
Barry Rappaport & Wil Tirion

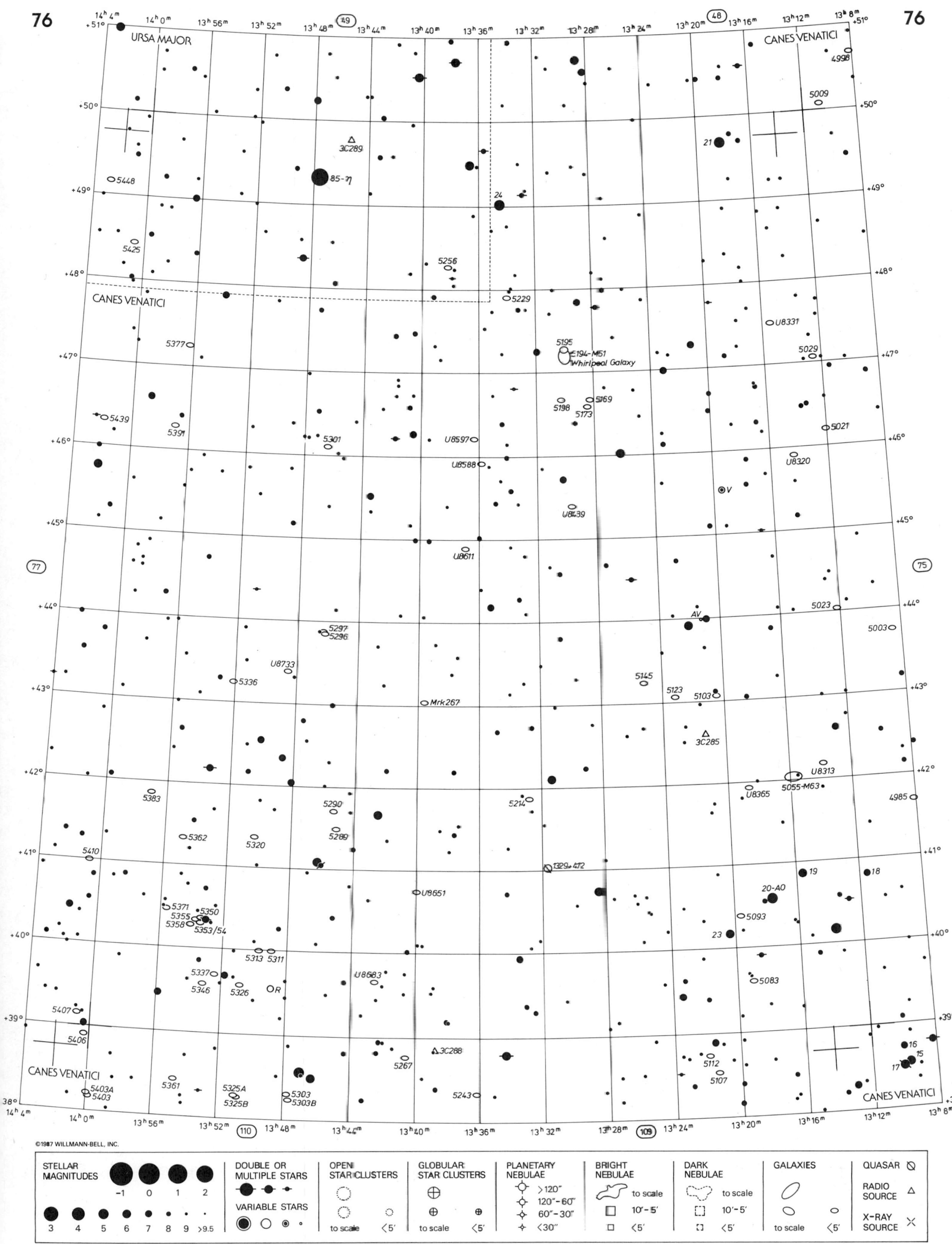

URSA MAJOR
CANES VENATICI
14h 4m
14h 0m
13h 56m
13h 52m
13h 48m
13h 44m
13h 40m
13h 36m
13h 32m
13h 28m
13h 24m
13h 20m
13h 16m
13h 12m
13h 8m
+51°
+50°
+49°
+48°
+47°
+46°
+45°
+44°
+43°
+42°
+41°
+40°
+39°
+38°
49
48
77
75
110
109
3C289
85-η
24
21
4998
5009
5448
5425
5256
5229
U8331
5377
5195
5194-M51
Whirlpool Galaxy
5029
5439
5391
5301
U8597
5198
5173
5169
5021
U8588
U8320
V
U8439
U8611
5023
5003
AV
5297
5296
U8733
5336
5145
5123
5103
Mrk267
3C285
U8313
5055-M63
U8365
4985
5383
5290
5214
5362
5320
5289
5410
1329+412
19
18
U8651
20-AO
5371
5355
5350
5358
5353/54
5093
23
5313
5311
5337
5346
5326
R
U8683
5083
5407
5406
3C288
5267
16
15
17
5112
5107
5361
5325A
5325B
5303
5303B
5243
5403A
5403
STELLAR MAGNITUDES
-1 0 1 2
3 4 5 6 7 8 9 >9.5
DOUBLE OR MULTIPLE STARS
VARIABLE STARS
OPEN STAR CLUSTERS
to scale <5'
GLOBULAR STAR CLUSTERS
to scale <5'
PLANETARY NEBULAE
>120"
120"-60"
60"-30"
<30"
BRIGHT NEBULAE
to scale
10'-5'
<5'
DARK NEBULAE
to scale
10'-5'
<5'
GALAXIES
to scale <5'
QUASAR
RADIO SOURCE
X-RAY SOURCE
Barry Rappaport & Wil Tirion

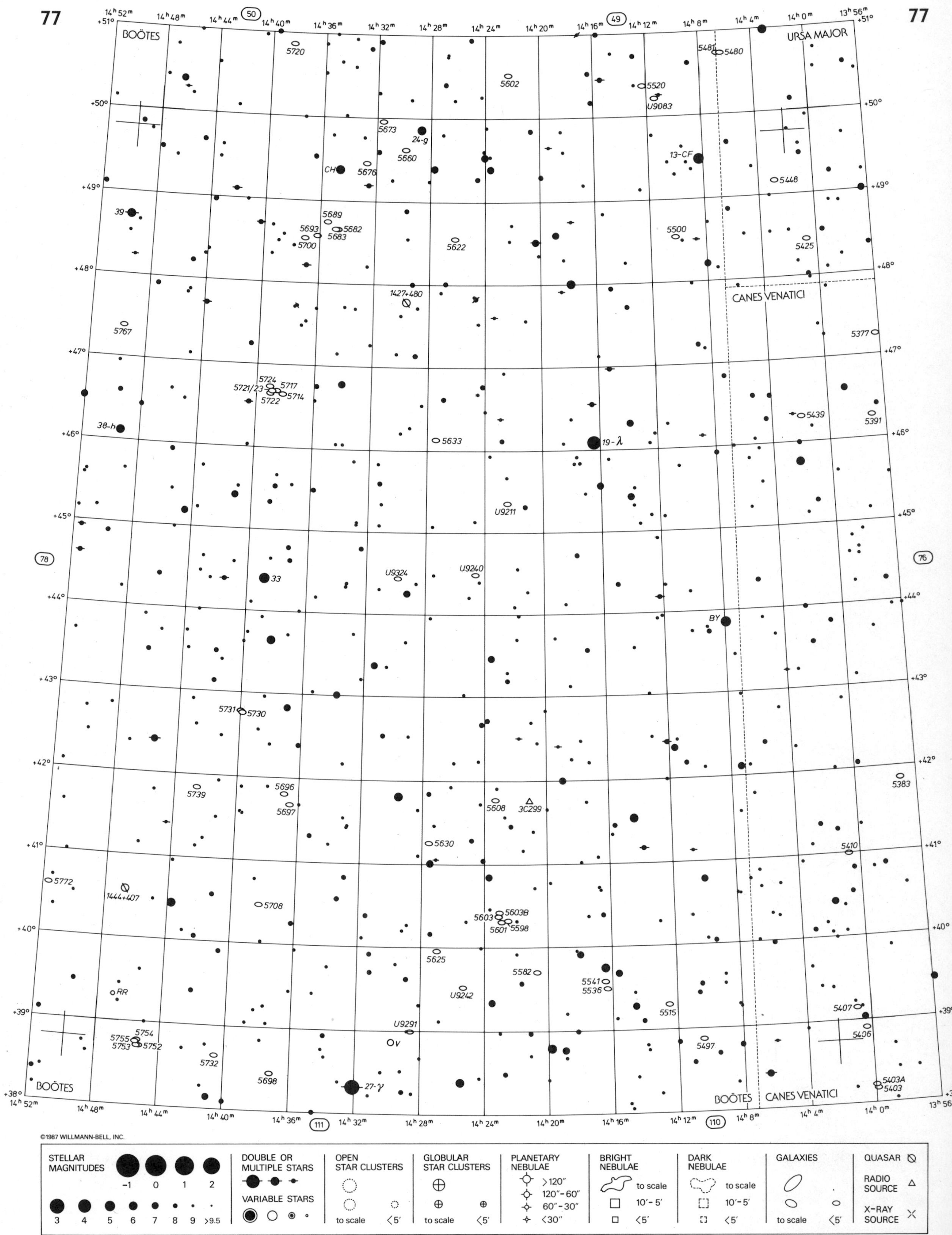
BOÖTES
URSA MAJOR
CANES VENATICI
BOÖTES
BOÖTES CANES VENATICI
14h 52m
14h 48m
14h 44m
14h 40m
14h 36m
14h 32m
14h 28m
14h 24m
14h 20m
14h 16m
14h 12m
14h 8m
14h 4m
14h 0m
13h 56m
+51°
+50°
+49°
+48°
+47°
+46°
+45°
+44°
+43°
+42°
+41°
+40°
+39°
+38°
50
49
78
76
111
110
5720
5602
5520
U9083
5481
5480
5673
24-g
5660
13-CF
CH
5676
5448
39
5689
5693
5682
5683
5700
5622
5500
5425
1427+480
5767
5377
5724
5721/23
5717
5714
5722
5439
5391
38-h
5633
19-λ
U9211
U9324
U9240
33
BY
5731
5730
5739
5696
5697
5608
3C299
5383
5630
5410
5772
1444+407
5708
5603
5603B
5601
5598
5625
5582
5541
5536
RR
U9242
5515
5407
U9291
V
5755
5754
5753
5752
5497
5406
5732
5698
27-γ
5403A
5403
STELLAR MAGNITUDES
-1 0 1 2
3 4 5 6 7 8 9 >9.5
DOUBLE OR MULTIPLE STARS
VARIABLE STARS
OPEN STAR CLUSTERS
to scale
<5'
GLOBULAR STAR CLUSTERS
to scale
<5'
PLANETARY NEBULAE
>120"
120"–60"
60"–30"
<30"
BRIGHT NEBULAE
to scale
10'–5'
<5'
DARK NEBULAE
to scale
10'–5'
<5'
GALAXIES
to scale
<5'
QUASAR
RADIO SOURCE
X-RAY SOURCE
Barry Rappaport & Wil Tirion

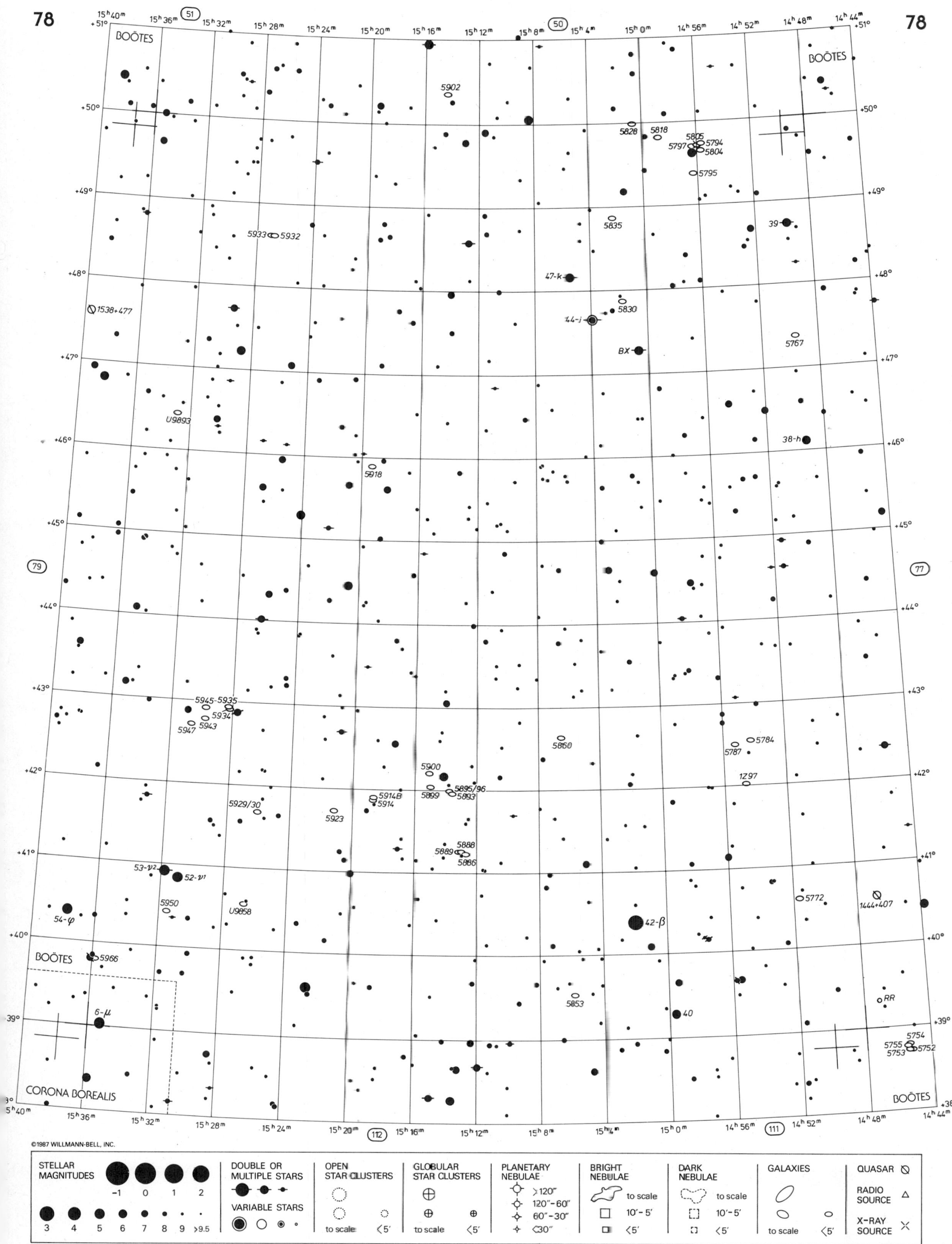

Barry Rappaport & Wil Tirion

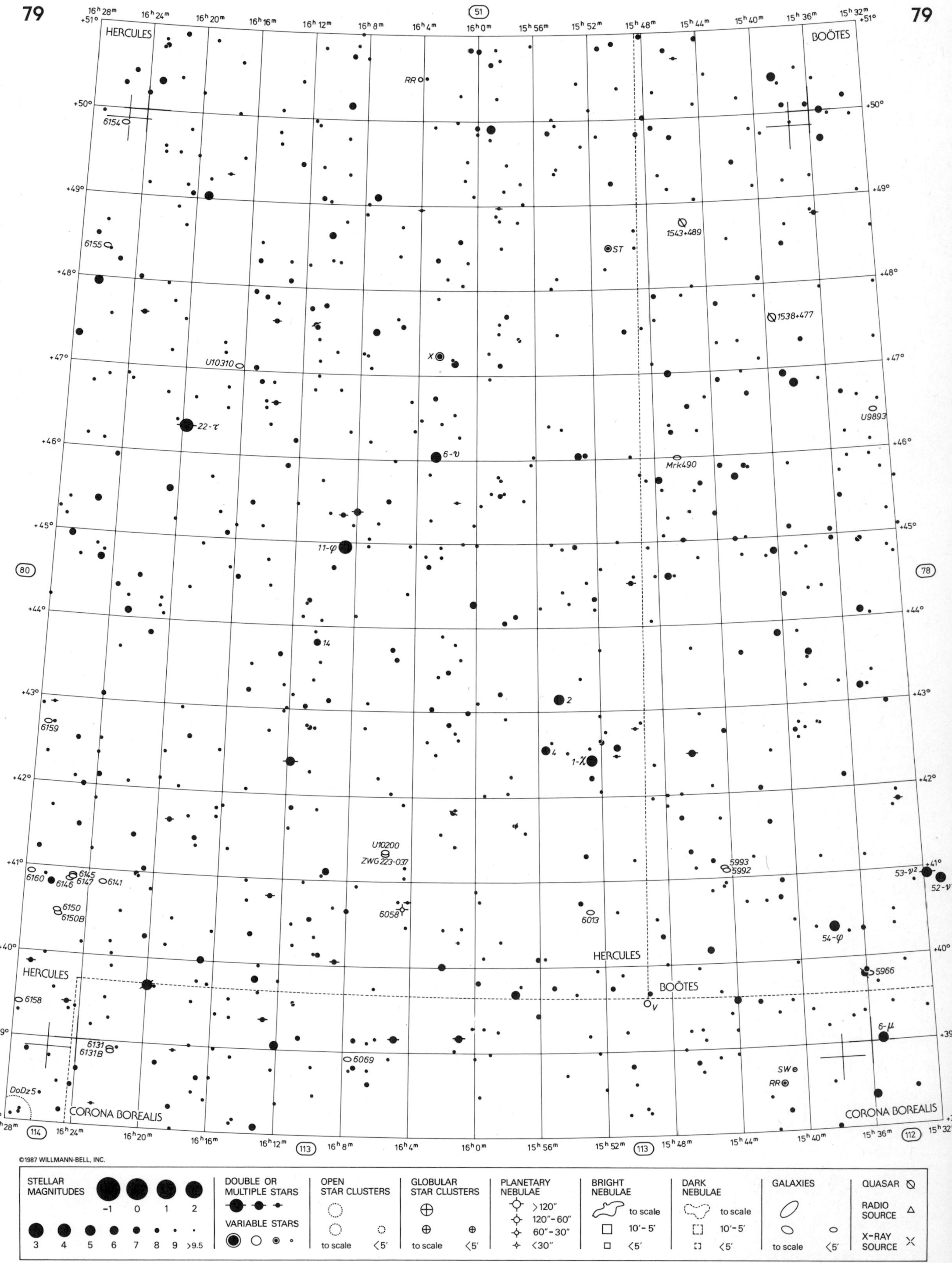

Barry Rappaport & Wil Tirion

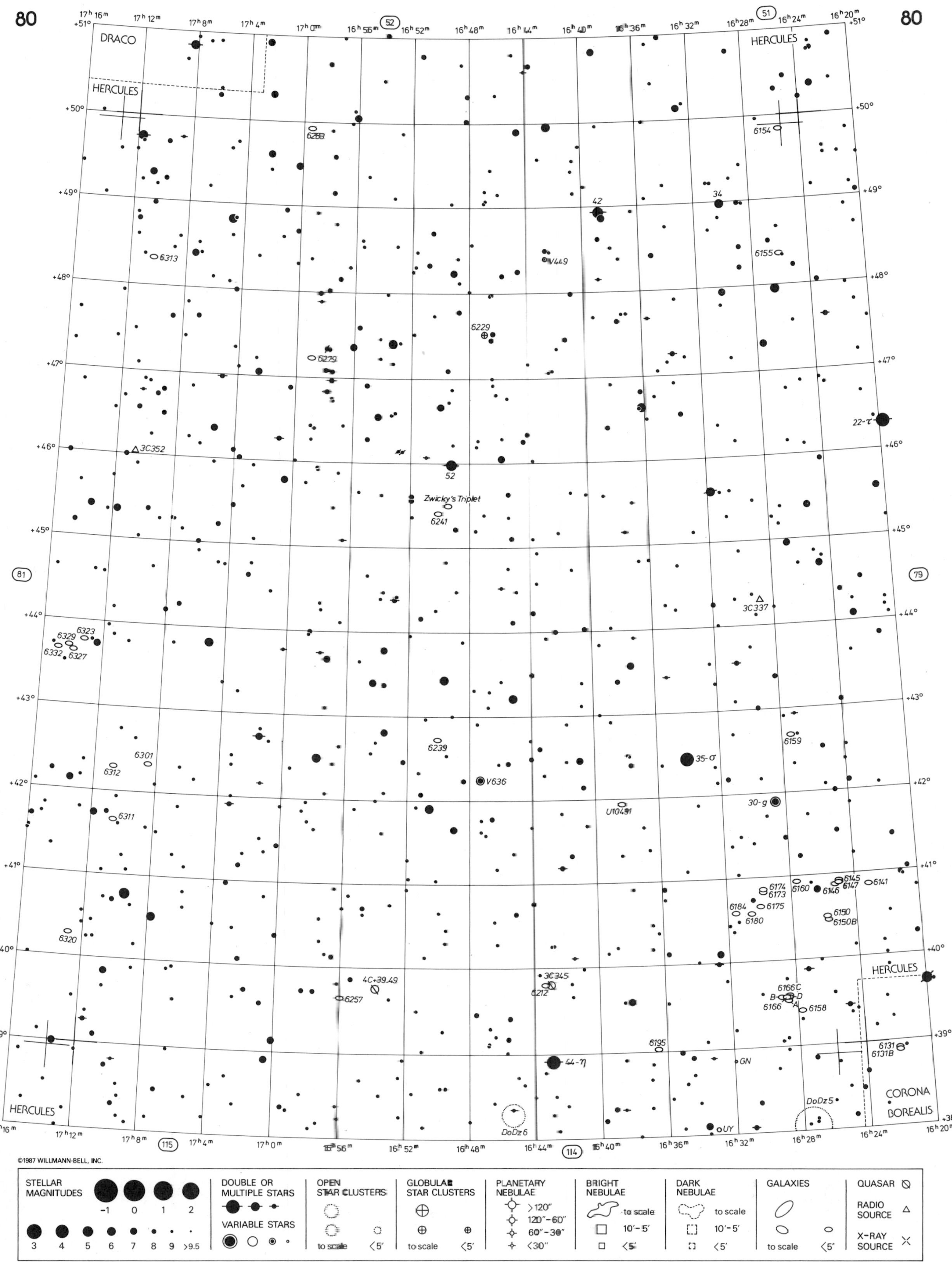
DRACO
HERCULES
CORONA BOREALIS
Zwicky's Triplet
6241
6229
6279
6283
6313
6154
6155
V449
42
34
52
22-τ
3C352
3C337
6323
6329
6332
6327
6301
6312
6311
6239
V636
35-σ
U10491
30-g
6159
6174
6173
6160
6145
6147
6146
6141
6184
6175
6180
6150
6150B
6320
4C+39.49
6257
3C345
6212
6166C
B
D
A
6166
6158
6195
44-η
GN
6131
6131B
DoDz5
DoDz6
UY
81
79
52
51
115
114
©1987 WILLMANN-BELL, INC.
STELLAR MAGNITUDES
DOUBLE OR MULTIPLE STARS
VARIABLE STARS
OPEN STAR CLUSTERS
GLOBULAR STAR CLUSTERS
PLANETARY NEBULAE
BRIGHT NEBULAE
DARK NEBULAE
GALAXIES
QUASAR
RADIO SOURCE
X-RAY SOURCE
to scale
Barry Rappaport & Wil Tirion

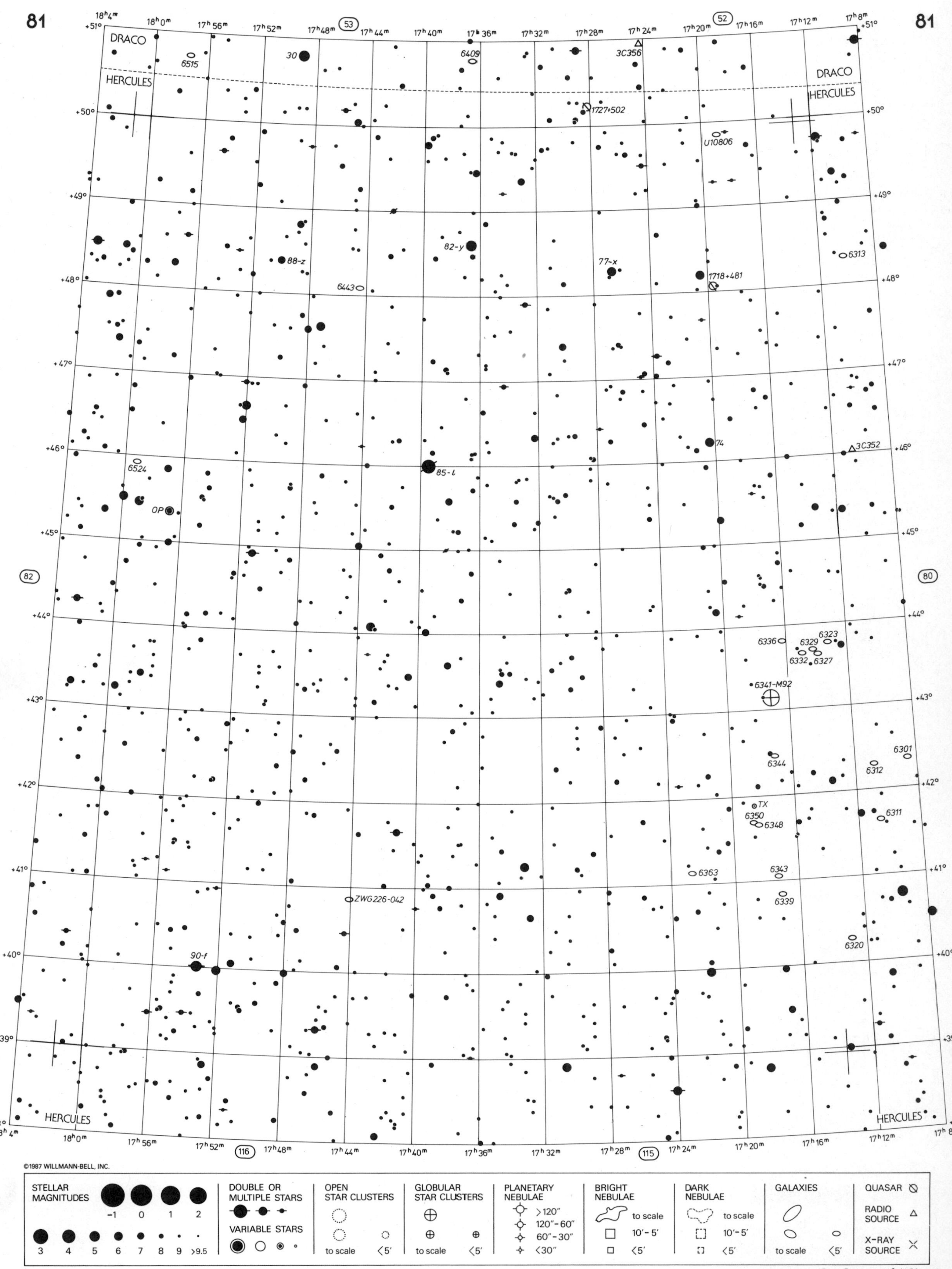

Barry Rappaport & Wil Tirion

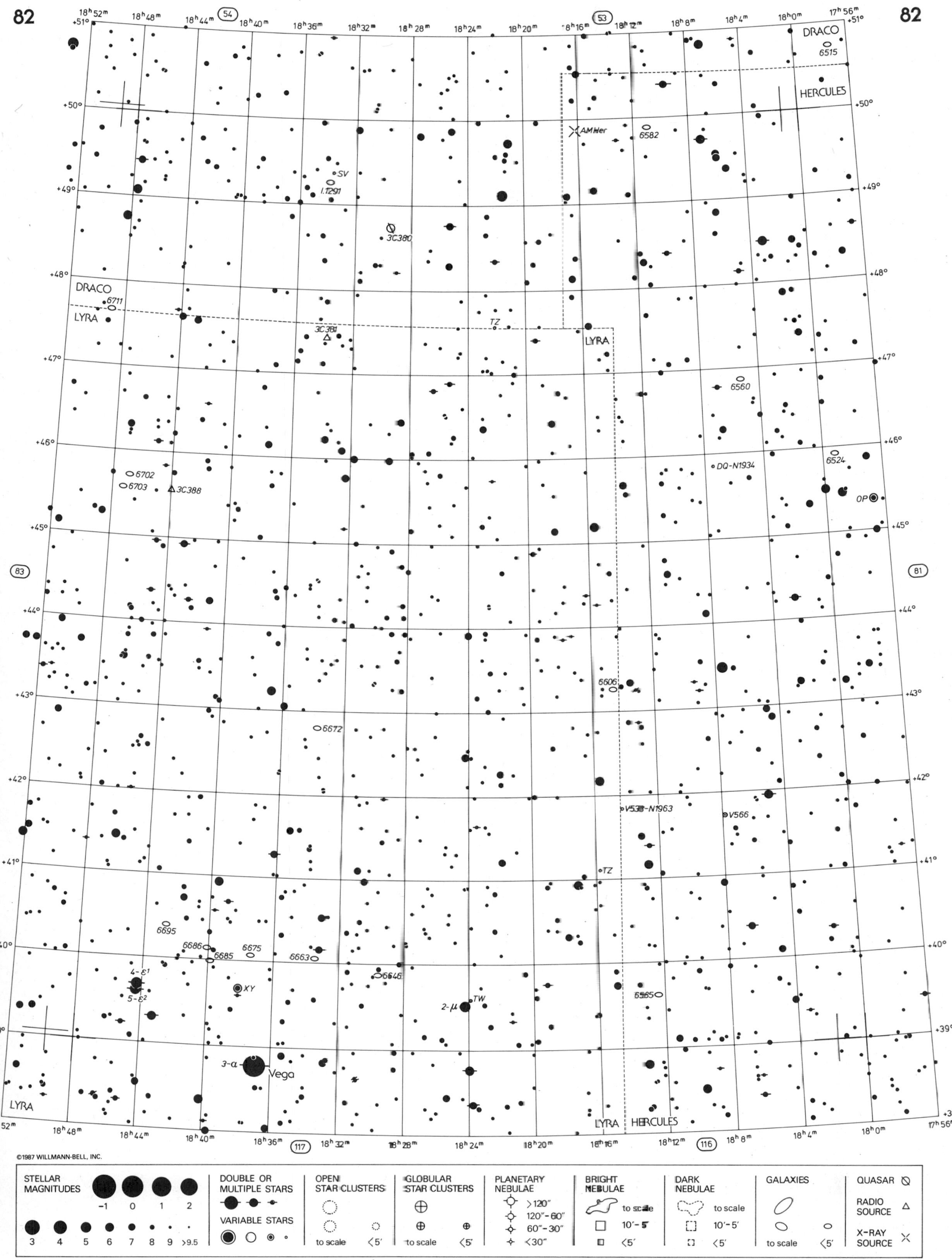

Barry Rappaport & Wil Tirion

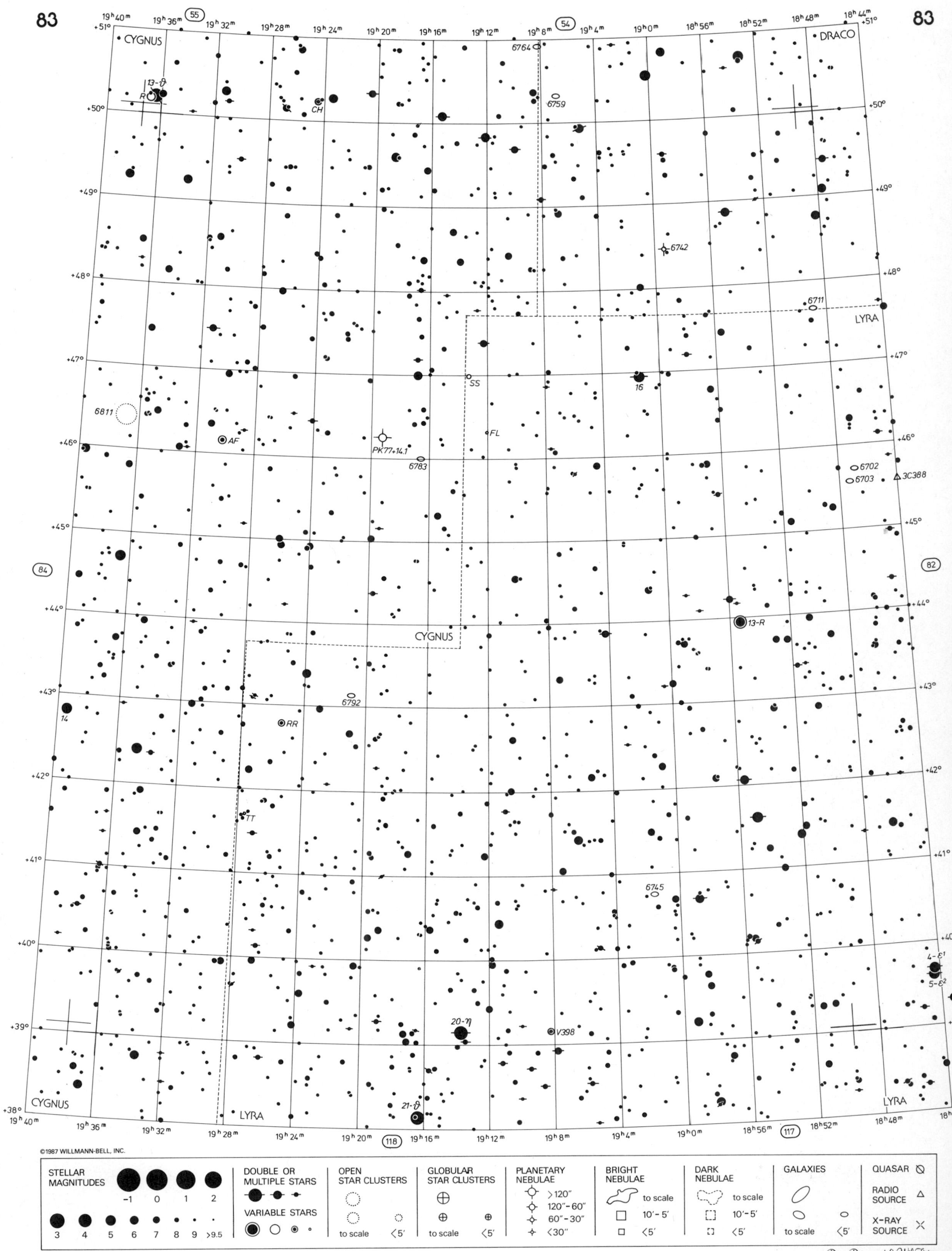

CYGNUS
DRACO
LYRA
13-θ
CH
6764
6759
6742
6711
SS
16
6811
AF
PK77+14.1
6783
FL
6702
6703
3C388
84
82
13-R
6792
RR
14
TT
6745
4-ε¹
5-ε²
20-η
V398
21-δ
55
54
118
117
STELLAR MAGNITUDES
DOUBLE OR MULTIPLE STARS
VARIABLE STARS
OPEN STAR CLUSTERS
GLOBULAR STAR CLUSTERS
PLANETARY NEBULAE
BRIGHT NEBULAE
DARK NEBULAE
GALAXIES
QUASAR
RADIO SOURCE
X-RAY SOURCE
to scale
Barry Rappaport & Wil Tirion

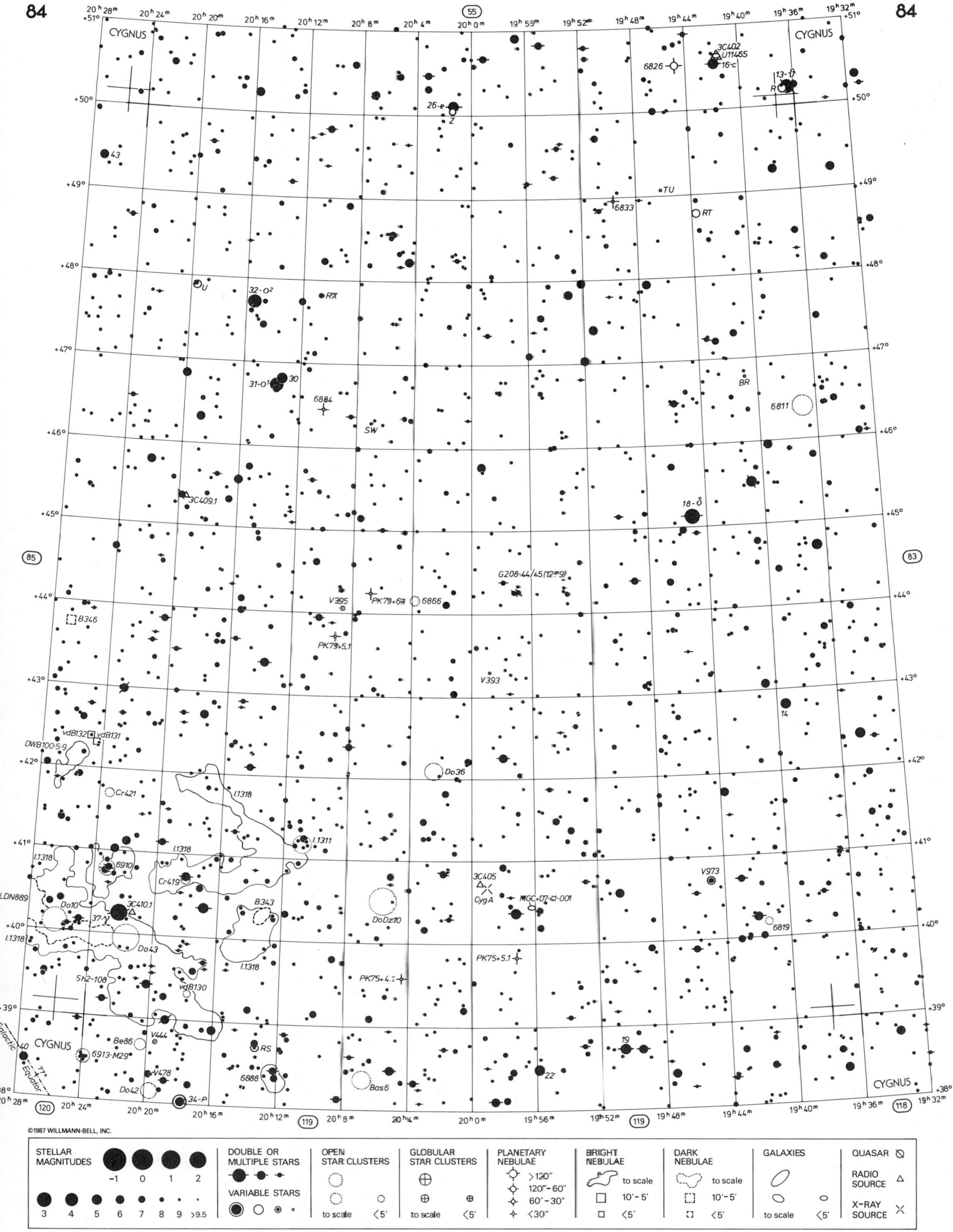

Barry Rappaport & Wil Tirion

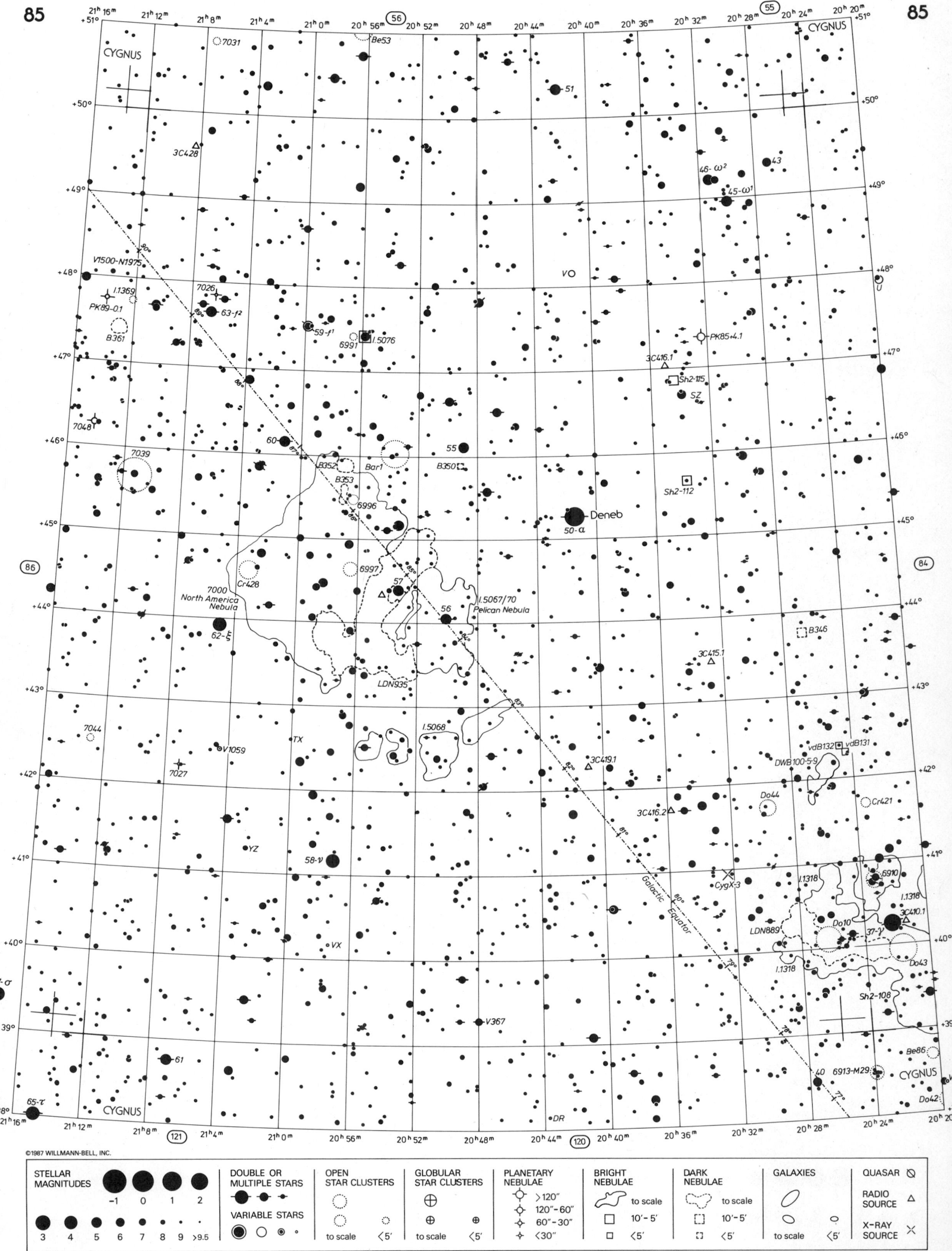

CYGNUS
21h 16m
21h 12m
21h 8m
21h 4m
21h 0m
20h 56m
20h 52m
20h 48m
20h 44m
20h 40m
20h 36m
20h 32m
20h 28m
20h 24m
20h 20m
56
55
120
121
86
84
+51°
+50°
+49°
+48°
+47°
+46°
+45°
+44°
+43°
+42°
+41°
+40°
+39°
+38°
7031
Be53
51
3C428
43
46-ω2
45-ω1
V1500-N1975
VO
U
I.1369
PK89-0.1
7026
63-f2
B361
59-f1
6991
I.5076
PK85+4.1
3C416.1
Sh2-115
SZ
7048
60
55
7039
B352
Bar1
B353
B350
Sh2-112
6996
Deneb
50-α
6997
Cr428
57
7000
North America
Nebula
I.5067/70
Pelican Nebula
56
62-ξ
B346
3C415.1
LDN935
7044
I.5068
TX
V1059
3C419.1
vdB132
vdB131
DWB 100-5-9
7027
Do44
Cr421
3C416.2
YZ
58-ν
6910
I.1318
CygX-3
Galactic Equator
3C410.1
Do10
LDN889
37-γ
VX
67-σ
Do43
Sh2-108
V367
61
Be86
40
6913-M29
V478
Do42
65-τ
DR
©1987 WILLMANN-BELL, INC.
STELLAR MAGNITUDES
-1 0 1 2
3 4 5 6 7 8 9 >9.5
DOUBLE OR MULTIPLE STARS
VARIABLE STARS
OPEN STAR CLUSTERS
to scale <5'
GLOBULAR STAR CLUSTERS
to scale <5'
PLANETARY NEBULAE
>120"
120"-60"
60"-30"
<30"
BRIGHT NEBULAE
to scale
10'-5'
<5'
DARK NEBULAE
to scale
10'-5'
<5'
GALAXIES
to scale <5'
QUASAR
RADIO SOURCE
X-RAY SOURCE
Barry Rappaport & Wil Tirion

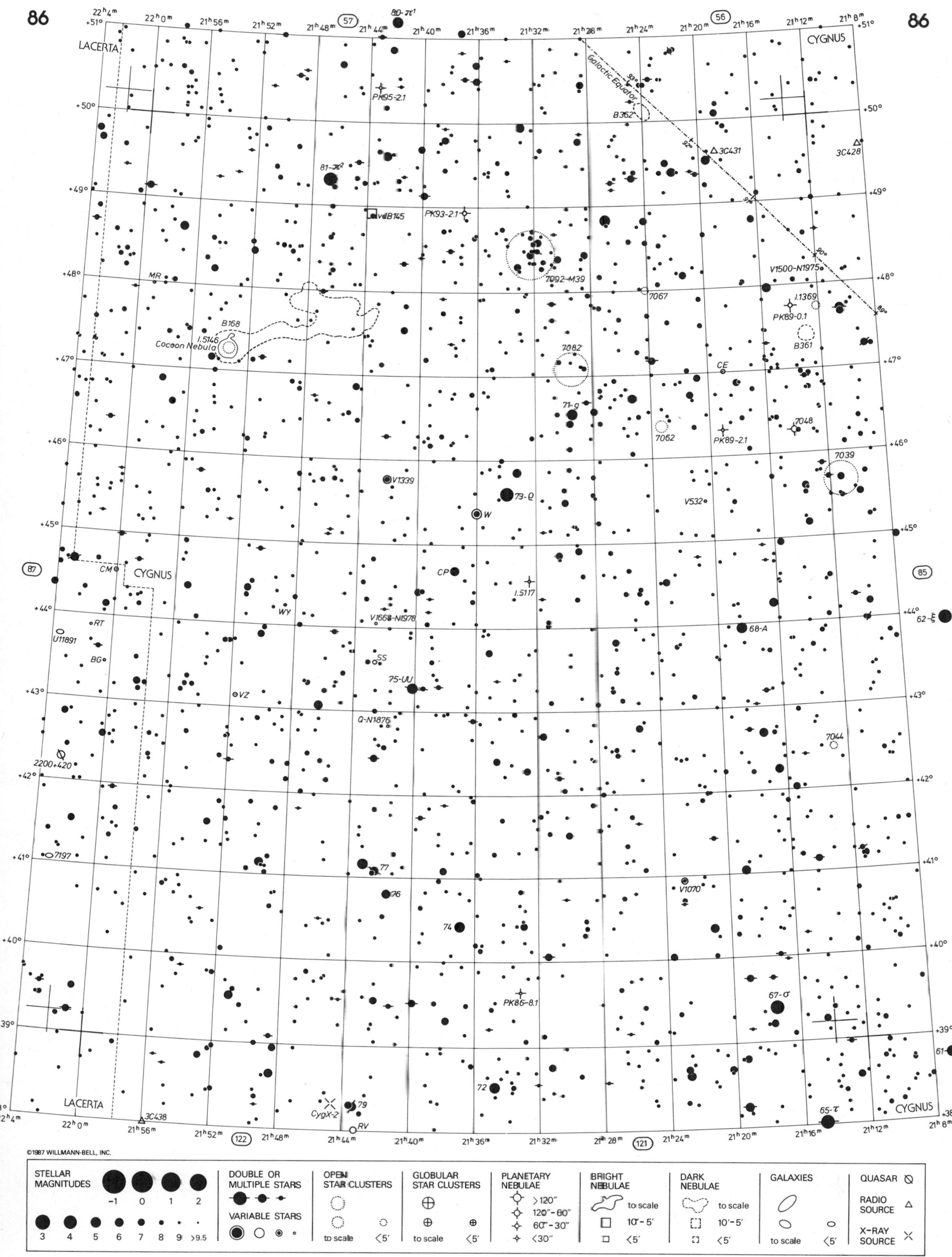

Barry Rappaport & Wil Tirion

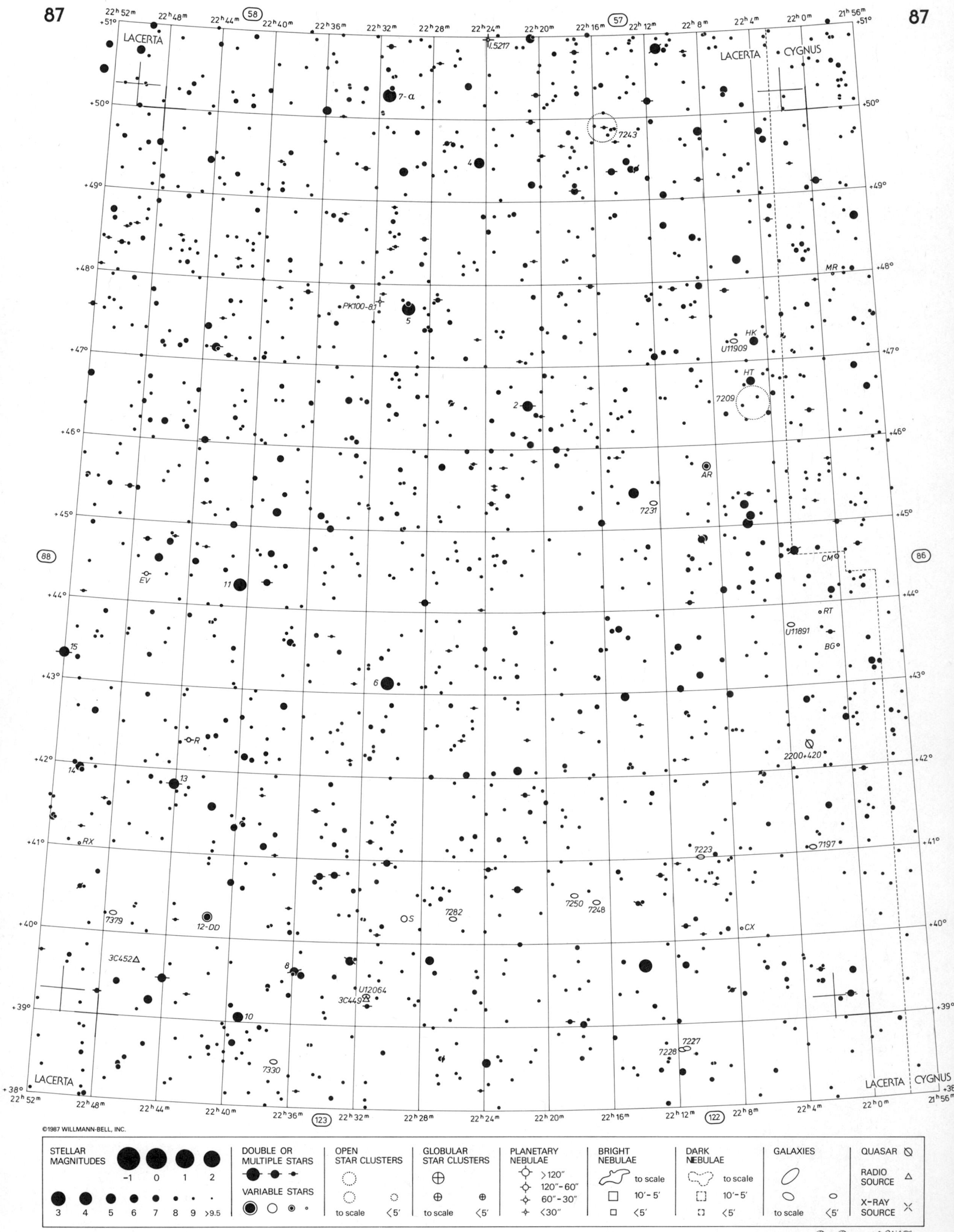
LACERTA
CYGNUS
7-α
I.5217
7243
PK100-8.1
U11909
HK
HT
7209
AR
7231
EV
CM
RT
U11891
BG
MR
2200+420
R
RX
7197
7223
7250
7248
7282
S
7379
12-DD
CX
3C452
U12064
3C449
7227
7228
7330
©1987 WILLMANN-BELL, INC.
STELLAR MAGNITUDES
-1 0 1 2
3 4 5 6 7 8 9 >9.5
DOUBLE OR MULTIPLE STARS
VARIABLE STARS
OPEN STAR CLUSTERS
to scale <5′
GLOBULAR STAR CLUSTERS
to scale <5′
PLANETARY NEBULAE
>120″
120″-60″
60″-30″
<30″
BRIGHT NEBULAE
to scale
10′-5′
<5′
DARK NEBULAE
to scale
10′-5′
<5′
GALAXIES
to scale <5′
QUASAR
RADIO SOURCE
X-RAY SOURCE
Barry Rappaport & Wil Tirion

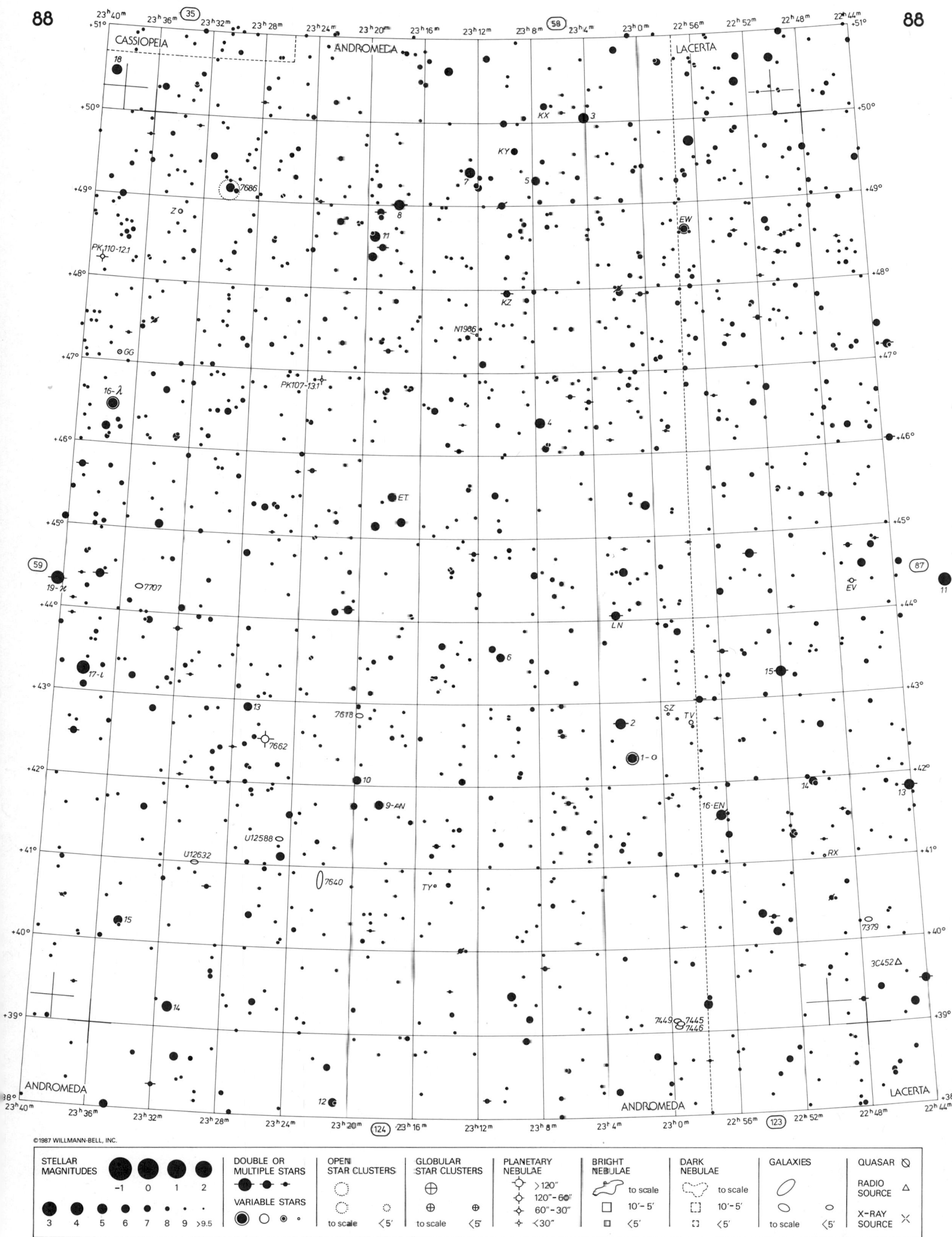

Barry Rappaport & Wil Tirion

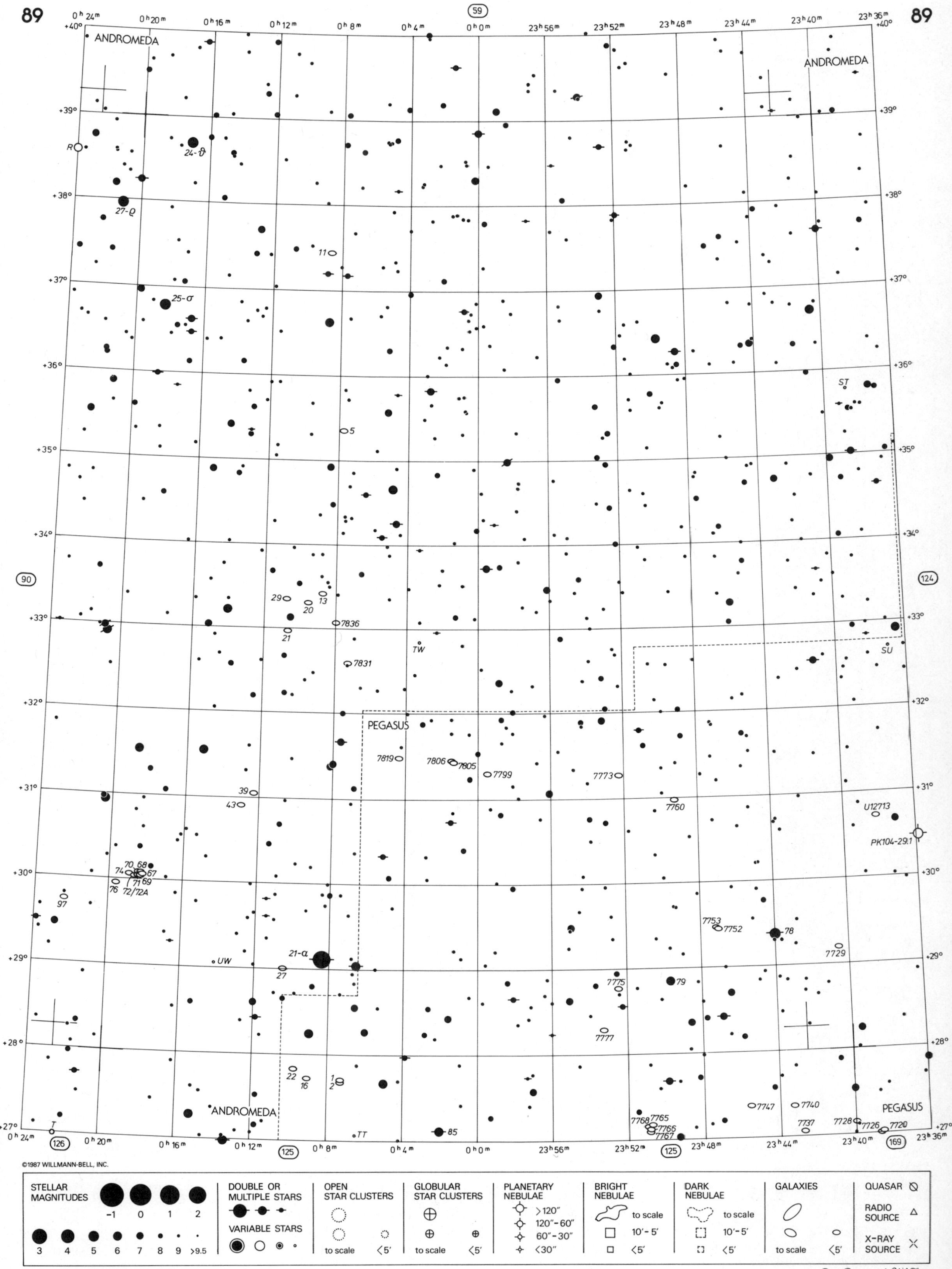

Barry Rappaport & Wil Tirion

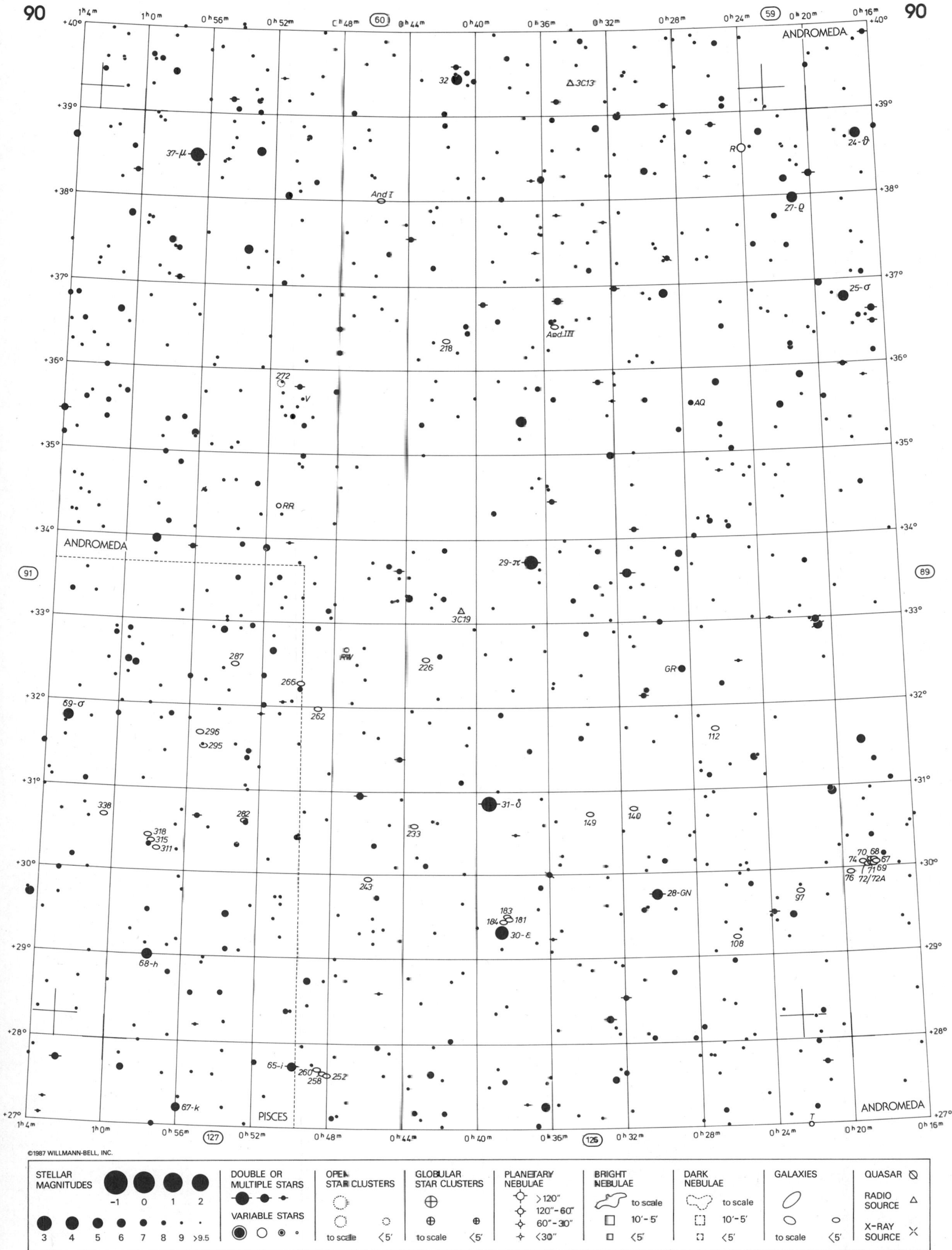
ANDROMEDA
PISCES
+40°
+39°
+38°
+37°
+36°
+35°
+34°
+33°
+32°
+31°
+30°
+29°
+28°
+27°
1h4m
1h0m
0h56m
0h52m
0h48m
0h44m
0h40m
0h36m
0h32m
0h28m
0h24m
0h20m
0h16m
60
59
91
89
127
126
32
3C13
37-μ
24-ϑ
R
And I
27-ρ
25-σ
And III
218
272
V
AQ
RR
29-π
3C19
287
RW
226
GR
266
262
69-σ
296
295
112
338
282
31-δ
149
140
318
315
311
233
70
68
74
67
71
69
76
72/72A
243
28-GN
97
183
184
181
30-ε
108
68-h
65-i
260
258
252
67-k
T
STELLAR MAGNITUDES
-1
0
1
2
3
4
5
6
7
8
9
>9.5
DOUBLE OR MULTIPLE STARS
VARIABLE STARS
OPEN STAR CLUSTERS
GLOBULAR STAR CLUSTERS
PLANETARY NEBULAE
BRIGHT NEBULAE
DARK NEBULAE
GALAXIES
QUASAR
RADIO SOURCE
X-RAY SOURCE
to scale
<5'
>120"
120"-60"
60"-30"
<30"
10'-5'
Barry Rappaport & Wil Tirion

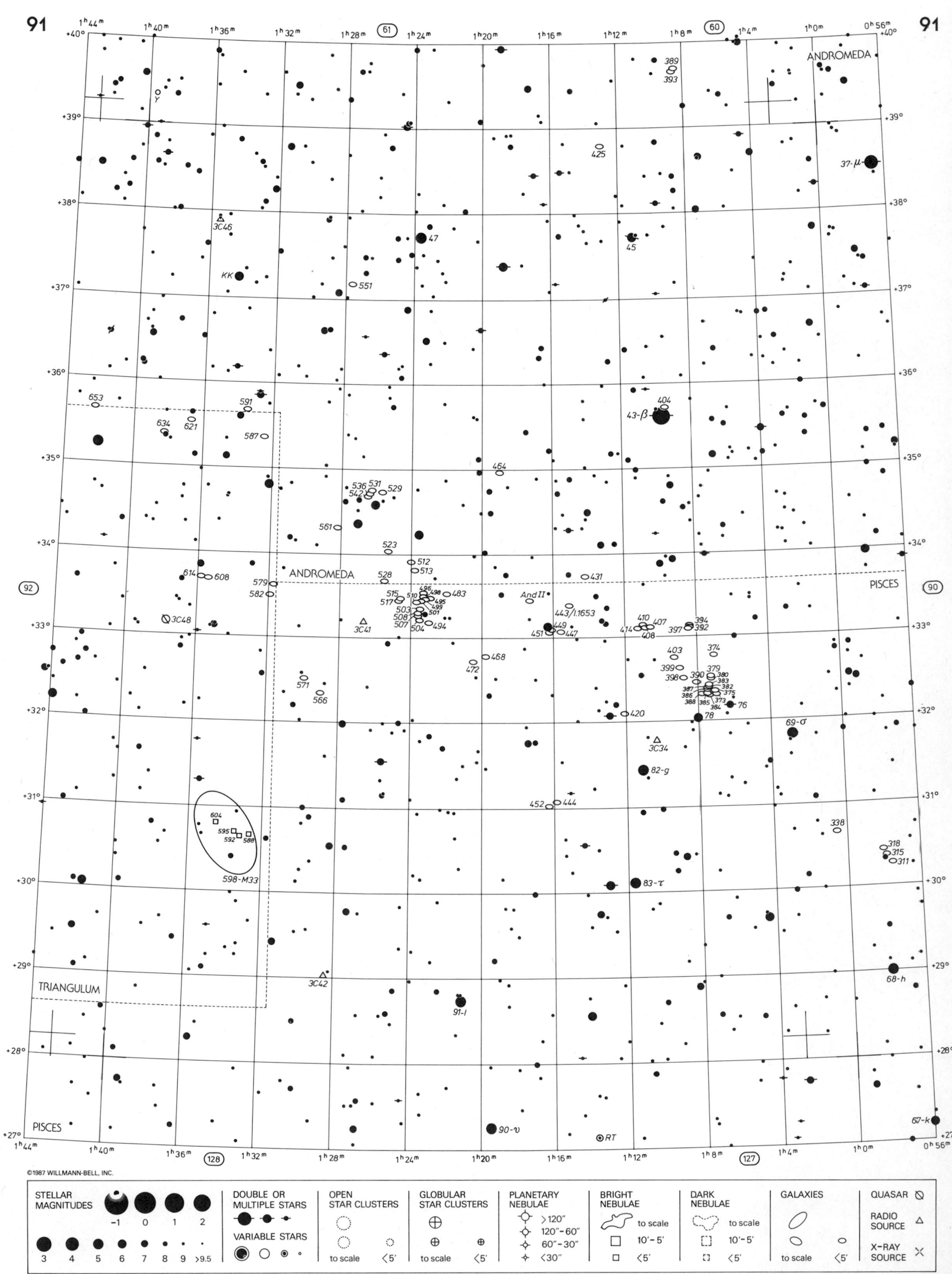

Barry Rappaport & Wil Tirion

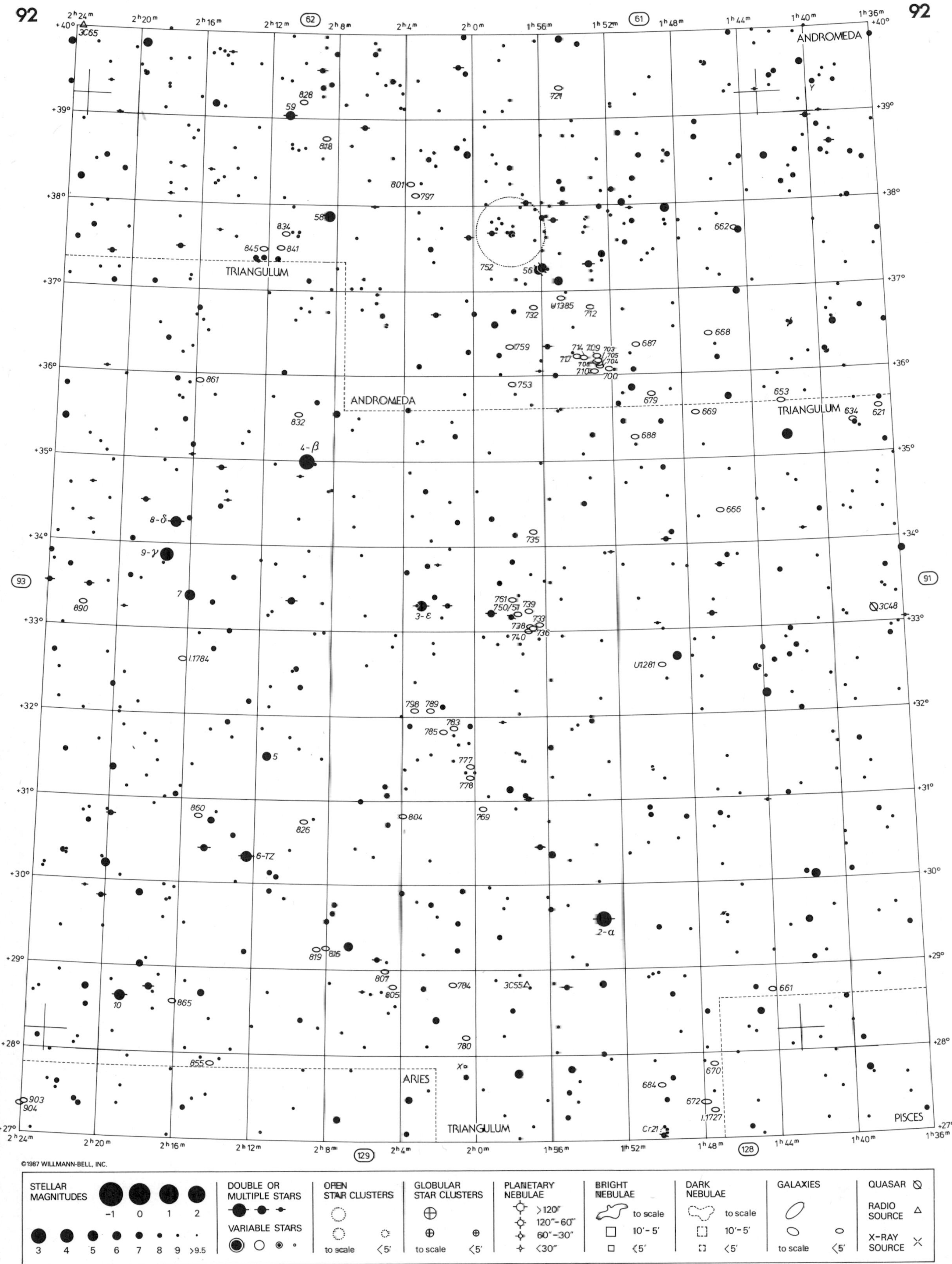

2h24m
2h20m
2h16m
2h12m
62
2h8m
2h4m
2h0m
1h56m
1h52m
61
1h48m
1h44m
1h40m
1h36m
+40°
+39°
+38°
+37°
+36°
+35°
+34°
+33°
+32°
+31°
+30°
+29°
+28°
+27°
93
91
129
128
3C65
ANDROMEDA
TRIANGULUM
ANDROMEDA
TRIANGULUM
ARIES
TRIANGULUM
PISCES
59
58
56
4-β
8-δ
9-γ
7
3-ε
5
6-TZ
2-α
10
3C48
3C55
X
Cr21
U1385
U1281
I.1784
I.1727
©1987 WILLMANN-BELL, INC.
STELLAR MAGNITUDES
-1
0
1
2
3
4
5
6
7
8
9
>9.5
DOUBLE OR MULTIPLE STARS
VARIABLE STARS
OPEN STAR CLUSTERS
to scale
<5'
GLOBULAR STAR CLUSTERS
to scale
<5'
PLANETARY NEBULAE
>120"
120"–60"
60"–30"
<30"
BRIGHT NEBULAE
to scale
10'–5'
<5'
DARK NEBULAE
to scale
10'–5'
<5'
GALAXIES
to scale
<5'
QUASAR
RADIO SOURCE
X-RAY SOURCE

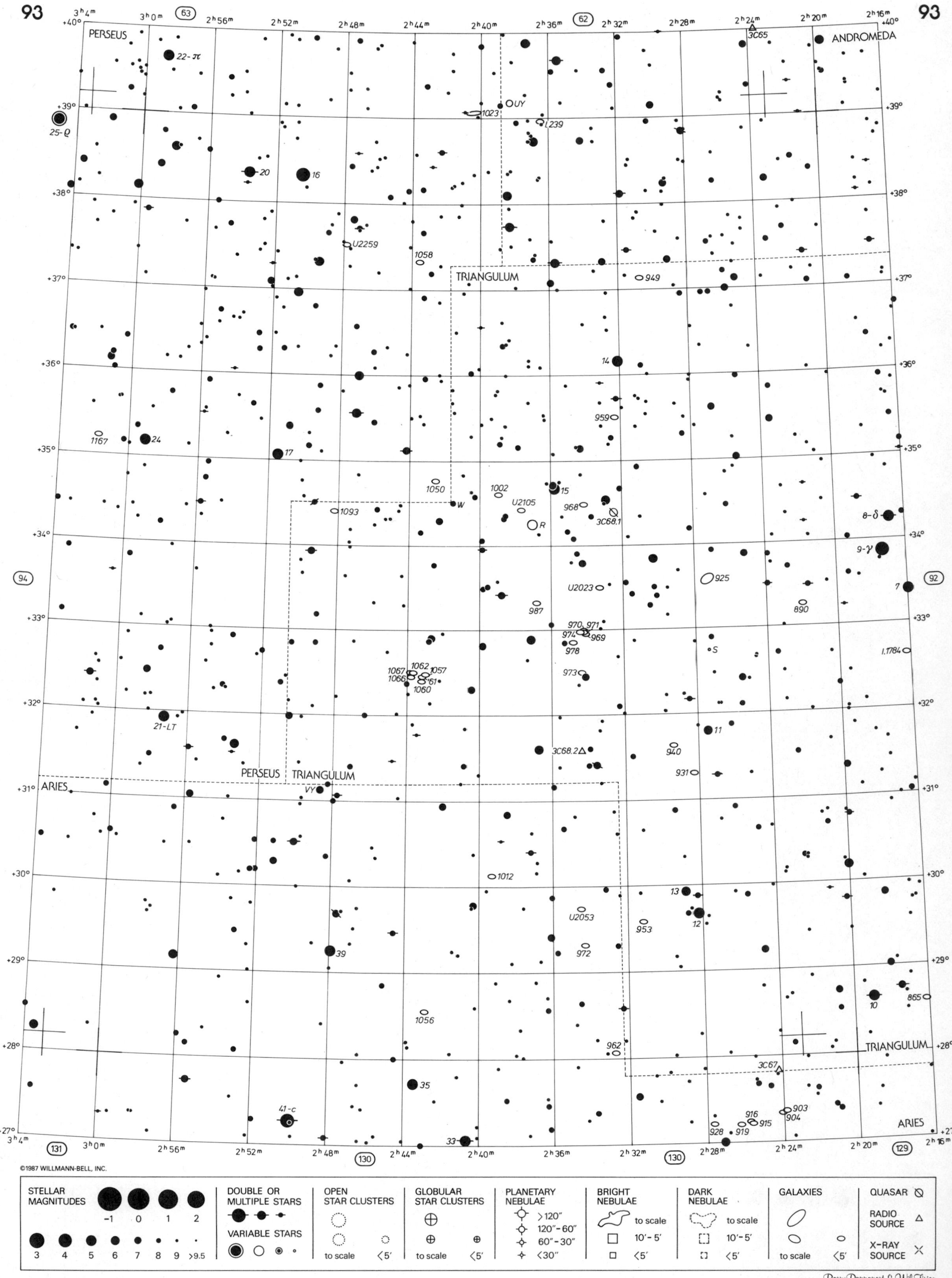

Barry Rappaport & Wil Tirion

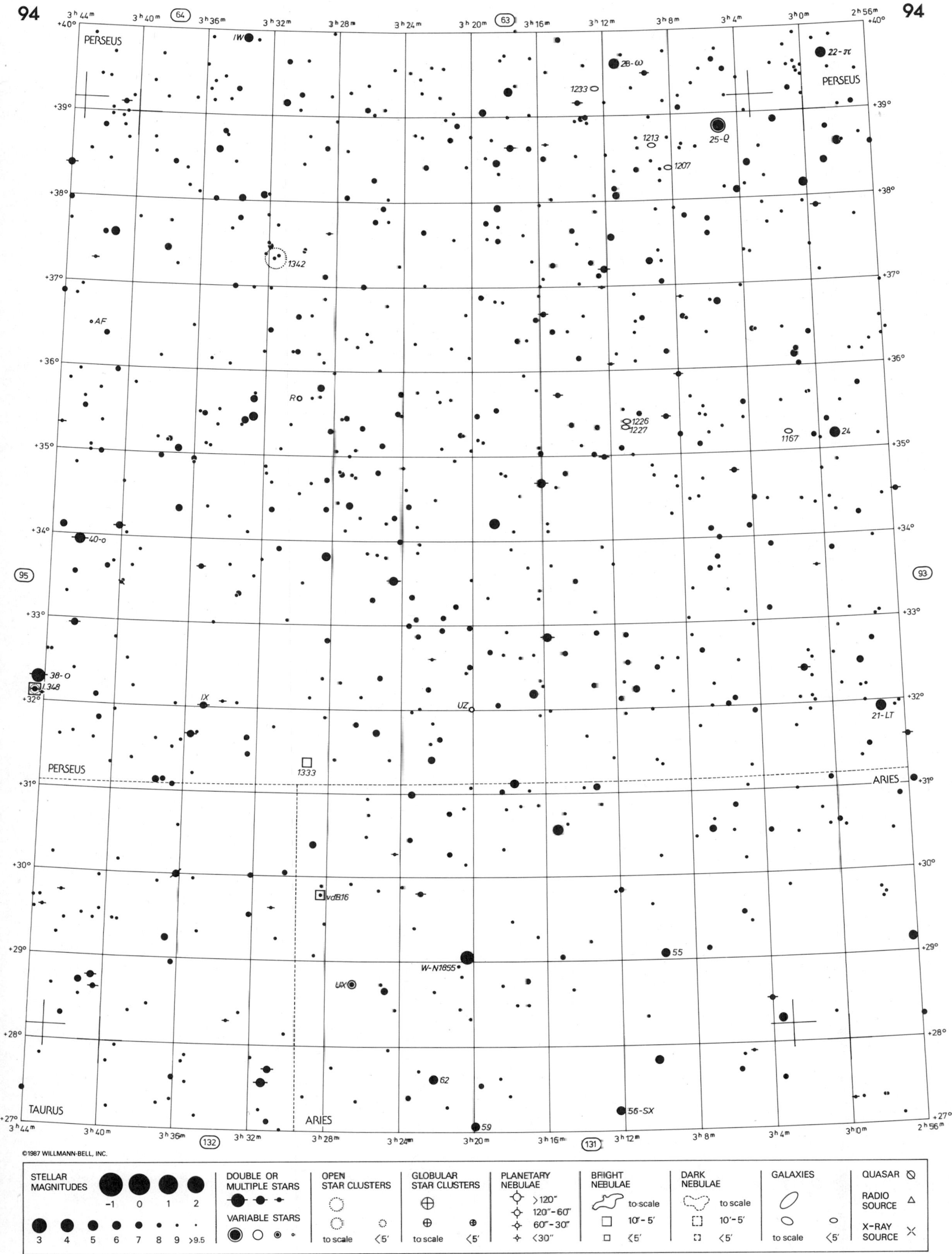

Barry Rappaport & Wil Tirion

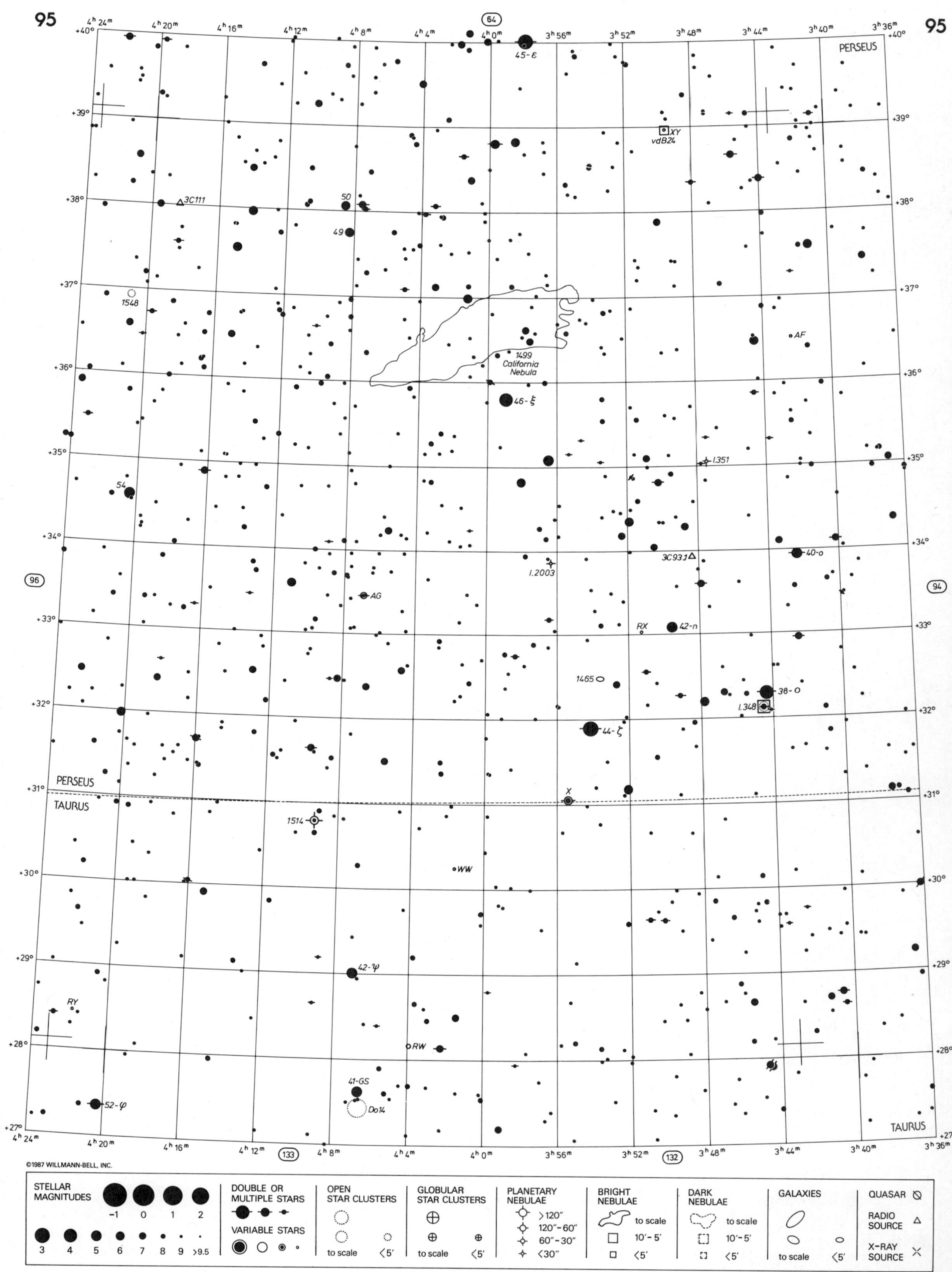

95
PERSEUS
TAURUS
64
96
94
133
132
45-ε
XY
vdB24
3C111
50
49
1548
1499
California
Nebula
AF
46-ξ
I.351
54
40-o
3C93.1
I.2003
AG
RX
42-n
1465
38-o
I.348
44-ζ
X
1514
WW
42-ψ
RY
RW
41-GS
52-φ
Do14
+40°
+39°
+38°
+37°
+36°
+35°
+34°
+33°
+32°
+31°
+30°
+29°
+28°
+27°
4h 24m
4h 20m
4h 16m
4h 12m
4h 8m
4h 4m
4h 0m
3h 56m
3h 52m
3h 48m
3h 44m
3h 40m
3h 36m
©1987 WILLMANN-BELL, INC.
STELLAR MAGNITUDES
-1 0 1 2
3 4 5 6 7 8 9 >9.5
DOUBLE OR MULTIPLE STARS
VARIABLE STARS
OPEN STAR CLUSTERS
to scale
<5′
GLOBULAR STAR CLUSTERS
to scale
<5′
PLANETARY NEBULAE
>120″
120″–60″
60″–30″
<30″
BRIGHT NEBULAE
to scale
10′–5′
<5′
DARK NEBULAE
to scale
10′–5′
<5′
GALAXIES
to scale
<5′
QUASAR
RADIO SOURCE
X-RAY SOURCE
Barry Rappaport & Wil Tirion

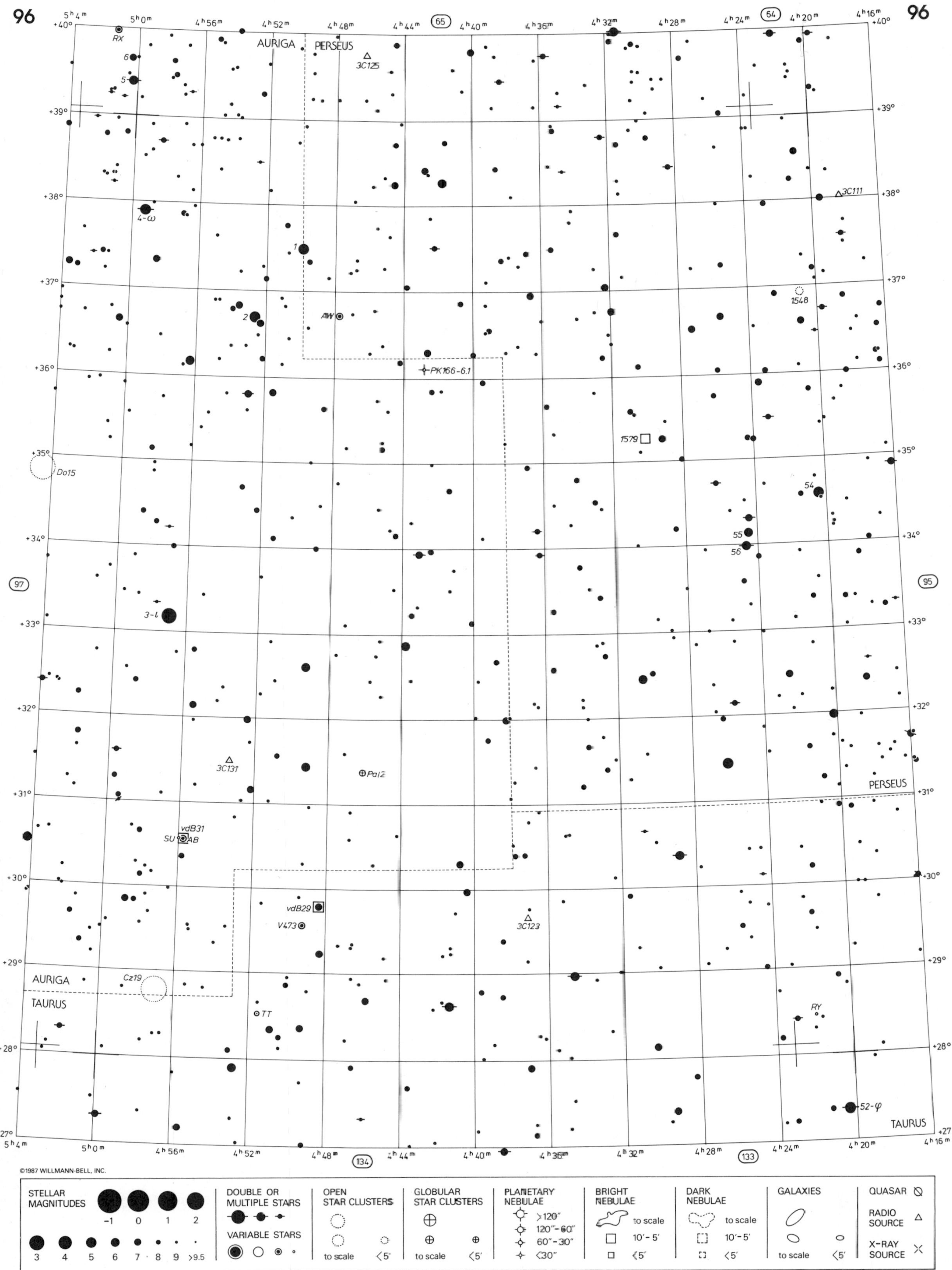
AURIGA
PERSEUS
TAURUS
RX
3C125
3C111
3C131
3C123
4-ω
3-ι
52-φ
1548
1579
PK166-6.1
Do15
Cz19
Pal2
vdB31
SU AB
vdB29
V473
TT
RY
AW
54
55
56
STELLAR MAGNITUDES
DOUBLE OR MULTIPLE STARS
VARIABLE STARS
OPEN STAR CLUSTERS
GLOBULAR STAR CLUSTERS
PLANETARY NEBULAE
BRIGHT NEBULAE
DARK NEBULAE
GALAXIES
QUASAR
RADIO SOURCE
X-RAY SOURCE
to scale
©1987 WILLMANN-BELL, INC.
Barry Rappaport & Wil Tirion

96

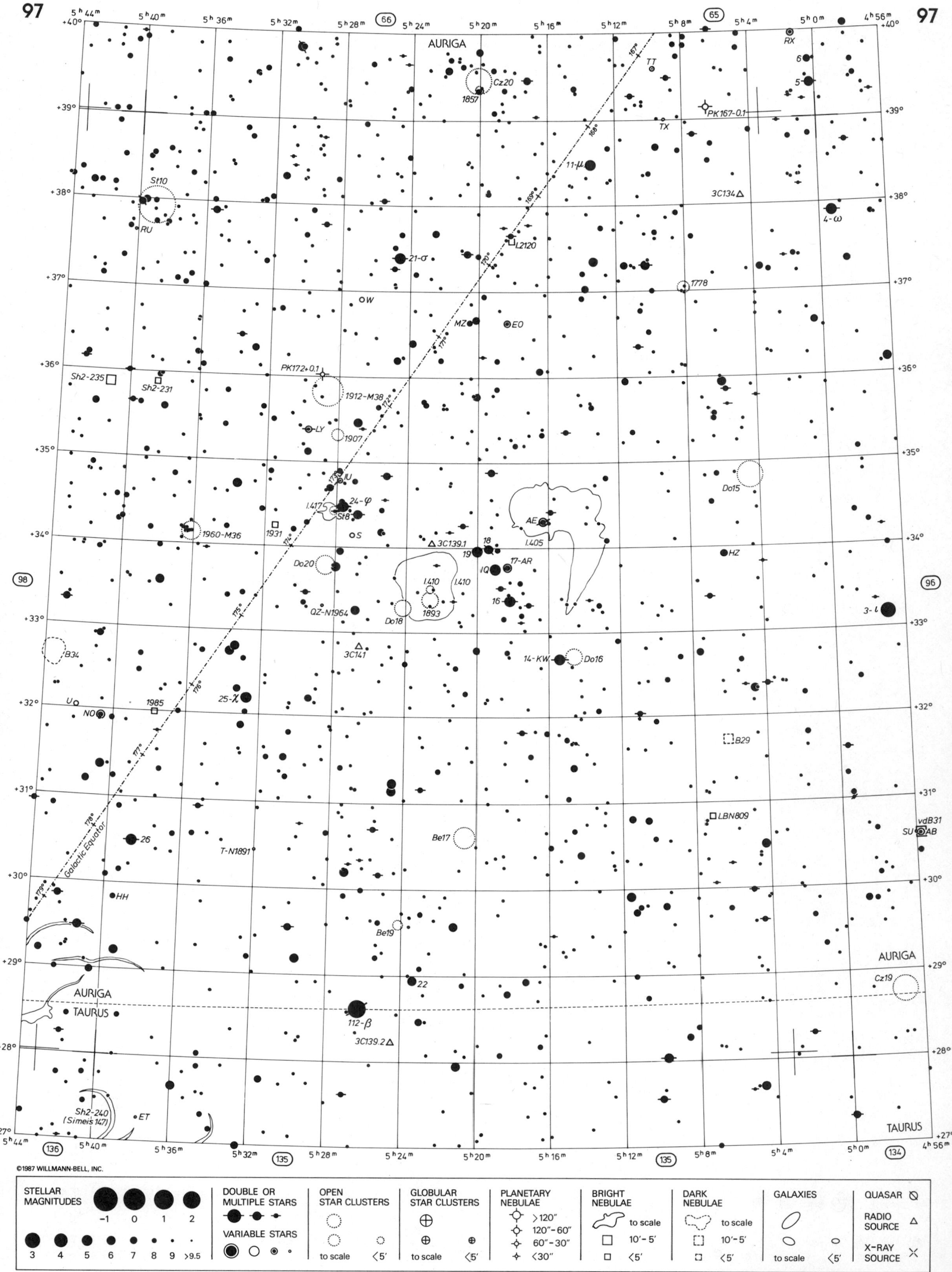

Barry Rappaport & Wil Tirion

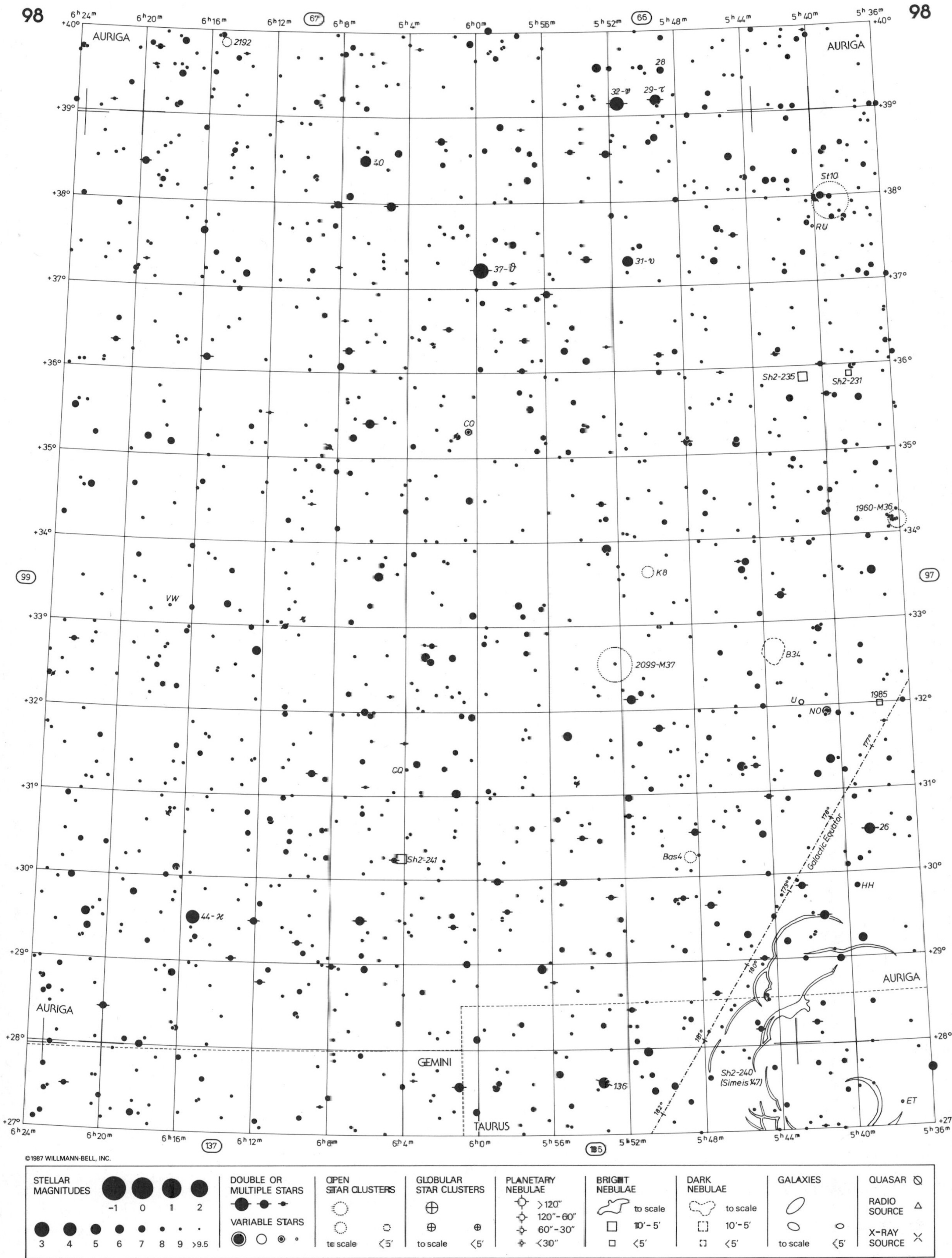

Barry Rappaport & Wil Tirion

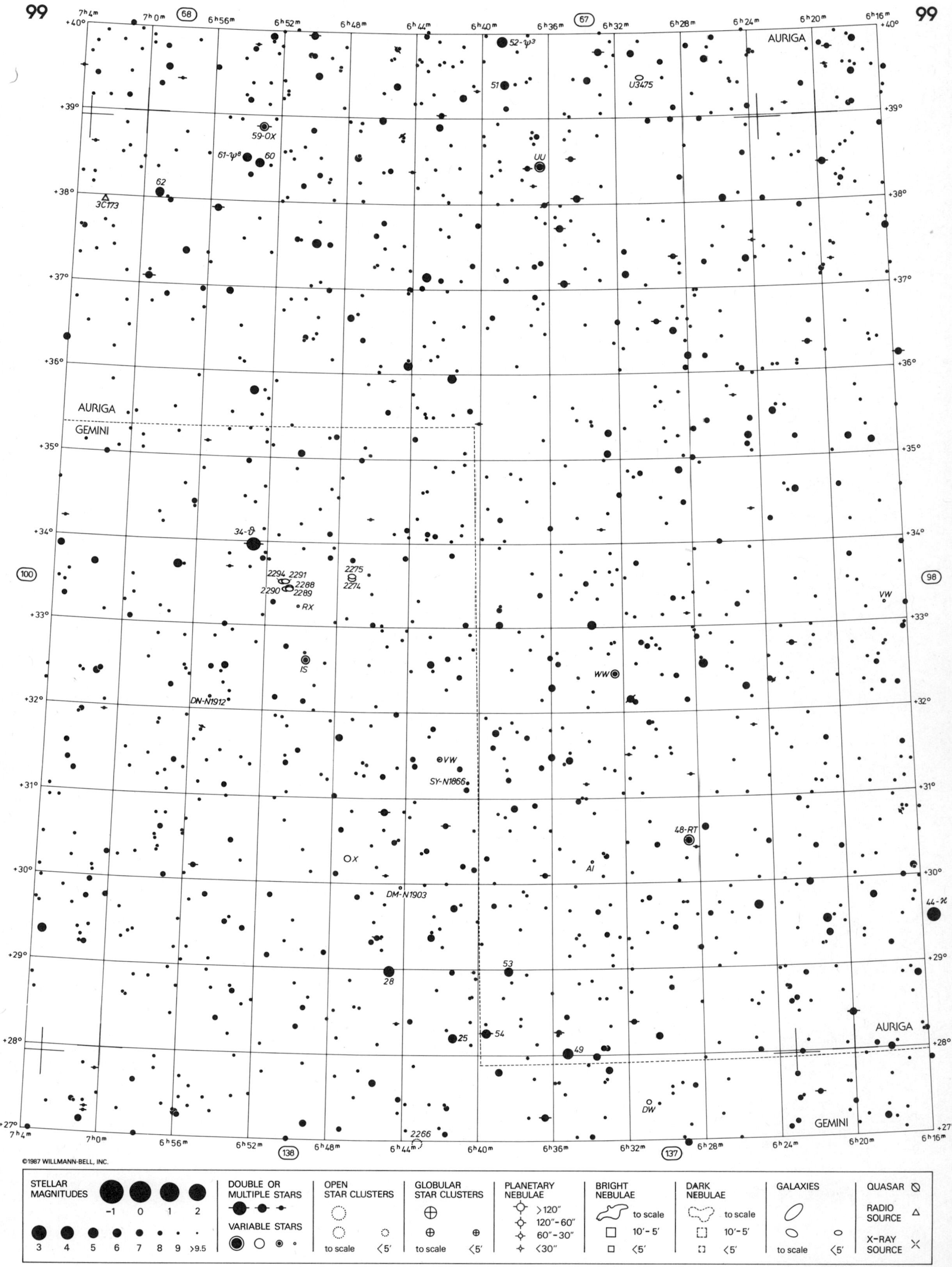

Barry Rappaport & Wil Tirion

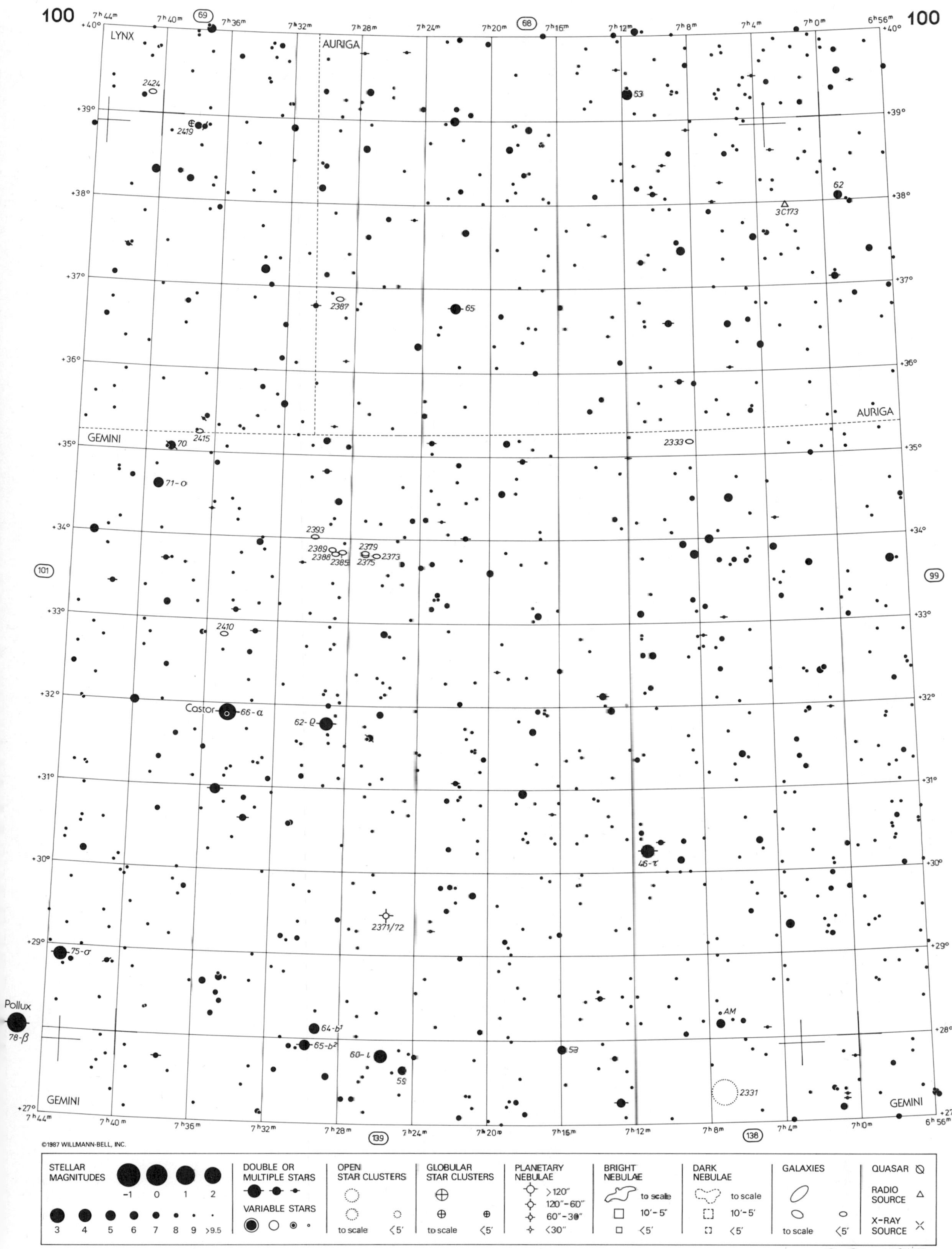

Barry Rappaport & Wil Tirion

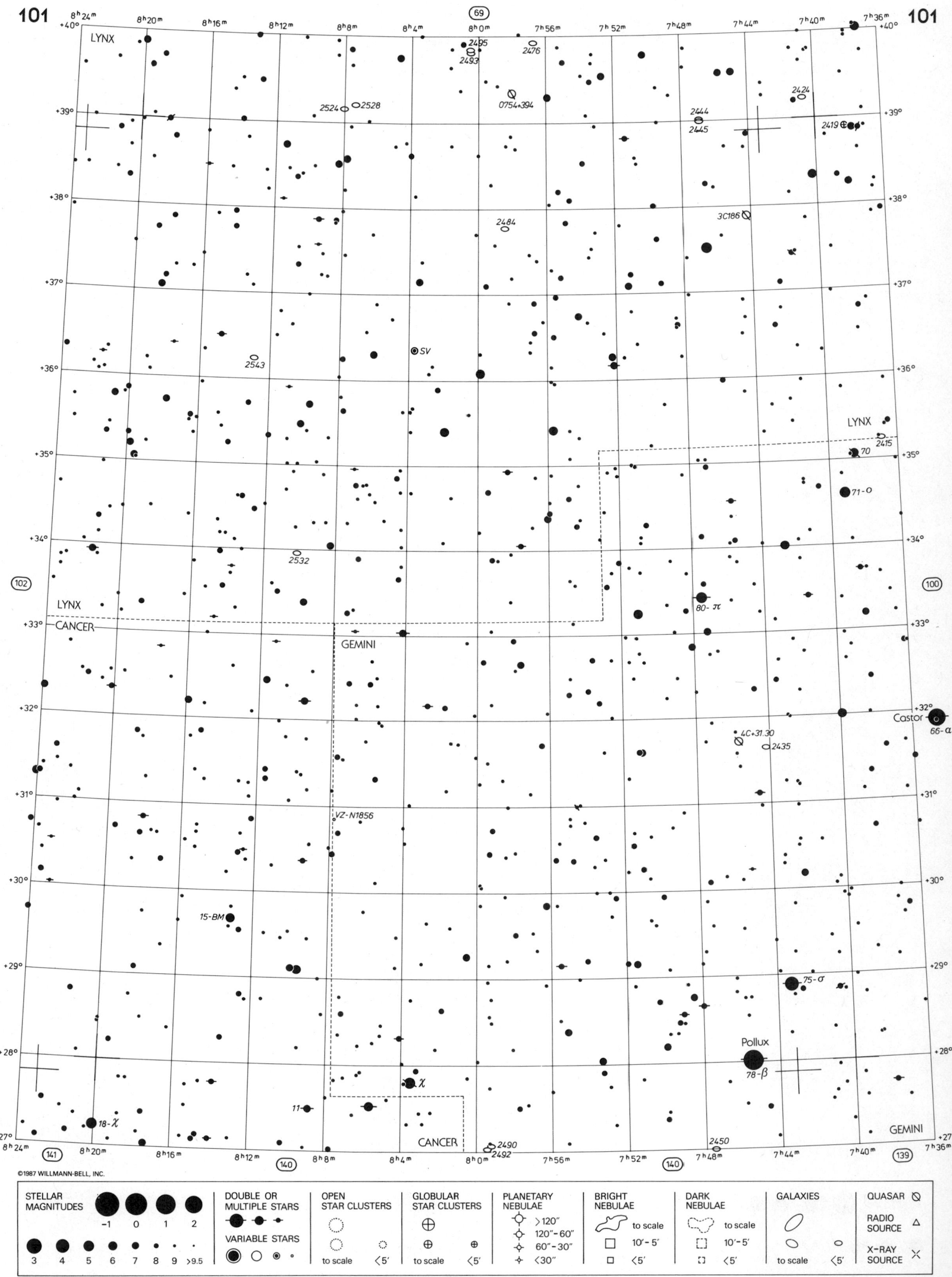

Barry Rappaport & Wil Tirion

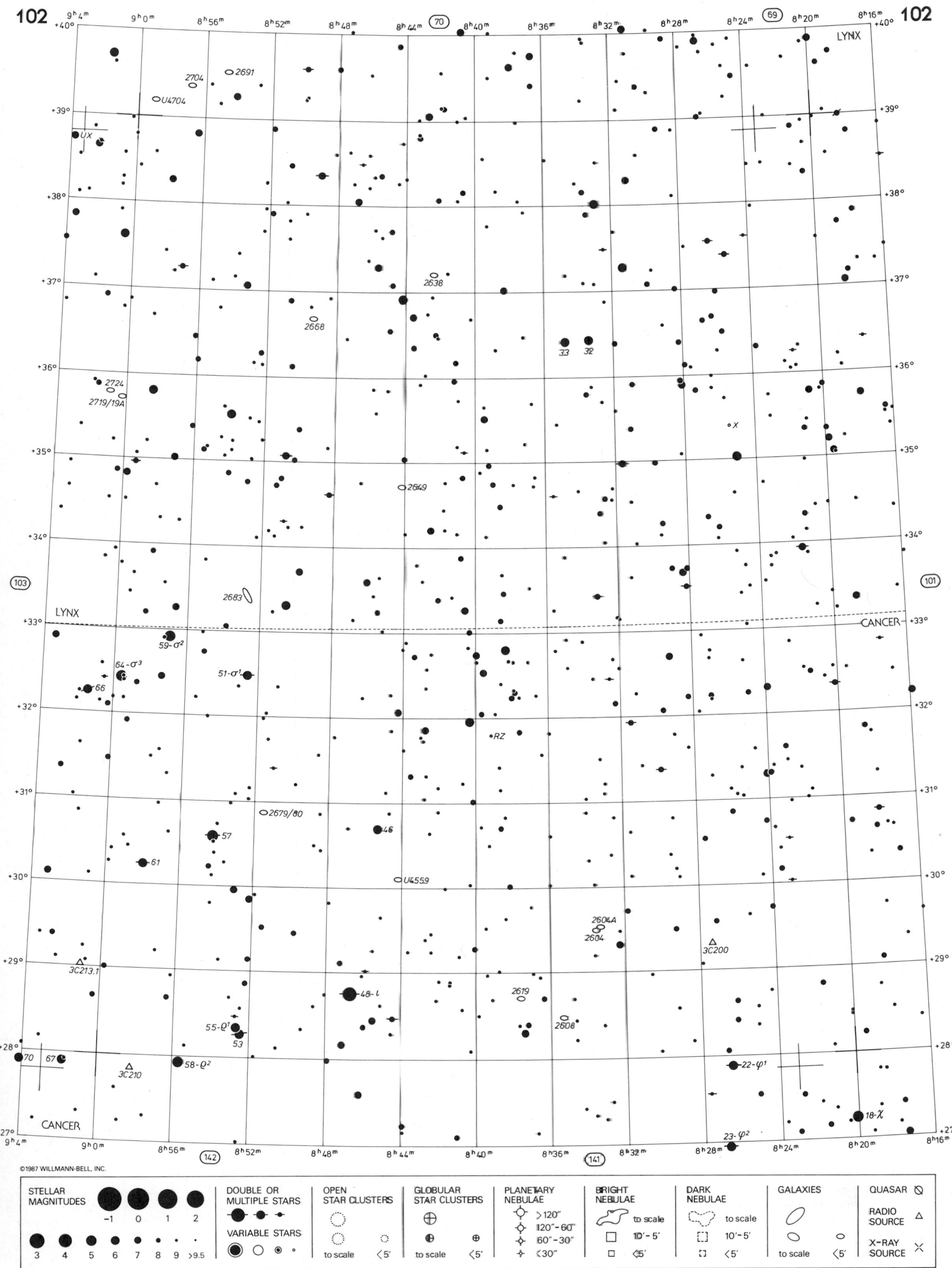

Barry Rappaport & Wil Tirion

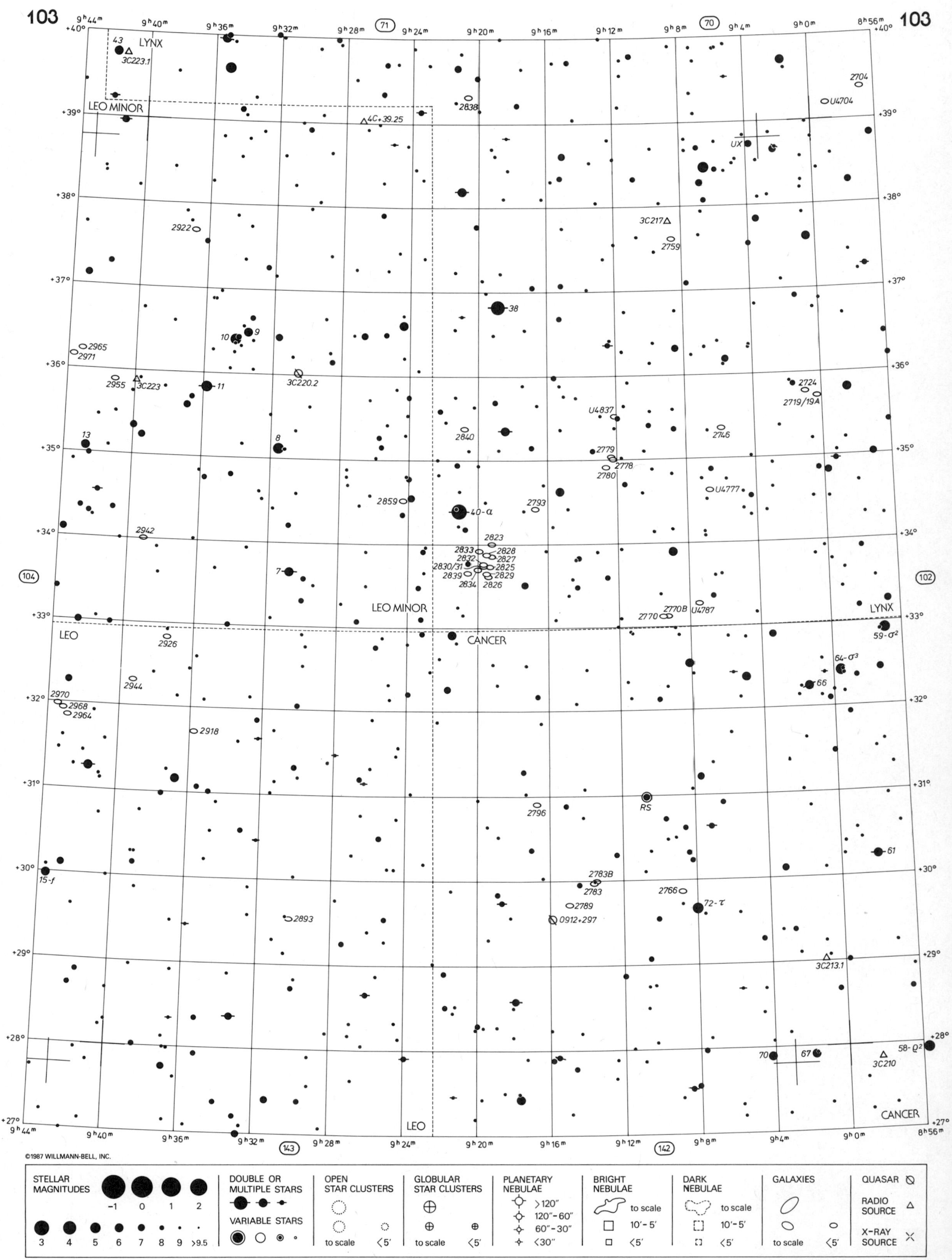
9h44m 9h40m 9h36m 9h32m 9h28m 9h24m 9h20m 9h16m 9h12m 9h8m 9h4m 9h0m 8h56m
+40° +39° +38° +37° +36° +35° +34° +33° +32° +31° +30° +29° +28° +27°
71
70
104
102
143
142
LYNX
LEO MINOR
LEO
CANCER
43
3C223.1
4C+39.25
2838
2704
U4704
UX
2922
3C217
2759
38
2965
2971
10
9
2955
3C223
11
3C220.2
2724
2719/19A
U4837
2840
2746
13
8
2779
2778
2780
2859
40-α
2793
U4777
2942
2823
2833
2828
2832
2827
2830/31
2825
2839
2829
2834
2826
7
2770
2770B
U4787
2926
59-σ2
64-σ3
2944
66
2970
2968
2964
2918
RS
2796
61
15-f
2783B
2783
2766
2789
72-τ
0912+297
2893
3C213.1
70
67
58-ρ2
3C210
©1987 WILLMANN-BELL, INC.
STELLAR MAGNITUDES
-1 0 1 2
3 4 5 6 7 8 9 >9.5
DOUBLE OR MULTIPLE STARS
VARIABLE STARS
OPEN STAR CLUSTERS
to scale
<5′
GLOBULAR STAR CLUSTERS
to scale
<5′
PLANETARY NEBULAE
>120″
120″-60″
60″-30″
<30″
BRIGHT NEBULAE
to scale
10′-5′
<5′
DARK NEBULAE
to scale
10′-5′
<5′
GALAXIES
to scale
<5′
QUASAR
RADIO SOURCE
X-RAY SOURCE
Barry Rappaport & Wil Tirion

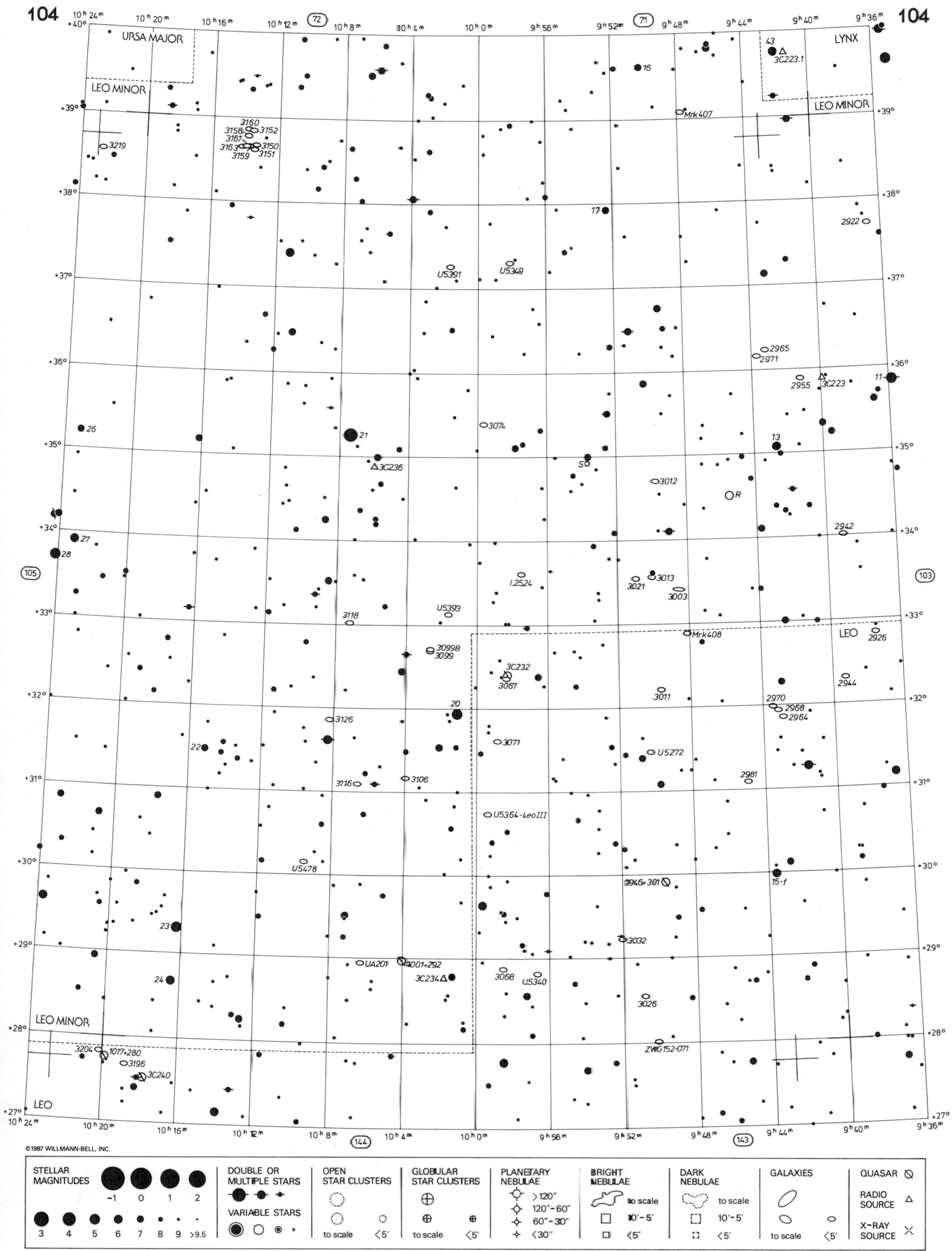
104
104
URSA MAJOR
LEO MINOR
LYNX
LEO
72
71
105
103
144
143
10h 24m
10h 20m
10h 16m
10h 12m
10h 8m
10h 4m
10h 0m
9h 56m
9h 52m
9h 48m
9h 44m
9h 40m
9h 36m
+40°
+39°
+38°
+37°
+36°
+35°
+34°
+33°
+32°
+31°
+30°
+29°
+28°
+27°
43
3C223.1
15
Mrk407
3160
3158
3152
3161
3150
3163
3151
3159
3219
17
2922
U5391
U5349
2965
2971
2955
3C223
11
26
3074
21
13
3C236
S
3012
R
27
28
2942
I.2524
3021
3013
3003
3118
U5393
Mrk408
2926
3099B
3099
3C232
3067
2944
3011
20
2970
2968
2964
3126
22
3071
U5272
3116
3106
2981
U5364-Leo III
U5478
0946+301
15-f
23
3032
UA201
1001+292
3C234
3068
U5340
24
3026
ZWG152-071
3204
1017+280
3196
3C240
©1987 WILLMANN-BELL, INC.
STELLAR MAGNITUDES
-1
0
1
2
3
4
5
6
7
8
9
>9.5
DOUBLE OR MULTIPLE STARS
VARIABLE STARS
OPEN STAR CLUSTERS
to scale
<5′
GLOBULAR STAR CLUSTERS
to scale
<5′
PLANETARY NEBULAE
>120″
120″-60″
60″-30″
<30″
BRIGHT NEBULAE
to scale
10′-5′
<5′
DARK NEBULAE
to scale
10′-5′
<5′
GALAXIES
to scale
<5′
QUASAR
RADIO SOURCE
X-RAY SOURCE
Barry Rappaport & Wil Tirion

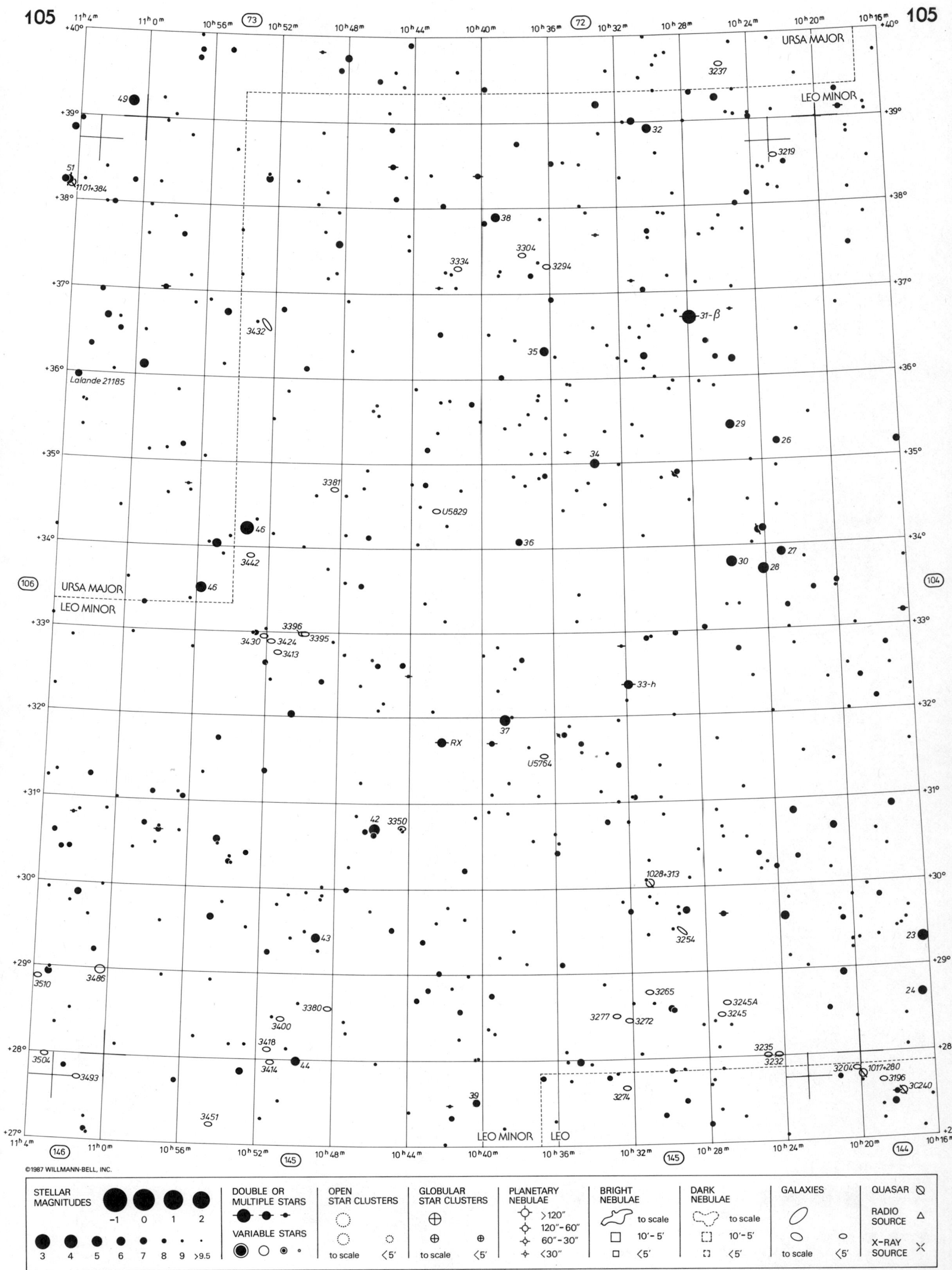

105
URSA MAJOR
LEO MINOR
LEO
31-β
33-h
RX
Lalande 21185
1101+384
1028+313
1017+280
3C240
U5829
U5764
STELLAR MAGNITUDES
DOUBLE OR MULTIPLE STARS
VARIABLE STARS
OPEN STAR CLUSTERS
GLOBULAR STAR CLUSTERS
PLANETARY NEBULAE
BRIGHT NEBULAE
DARK NEBULAE
GALAXIES
QUASAR
RADIO SOURCE
X-RAY SOURCE
to scale
©1987 WILLMANN-BELL, INC.
Barry Rappaport & Wil Tirion

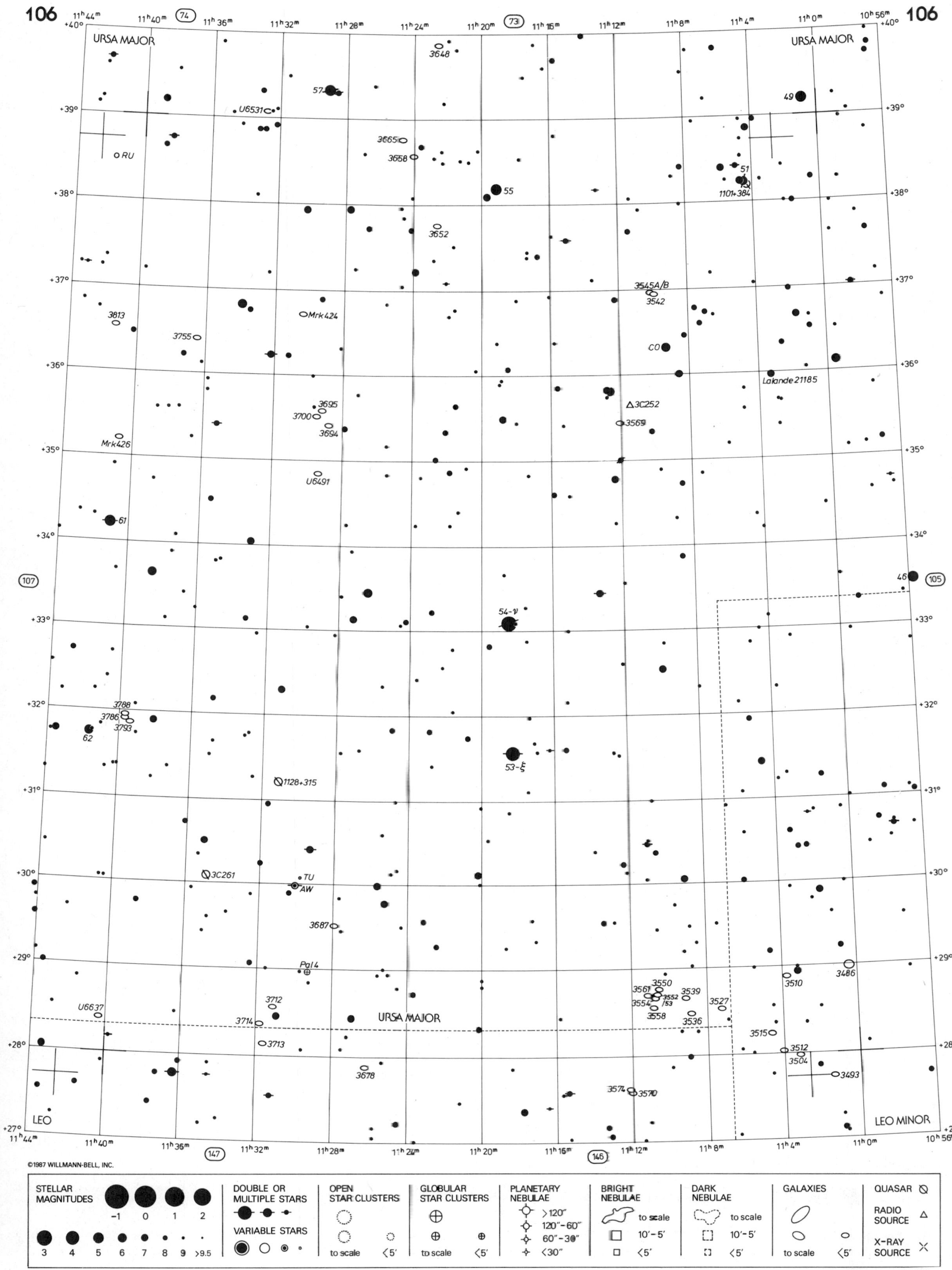

Barry Rappaport & Wil Tirion

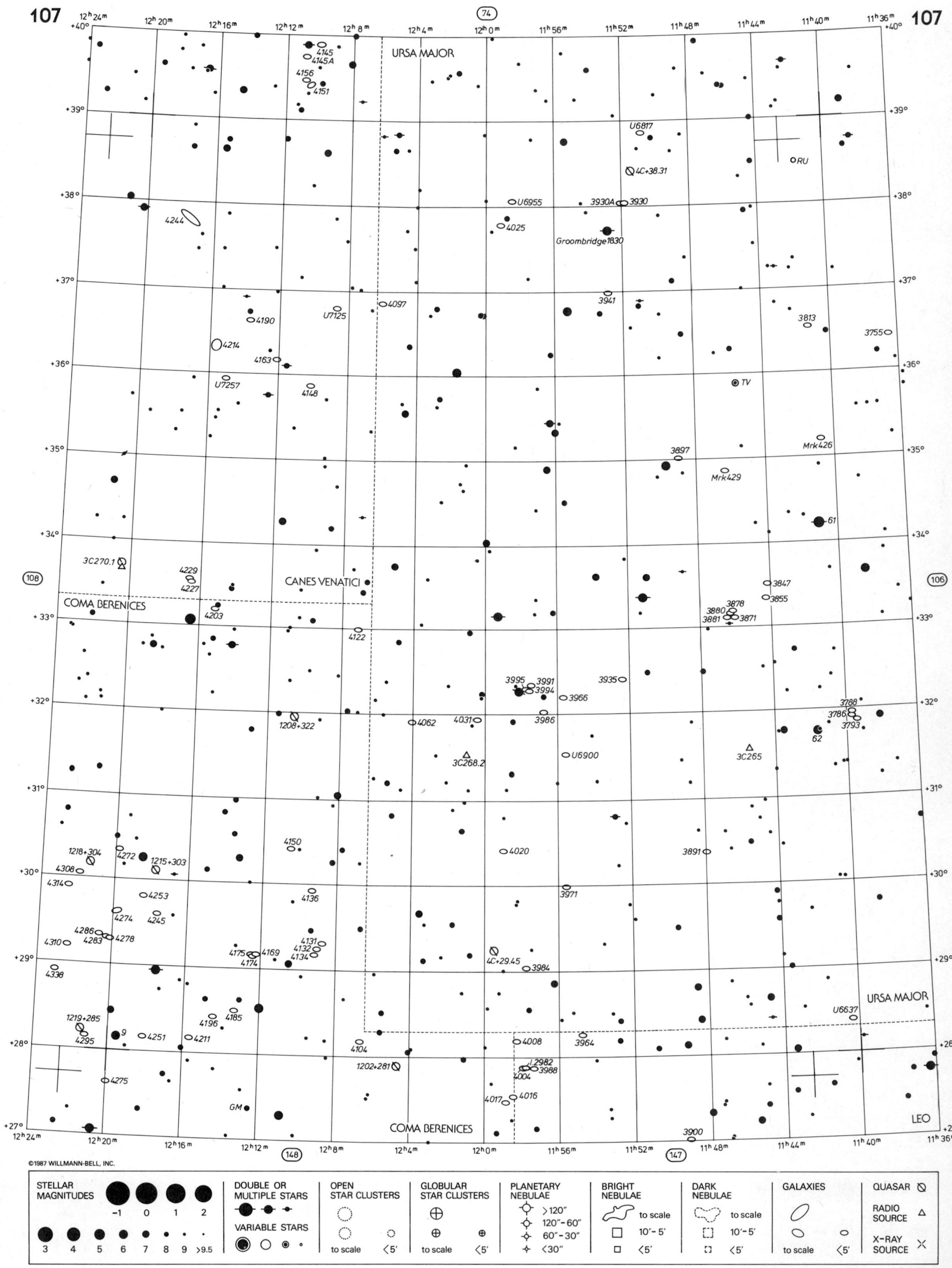

Barry Rappaport & Wil Tirion

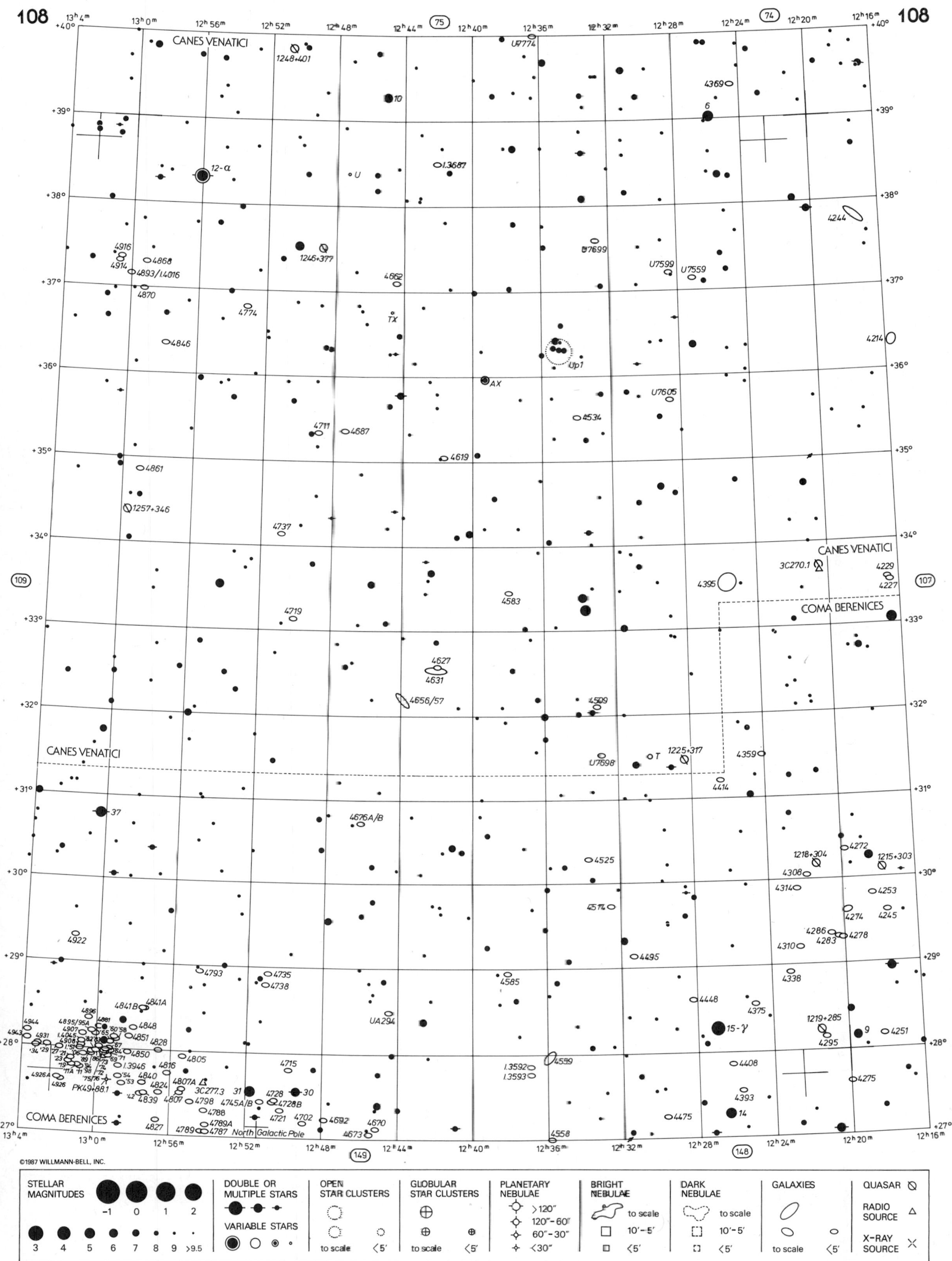

Barry Rappaport & Wil Tirion

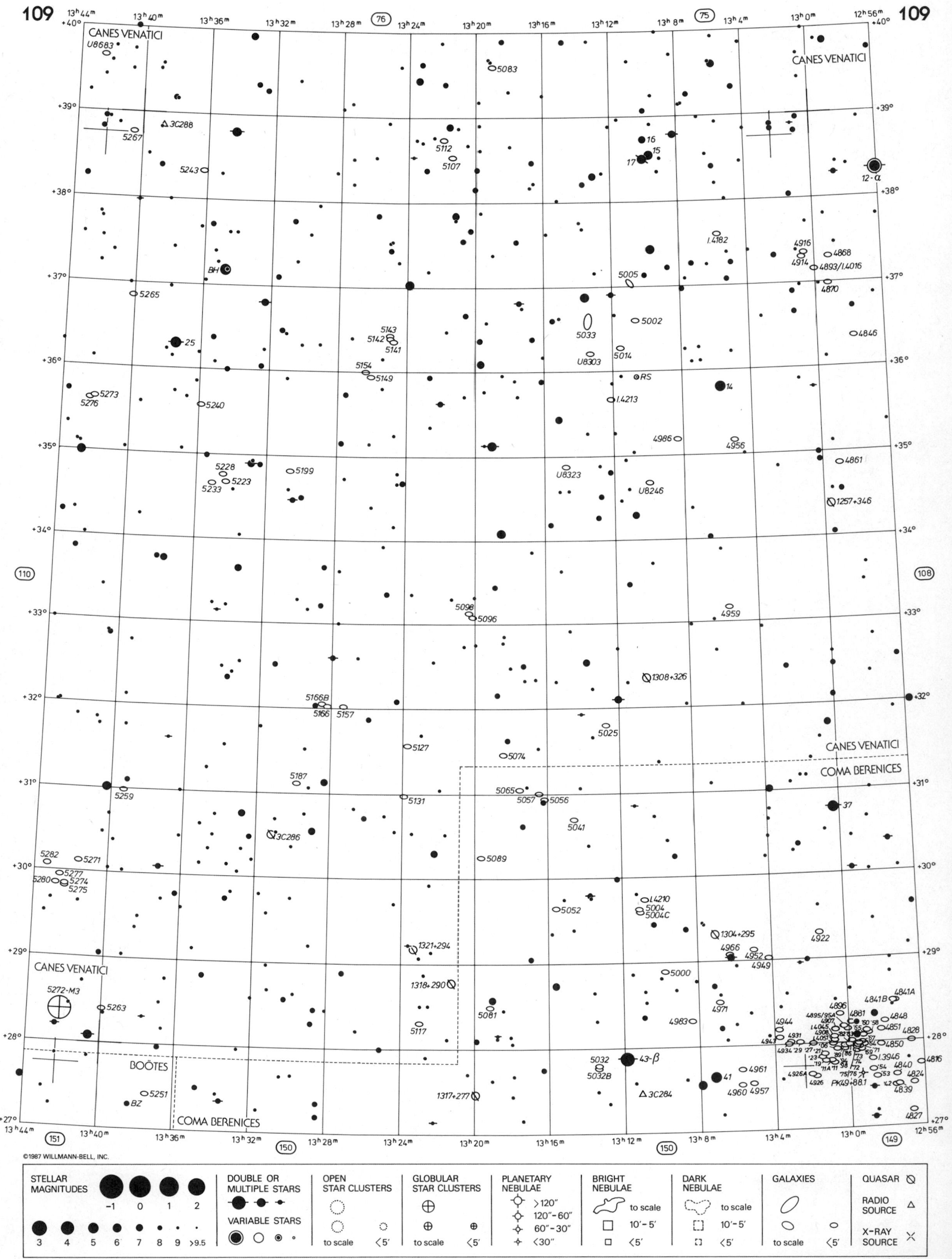
CANES VENATICI
COMA BERENICES
BOÖTES
13h44m
13h40m
13h36m
13h32m
13h28m
13h24m
13h20m
13h16m
13h12m
13h8m
13h4m
13h0m
12h56m
+40°
+39°
+38°
+37°
+36°
+35°
+34°
+33°
+32°
+31°
+30°
+29°
+28°
+27°
76
75
110
108
151
150
149
U8683
3C288
5267
5243
5083
5112
5107
16
15
17
12-α
I.4182
4916
4868
4914
4893/I.4016
4870
BH
5005
5265
5002
5033
4846
5143
5142
5141
25
5014
U8303
5154
5149
RS
14
5273
5276
5240
I.4213
4986
4956
4861
5228
5199
U8323
5223
5233
U8246
1257+346
5098
5096
4959
1308+326
5166B
5166
5157
5025
5127
5074
5187
5065
5057
5056
5259
5131
37
3C286
5041
5282
5271
5089
5277
5280
5274
5275
I.4210
5004
5004C
5052
1304+295
4922
1321+294
4966
4952
4949
5000
1318+290
5272-M3
5263
4971
4841B
4841A
4896
4881
4848
5081
4983
4851
4828
5117
4944
4850
5032
5032B
43-β
4961
41
I.3946
4840
4816
4824
4839
5251
BZ
1317+277
3C284
4960
4957
4926A
4926
PK49+88.1
4827
©1987 WILLMANN-BELL, INC.
STELLAR MAGNITUDES
−1
0
1
2
3
4
5
6
7
8
9
>9.5
DOUBLE OR MULTIPLE STARS
VARIABLE STARS
OPEN STAR CLUSTERS
to scale
<5'
GLOBULAR STAR CLUSTERS
to scale
<5'
PLANETARY NEBULAE
>120"
120"–60"
60"–30"
<30"
BRIGHT NEBULAE
to scale
10'–5'
<5'
DARK NEBULAE
to scale
10'–5'
<5'
GALAXIES
to scale
<5'
QUASAR
RADIO SOURCE
X-RAY SOURCE
Barry Rappaport & Wil Tirion

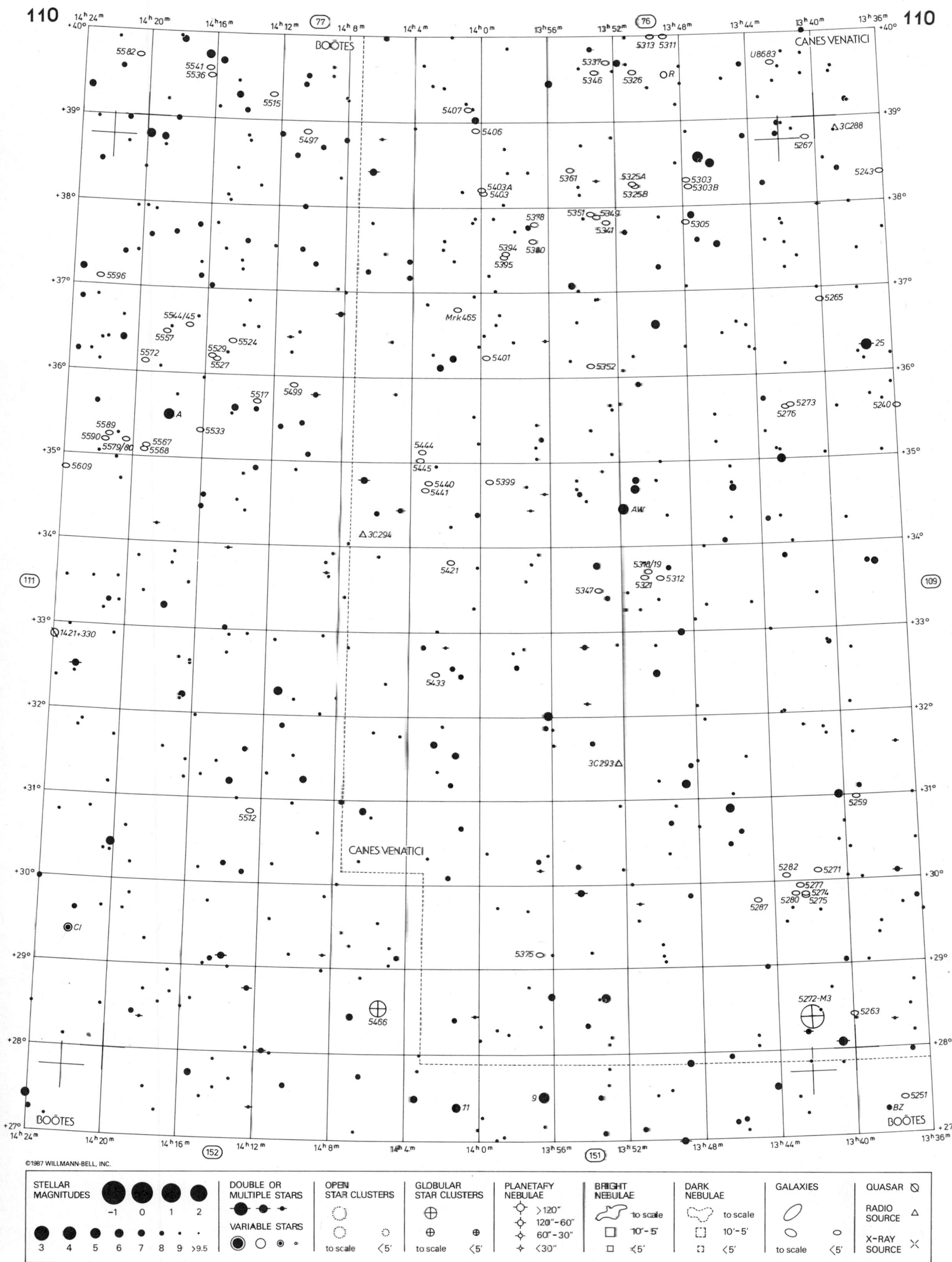

STELLAR MAGNITUDES: −1, 0, 1, 2, 3, 4, 5, 6, 7, 8, 9, >9.5
DOUBLE OR MULTIPLE STARS
VARIABLE STARS
OPEN STAR CLUSTERS: to scale, <5′
GLOBULAR STAR CLUSTERS: to scale, <5′
PLANETARY NEBULAE: >120″, 120″–60″, 60″–30″, <30″
BRIGHT NEBULAE: to scale, 10′–5′, <5′
DARK NEBULAE: to scale, 10′–5′, <5′
GALAXIES: to scale, <5′
QUASAR
RADIO SOURCE
X-RAY SOURCE

Barry Rappaport & Wil Tirion

15h 4m · 15h 0m · 14h 56m · (78) · 14h 52m · 14h 48m · 14h 44m · 14h 40m · 14h 36m · (77) · 14h 32m · 14h 28m · 14h 24m · 14h 20m · 14h 16m

111

42-β

BOÖTES

+40° +39° +38° +37° +36° +35° +34° +33° +32° +31° +30° +29° +28° +27°

40 · RR · 5755 · 5754 · 5753 · 5752 · 5732 · 5698 · 27-γ · 5625 · U9242 · U9291 · V · 5582 · 5541 · 5536

5596 · CP · 5695 · 5684 · 5686 · 5675 · 5654 · 5616 · 5544/45 · 5557 · 5572

U9560 · U9562 · BW · 5646 · 5656 · A · 5533 · 5589 · 5590 · 5579/80 · 5567 · 5568 · 5613 · 5615 · 5614 · 5609

5727 · (112) · (110) · 5623 · 5611 · 1421+330 · RV

5672 · RW · 5653 · 5706 · 5709 · 25-ρ · 5639 · 5639B · 5642 · 5789 · 5798 · 5771 · 5773 · 5685 · 28-σ · CI

5657 · 5780 · 5735 · 5641 · 5635 · 36-ε

BOÖTES · BOÖTES

15h 4m · 15h 0m · 14h 56m · 14h 52m · (153) · 14h 48m · 14h 44m · 14h 40m · 14h 36m · 14h 32m · (152) · 14h 28m · 14h 24m · 14h 20m · 14h 16m

STELLAR MAGNITUDES	DOUBLE OR MULTIPLE STARS	OPEN STAR CLUSTERS	GLOBULAR STAR CLUSTERS	PLANETARY NEBULAE	BRIGHT NEBULAE	DARK NEBULAE	GALAXIES	
-1 0 1 2	VARIABLE STARS	to scale <5′	to scale <5′	>120″ 120″-60″ 60″-30″ <30″	to scale 10′-5′ <5′	to scale 10′-5′ <5′	to scale <5′	QUASAR RADIO SOURCE X-RAY SOURCE
3 4 5 6 7 8 9 >9.5								

Barry Rappaport & Wil Tirion

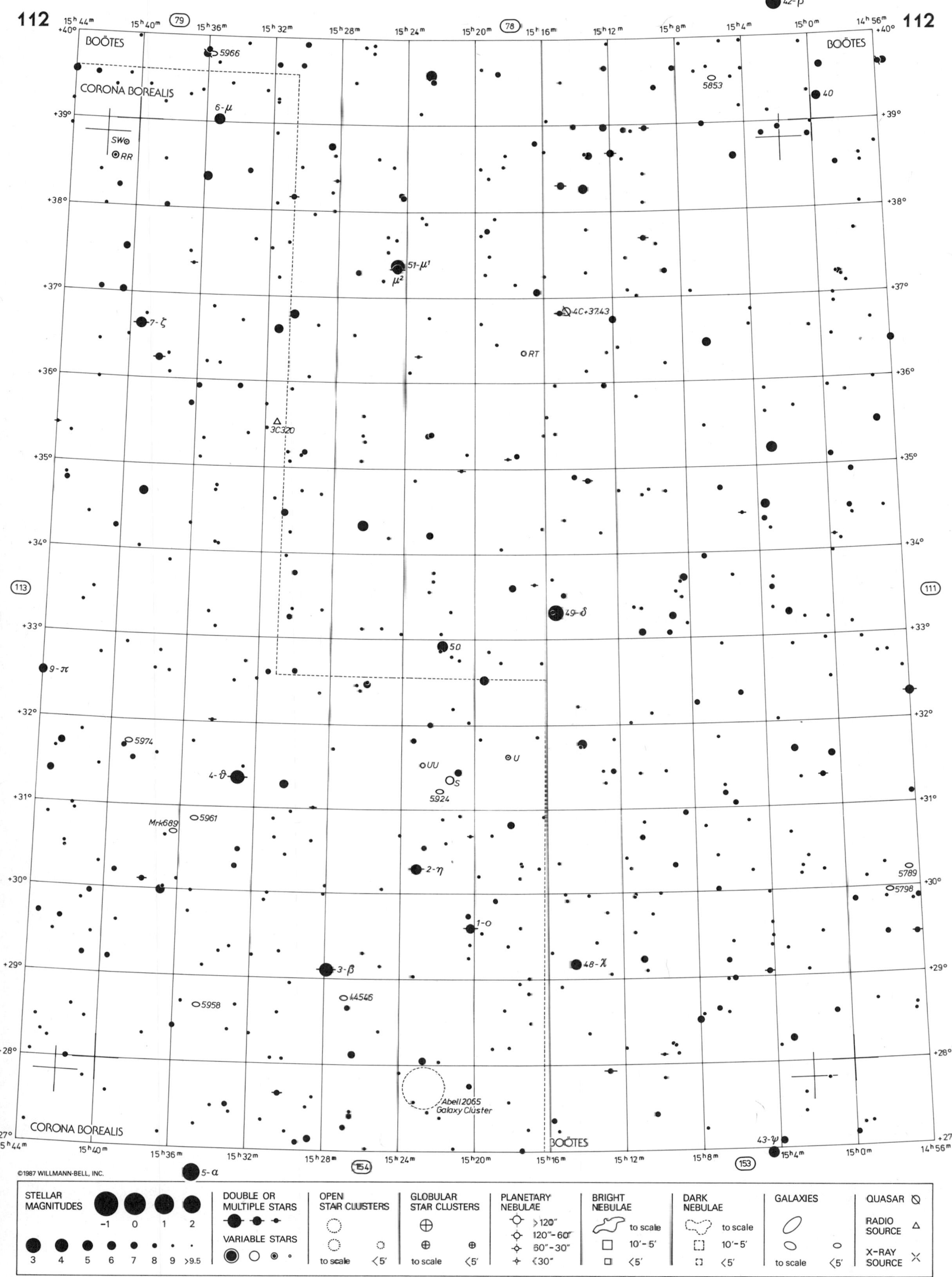

Barry Rappaport & Wil Tirion

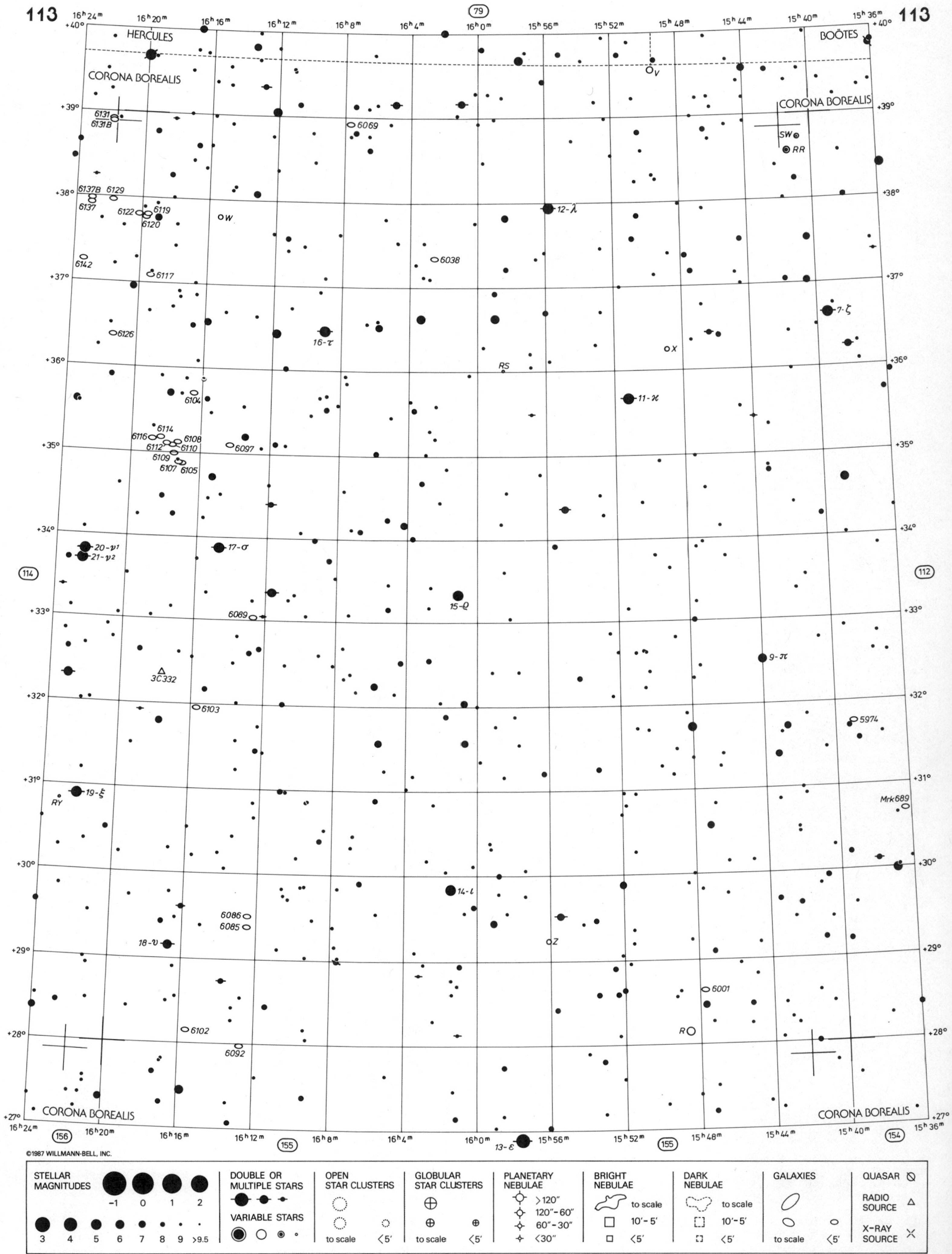

113
HERCULES
BOÖTES
CORONA BOREALIS
16h24m
16h20m
16h16m
16h12m
16h8m
16h4m
16h0m
15h56m
15h52m
15h48m
15h44m
15h40m
15h36m
+40°
+39°
+38°
+37°
+36°
+35°
+34°
+33°
+32°
+31°
+30°
+29°
+28°
+27°
79
114
112
156
155
154
6131
6131B
6069
6137B
6129
6137
6122
6119
6120
W
12-λ
V
SW
RR
6142
6038
6117
7-ζ
6126
16-τ
X
RS
6104
11-κ
6114
6116
6108
6112
6110
6097
6109
6107
6105
20-ν¹
21-ν²
17-σ
15-ρ
6089
9-π
3C332
6103
5974
19-ξ
RY
Mrk689
14-ι
6086
6085
18-υ
Z
6001
6102
R
6092
13-ε
©1987 WILLMANN-BELL, INC.
STELLAR MAGNITUDES
-1
0
1
2
3
4
5
6
7
8
9
>9.5
DOUBLE OR MULTIPLE STARS
VARIABLE STARS
OPEN STAR CLUSTERS
to scale
<5'
GLOBULAR STAR CLUSTERS
to scale
<5'
PLANETARY NEBULAE
>120"
120"–60"
60"–30"
<30"
BRIGHT NEBULAE
to scale
10'–5'
<5'
DARK NEBULAE
to scale
10'–5'
<5'
GALAXIES
to scale
<5'
QUASAR
RADIO SOURCE
X-RAY SOURCE
Barry Rappaport & Wil Tirion

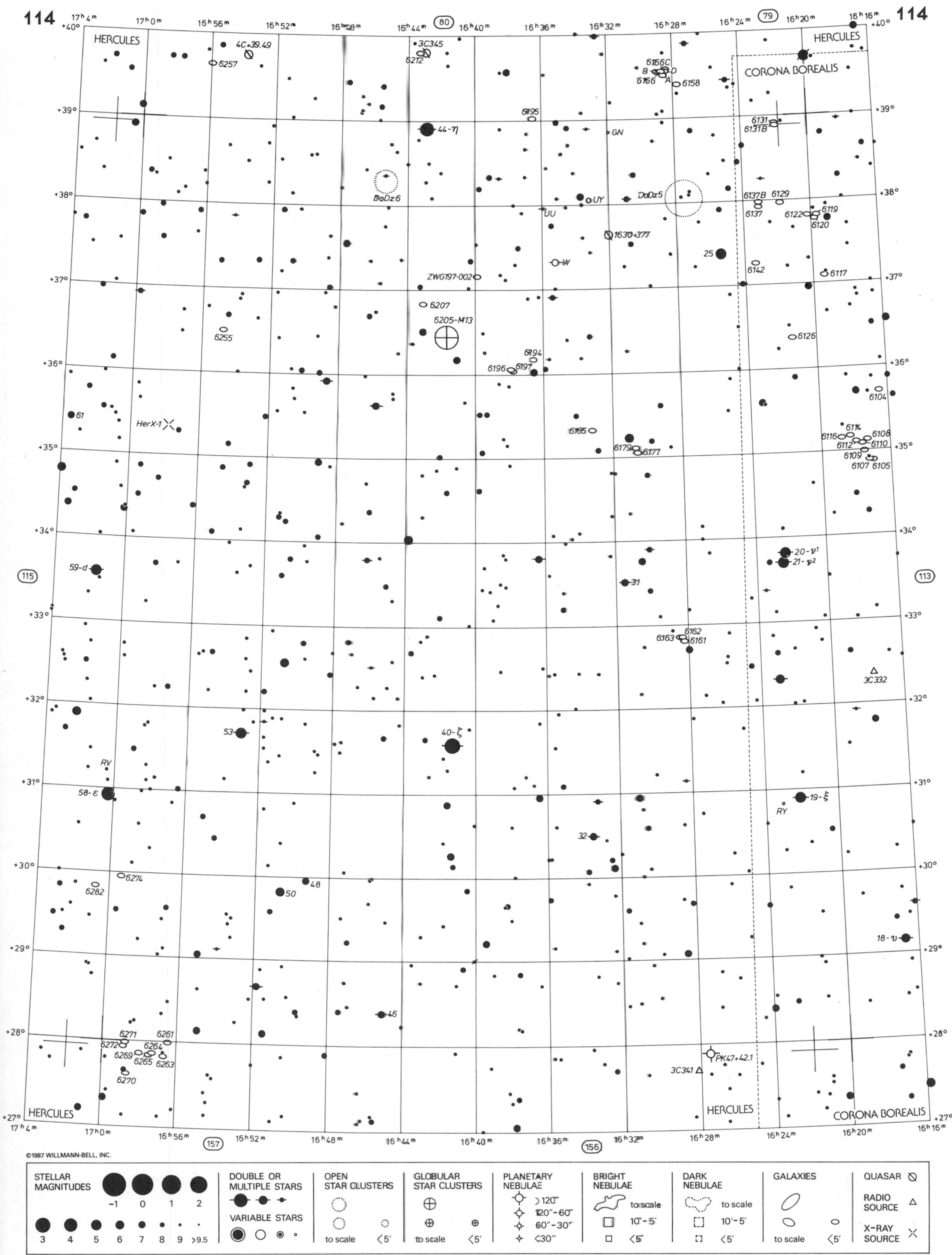

STELLAR MAGNITUDES -1 0 1 2 3 4 5 6 7 8 9 >9.5 | DOUBLE OR MULTIPLE STARS | VARIABLE STARS | OPEN STAR CLUSTERS to scale <5' | GLOBULAR STAR CLUSTERS to scale <5' | PLANETARY NEBULAE >120" 120"-60" 60"-30" <30" | BRIGHT NEBULAE to scale 10'-5' <5' | DARK NEBULAE to scale 10'-5' <5' | GALAXIES to scale <5' | QUASAR | RADIO SOURCE | X-RAY SOURCE

Barry Rappaport & Wil Tirion

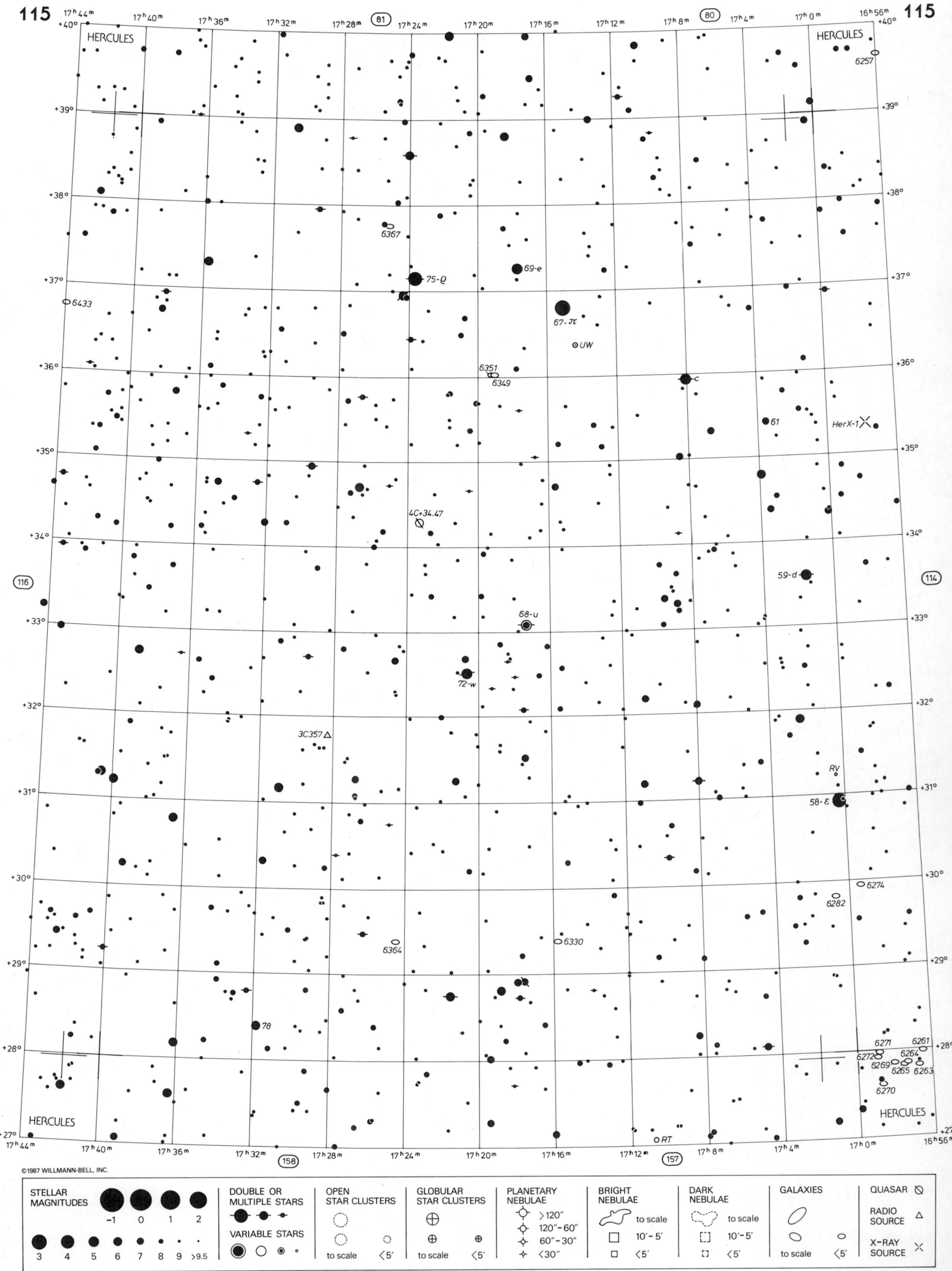

Barry Rappaport & Wil Tirion

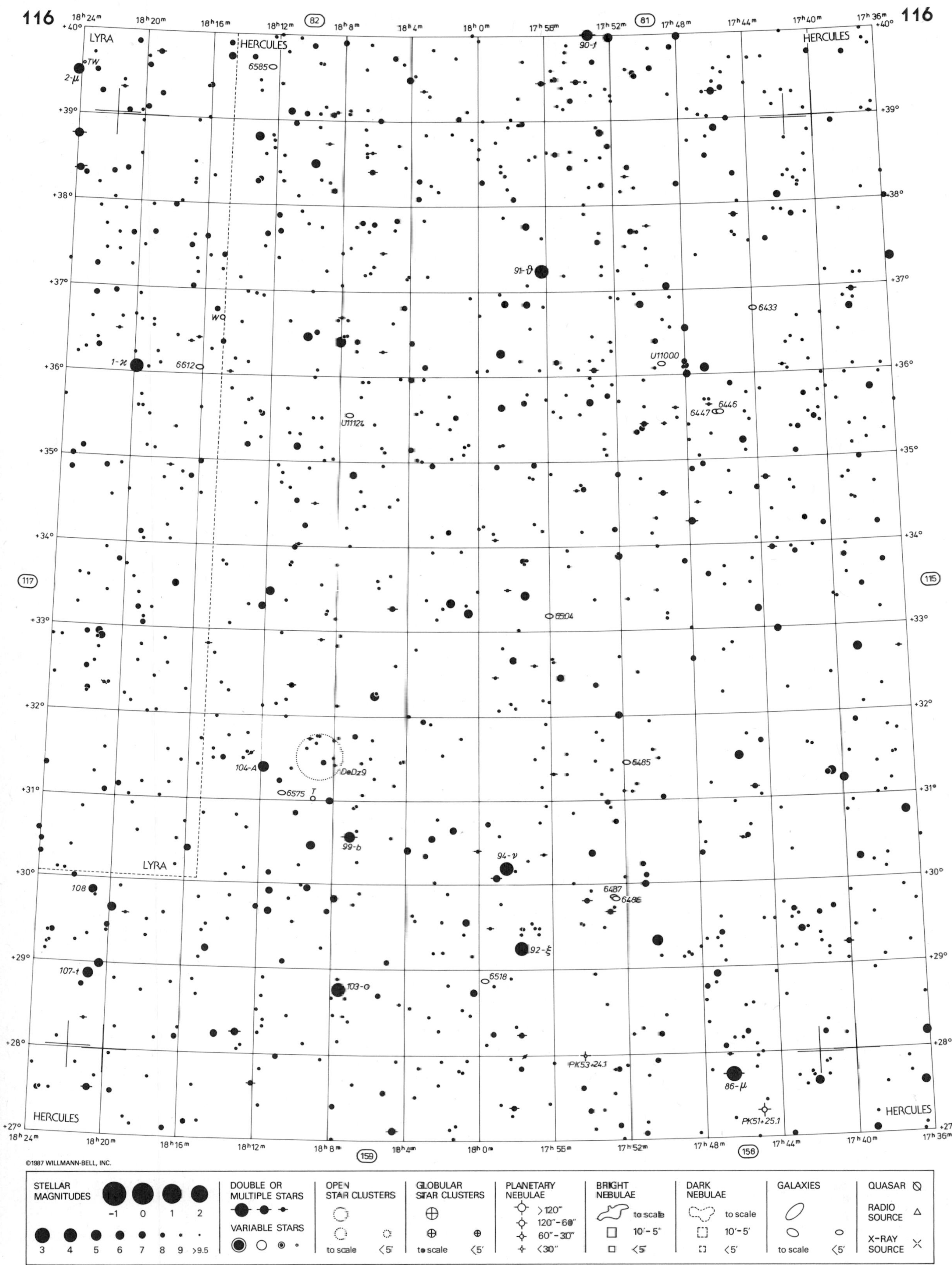

Barry Rappaport & Wil Tirion

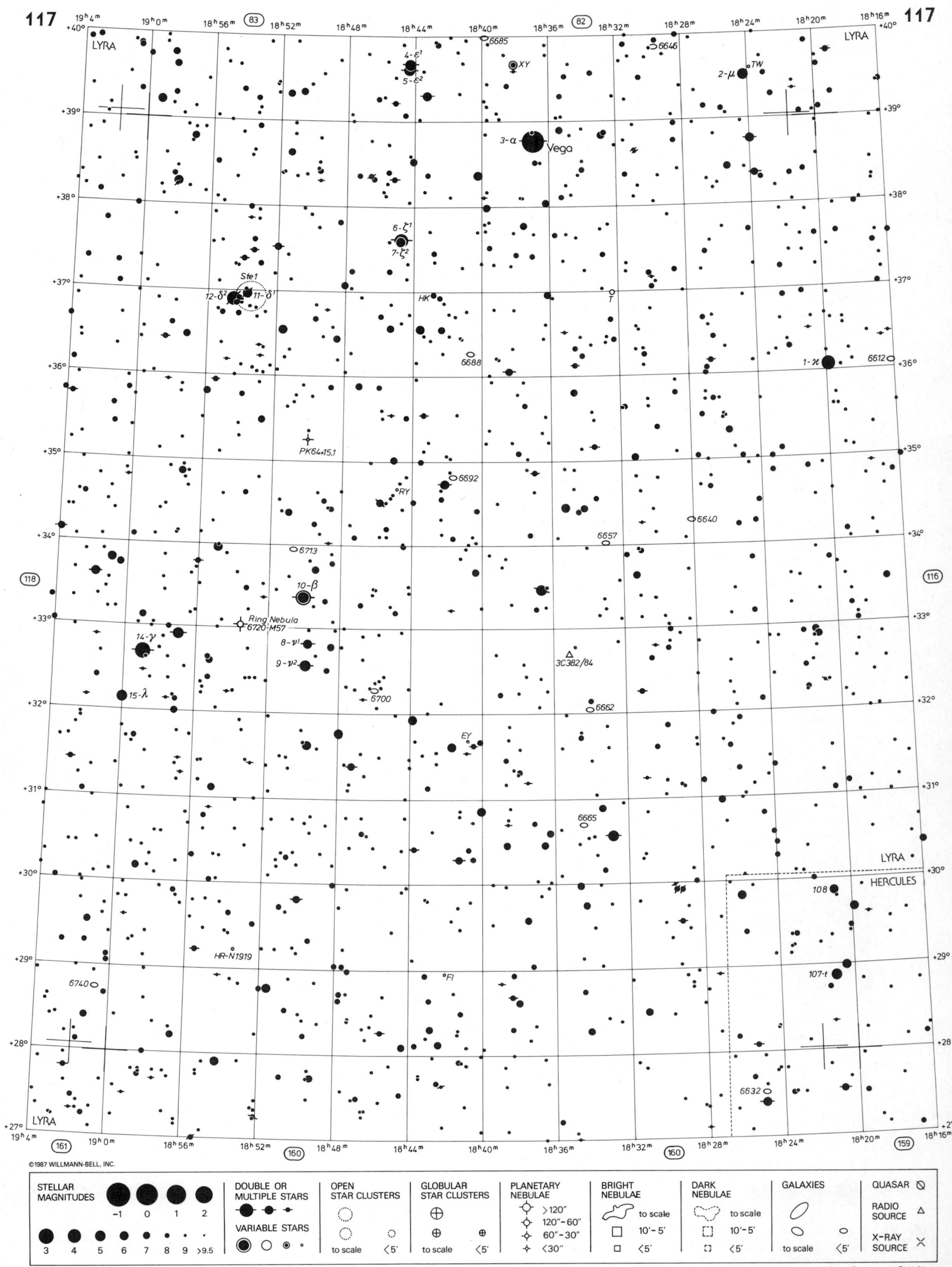

Barry Rappaport & Wil Tirion

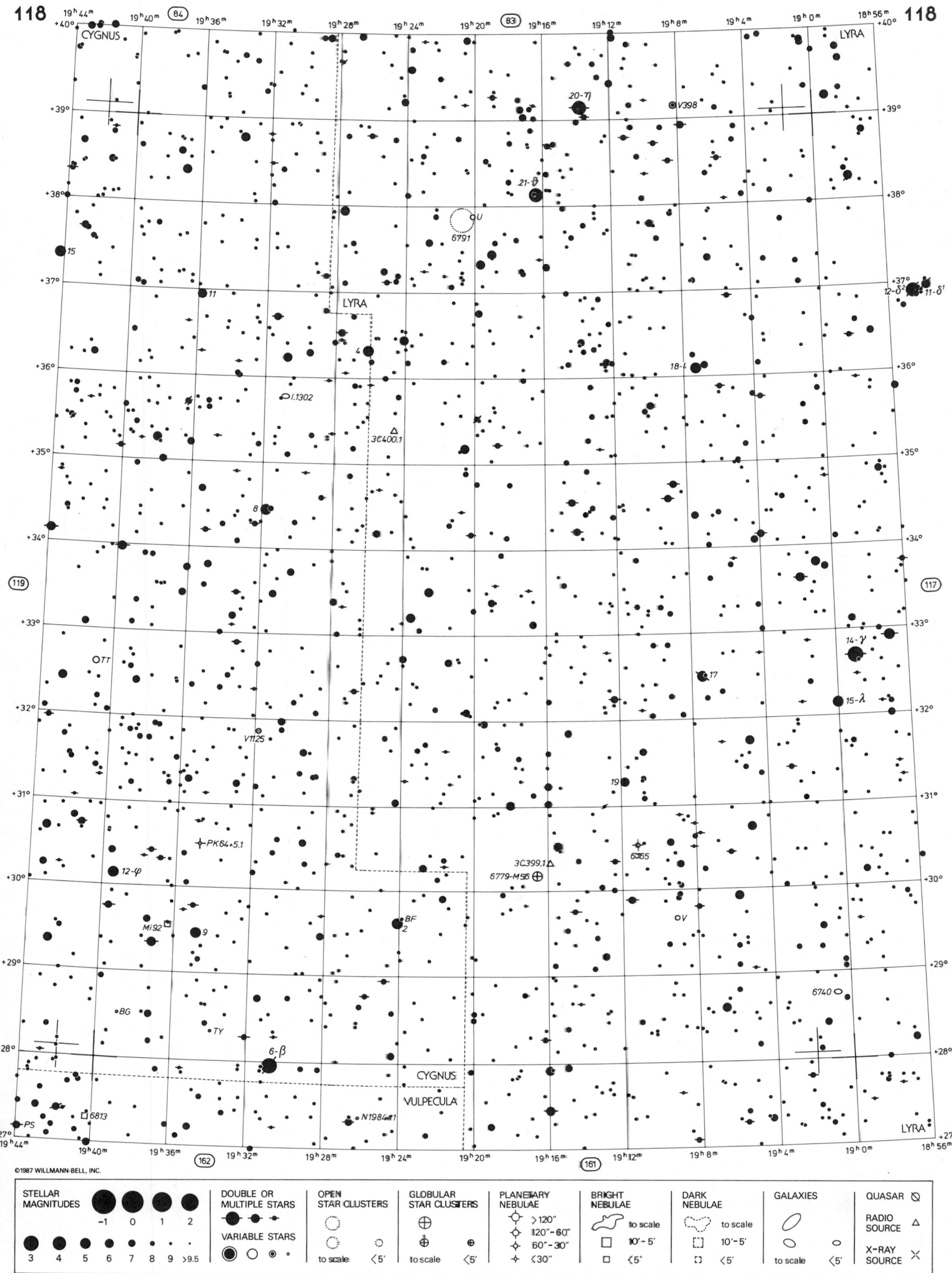
19h44m
19h40m
19h36m
19h32m
19h28m
19h24m
19h20m
19h16m
19h12m
19h8m
19h4m
19h0m
18h56m
84
83
+40°
+39°
+38°
+37°
+36°
+35°
+34°
+33°
+32°
+31°
+30°
+29°
+28°
+27°
119
117
162
161
CYGNUS
LYRA
VULPECULA
20-η
V398
21-ϑ
U
6791
15
11
12-δ²
11-δ¹
4
18-ι
I.1302
3C400.1
8
14-γ
TT
17
15-λ
V1125
19
PK64+5.1
3C399.1
6779-M56
6765
12-φ
M192
9
BF
2
V
6740
BG
TY
6-β
6813
PS
N1984-41
STELLAR MAGNITUDES
-1
0
1
2
3
4
5
6
7
8
9
>9.5
DOUBLE OR MULTIPLE STARS
VARIABLE STARS
OPEN STAR CLUSTERS
to scale
<5′
GLOBULAR STAR CLUSTERS
to scale
<5′
PLANETARY NEBULAE
>120″
120″-60″
60″-30″
<30″
BRIGHT NEBULAE
to scale
10′-5′
<5′
DARK NEBULAE
to scale
10′-5′
<5′
GALAXIES
to scale
<5′
QUASAR
RADIO SOURCE
X-RAY SOURCE
Barry Rappaport & Wil Tirion

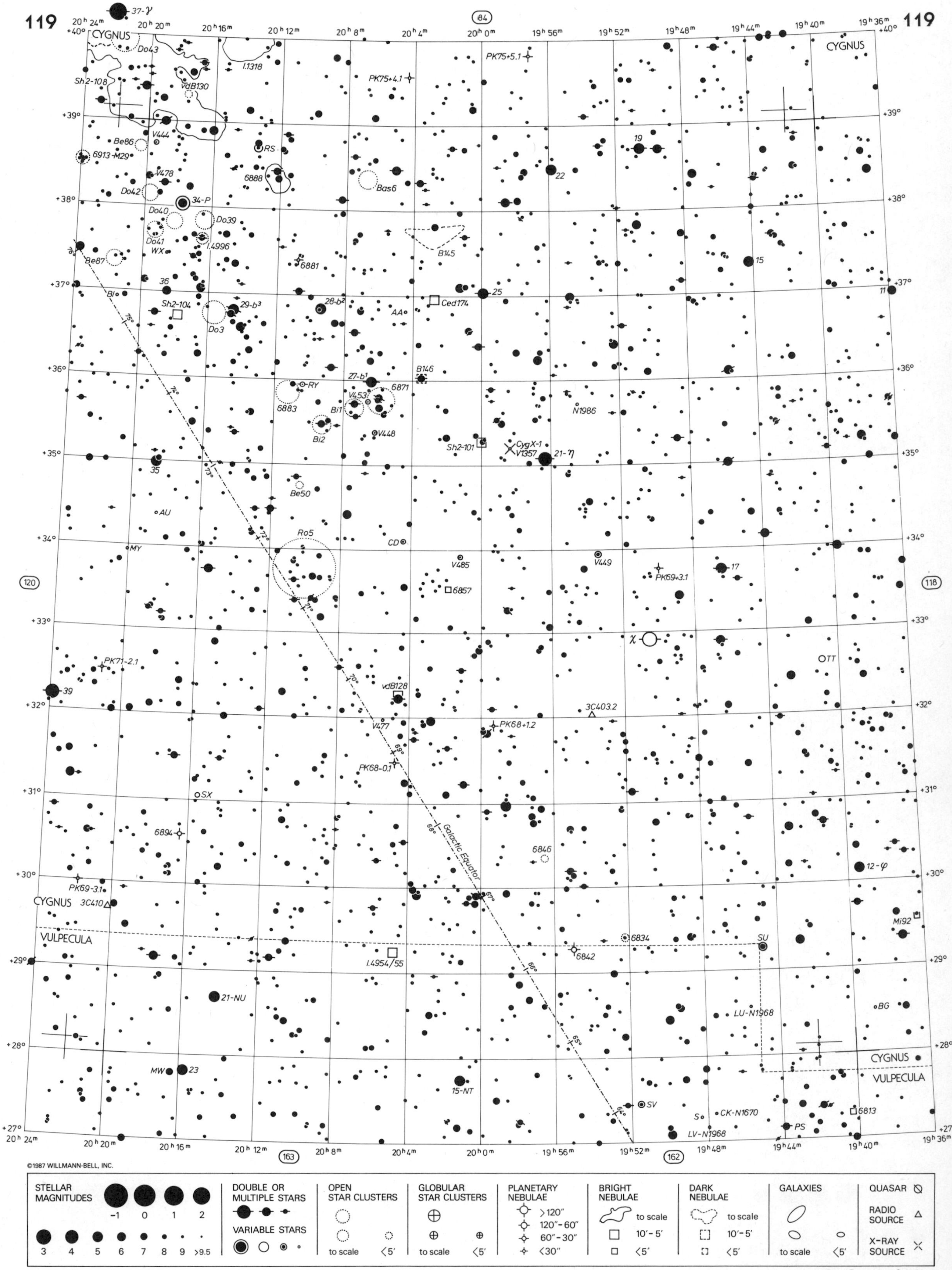

Barry Rappaport & Wil Tirion

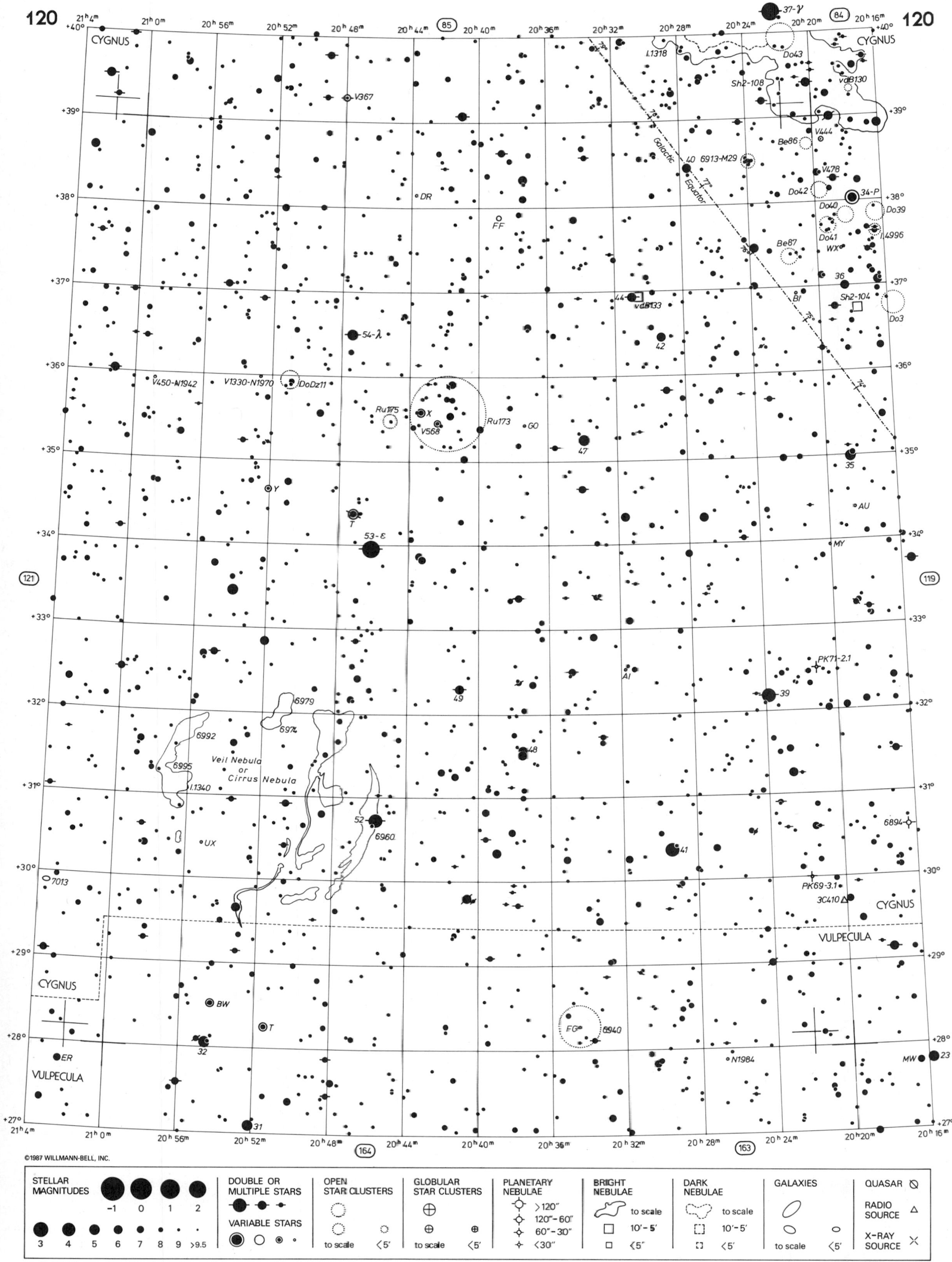
CYGNUS
VULPECULA
37-γ
34-P
53-ε
54-λ
Galactic Equator
6913-M29
Veil Nebula or Cirrus Nebula
6992
6995
6979
6974
6960
I.1340
I.1318
I.4996
Sh2-108
Sh2-104
vdB130
vdB133
Be86
Be87
Do43
Do42
Do40
Do41
Do39
Do3
DoDz11
Ru175
Ru173
V367
V444
V478
V568
V450-N1942
V1330-N1970
N1984
PK71-2.1
PK69-3.1
3C410
6894
6940
7013
STELLAR MAGNITUDES
DOUBLE OR MULTIPLE STARS
VARIABLE STARS
OPEN STAR CLUSTERS
GLOBULAR STAR CLUSTERS
PLANETARY NEBULAE
BRIGHT NEBULAE
DARK NEBULAE
GALAXIES
QUASAR
RADIO SOURCE
X-RAY SOURCE
to scale
Barry Rappaport & Wil Tirion

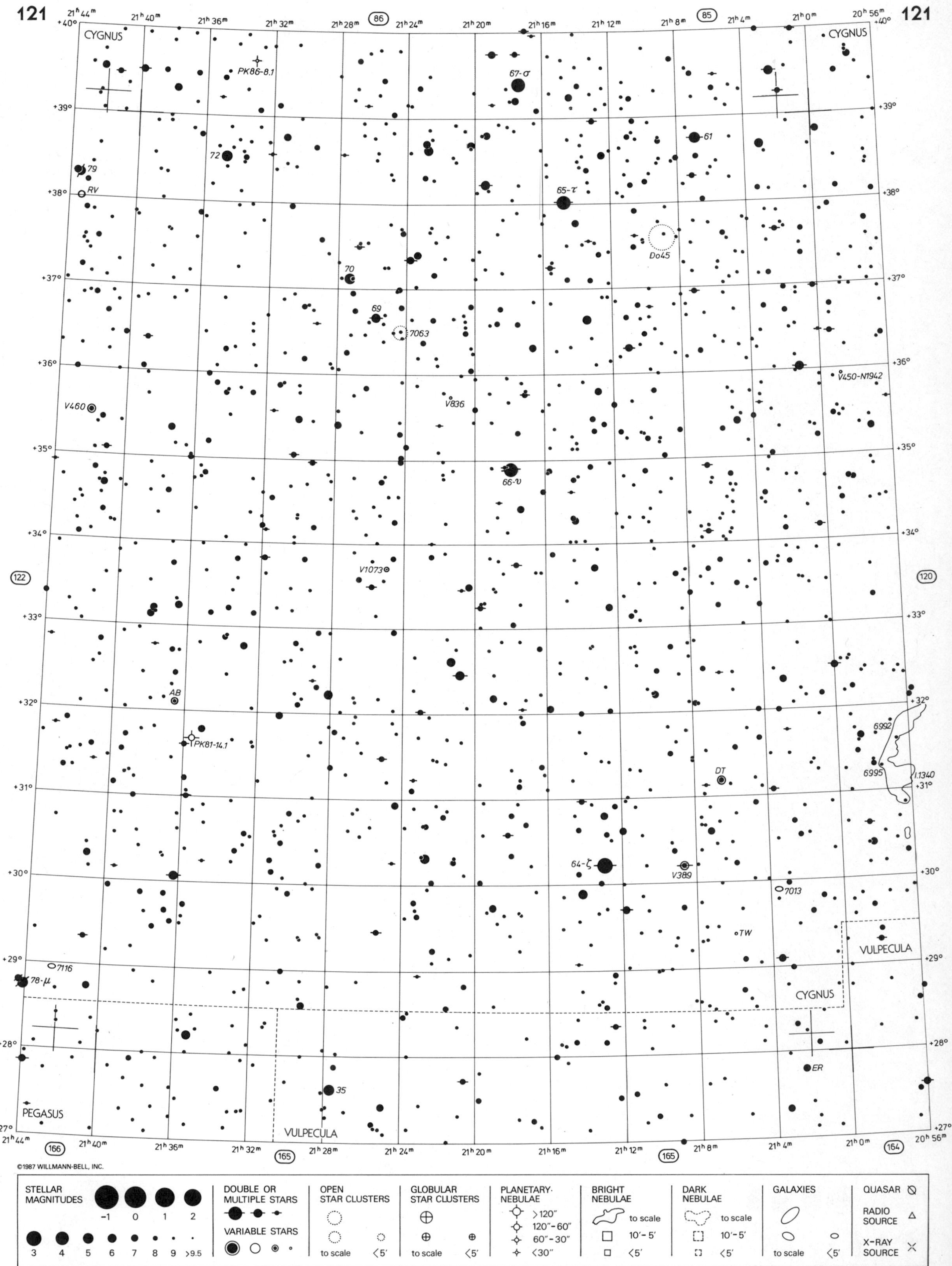

CYGNUS
PK86-8.1
67-σ
72
61
79
RV
65-τ
Do45
70
69
7063
V450-N1942
V460
V836
66-υ
V1073
AB
PK81-14.1
DT
6992
6995
I.1340
64-ζ
V389
7013
TW
VULPECULA
7116
78-μ
ER
35
PEGASUS
86
85
122
120
166
165
164
STELLAR MAGNITUDES
DOUBLE OR MULTIPLE STARS
VARIABLE STARS
OPEN STAR CLUSTERS
GLOBULAR STAR CLUSTERS
PLANETARY NEBULAE
BRIGHT NEBULAE
DARK NEBULAE
GALAXIES
QUASAR
RADIO SOURCE
X-RAY SOURCE
to scale
Barry Rappaport & Wil Tirion

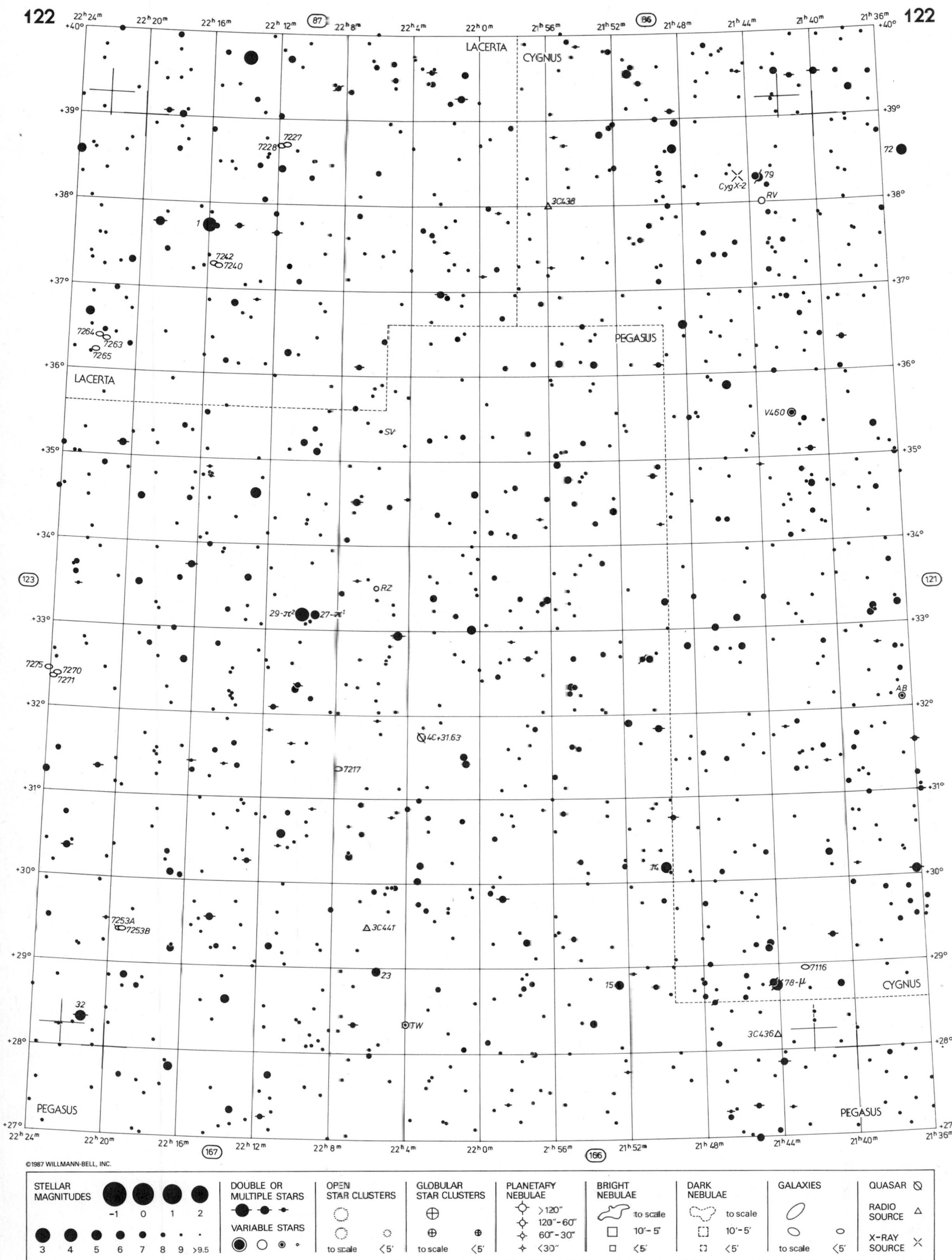

LACERTA
CYGNUS
PEGASUS
LACERTA
PEGASUS
CYGNUS
PEGASUS
Cyg X-2
79
72
RV
3C438
V460
SV
RZ
29-π²
27-π¹
AB
4C+31.63
7217
7227
7228
7242
7240
7264
7263
7265
7275
7270
7271
7253A
7253B
3C441
23
32
TW
15
14
7116
78-μ
3C436
STELLAR MAGNITUDES
DOUBLE OR MULTIPLE STARS
VARIABLE STARS
OPEN STAR CLUSTERS
GLOBULAR STAR CLUSTERS
PLANETARY NEBULAE
BRIGHT NEBULAE
DARK NEBULAE
GALAXIES
QUASAR
RADIO SOURCE
X-RAY SOURCE
to scale
Barry Rappaport & Wil Tirion

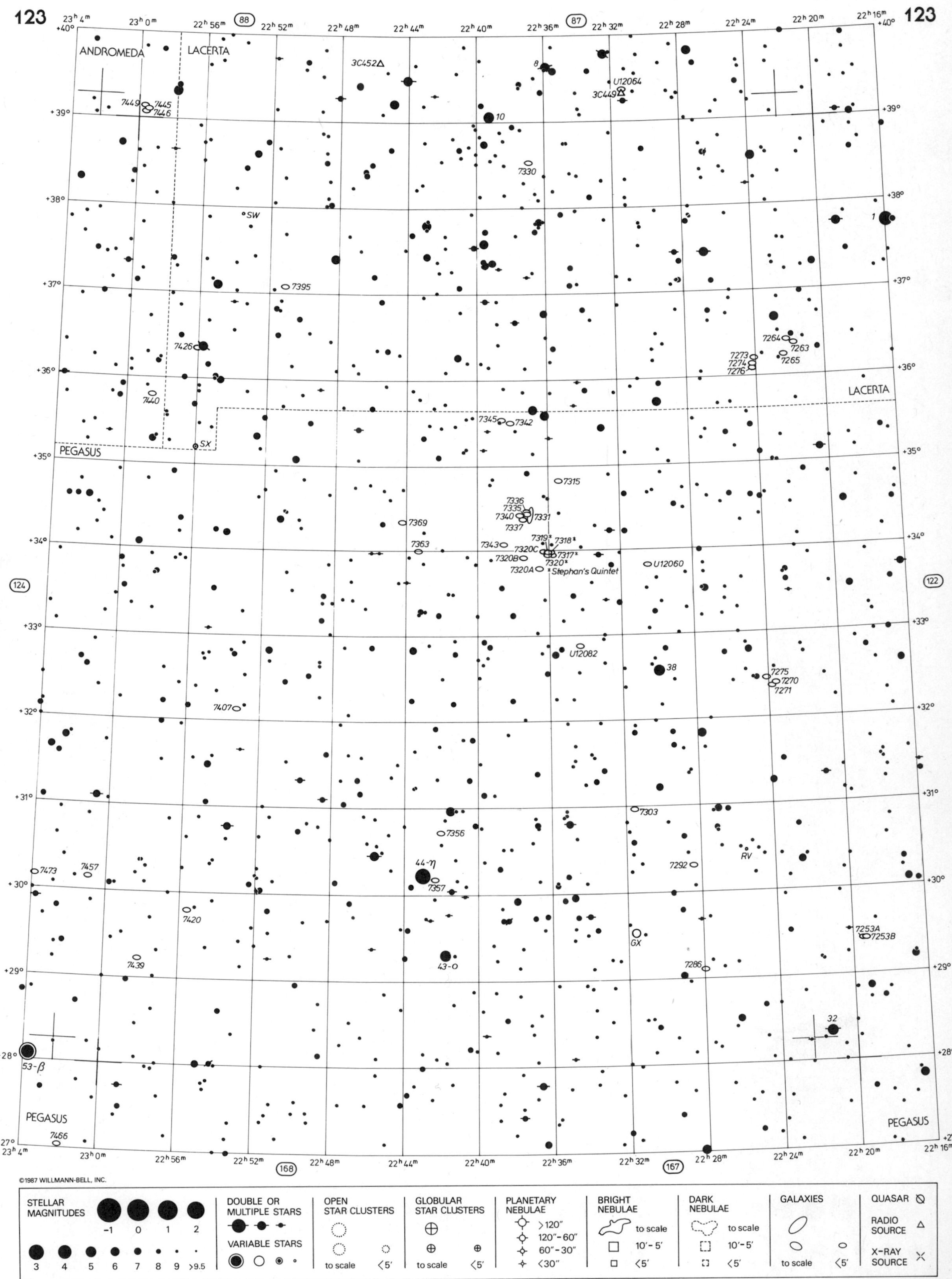

Barry Rappaport & Wil Tirion

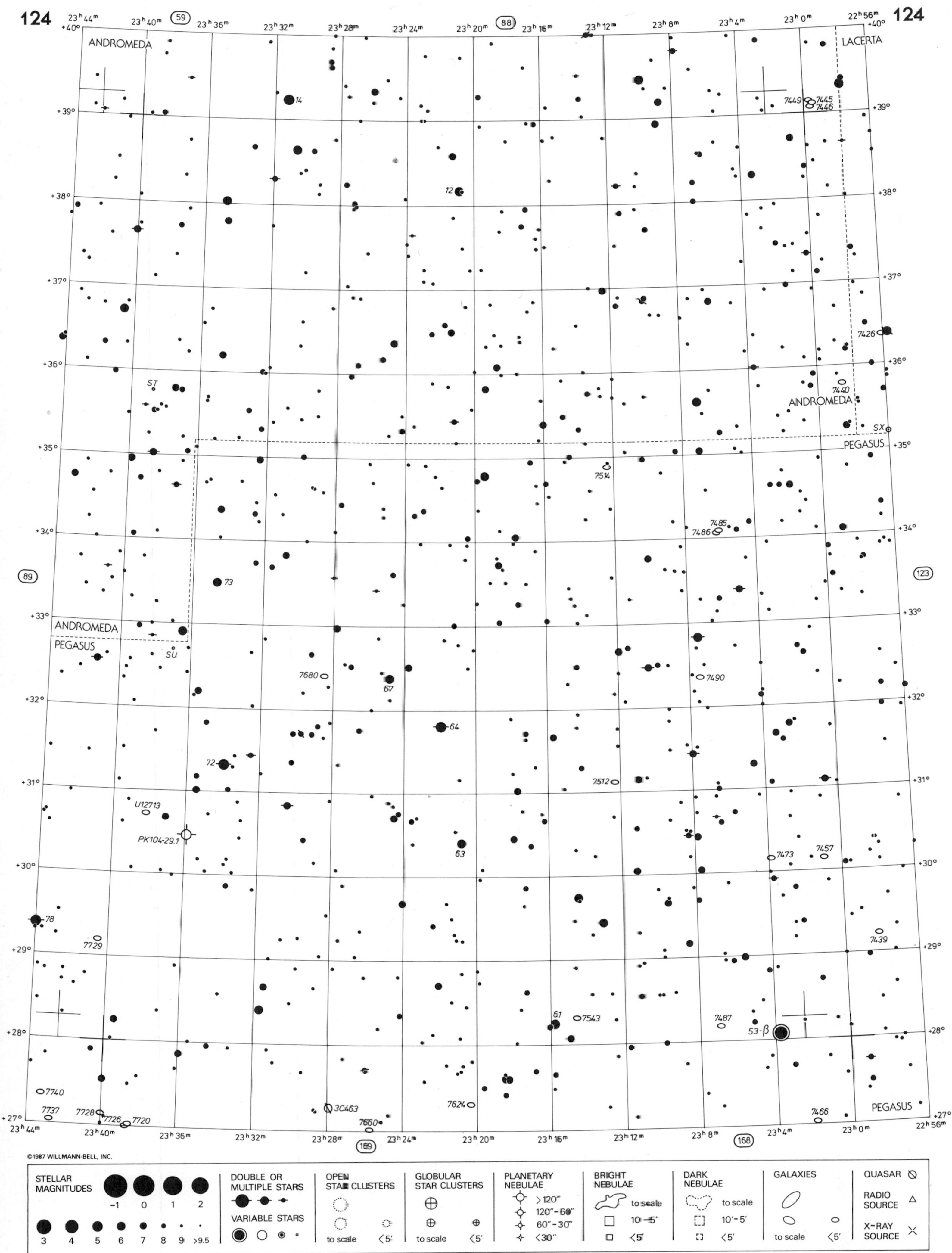

Barry Rappaport & Wil Tirion

124

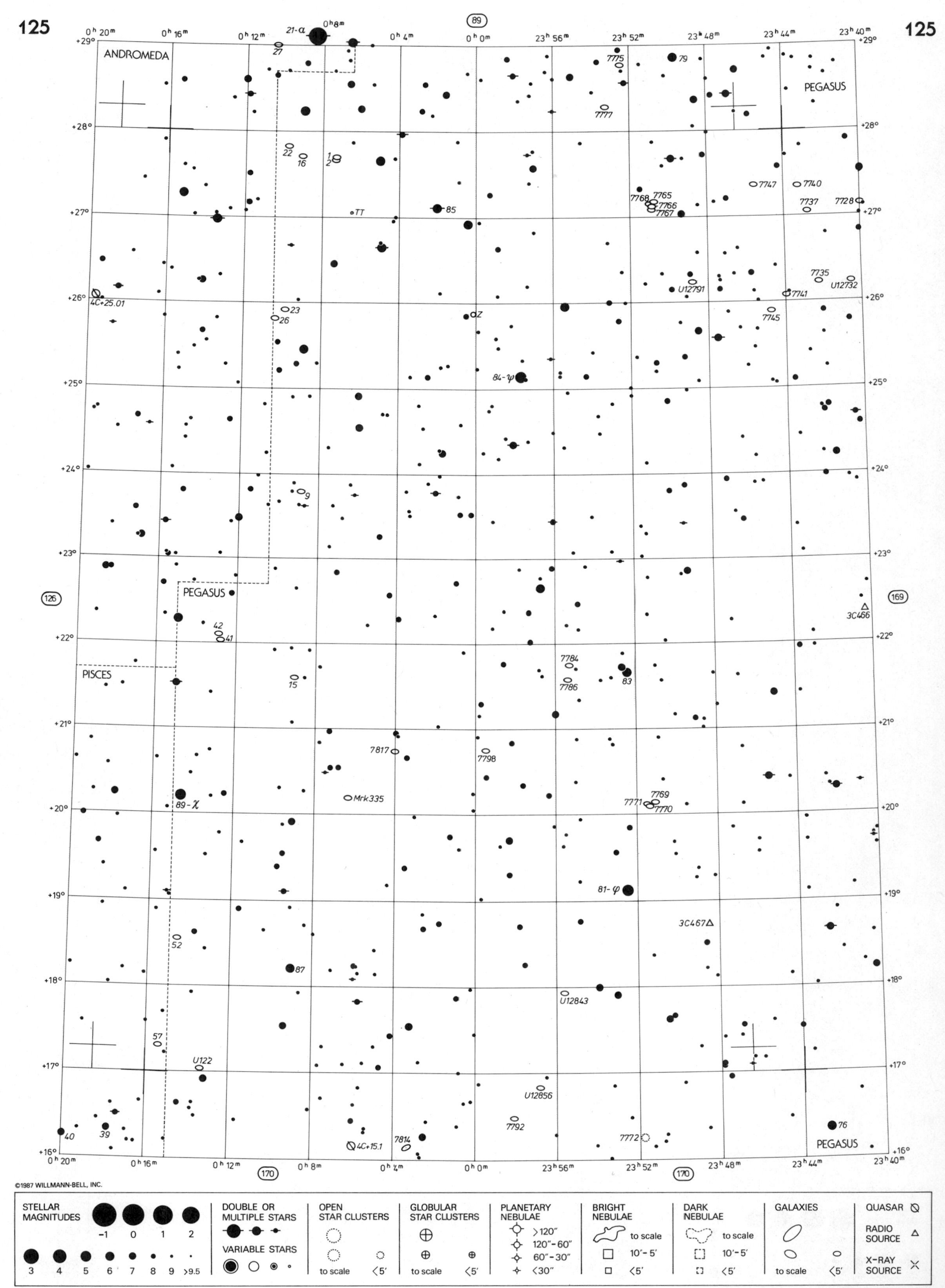

125
89
0h 20m
0h 16m
0h 12m
0h 8m
0h 4m
0h 0m
23h 56m
23h 52m
23h 48m
23h 44m
23h 40m
+29°
+28°
+27°
+26°
+25°
+24°
+23°
+22°
+21°
+20°
+19°
+18°
+17°
+16°
ANDROMEDA
PEGASUS
PISCES
21-α
27
7775
79
7777
22
16
1
2
7747
7740
7768
7765
7766
7767
7737
7728
TT
85
7735
U12732
U12791
7741
4C+25.01
23
26
Z
7745
84-ψ
9
126
169
3C466
42
41
7784
7786
83
15
7817
7798
89-χ
Mrk335
7769
7771
7770
81-φ
3C467
52
87
U12843
57
U122
U12856
7792
40
39
4C+15.1
7814
7772
76
170
©1987 WILLMANN-BELL, INC.
STELLAR MAGNITUDES
-1
0
1
2
3
4
5
6
7
8
9
>9.5
DOUBLE OR MULTIPLE STARS
VARIABLE STARS
OPEN STAR CLUSTERS
to scale
<5'
GLOBULAR STAR CLUSTERS
to scale
<5'
PLANETARY NEBULAE
>120"
120"-60"
60"-30"
<30"
BRIGHT NEBULAE
to scale
10'-5'
<5'
DARK NEBULAE
to scale
10'-5'
<5'
GALAXIES
to scale
<5'
QUASAR
RADIO SOURCE
X-RAY SOURCE
Barry Rappaport & Wil Tirion

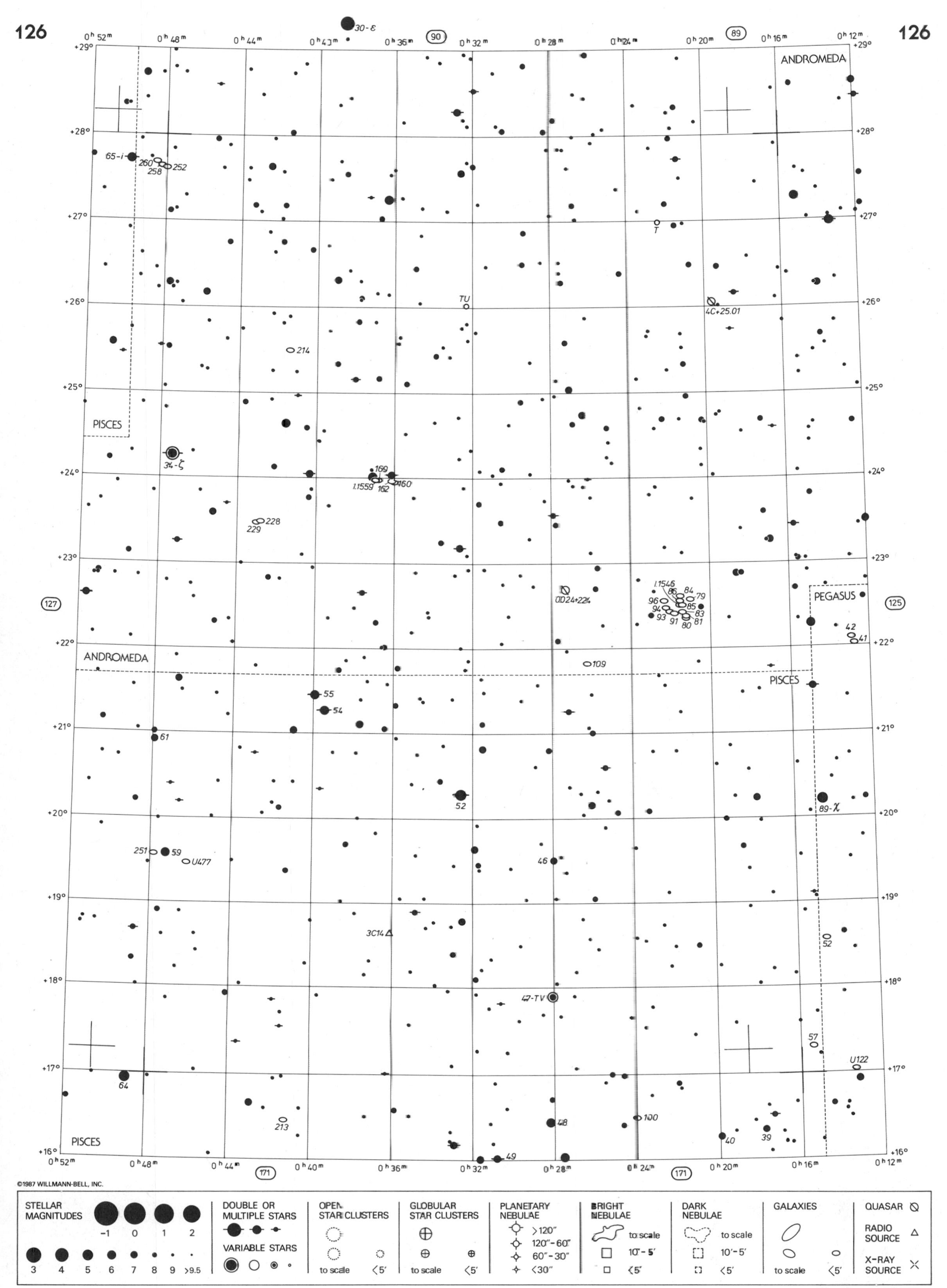

Barry Rappaport & Wil Tirion

1h 24m 1h 20m 1h 16m 1h 12m (91) 1h 8m 1h 4m 1h 0m 0h 56m (90) 0h 52m 0h 48m 0h 44m

+29° +28° +27° +26° +25° +24° +23° +22° +21° +20° +19° +18° +17° +16°

PISCES

91-i

90-υ

RT

Z

85-φ

0109+224

X

0117+213

84-χ

79-ψ2

74-ψ1

81-ψ3

459

473

463

476

87

68-h

67-k

65-i

260 258 252

326

3C28

0052+251

ANDROMEDA

280

304

34-ζ

36

38-η

354

LGS3

61

251

59

U477

66

64

PISCES

(128) (126)

1h 24m 1h 20m 1h 16m (172) 1h 12m 1h 8m 1h 4m 1h 0m 0h 56m (172) 0h 52m 0h 48m 0h 44m

STELLAR MAGNITUDES: −1, 0, 1, 2, 3, 4, 5, 6, 7, 8, 9, >9.5

DOUBLE OR MULTIPLE STARS

VARIABLE STARS

OPEN STAR CLUSTERS: to scale, <5′

GLOBULAR STAR CLUSTERS: to scale, <5′

PLANETARY NEBULAE: >120″, 120″–60″, 60″–30″, <30″

BRIGHT NEBULAE: to scale, 10′–5′, <5′

DARK NEBULAE: to scale, 10′–5′, <5′

GALAXIES: to scale, <5′

QUASAR

RADIO SOURCE

X-RAY SOURCE

Barry Rappaport & Wil Tirion

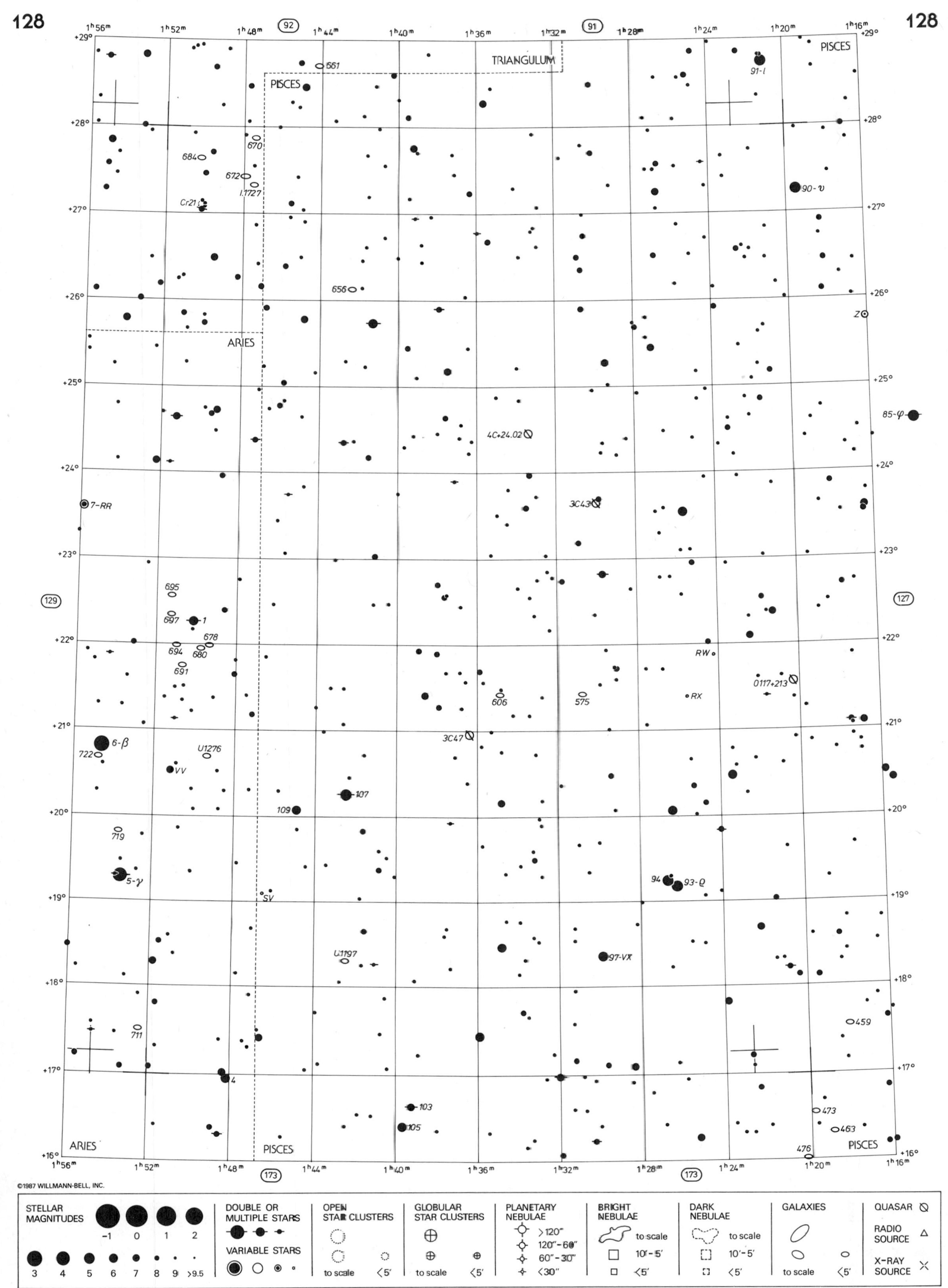

STELLAR MAGNITUDES	DOUBLE OR MULTIPLE STARS	OPEN STAR CLUSTERS	GLOBULAR STAR CLUSTERS	PLANETARY NEBULAE	BRIGHT NEBULAE	DARK NEBULAE	GALAXIES	QUASAR
−1 0 1 2	VARIABLE STARS	to scale <5′	to scale <5′	>120″ 120″–60″ 60″–30″ <30″	to scale 10′–5′ <5′	to scale 10′–5′ <5′	to scale <5′	RADIO SOURCE
3 4 5 6 7 8 9 >9.5								X-RAY SOURCE

Barry Rappaport & Wil Tirion

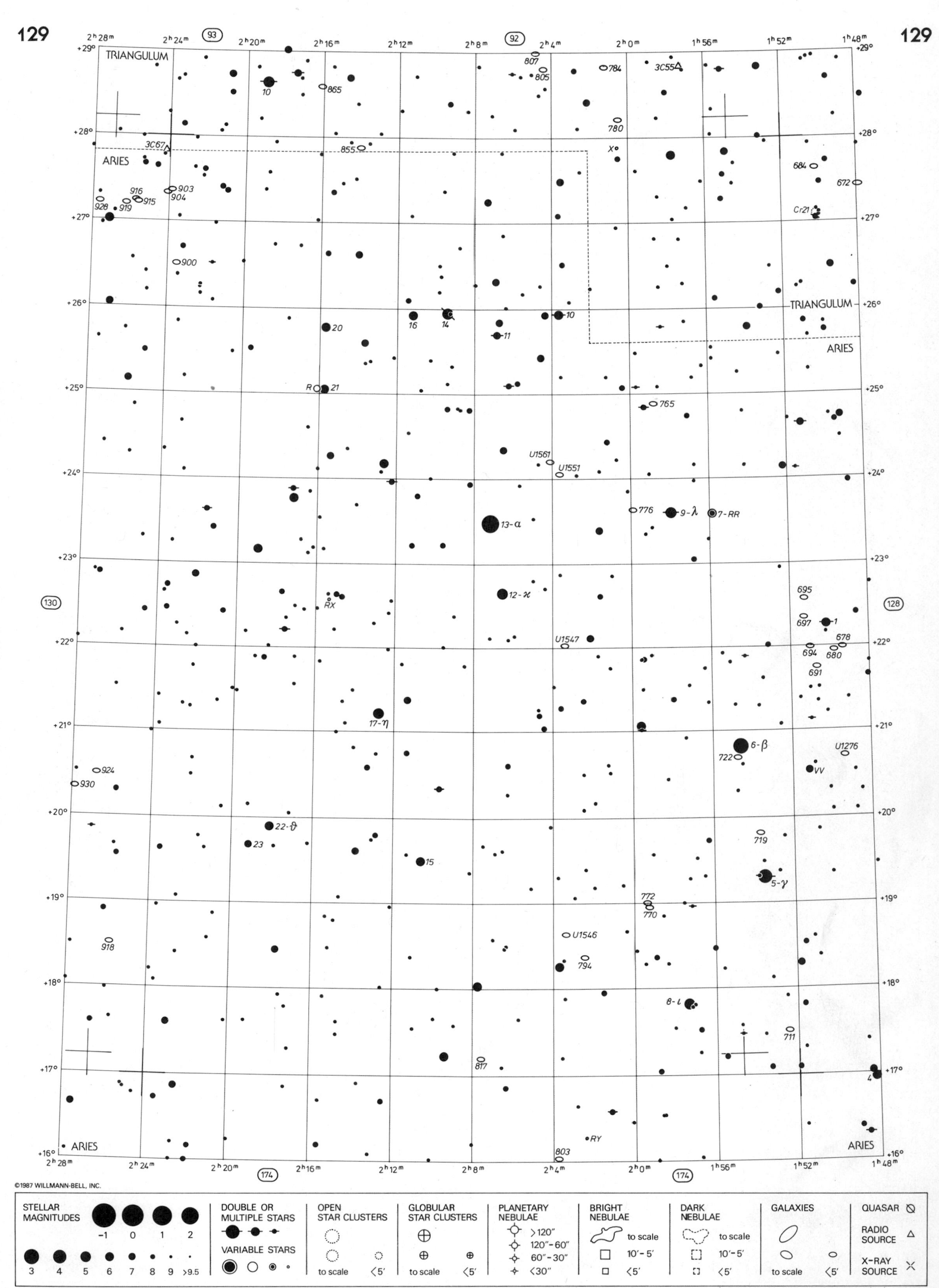
129
129
2h 28m
2h 24m
2h 20m
2h 16m
2h 12m
2h 8m
2h 4m
2h 0m
1h 56m
1h 52m
1h 48m
93
92
130
128
174
174
+29°
+28°
+27°
+26°
+25°
+24°
+23°
+22°
+21°
+20°
+19°
+18°
+17°
+16°
TRIANGULUM
ARIES
TRIANGULUM
ARIES
ARIES
ARIES
3C67
3C55
X
Cr21
10
865
855
807
805
784
780
684
672
916
903
904
915
926
919
900
20
16
14
10
11
R
21
765
U1561
U1551
776
9-λ
7-RR
13-α
12-κ
RX
695
697
1
U1547
678
694
680
691
17-η
6-β
722
U1276
924
930
VV
22-ϑ
23
15
719
5-γ
772
770
U1546
918
794
8-ι
711
817
4
RY
803
©1987 WILLMANN-BELL, INC.
STELLAR MAGNITUDES
-1
0
1
2
3
4
5
6
7
8
9
>9.5
DOUBLE OR MULTIPLE STARS
VARIABLE STARS
OPEN STAR CLUSTERS
to scale
<5'
GLOBULAR STAR CLUSTERS
to scale
<5'
PLANETARY NEBULAE
>120"
120"-60"
60"-30"
<30"
BRIGHT NEBULAE
to scale
10'-5'
<5'
DARK NEBULAE
to scale
10'-5'
<5'
GALAXIES
to scale
<5'
QUASAR
RADIO SOURCE
X-RAY SOURCE
Barry Rappaport & Wil Tirion

130

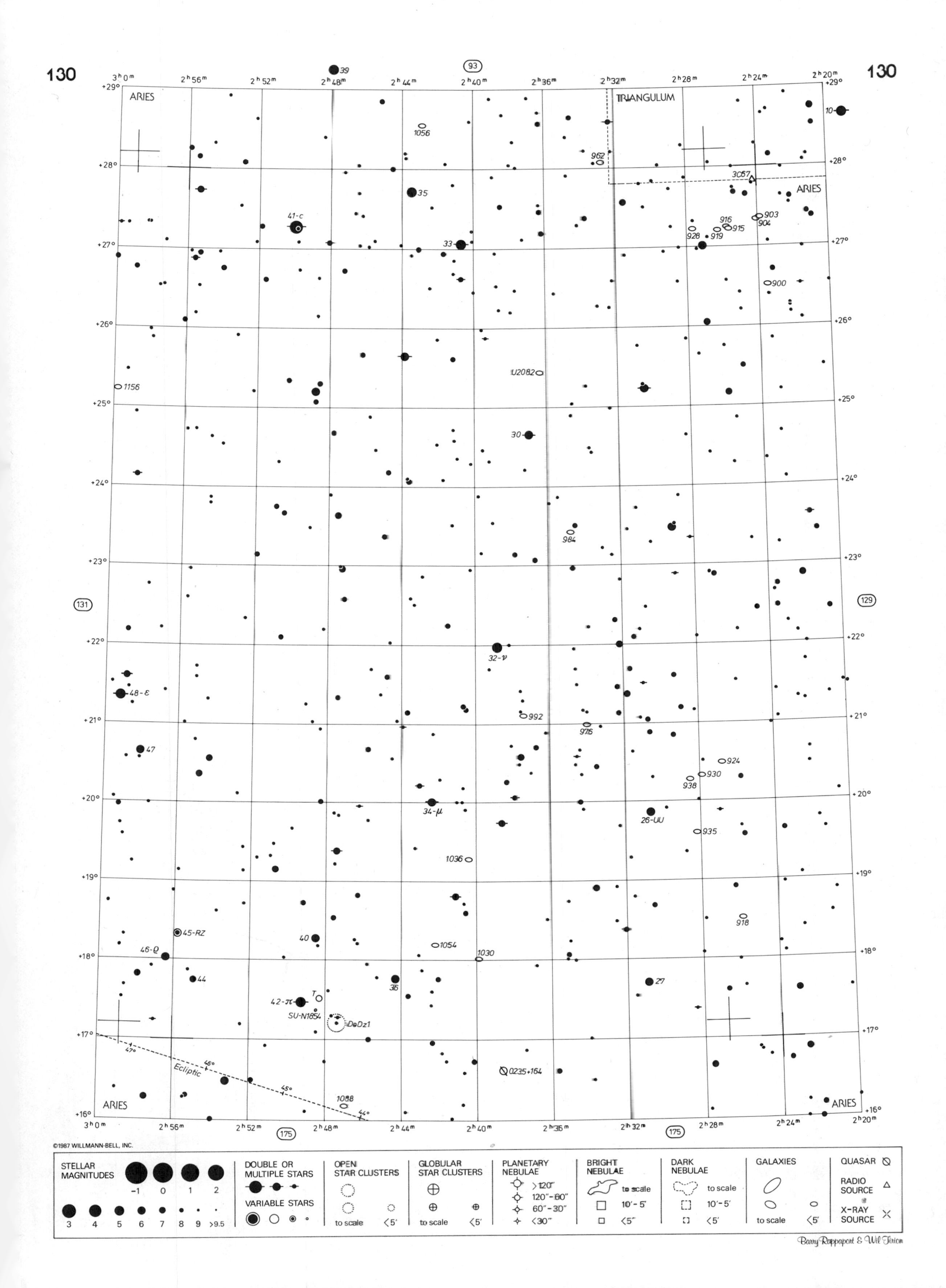

STELLAR MAGNITUDES	DOUBLE OR MULTIPLE STARS	OPEN STAR CLUSTERS	GLOBULAR STAR CLUSTERS	PLANETARY NEBULAE	BRIGHT NEBULAE	DARK NEBULAE	GALAXIES	QUASAR
−1 0 1 2	VARIABLE STARS	to scale <5′	to scale <5′	>120″ 120″−60″ 60″−30″ <30″	to scale 10′−5′ <5′	to scale 10′−5′ <5′	to scale <5′	RADIO SOURCE
3 4 5 6 7 8 9 >9.5								X-RAY SOURCE

Barry Rappaport & Wil Tirion

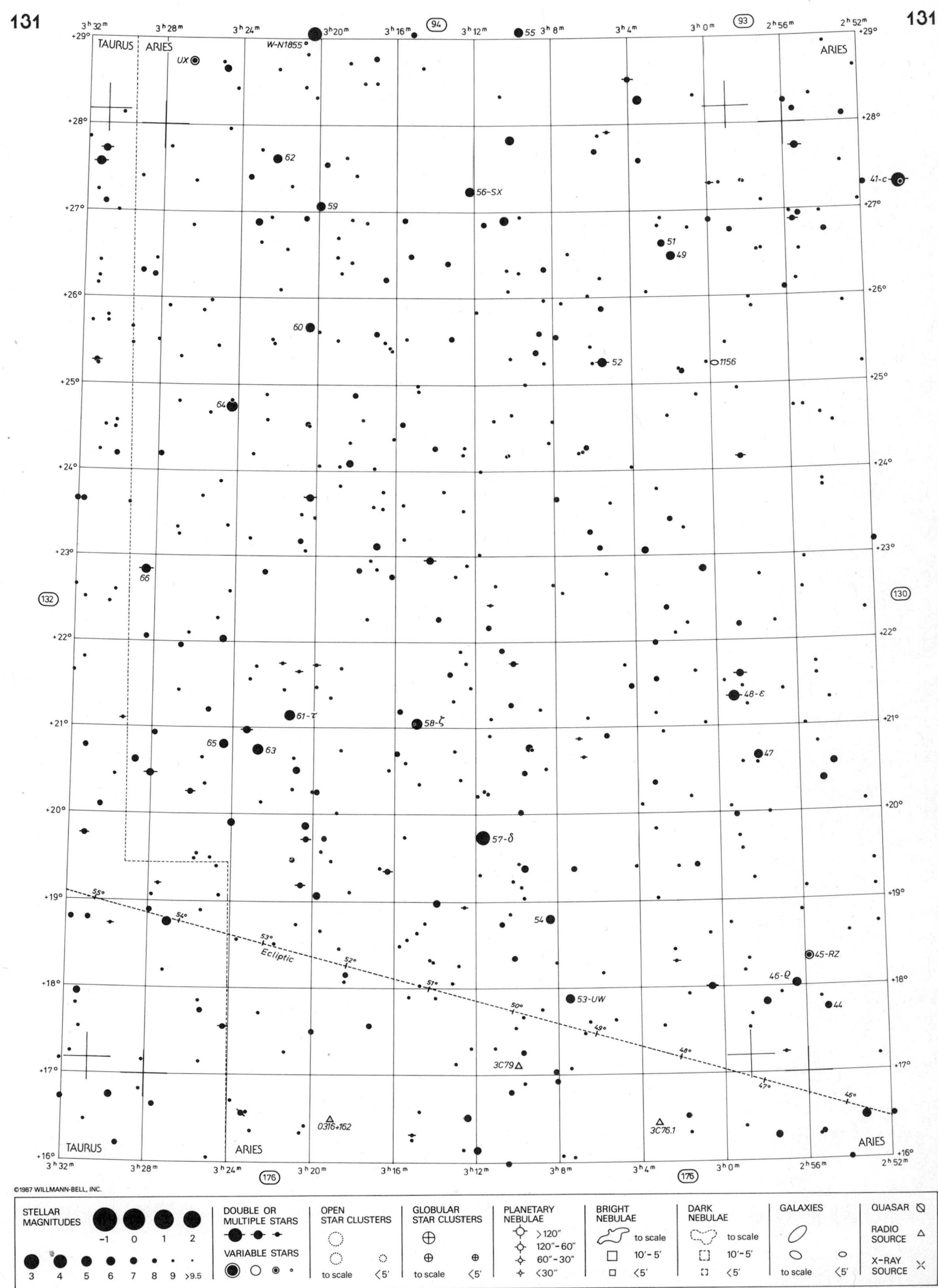

Barry Rappaport & Wil Tirion

STELLAR MAGNITUDES -1 0 1 2 3 4 5 6 7 8 9 >9.5
DOUBLE OR MULTIPLE STARS
VARIABLE STARS
OPEN STAR CLUSTERS to scale <5′
GLOBULAR STAR CLUSTERS to scale <5′
PLANETARY NEBULAE >120″ 120″-60″ 60″-30″ <30″
BRIGHT NEBULAE to scale 10′-5′ <5′
DARK NEBULAE to scale 10′-5′ <5′
GALAXIES to scale <5′
QUASAR
RADIO SOURCE
X-RAY SOURCE

Barry Rappaport & Wil Tirion

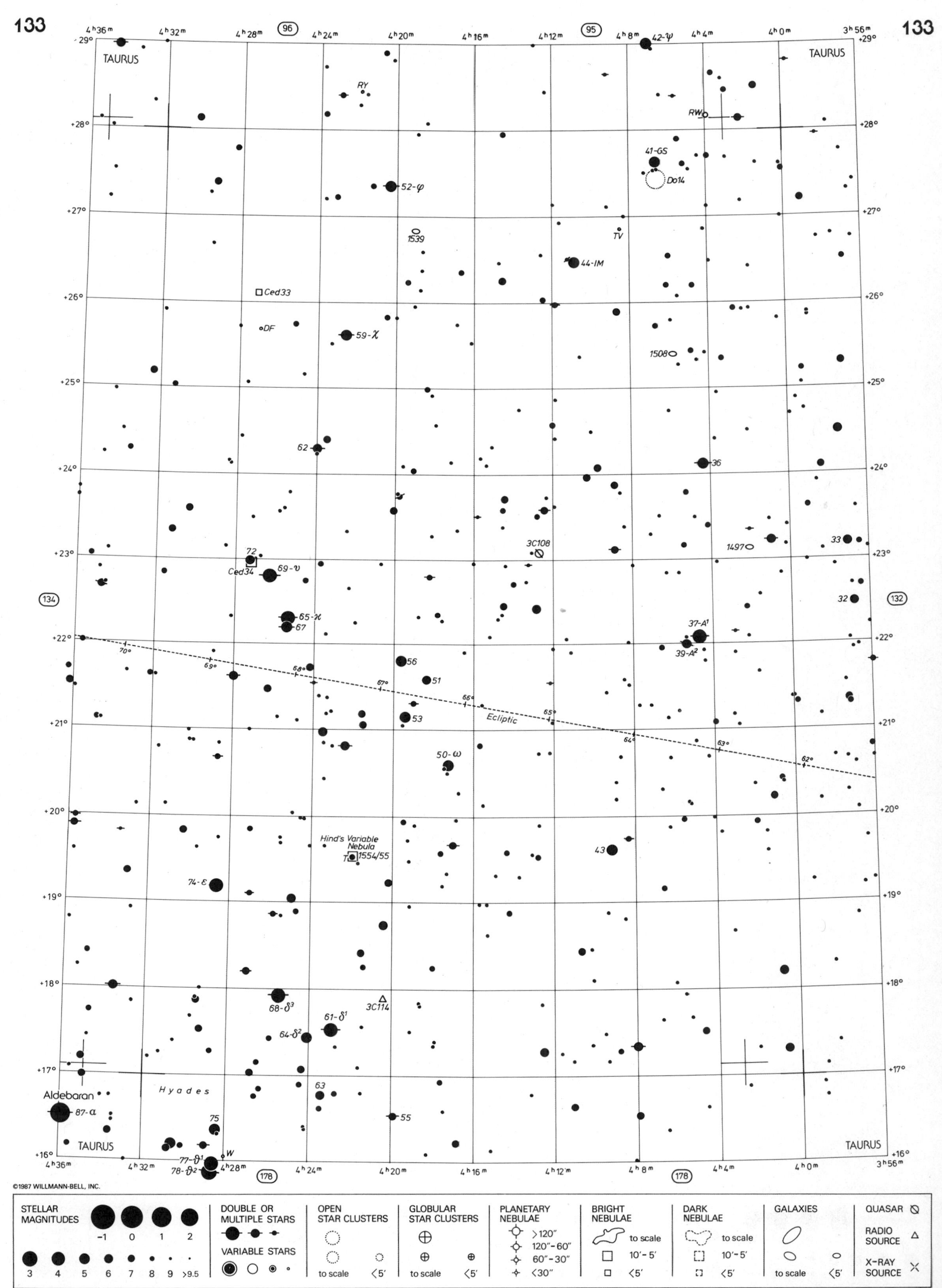
TAURUS
Aldebaran
87-α
Hyades
74-ε
68-δ3
61-δ1
64-δ2
65-κ
67
69-υ
Ced34
Ced33
DF
59-χ
52-φ
44-IM
42-ψ
41-GS
Do14
37-A1
39-A2
50-ω
3C108
3C114
Hind's Variable Nebula
1554/55
1539
1508
1497
Ecliptic
77-θ1
78-θ2
RY
RW
TV
W
STELLAR MAGNITUDES
DOUBLE OR MULTIPLE STARS
VARIABLE STARS
OPEN STAR CLUSTERS
GLOBULAR STAR CLUSTERS
PLANETARY NEBULAE
BRIGHT NEBULAE
DARK NEBULAE
GALAXIES
QUASAR
RADIO SOURCE
X-RAY SOURCE
to scale
Barry Rappaport & Wil Tirion

134

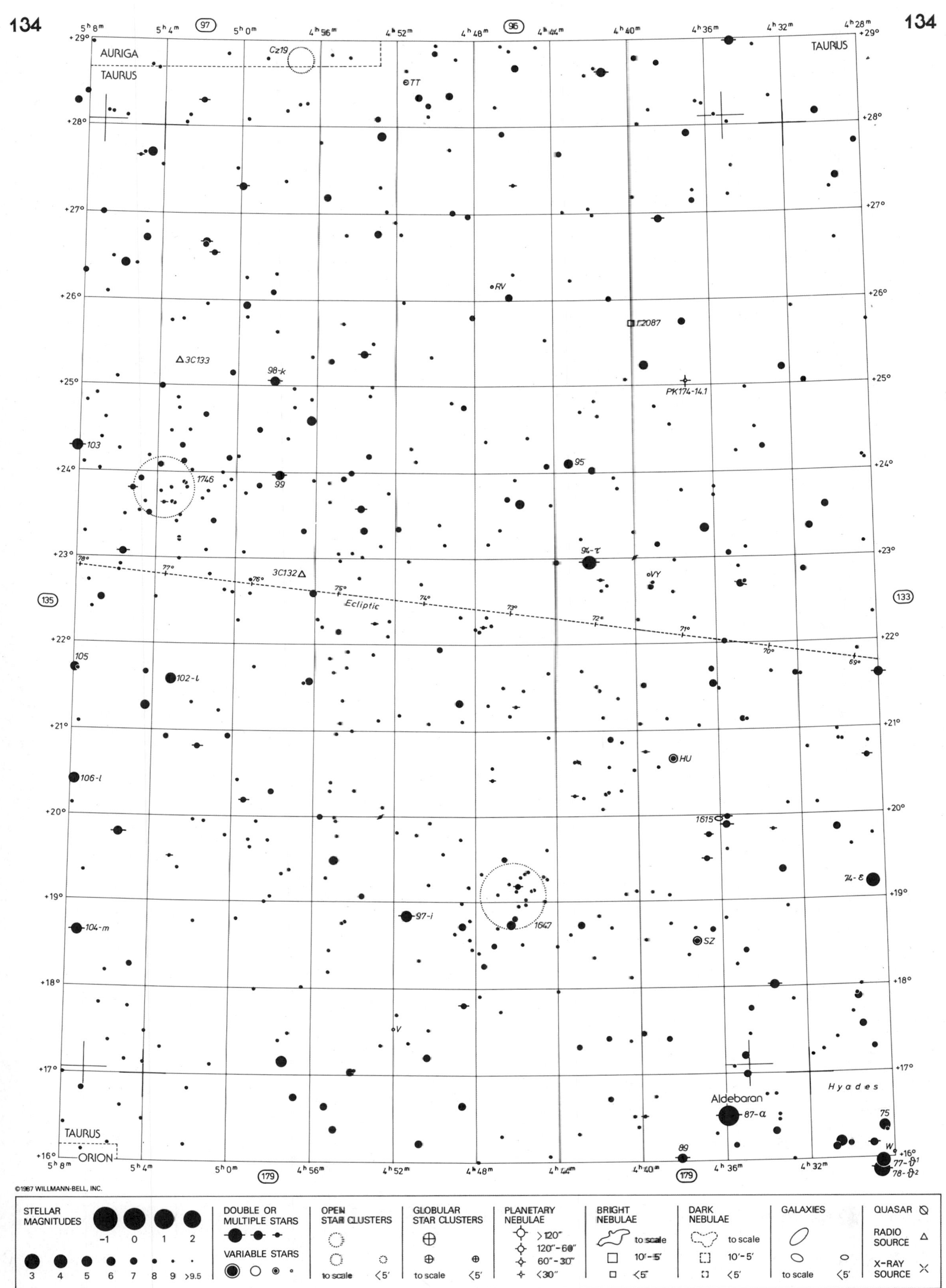

AURIGA
TAURUS
Cz19
TT
TAURUS
RV
Γ2087
3C133
98-k
PK174-14.1
103
95
1746
99
94-τ
3C132
VY
Ecliptic
105
102-ι
HU
106-l
1615
74-ε
97-i
1647
104-m
SZ
V
Hyades
Aldebaran
87-α
75
89
W
77-ϑ1
78-ϑ2
TAURUS
ORION
STELLAR MAGNITUDES
DOUBLE OR MULTIPLE STARS
VARIABLE STARS
OPEN STAR CLUSTERS
GLOBULAR STAR CLUSTERS
PLANETARY NEBULAE
BRIGHT NEBULAE
DARK NEBULAE
GALAXIES
QUASAR
RADIO SOURCE
X-RAY SOURCE
to scale
Barry Rappaport & Wil Tirion

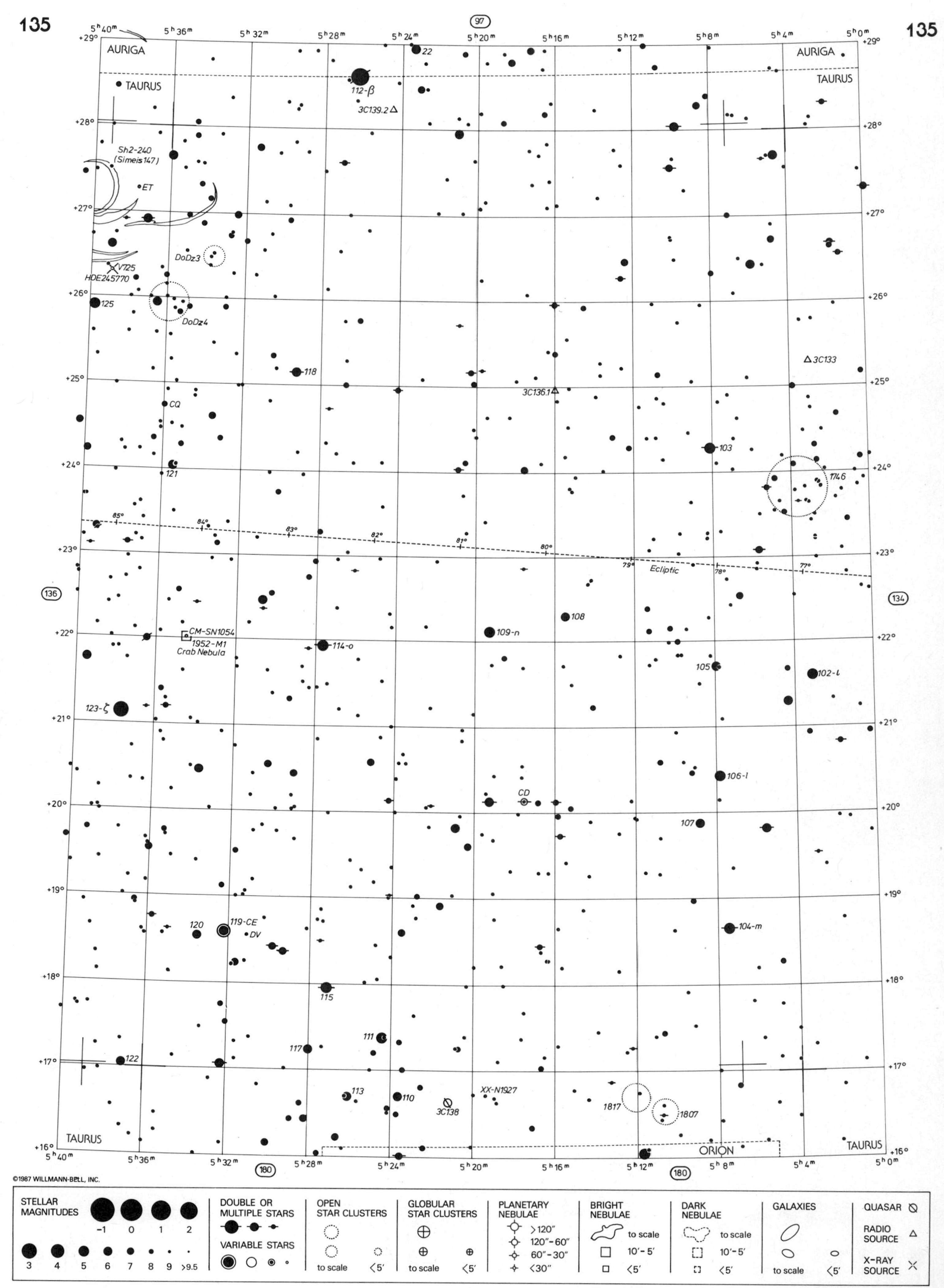
AURIGA
TAURUS
112-β
3C139.2
22
Sh2-240
(Simeis 147)
ET
DoDz3
V725
HDE245770
125
DoDz4
118
3C133
3C136.1
CQ
103
1746
121
Ecliptic
136
134
108
109-n
CM-SN1054
1952-M1
Crab Nebula
114-o
105
102-ι
123-ζ
106-l
CD
107
120
119-CE
DV
104-m
115
111
117
122
113
110
3C138
XX-N1927
1817
1807
ORION
97
180
©1987 WILLMANN-BELL, INC.
STELLAR MAGNITUDES
DOUBLE OR MULTIPLE STARS
VARIABLE STARS
OPEN STAR CLUSTERS
GLOBULAR STAR CLUSTERS
PLANETARY NEBULAE
BRIGHT NEBULAE
DARK NEBULAE
GALAXIES
QUASAR
RADIO SOURCE
X-RAY SOURCE
to scale
Barry Rappaport & Wil Tirion

STELLAR MAGNITUDES	DOUBLE OR MULTIPLE STARS	OPEN STAR CLUSTERS	GLOBULAR STAR CLUSTERS	PLANETARY NEBULAE	BRIGHT NEBULAE	DARK NEBULAE	GALAXIES	QUASAR
−1 0 1 2	VARIABLE STARS	to scale ‹5′	to scale ‹5′	>120″ 120″–60″ 60″–30″ ‹30″	to scale 10′–5′ ‹5′	to scale 10′–5′ ‹5′	to scale ‹5′	RADIO SOURCE
3 4 5 6 7 8 9 >9.5								X-RAY SOURCE

Barry Rappaport & Wil Tirion

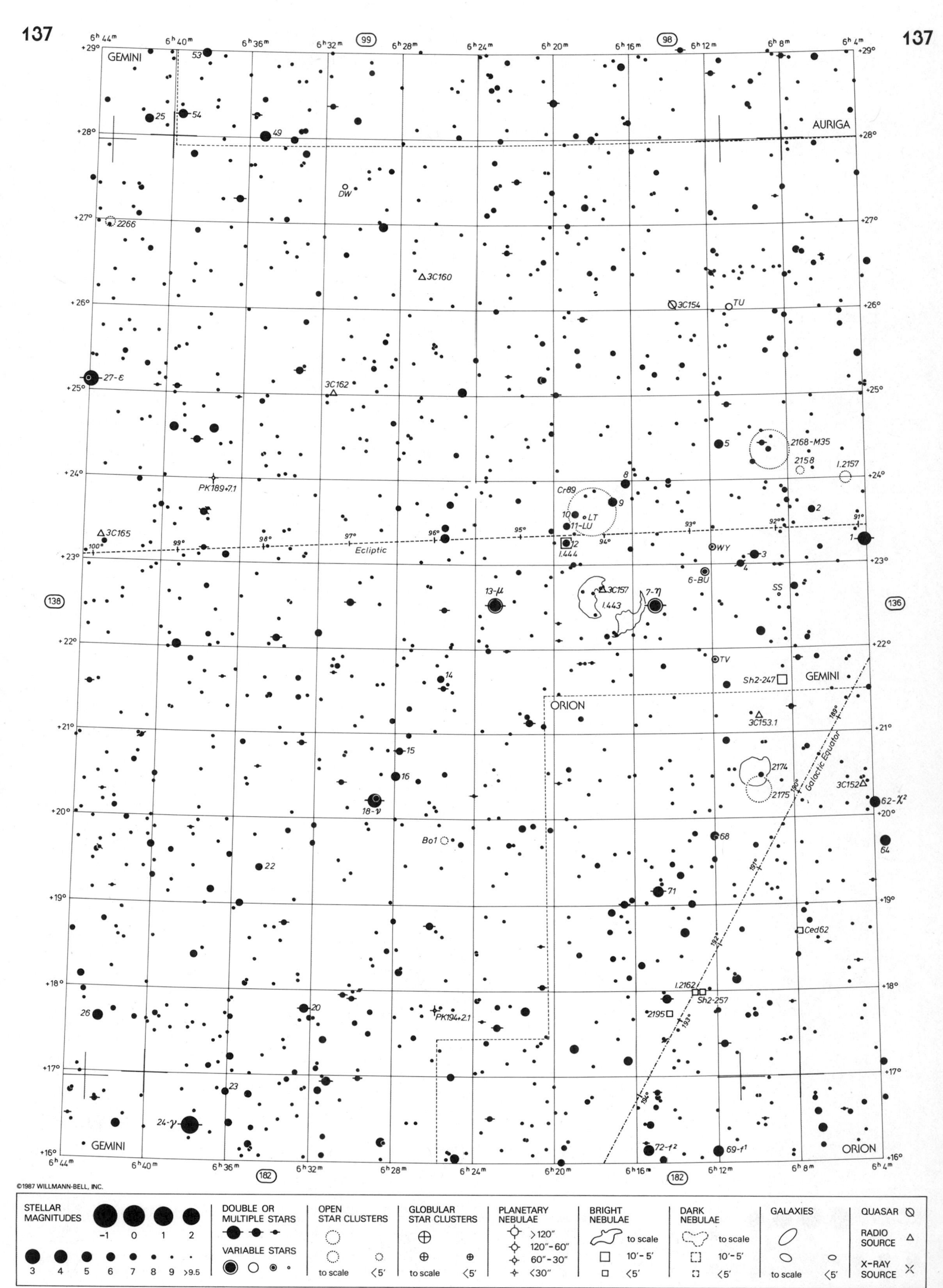

Barry Rappaport & Wil Tirion

138

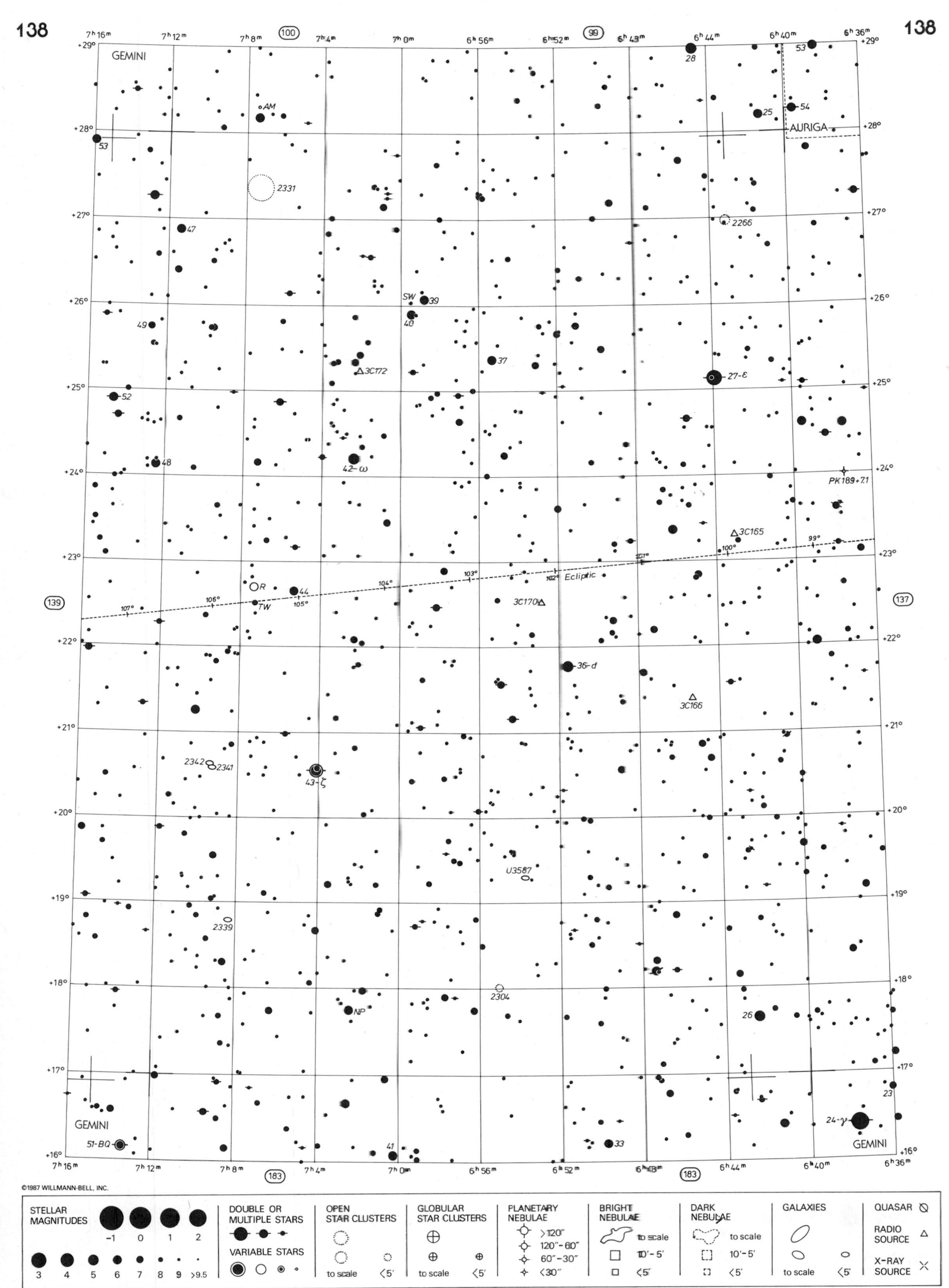

Barry Rappaport & Wil Tirion

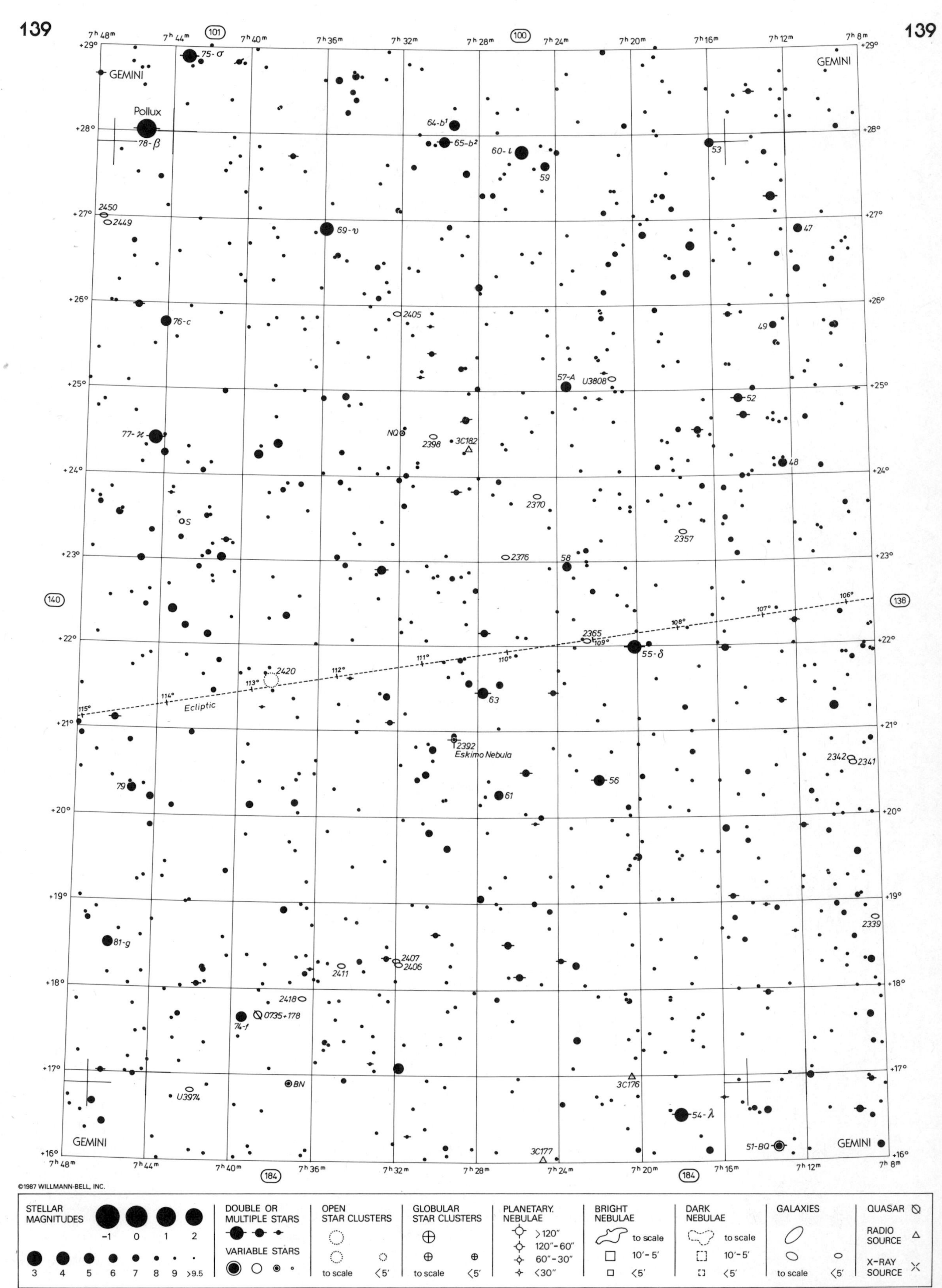
139
GEMINI
Pollux
78-β
75-σ
64-b¹
65-b²
60-ι
59
53
69-υ
47
76-c
49
57-A
U3808
52
77-κ
NQ
2398
3C182
48
2370
S
2357
2376
58
2365
55-δ
2420
Ecliptic
63
2392
Eskimo Nebula
2342
2341
56
79
61
81-g
2339
2407
2406
2411
2418
74-f
0735+178
BN
U3974
3C176
54-λ
51-BQ
3C177
2450
2449
2405
©1987 WILLMANN-BELL, INC.
STELLAR MAGNITUDES
DOUBLE OR MULTIPLE STARS
VARIABLE STARS
OPEN STAR CLUSTERS
GLOBULAR STAR CLUSTERS
PLANETARY NEBULAE
BRIGHT NEBULAE
DARK NEBULAE
GALAXIES
QUASAR
RADIO SOURCE
X-RAY SOURCE
to scale
Barry Rappaport & Wil Tirion

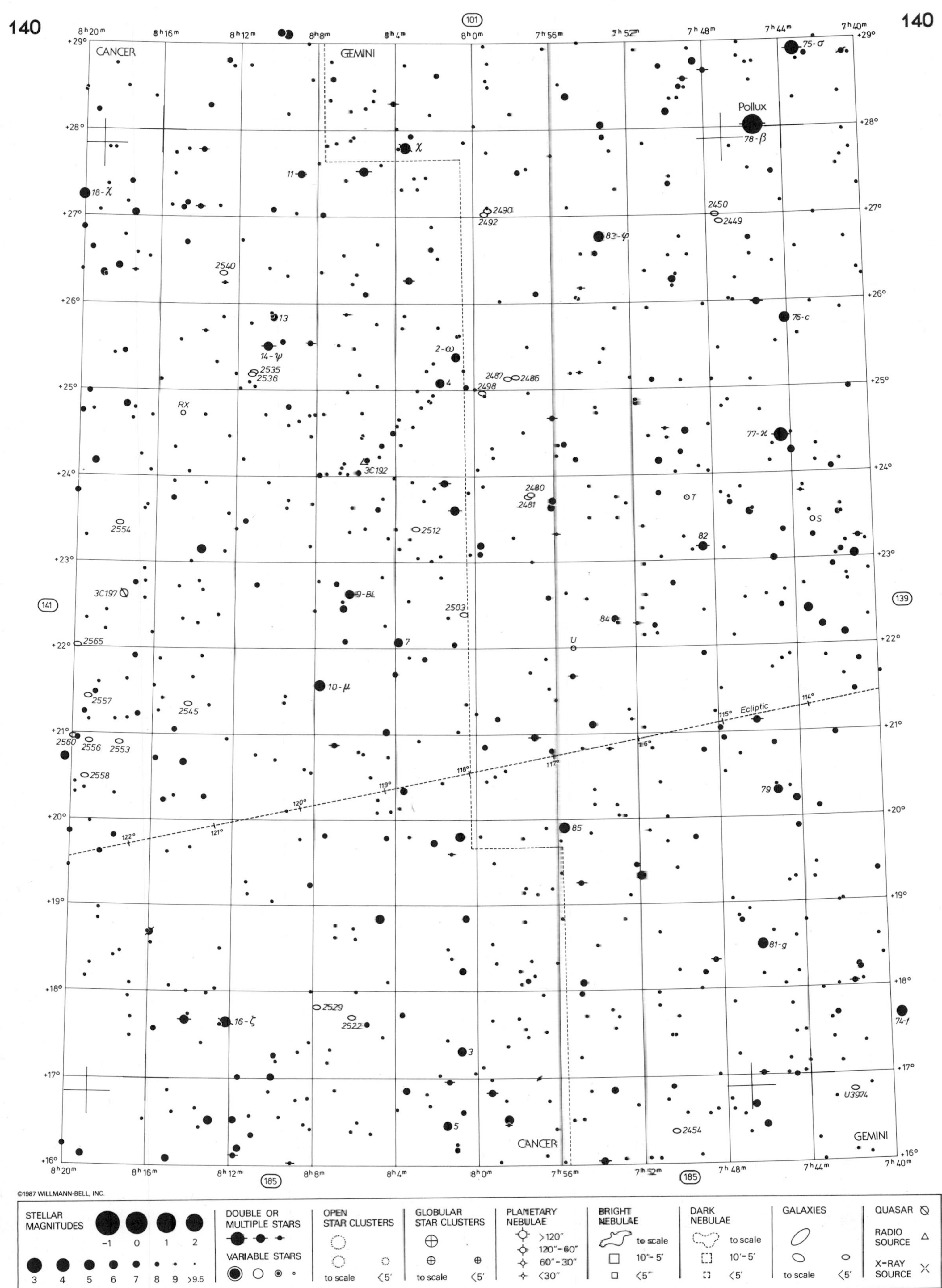

CANCER
GEMINI
Pollux
78-β
75-σ
76-c
77-κ
83-φ
82
84
85
79
81-g
74-f
18-χ
11
χ
13
14-ψ
2-ω
4
7
10-μ
9-BL
3
5
16-ζ
RX
T
S
U
3C192
3C197
2490
2492
2450
2449
2540
2535
2536
2487
2486
2498
2480
2481
2512
2554
2503
2565
2557
2545
2560
2556
2553
2558
2529
2522
2454
U3974
Ecliptic
STELLAR MAGNITUDES
DOUBLE OR MULTIPLE STARS
VARIABLE STARS
OPEN STAR CLUSTERS
GLOBULAR STAR CLUSTERS
PLANETARY NEBULAE
BRIGHT NEBULAE
DARK NEBULAE
GALAXIES
QUASAR
RADIO SOURCE
X-RAY SOURCE
to scale
©1987 WILLMANN-BELL, INC.
Barry Rappaport & Wil Tirion

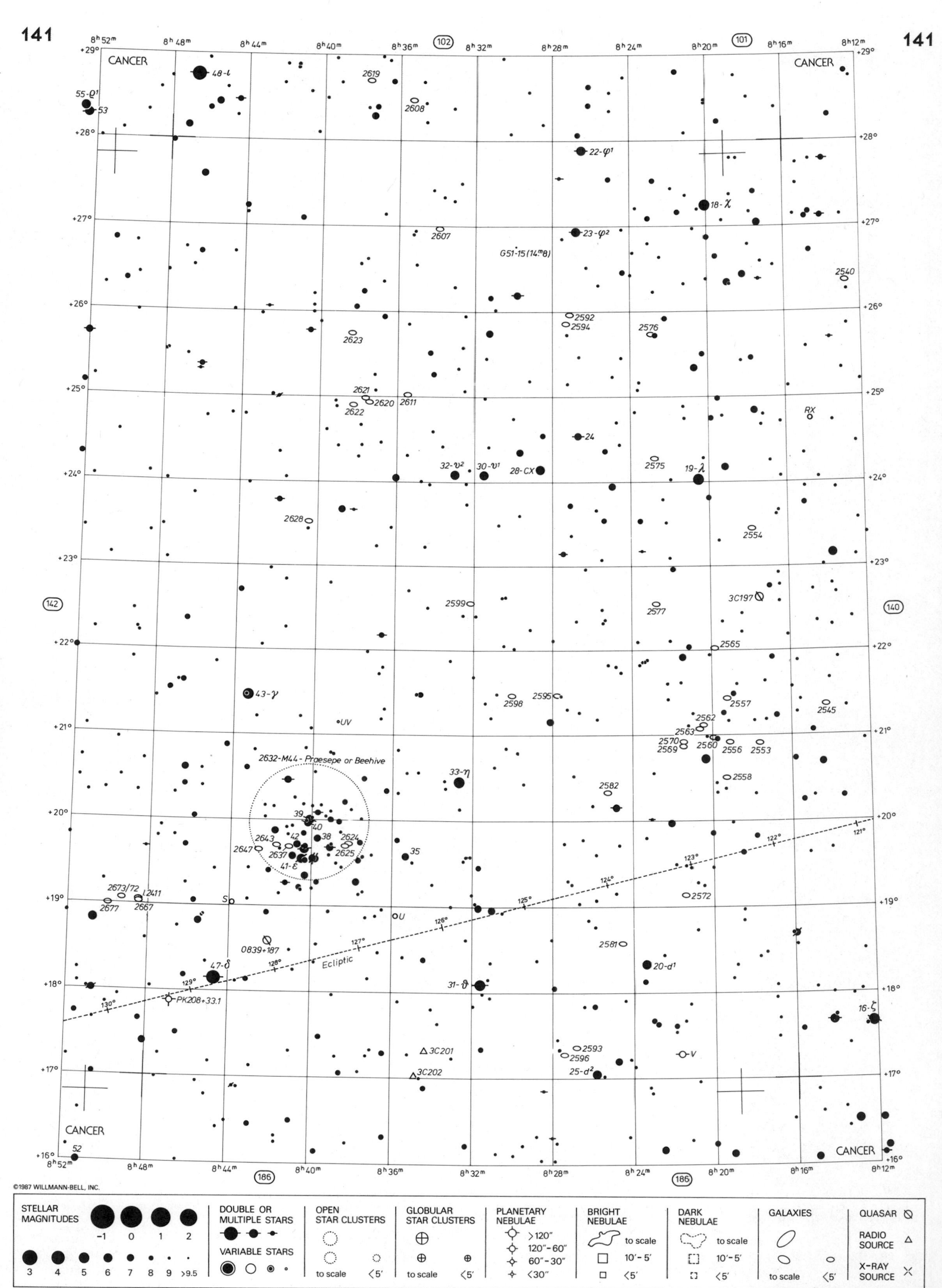

CANCER
48-ι
55-ρ¹
53
22-φ¹
18-χ
23-φ²
G51-15 (14ᵐ8)
2619
2608
2607
2540
2592
2594
2576
2623
2621
2620
2611
2622
RX
24
32-υ²
30-υ¹
28-CX
2575
19-λ
2628
2554
2599
2577
3C197
2565
43-γ
2598
2595
2557
2545
UV
2562
2563
2570
2569
2560
2556
2553
2632-M44 - Praesepe or Beehive
33-η
2558
2582
39
40
38
2643
42
2624
2647
2637
2625
41-ε
35
2673/72
I 2411
2677
2667
S
2572
U
0839+187
2581
47-δ
Ecliptic
20-d¹
31-ϑ
PK208+33.1
16-ζ
3C201
2593
2596
V
3C202
25-d²
52
©1987 WILLMANN-BELL, INC.
STELLAR MAGNITUDES
DOUBLE OR MULTIPLE STARS
VARIABLE STARS
OPEN STAR CLUSTERS
GLOBULAR STAR CLUSTERS
PLANETARY NEBULAE
BRIGHT NEBULAE
DARK NEBULAE
GALAXIES
QUASAR
RADIO SOURCE
X-RAY SOURCE
to scale
<5′
>120″
120″–60″
60″–30″
<30″
10′–5′
Barry Rappaport & Wil Tirion

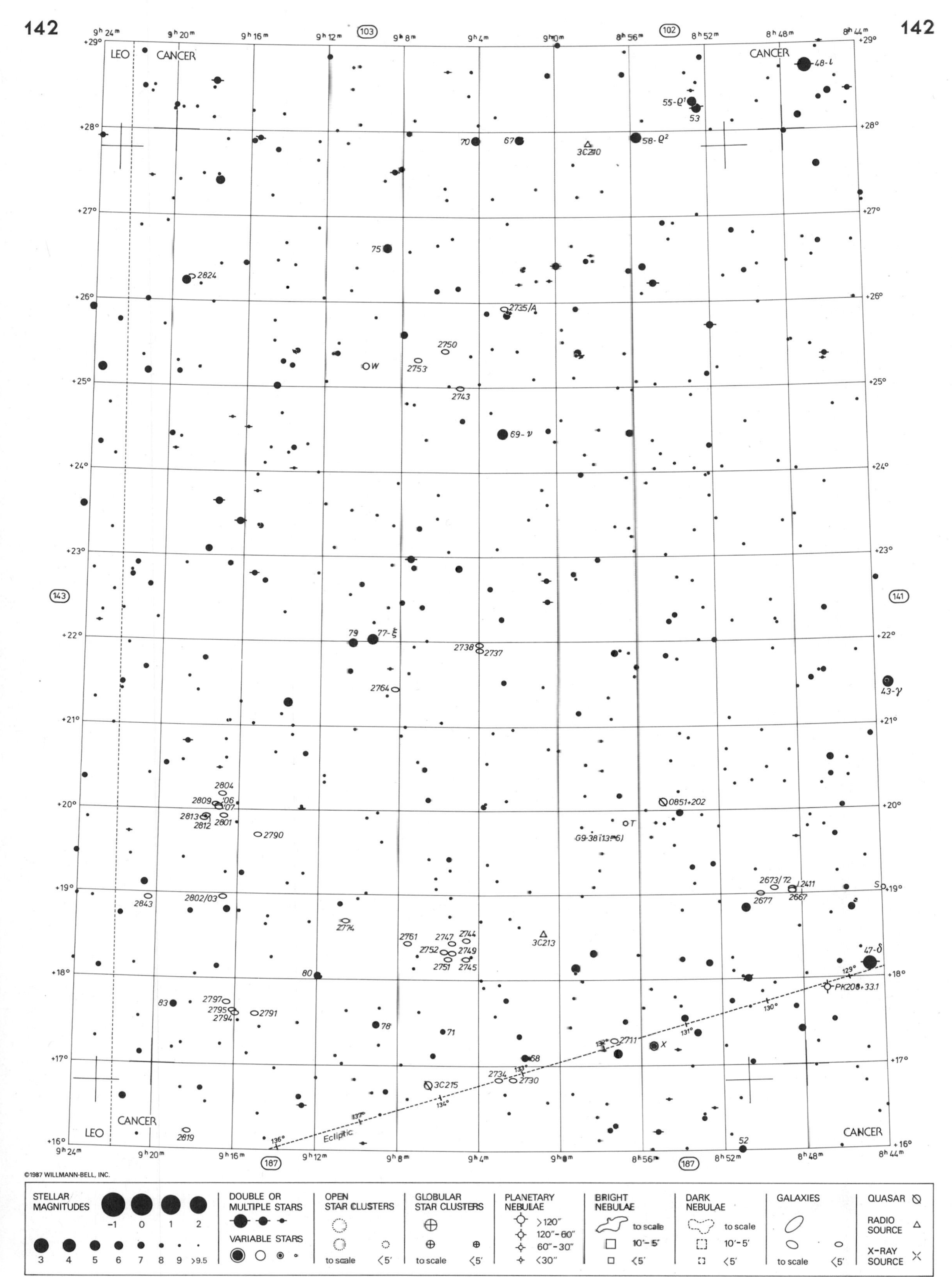

Barry Rappaport & Wil Tirion

142

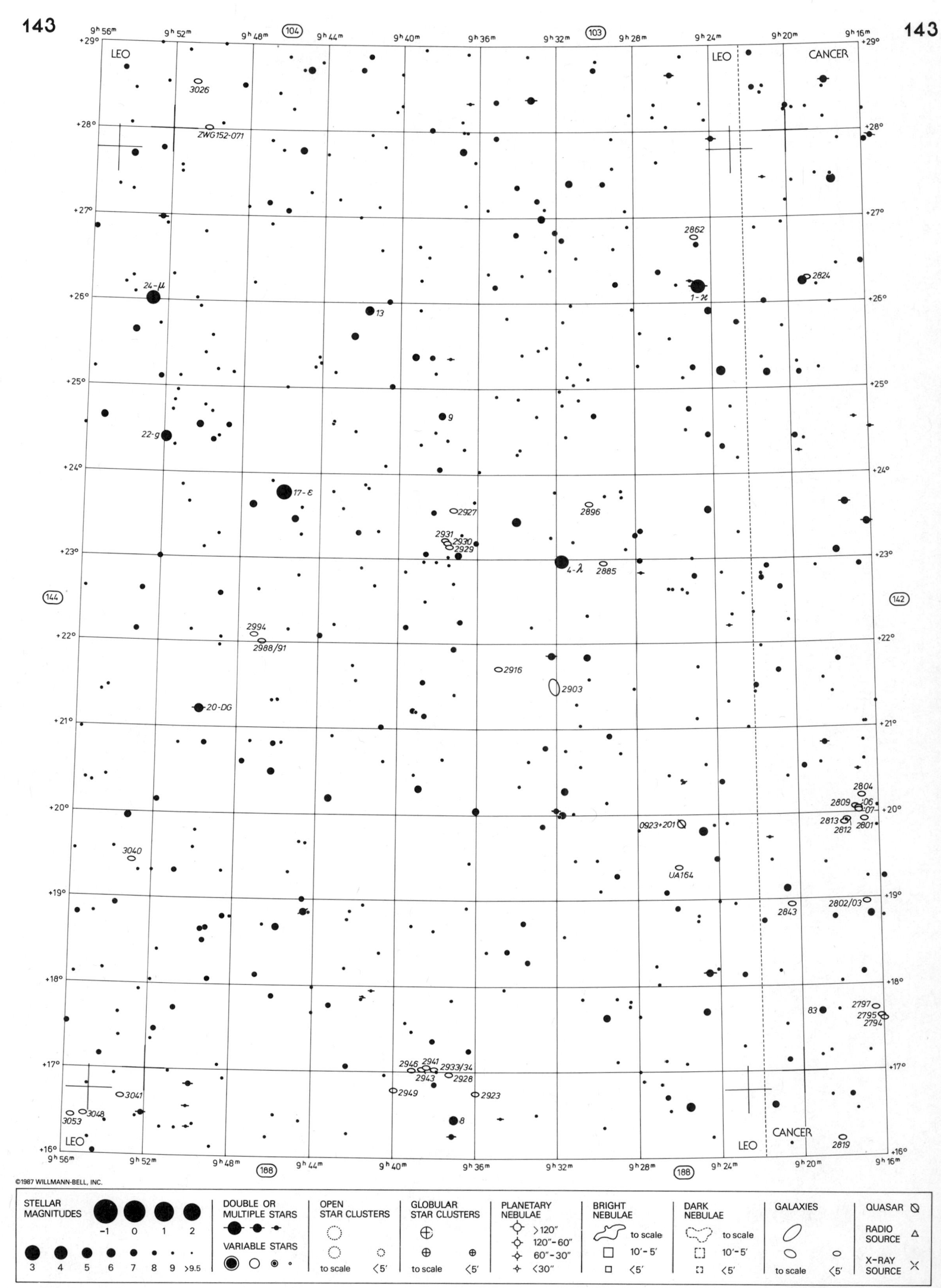

Barry Rappaport & Wil Tirion

10ʰ 28ᵐ 10ʰ 24ᵐ (105) 10ʰ 20ᵐ 10ʰ 16ᵐ 10ʰ 12ᵐ 10ʰ 8ᵐ (104) 10ʰ 4ᵐ 10ʰ 0ᵐ 9ʰ 56ᵐ 9ʰ 52ᵐ 9ʰ 48ᵐ

144

LEO MINOR

LEO

+29° +28° +27° +26° +25° +24° +23° +22° +21° +20° +19° +18° +17° +16°

3245A
3245
3235 3232
3204 1017+280
3196
3C240
24
UA201 1001+292
3C234
3068 U5340
3026
ZWG152-071
24-μ
3209
U5588
1011+250
3098
22-g
17-ε
3216
35
36-ζ
39
3248
3162
(145)
(143)
3088A/B
3C241
3193 3187
3190 3185
3221
3177
V
20-DG
3C242
3226
3227 3222
AD
41-γ
3213
40
3040
3131
3239
3154
30-η
3060
3041
3048
3053

LEO LEO

10ʰ 28ᵐ 10ʰ 24ᵐ 10ʰ 20ᵐ (189) 10ʰ 16ᵐ 10ʰ 12ᵐ 10ʰ 8ᵐ 10ʰ 4ᵐ 10ʰ 0ᵐ (189) 9ʰ 56ᵐ 9ʰ 52ᵐ 9ʰ 48ᵐ

STELLAR MAGNITUDES	DOUBLE OR MULTIPLE STARS	OPEN STAR CLUSTERS	GLOBULAR STAR CLUSTERS	PLANETARY NEBULAE	BRIGHT NEBULAE	DARK NEBULAE	GALAXIES	
-1 0 1 2	VARIABLE STARS	to scale <5′	to scale <5′	>120″ 120″-60″ 60″-30″ <30″	to scale 10′-5′ <5′	to scale 10′-5′ <5′	to scale <5′	QUASAR RADIO SOURCE X-RAY SOURCE
3 4 5 6 7 8 9 >9.5								

Barry Rappaport & Wil Tirion

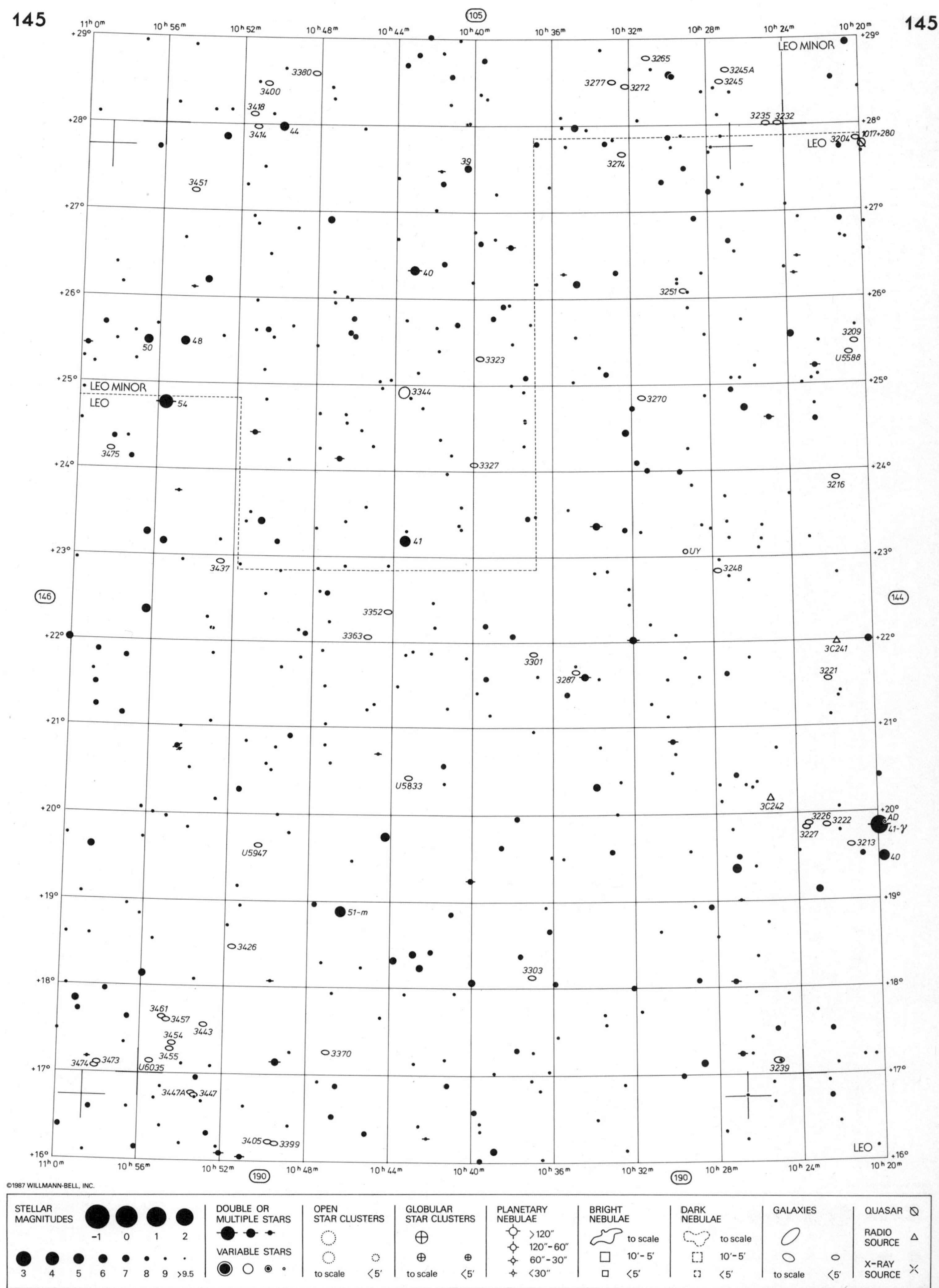

Barry Rappaport & Wil Tirion

11h 32m · 11h 28m · 11h 24m · 11h 20m · 11h 16m · (106) · 11h 12m · 11h 8m · 11h 4m · 11h 0m · (105) · 10h 56m · 10h 52m

+29° · +28° · +27° · +26° · +25° · +24° · +23° · +22° · +21° · +20° · +19° · +18° · +17° · +16°

URSA MAJOR · LEO · LEO MINOR

Pal4 · 3712 · 3714 · 3713 · 3678 · 3561 · 3550 · 3539 · 3554 · 3552/53 · 3558 · 3536 · 3527 · 3510 · 3486 · 3515 · 3512 · 3504 · 3493 · 3574 · 3570 · 3451 · 3629 · 3563A/B · 3612 · 3609 · 3534 · 3534B · 3689 · 50 · 48 · 51 · 3C250 · 67 · 54 · 3701 · 3651 · 3653 · 3475 · 3670 · 3618 · 3615 · 3C256 · 64 · 72 · 3710 · 3437 · (147) · (145) · U6253-Leo II · 3555 · 3551 · 1116+215 · 3697A · 3697C · 3697B · 3650 · 68-δ · 3588 · 3649 · 3646 · 3522 · 60-b · 3C258 · U6320 · 86 · 3639 · 3626 · U6171 · 3608 · 3607 · 3599 · 3605 · 3507 · 3501 · 3659 · 3461 · 3457 · 3443 · 3487 · 71 · 3602 · U6157 · 3454 · 3686 · 3598 · 3592 · 3455 · 3684 · 3474 · 3473 · U6035 · 3691 · 3681 · 3447A · 3447 · U6112 · 3655 · 81 · 4C+16.30

11h 32m · 11h 28m · 11h 24m · (191) · 11h 20m · 11h 16m · 11h 12m · 11h 8m · 11h 4m · (191) · 11h 0m · 10h 56m · 10h 52m

STELLAR MAGNITUDES	DOUBLE OR MULTIPLE STARS	OPEN STAR CLUSTERS	GLOBULAR STAR CLUSTERS	PLANETARY NEBULAE	BRIGHT NEBULAE	DARK NEBULAE	GALAXIES	QUASAR
−1 0 1 2	VARIABLE STARS	to scale	to scale	>120″	to scale	to scale	to scale	RADIO SOURCE
3 4 5 6 7 8 9 >9.5		<5′	<5′	120″–60″	10′–5′	10′–5′	<5′	X-RAY SOURCE
				60″–30″	<5′	<5′		
				<30″				

Barry Rappaport & Wil Tirion

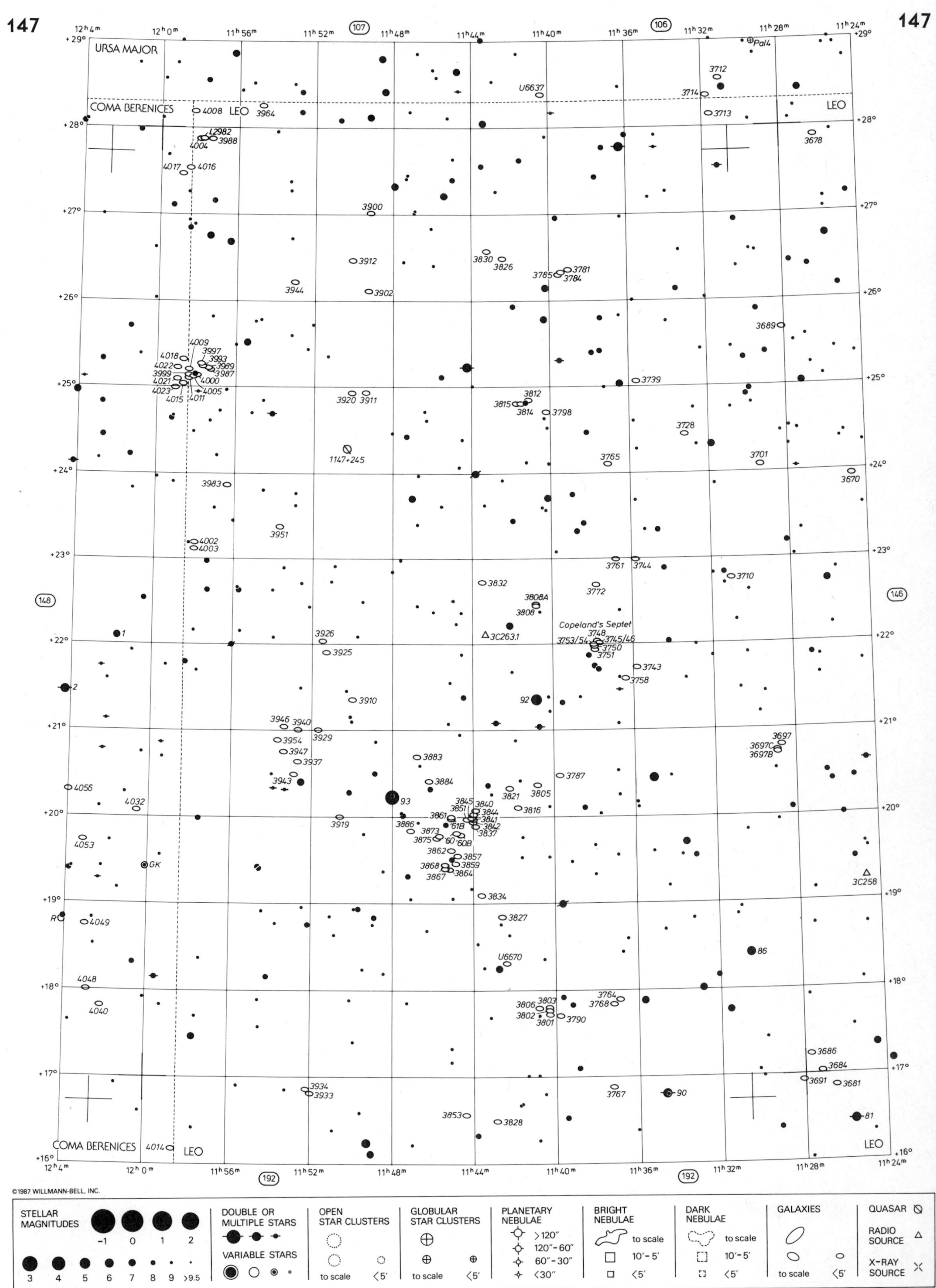

Barry Rappaport & Wil Tirion

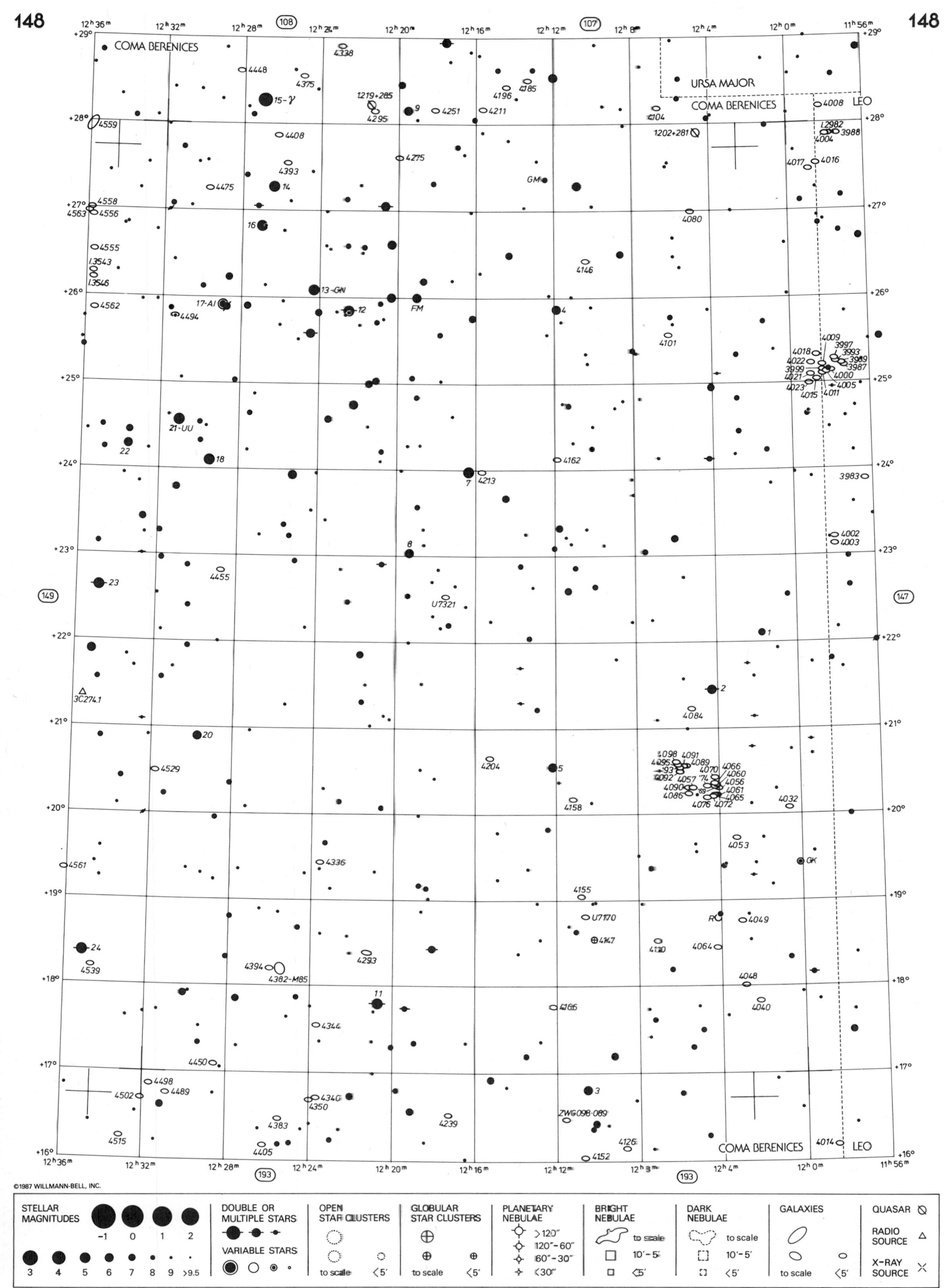

Barry Rappaport & Wil Tirion

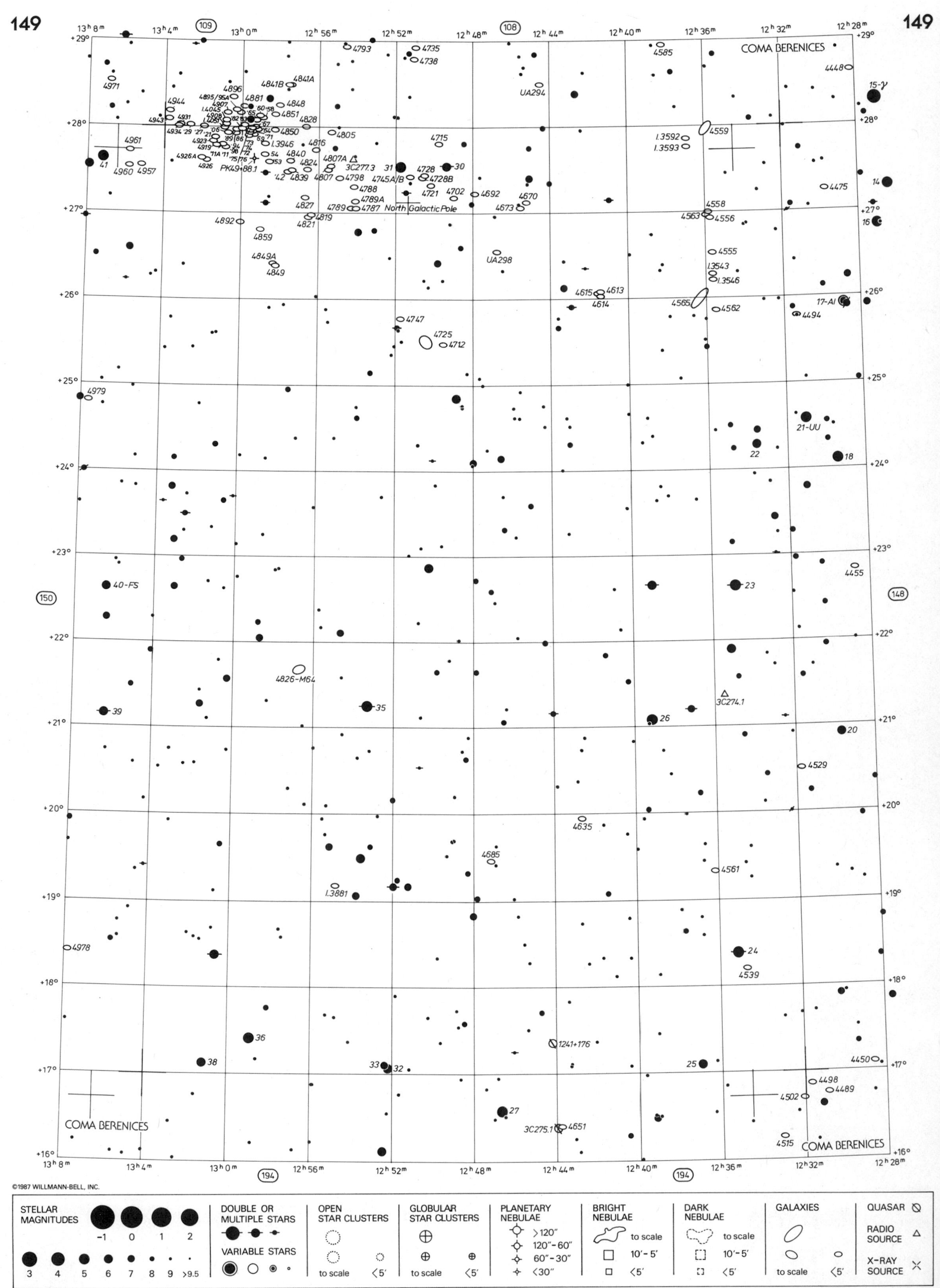

COMA BERENICES
13h 8m
13h 4m
13h 0m
12h 56m
12h 52m
12h 48m
12h 44m
12h 40m
12h 36m
12h 32m
12h 28m
+29°
+28°
+27°
+26°
+25°
+24°
+23°
+22°
+21°
+20°
+19°
+18°
+17°
+16°
109
108
150
148
194
North Galactic Pole
4826-M64
3C274.1
3C275.1
1241+176
PK49+88.1
3C277.3
15-γ
17-AI
21-UU
40-FS
©1987 WILLMANN-BELL, INC.
STELLAR MAGNITUDES
-1 0 1 2
3 4 5 6 7 8 9 >9.5
DOUBLE OR MULTIPLE STARS
VARIABLE STARS
OPEN STAR CLUSTERS
to scale <5'
GLOBULAR STAR CLUSTERS
to scale <5'
PLANETARY NEBULAE
>120"
120"-60"
60"-30"
<30"
BRIGHT NEBULAE
to scale
10'-5'
<5'
DARK NEBULAE
to scale
10'-5'
<5'
GALAXIES
to scale <5'
QUASAR
RADIO SOURCE
X-RAY SOURCE
Barry Rappaport & Wil Tirion

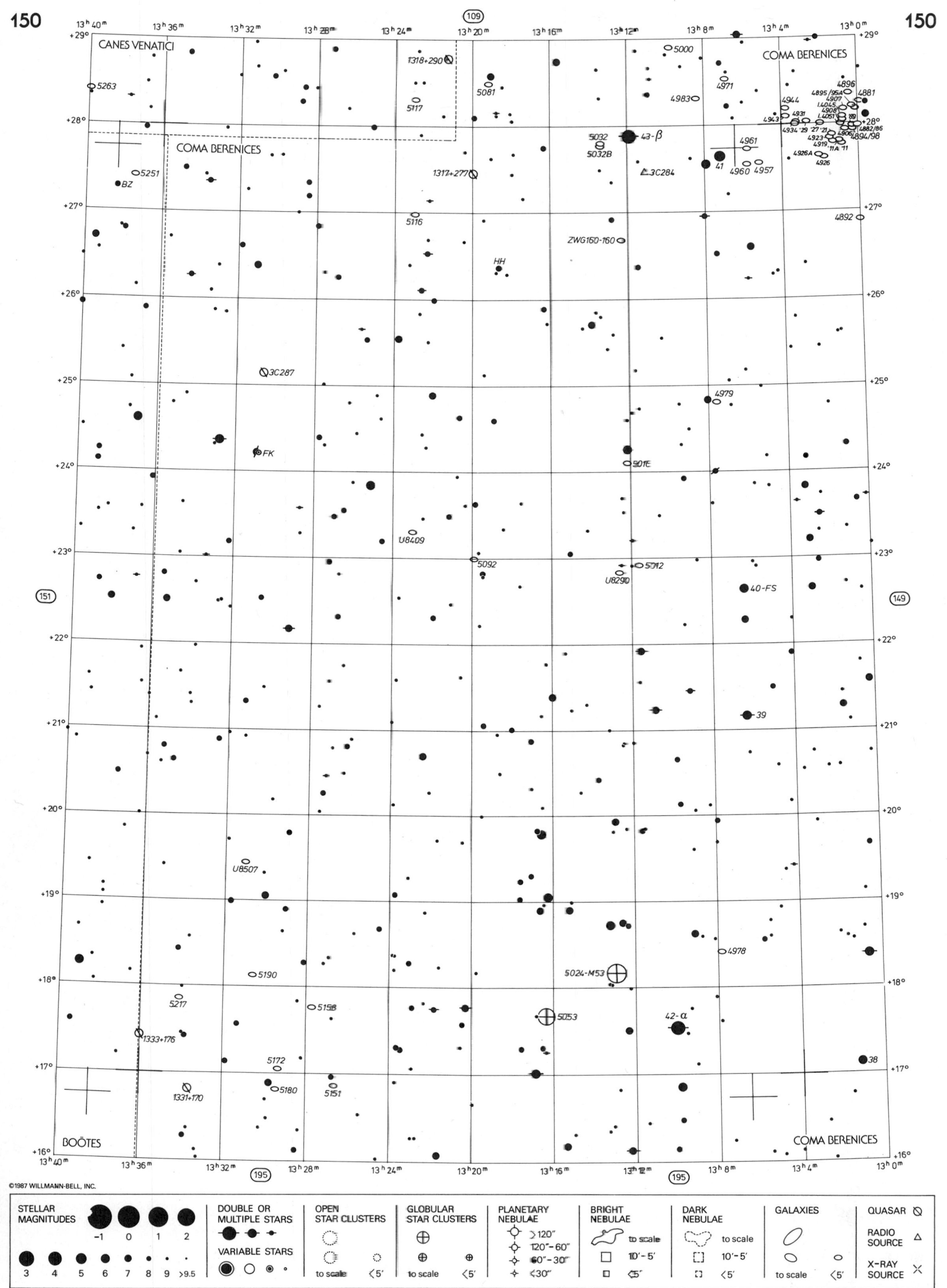
150
109
13h 40m
13h 36m
13h 32m
13h 28m
13h 24m
13h 20m
13h 16m
13h 12m
13h 8m
13h 4m
13h 0m
+29°
+28°
+27°
+26°
+25°
+24°
+23°
+22°
+21°
+20°
+19°
+18°
+17°
+16°
151
149
195
CANES VENATICI
COMA BERENICES
BOÖTES
5263
5251
BZ
1318+290
5117
5081
5000
4971
4983
4896
4881
4944
4943
43-β
5032
5032B
3C284
1317+277
41
4961
4960
4957
4892
5116
ZWG160-160
HH
3C287
4979
FK
5016
U8409
5092
5012
U8290
40-FS
39
U8507
4978
5190
5024-M53
5217
5158
5053
42-α
1333+176
5172
38
5180
5151
1331+170
©1987 WILLMANN-BELL, INC.
STELLAR MAGNITUDES
-1 0 1 2
3 4 5 6 7 8 9 >9.5
DOUBLE OR MULTIPLE STARS
VARIABLE STARS
OPEN STAR CLUSTERS
to scale <5'
GLOBULAR STAR CLUSTERS
to scale <5'
PLANETARY NEBULAE
>120"
120"-60"
60"-30"
<30"
BRIGHT NEBULAE
to scale
10'-5'
<5'
DARK NEBULAE
to scale
10'-5'
<5'
GALAXIES
to scale <5'
QUASAR
RADIO SOURCE
X-RAY SOURCE
Barry Rappaport & Wil Tirion

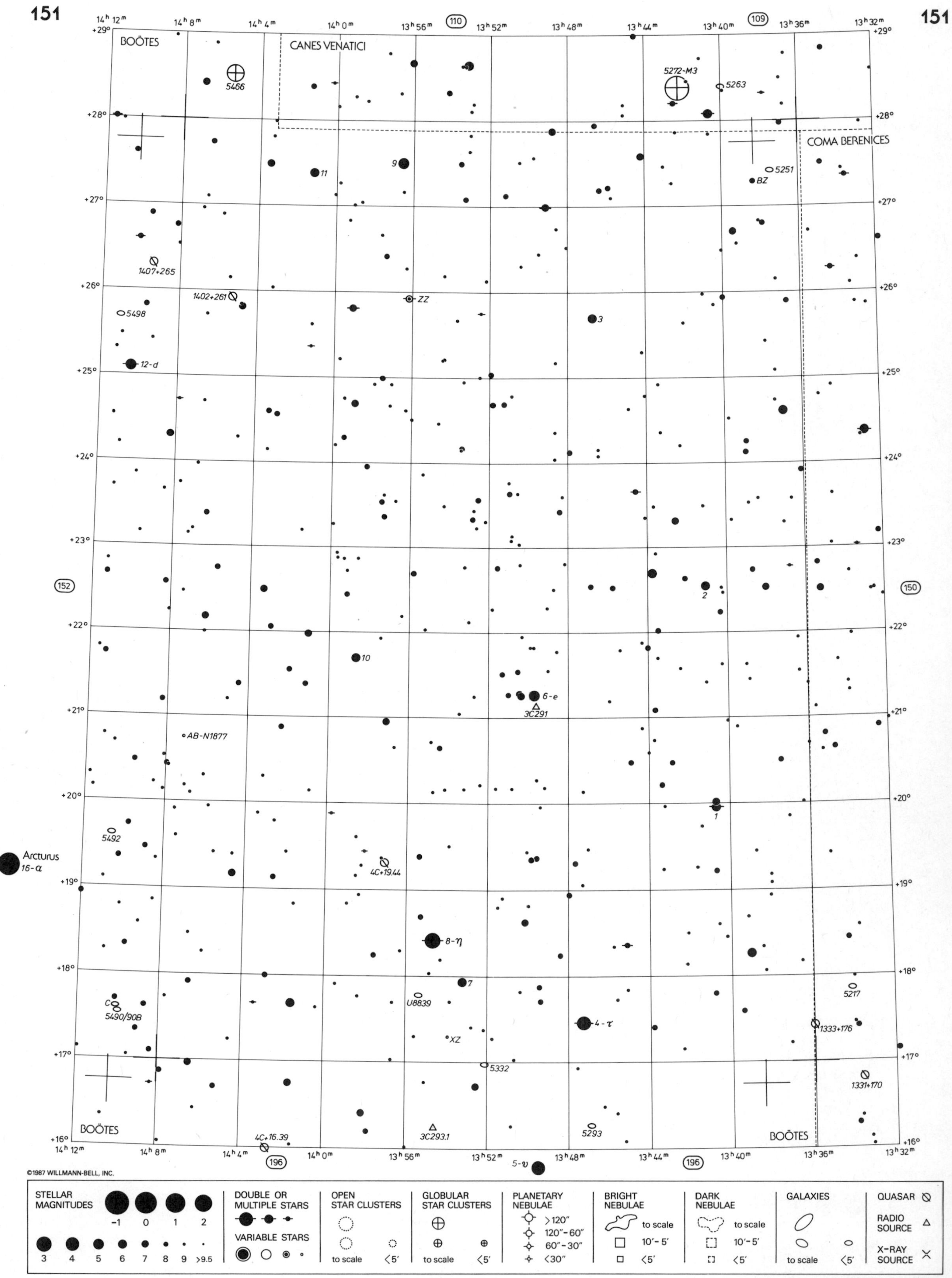
BOÖTES
CANES VENATICI
COMA BERENICES
Arcturus 16-α
5466
5272-M3
5263
5251
BZ
1407+265
1402+261
5498
ZZ
12-d
3
11
9
2
10
6-e
3C291
AB-N1877
1
5492
4C+19.44
8-η
7
U8839
5217
C
5490/90B
4-τ
1333+176
XZ
5332
1331+170
3C293.1
5293
4C+16.39
5-υ
14h 12m
14h 8m
14h 4m
14h 0m
13h 56m
13h 52m
13h 48m
13h 44m
13h 40m
13h 36m
13h 32m
+29°
+28°
+27°
+26°
+25°
+24°
+23°
+22°
+21°
+20°
+19°
+18°
+17°
+16°
110
109
152
150
196
©1987 WILLMANN-BELL, INC.
STELLAR MAGNITUDES
-1 0 1 2
3 4 5 6 7 8 9 >9.5
DOUBLE OR MULTIPLE STARS
VARIABLE STARS
OPEN STAR CLUSTERS
to scale <5'
GLOBULAR STAR CLUSTERS
to scale <5'
PLANETARY NEBULAE
>120"
120"-60"
60"-30"
<30"
BRIGHT NEBULAE
to scale
10'-5'
<5'
DARK NEBULAE
to scale
10'-5'
<5'
GALAXIES
to scale <5'
QUASAR
RADIO SOURCE
X-RAY SOURCE
Barry Rappaport & Wil Tirion

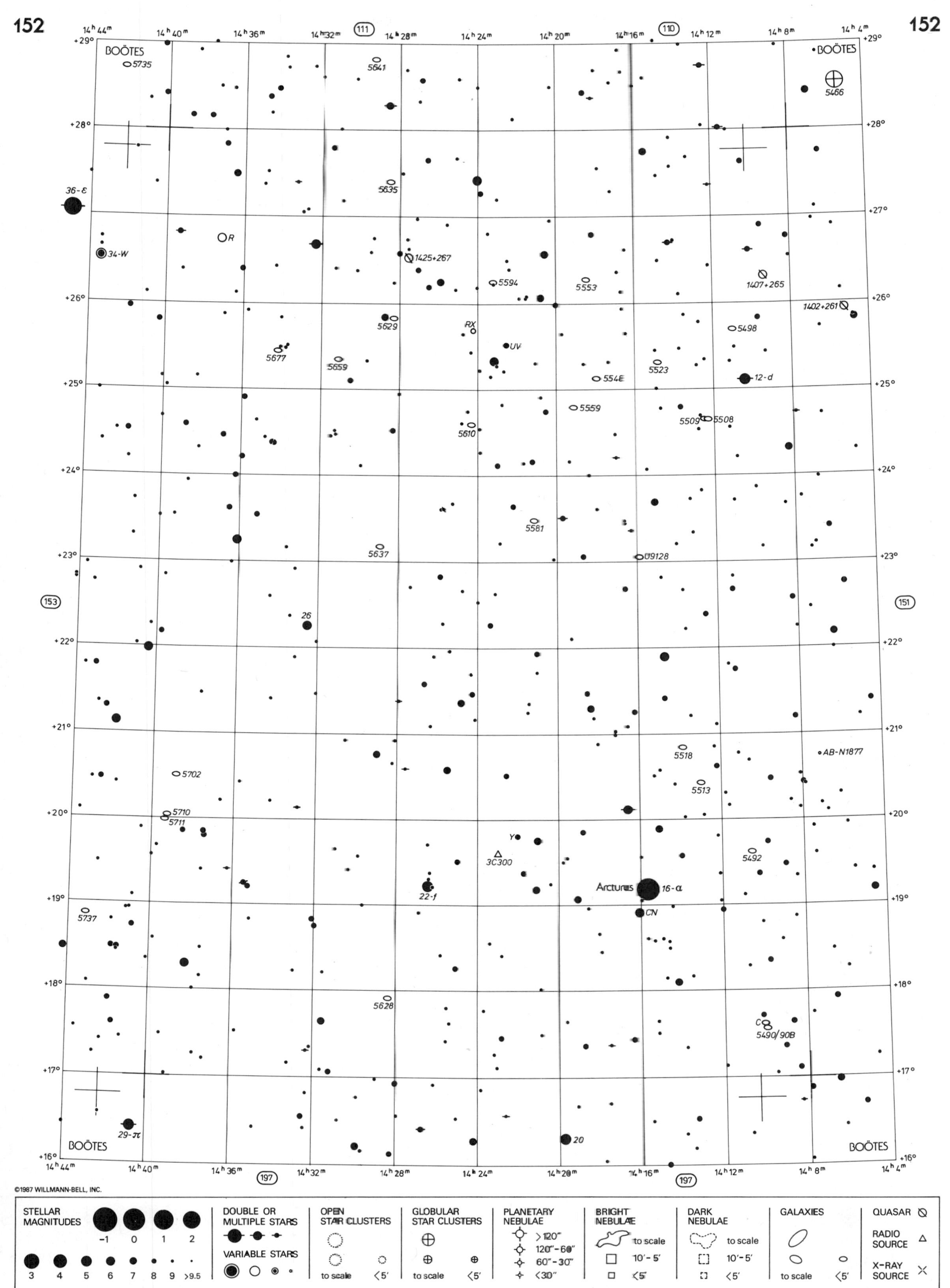

STELLAR MAGNITUDES −1 0 1 2 3 4 5 6 7 8 9 >9.5
DOUBLE OR MULTIPLE STARS
VARIABLE STARS
OPEN STAR CLUSTERS to scale <5'
GLOBULAR STAR CLUSTERS to scale <5'
PLANETARY NEBULAE >120" 120"-60" 60"-30" <30"
BRIGHT NEBULAE to scale 10'-5' <5'
DARK NEBULAE to scale 10'-5' <5'
GALAXIES to scale <5'
QUASAR
RADIO SOURCE
X-RAY SOURCE

Barry Rappaport & Wil Tirion

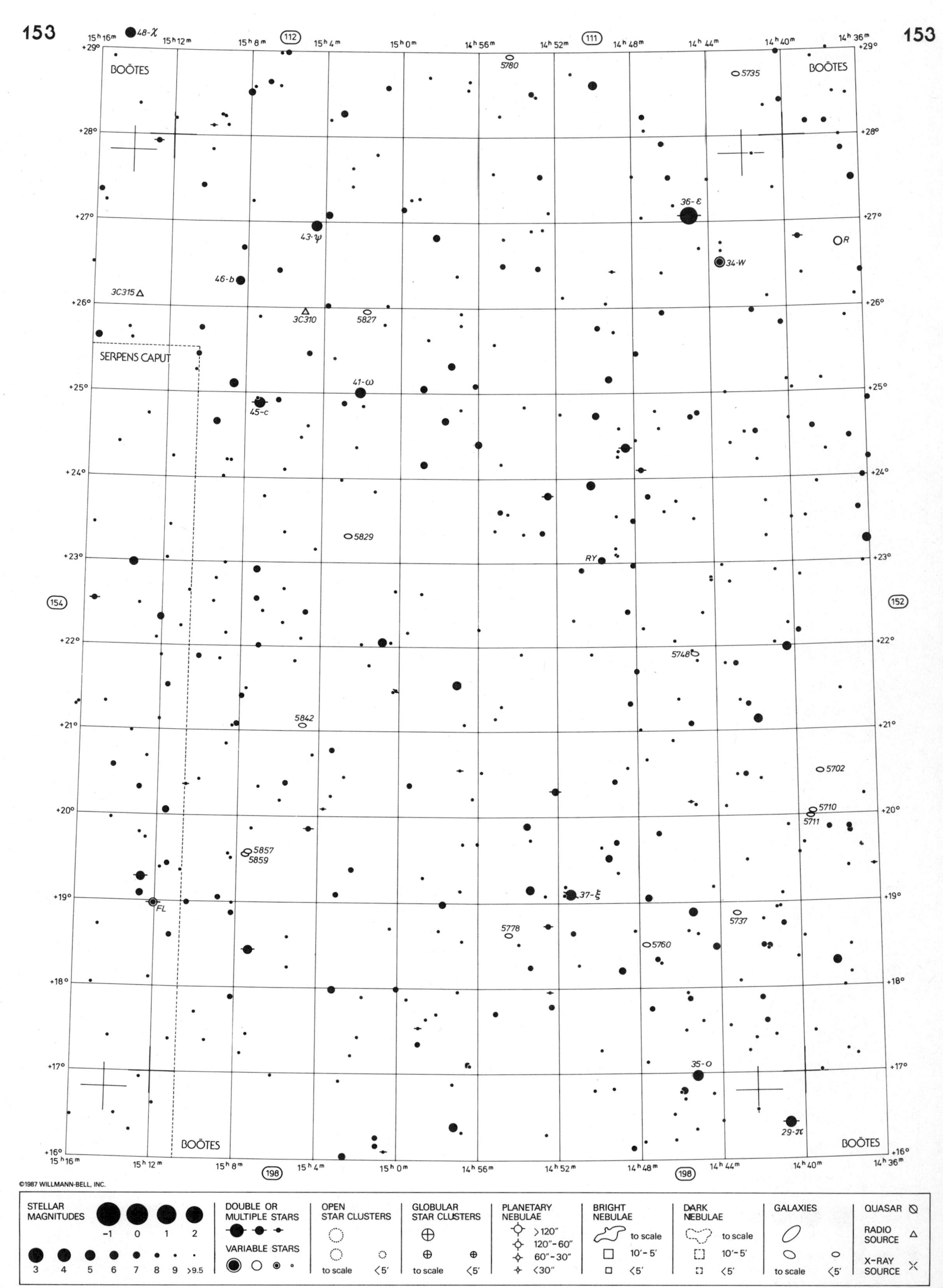

153
15h16m
15h12m
15h8m
15h4m
15h0m
14h56m
14h52m
14h48m
14h44m
14h40m
14h36m
112
111
198
154
152
+29°
+28°
+27°
+26°
+25°
+24°
+23°
+22°
+21°
+20°
+19°
+18°
+17°
+16°
BOÖTES
SERPENS CAPUT
48-χ
43-ψ
46-b
41-ω
45-c
36-ε
34-W
37-ξ
35-O
29-π
R
RY
FL
3C315
3C310
5780
5735
5827
5829
5748
5842
5702
5710
5711
5857
5859
5778
5760
5737
©1987 WILLMANN-BELL, INC.
STELLAR MAGNITUDES
-1 0 1 2
3 4 5 6 7 8 9 >9.5
DOUBLE OR MULTIPLE STARS
VARIABLE STARS
OPEN STAR CLUSTERS
to scale <5'
GLOBULAR STAR CLUSTERS
to scale <5'
PLANETARY NEBULAE
>120"
120"-60"
60"-30"
<30"
BRIGHT NEBULAE
to scale
10'-5'
<5'
DARK NEBULAE
to scale
10'-5'
<5'
GALAXIES
to scale <5'
QUASAR
RADIO SOURCE
X-RAY SOURCE
Barry Rappaport & Wil Tirion

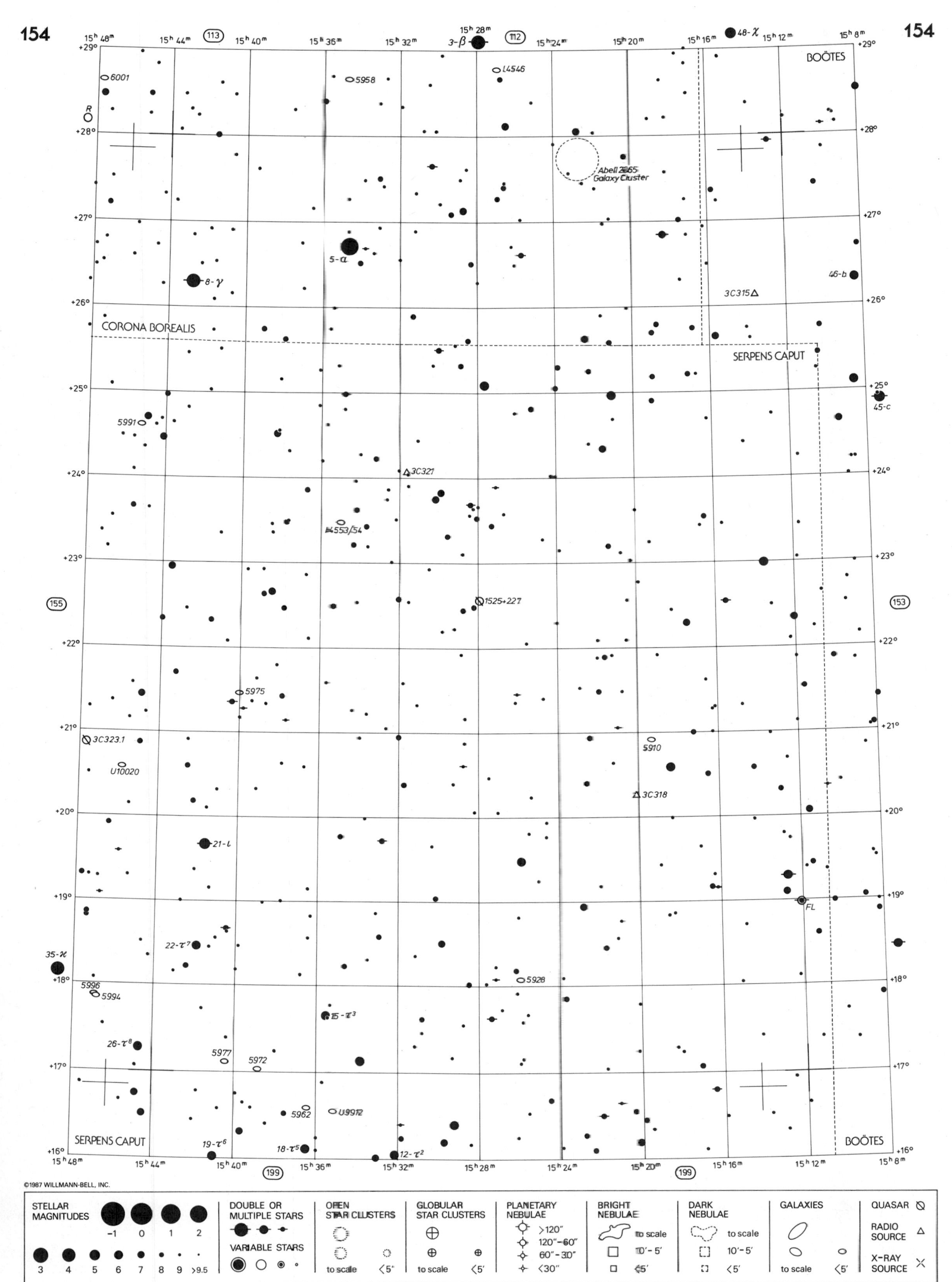
BOÖTES
CORONA BOREALIS
SERPENS CAPUT
Abell 2065 Galaxy Cluster
5-α
8-γ
3-β
48-χ
46-b
45-c
21-ι
35-κ
FL
3C315
3C321
3C318
3C323.1
1525+227
STELLAR MAGNITUDES
DOUBLE OR MULTIPLE STARS
VARIABLE STARS
OPEN STAR CLUSTERS
GLOBULAR STAR CLUSTERS
PLANETARY NEBULAE
BRIGHT NEBULAE
DARK NEBULAE
GALAXIES
QUASAR
RADIO SOURCE
X-RAY SOURCE
Barry Rappaport & Wil Tirion

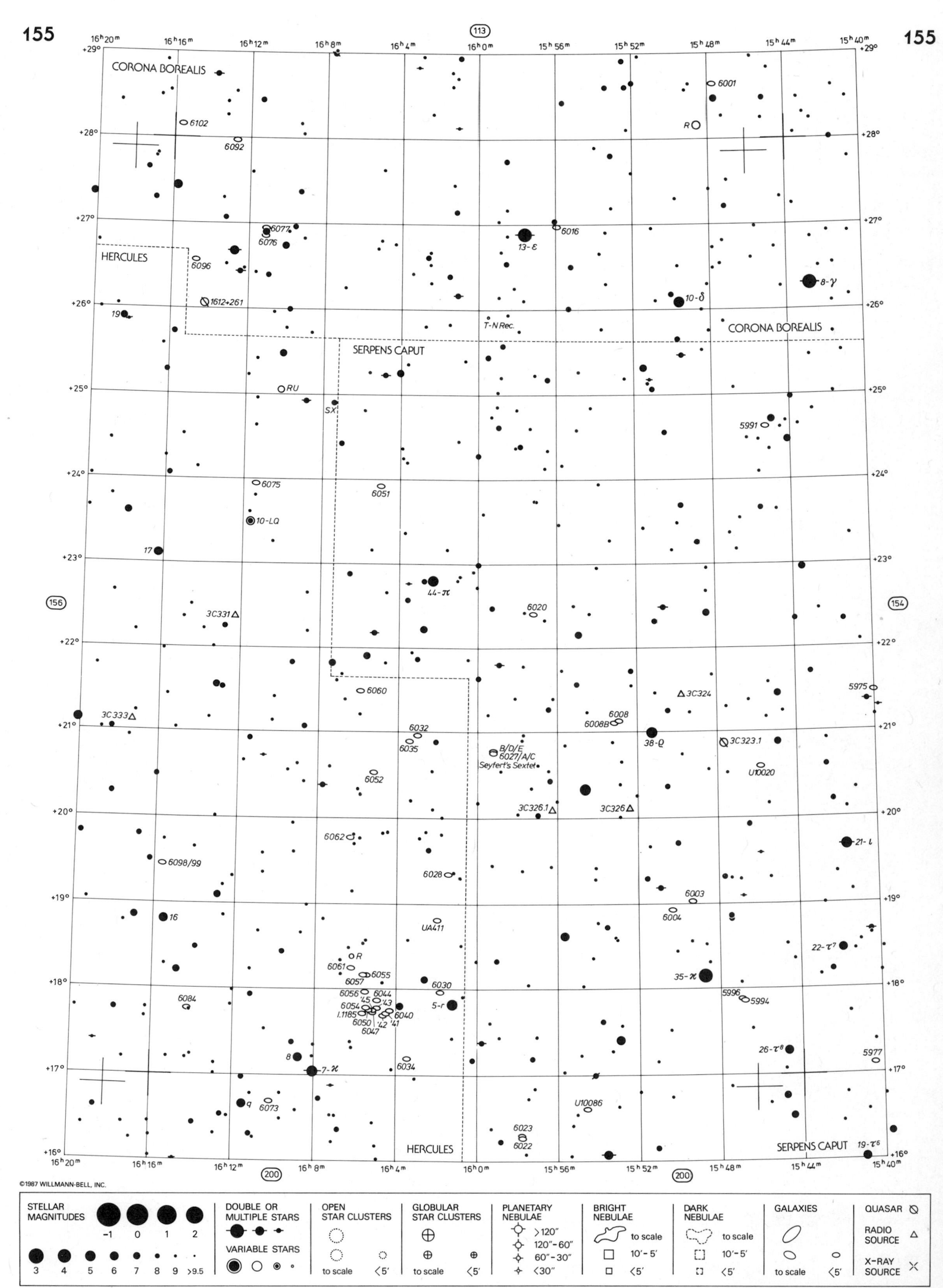

STELLAR MAGNITUDES	DOUBLE OR MULTIPLE STARS	OPEN STAR CLUSTERS	GLOBULAR STAR CLUSTERS	PLANETARY NEBULAE	BRIGHT NEBULAE	DARK NEBULAE	GALAXIES	QUASAR
−1 0 1 2	VARIABLE STARS	to scale ‹5′	to scale ‹5′	>120″ 120″–60″ 60″–30″ ‹30″	to scale 10′–5′ ‹5′	to scale 10′–5′ ‹5′	to scale ‹5′	RADIO SOURCE
3 4 5 6 7 8 9 >9.5								X-RAY SOURCE

Barry Rappaport & Wil Tirion

16h 52m 16h 48m 16h 44m 16h 40m 16h 36m (114) 16h 32m 16h 28m 16h 24m 16h 20m (113) 16h 16m 16h 12m

156

HERCULES

CORONA BOREALIS

HERCULES

HERCULES

HERCULES

+29° +28° +27° +26° +25° +24° +23° +22° +21° +20° +19° +18° +17° +16°

46, 39, 51, 27-β, s, 20-γ, 16, 17, 19, U, 10-LQ

6102, 6092, 6096, 6228, 6233, 6210, 6203, 6201, 6148, 6186, 6168, 6181, 6149, 6098/99, 6084

PK47+42.1, 3C341, 3C342, 1634+267A/B, 1612+261, 3C336, 3C340, 3C331, 3C333, 3C334, 3C346

(157) (155)

16h 52m 16h 48m 16h 44m (201) 16h 40m 16h 36m 16h 32m 16h 28m 16h 24m (201) 16h 20m 16h 16m 16h 12m

STELLAR MAGNITUDES −1 0 1 2 3 4 5 6 7 8 9 >9.5

DOUBLE OR MULTIPLE STARS

VARIABLE STARS

OPEN STAR CLUSTERS to scale <5′

GLOBULAR STAR CLUSTERS to scale <5′

PLANETARY NEBULAE >120″ 120″–60″ 60″–30″ <30″

BRIGHT NEBULAE to scale 10′–5′ <5′

DARK NEBULAE to scale 10′–5′ <5′

GALAXIES to scale <5′

QUASAR

RADIO SOURCE

X-RAY SOURCE

Barry Rappaport & Wil Tirion

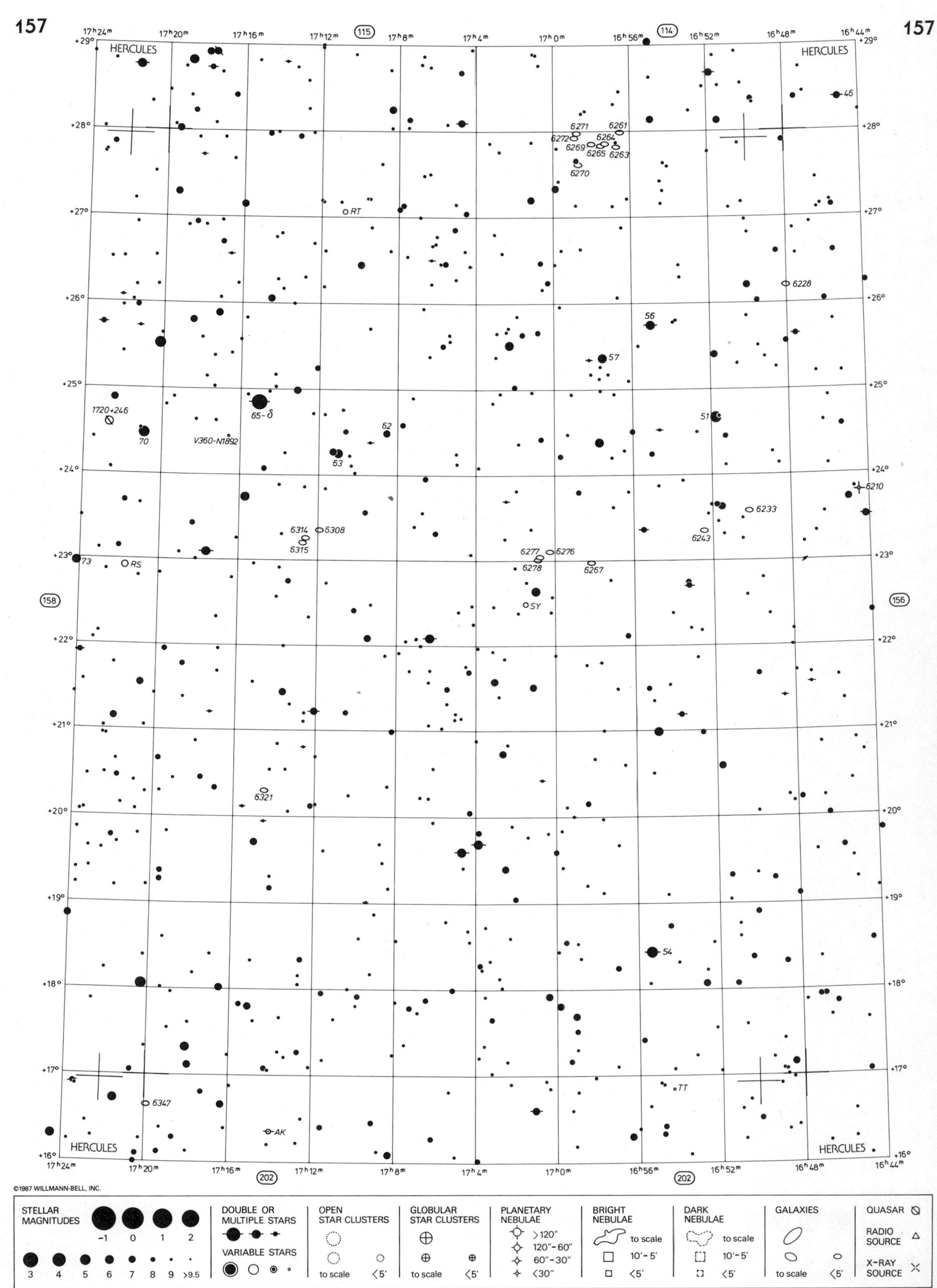

Barry Rappaport & Wil Tirion

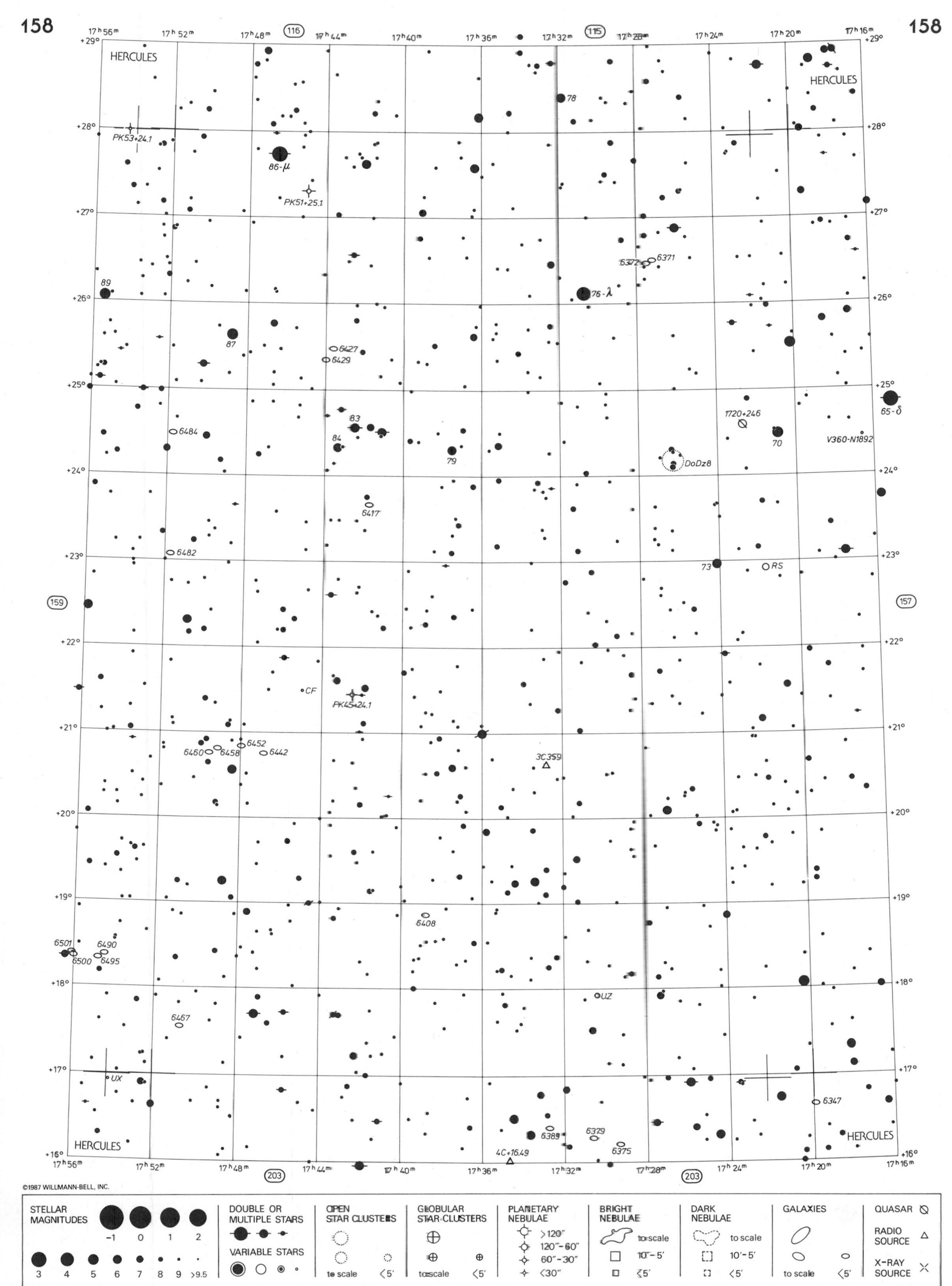

Barry Rappaport & Wil Tirion

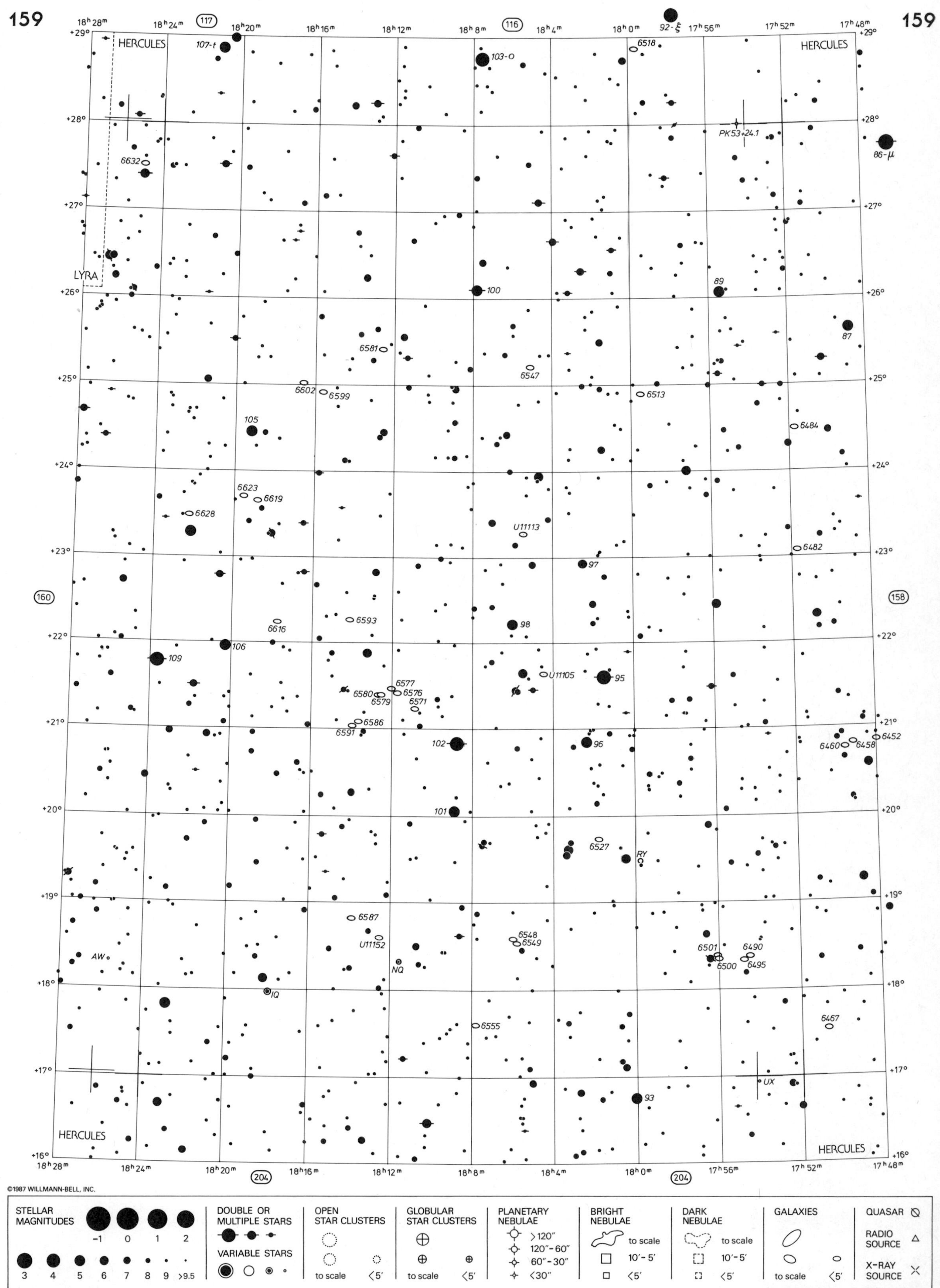
HERCULES
LYRA
18h28m
18h24m
18h20m
18h16m
18h12m
18h8m
18h4m
18h0m
17h56m
17h52m
17h48m
+29°
+28°
+27°
+26°
+25°
+24°
+23°
+22°
+21°
+20°
+19°
+18°
+17°
+16°
117
116
160
158
204
107-t
103-o
92-ξ
86-μ
6518
PK53+24.1
6632
100
89
87
6581
6547
6602
6599
6513
6484
105
6623
6619
6628
U11113
6482
97
6616
6593
98
109
106
U11105
95
6577
6580
6576
6579
6571
6586
6591
102
96
6460
6458
6452
101
6527
RY
6587
U11152
6548
6549
6501
6490
6500
6495
AW
NQ
IQ
6555
6467
UX
93
STELLAR MAGNITUDES
-1 0 1 2
3 4 5 6 7 8 9 >9.5
DOUBLE OR MULTIPLE STARS
VARIABLE STARS
OPEN STAR CLUSTERS
to scale
<5'
GLOBULAR STAR CLUSTERS
to scale
<5'
PLANETARY NEBULAE
>120"
120"-60"
60"-30"
<30"
BRIGHT NEBULAE
to scale
10'-5'
<5'
DARK NEBULAE
to scale
10'-5'
<5'
GALAXIES
to scale
<5'
QUASAR
RADIO SOURCE
X-RAY SOURCE
Barry Rappaport & Wil Tirion

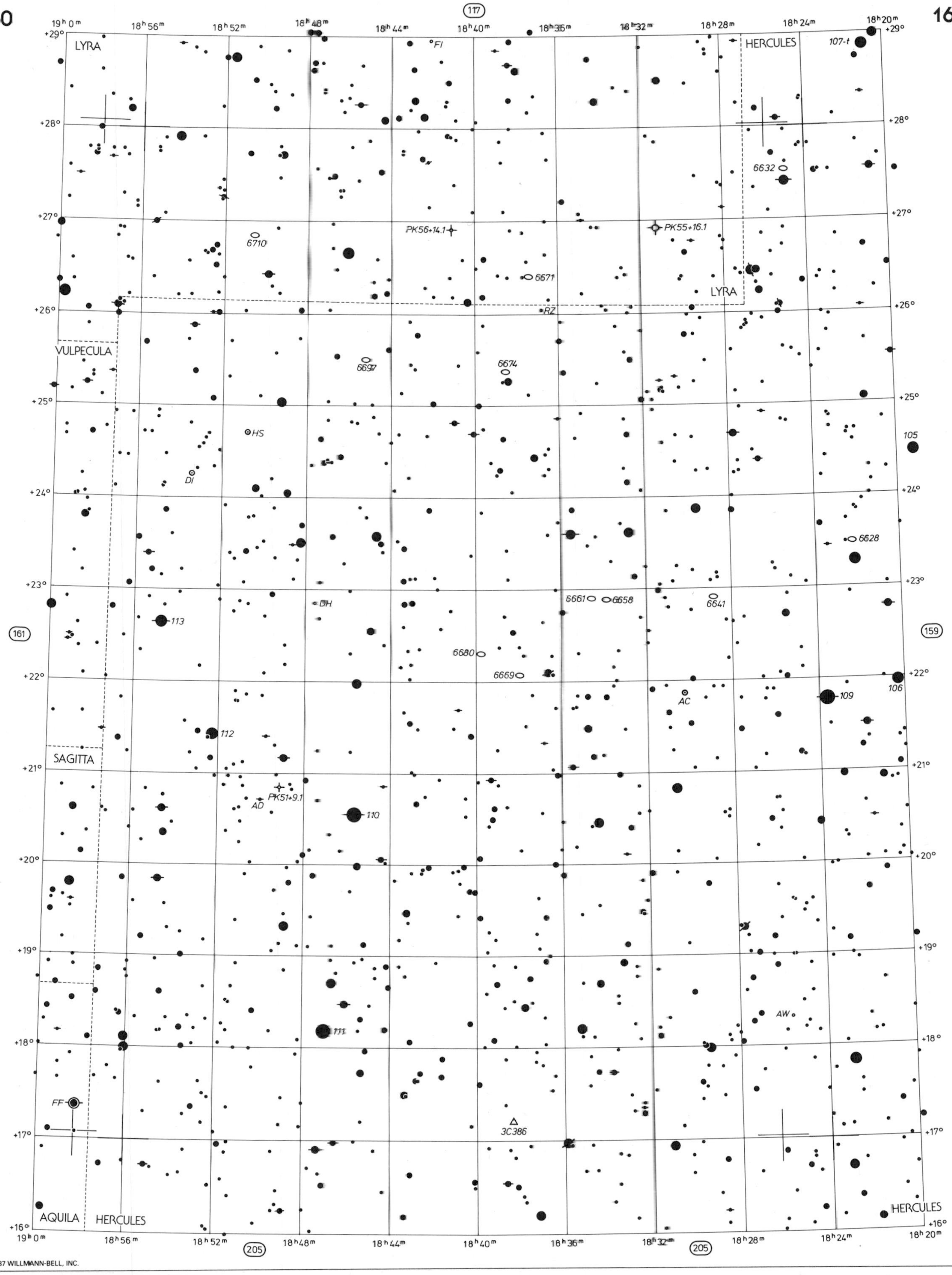

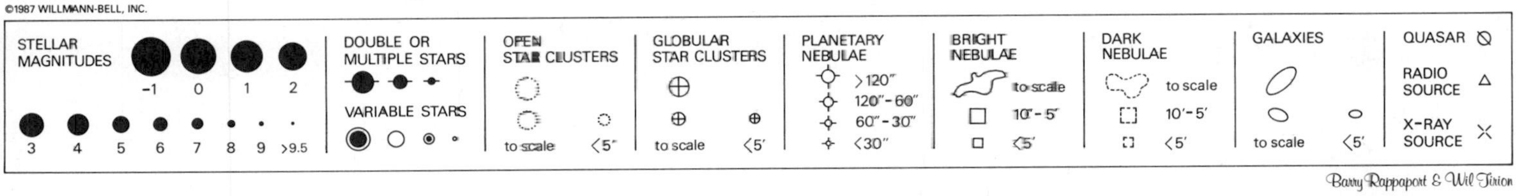

Barry Rappaport & Wil Tirion

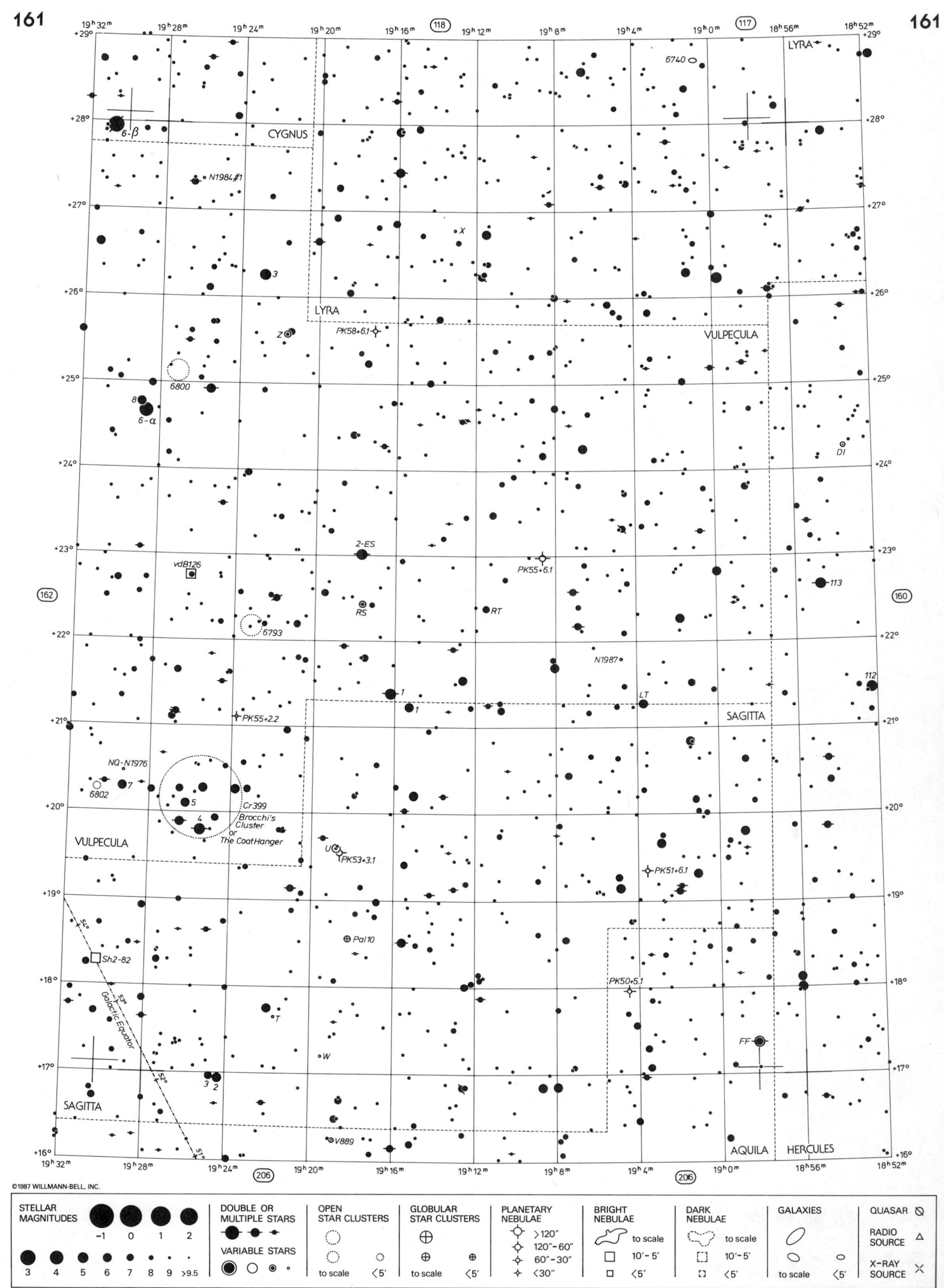

Barry Rappaport & Wil Tirion

20h 4m 20h 0m 19h 56m 19h 52m (119) 19h 48m 19h 44m 19h 40m 19h 36m (118) 19h 32m 19h 28m 19h 24m

+29° +28° +27° +26° +25° +24° +23° +22° +21° +20° +19° +18° +17° +16°

VULPECULA

CYGNUS

LU-N1968

BG

TY

15-NT

6-β

SV

S

CK-N1670

LV-N1968

PS

6813

N1984#1

Sh2-90

6815

PK61+3.1

X

PK62-0.1

10

Sh2-88

St1

6800

16

8

6-α

13

6823

6830

6820

14

vdB126

6853-M27

Dumbbell Nebula

12

(163)

(161)

PK59-1.1

PK60-4.1

4C+21.53

6827

Cr 399 - Brocchi's Cluster or The Coat Hanger

NQ-N1976

U

PK56-0.1

6802

7

5

4

9

VULPECULA

SAGITTA

12-γ

8-ζ

6838-M71

PK55-1.1

9

7-δ

Sh2-84

Sh2-82

H20

HS-N1977

5-α

Galactic Equator

WY-N1783

13-VZ

3C400.2

6-β

15

PK53-1.1

PK53-3.1

3 2

11

10-S

4-ε

SAGITTA

AQUILA

20h 4m 20h 0m 19h 56m (207) 19h 52m 19h 48m 19h 44m 19h 40m 19h 36m (207) 19h 32m 19h 28m 19h 24m

STELLAR MAGNITUDES	DOUBLE OR MULTIPLE STARS	OPEN STAR CLUSTERS	GLOBULAR STAR CLUSTERS	PLANETARY NEBULAE	BRIGHT NEBULAE	DARK NEBULAE	GALAXIES	QUASAR
-1 0 1 2	VARIABLE STARS	to scale <5'	to scale <5'	>120" 120"-60" 60"-30" <30"	to scale 10'-5' <5'	to scale 10'-5' <5'	to scale <5'	RADIO SOURCE X-RAY SOURCE
3 4 5 6 7 8 9 >9.5								

Barry Rappaport & Wil Tirion

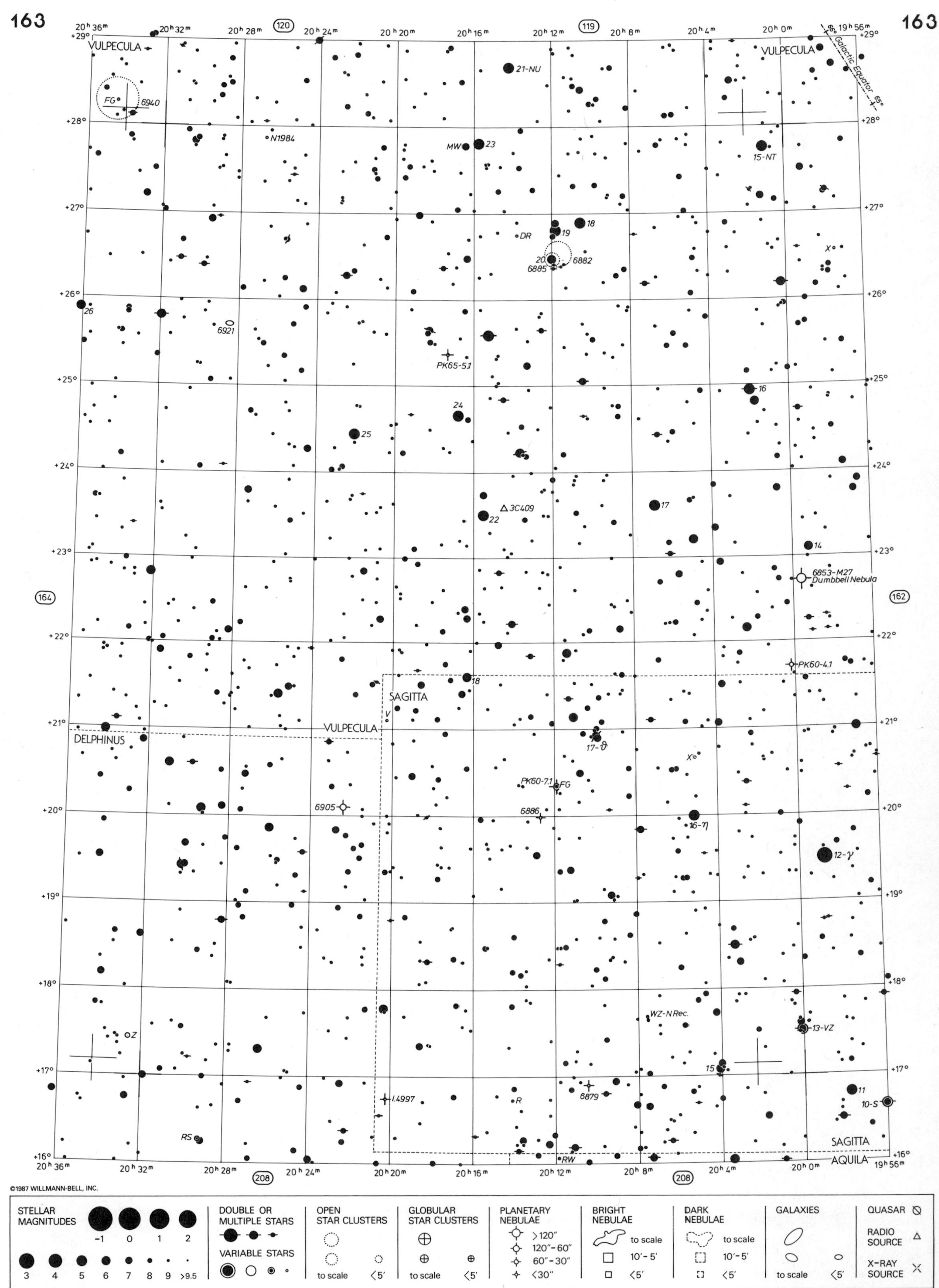
163
20h 36m
20h 32m
20h 28m
20h 24m
20h 20m
20h 16m
20h 12m
20h 8m
20h 4m
20h 0m
19h 56m
(120)
(119)
(164)
(162)
(208)
+29°
+28°
+27°
+26°
+25°
+24°
+23°
+22°
+21°
+20°
+19°
+18°
+17°
+16°
VULPECULA
DELPHINUS
SAGITTA
AQUILA
Galactic Equator
21-NU
FG
6940
N1984
MW
23
15-NT
DR
18
19
20
6885
6882
26
6921
PK65-5.1
16
24
25
3C409
22
17
14
6853-M27
Dumbbell Nebula
PK60-4.1
V
17-ϑ
X
PK60-7.1
6905
6886
16-η
12-γ
WZ-N Rec.
13-VZ
Z
15
I.4997
R
6879
11
10-S
RS
RW
©1987 WILLMANN-BELL, INC.
STELLAR MAGNITUDES
-1 0 1 2
3 4 5 6 7 8 9 >9.5
DOUBLE OR MULTIPLE STARS
VARIABLE STARS
OPEN STAR CLUSTERS
to scale
<5′
GLOBULAR STAR CLUSTERS
to scale
<5′
PLANETARY NEBULAE
>120″
120″-60″
60″-30″
<30″
BRIGHT NEBULAE
to scale
10′-5′
<5′
DARK NEBULAE
to scale
10′-5′
<5′
GALAXIES
to scale
<5′
QUASAR
RADIO SOURCE
X-RAY SOURCE
Barry Rappaport & Wil Tirion

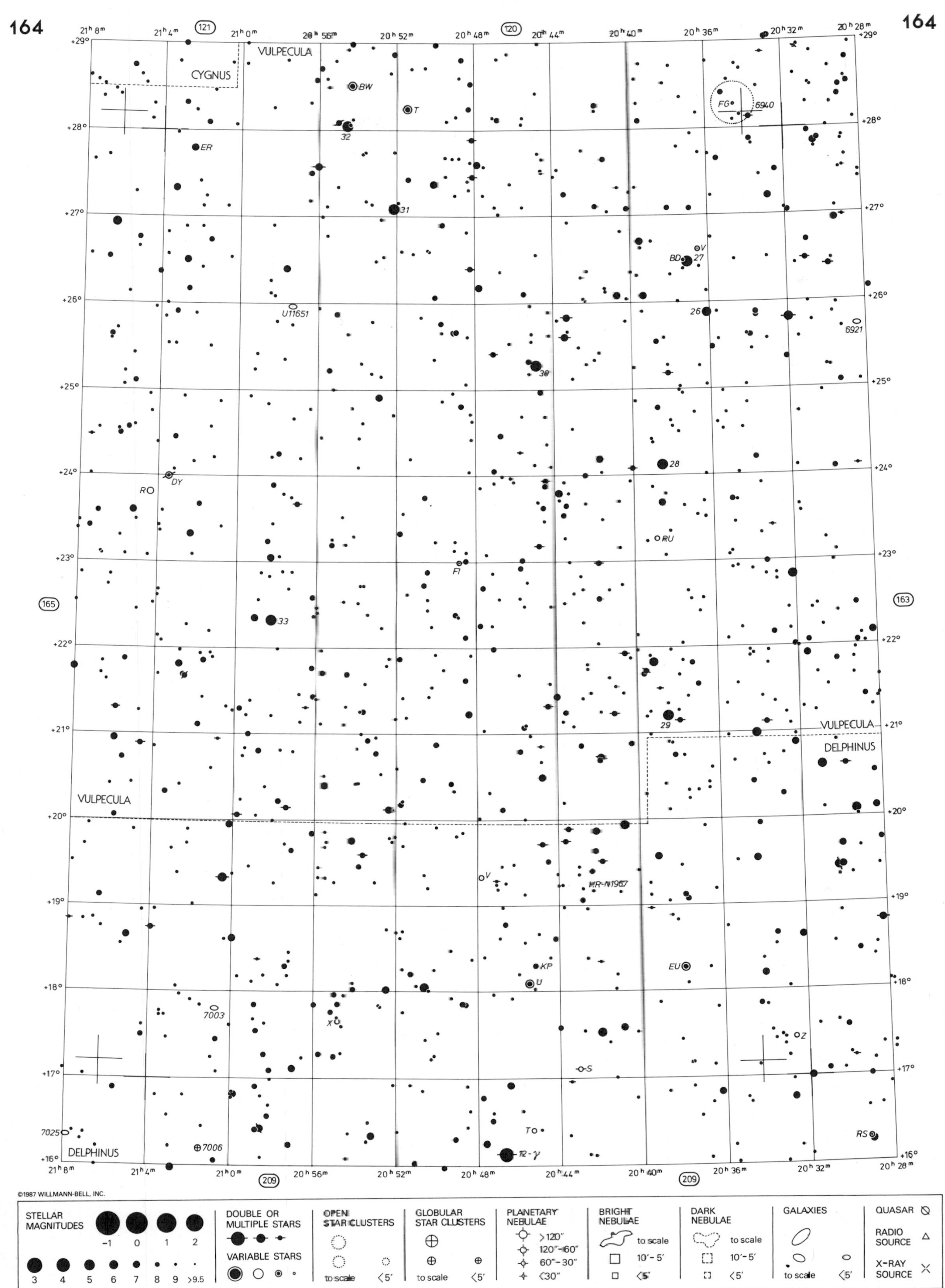

121
120
209
165
163
VULPECULA
CYGNUS
DELPHINUS
21h 8m
21h 4m
21h 0m
20h 56m
20h 52m
20h 48m
20h 44m
20h 40m
20h 36m
20h 32m
20h 28m
+29°
+28°
+27°
+26°
+25°
+24°
+23°
+22°
+21°
+20°
+19°
+18°
+17°
+16°
BW
T
32
ER
31
FG
6940
V
BD
27
26
U11651
6921
30
28
DY
RO
RU
FI
33
29
V
MR-N1967
KP
U
EU
7003
X
Z
S
T
7025
7006
12-γ
RS
©1987 WILLMANN-BELL, INC.
STELLAR MAGNITUDES
-1 0 1 2
3 4 5 6 7 8 9 >9.5
DOUBLE OR MULTIPLE STARS
VARIABLE STARS
OPEN STAR CLUSTERS
to scale
<5′
GLOBULAR STAR CLUSTERS
to scale
<5′
PLANETARY NEBULAE
>120″
120″–60″
60″–30″
<30″
BRIGHT NEBULAE
to scale
10′–5′
<5′
DARK NEBULAE
to scale
10′–5′
<5′
GALAXIES
to scale
<5′
QUASAR
RADIO SOURCE
X-RAY SOURCE
Barry Rappaport & Wil Tirion

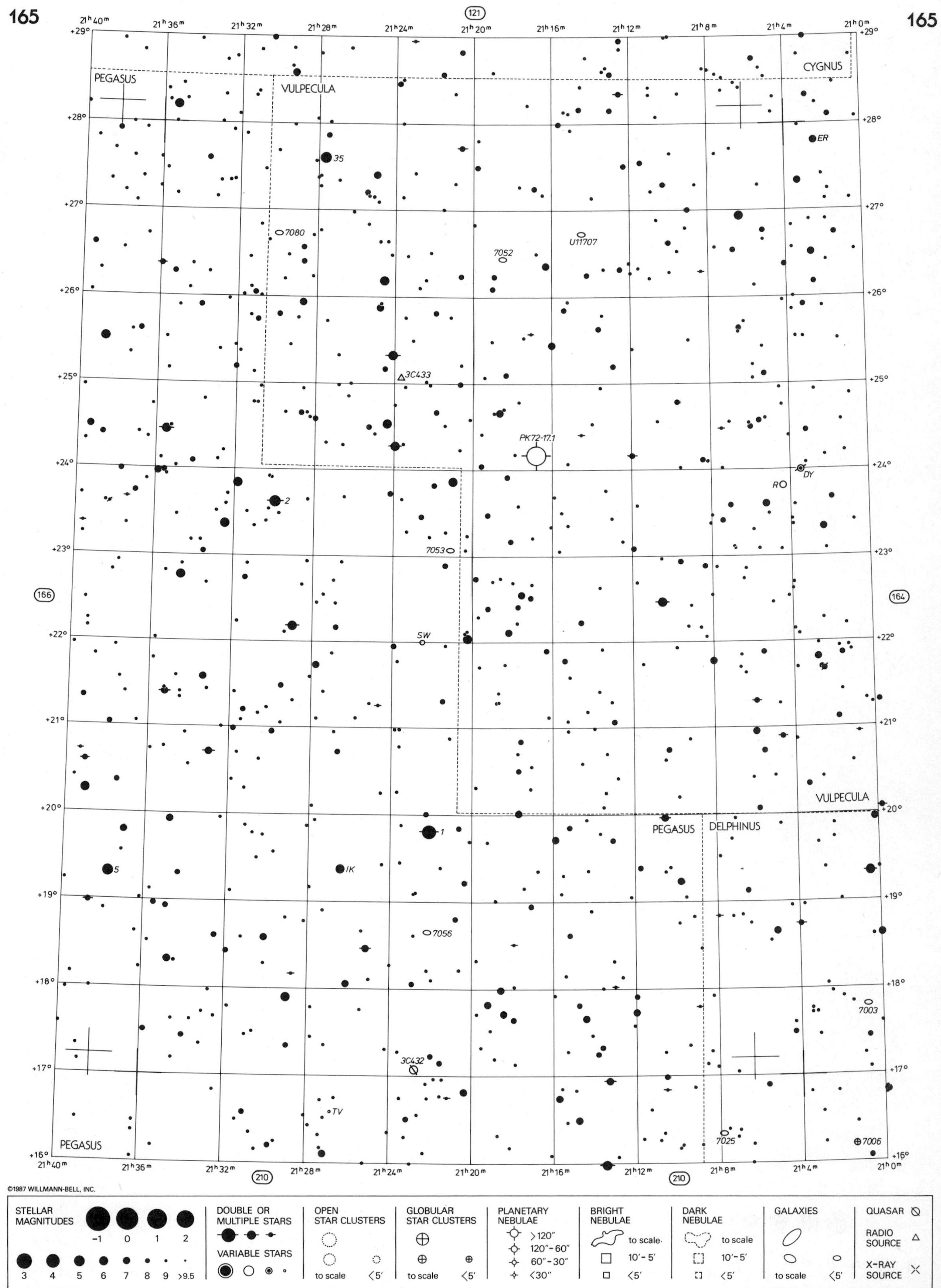

Barry Rappaport & Wil Tirion

166

22h 12m 22h 8m 22h 4m 22h 0m 21h 56m (122) 21h 52m 21h 48m 21h 44m 21h 40m (121) 21h 36m 21h 32m

+29° +28° +27° +26° +25° +24° +23° +22° +21° +20° +19° +18° +17° +16°

PEGASUS

CYGNUS

PEGASUS

23 TW 15 7116 78-μ 3C436

7224 16 10-ι 24-ι RR

AW DX

12 RX (167) (165) 7137

25 HO 28

I.1420 5

7177 2141+175 9 13

7206 7207

PEGASUS PEGASUS

22h 12m 22h 8m 22h 4m (211) 22h 0m 21h 56m 21h 52m 21h 48m 21h 44m (211) 21h 40m 21h 36m 21h 32m

STELLAR MAGNITUDES: -1 0 1 2 3 4 5 6 7 8 9 >9.5

DOUBLE OR MULTIPLE STARS

VARIABLE STARS

OPEN STAR CLUSTERS: to scale <5'

GLOBULAR STAR CLUSTERS: to scale <5'

PLANETARY NEBULAE: >120" 120"-60" 60"-30" <30"

BRIGHT NEBULAE: to scale 10'-5' <5'

DARK NEBULAE: to scale 10'-5' <5'

GALAXIES: to scale <5'

QUASAR

RADIO SOURCE

X-RAY SOURCE

Barry Rappaport & Wil Tirion

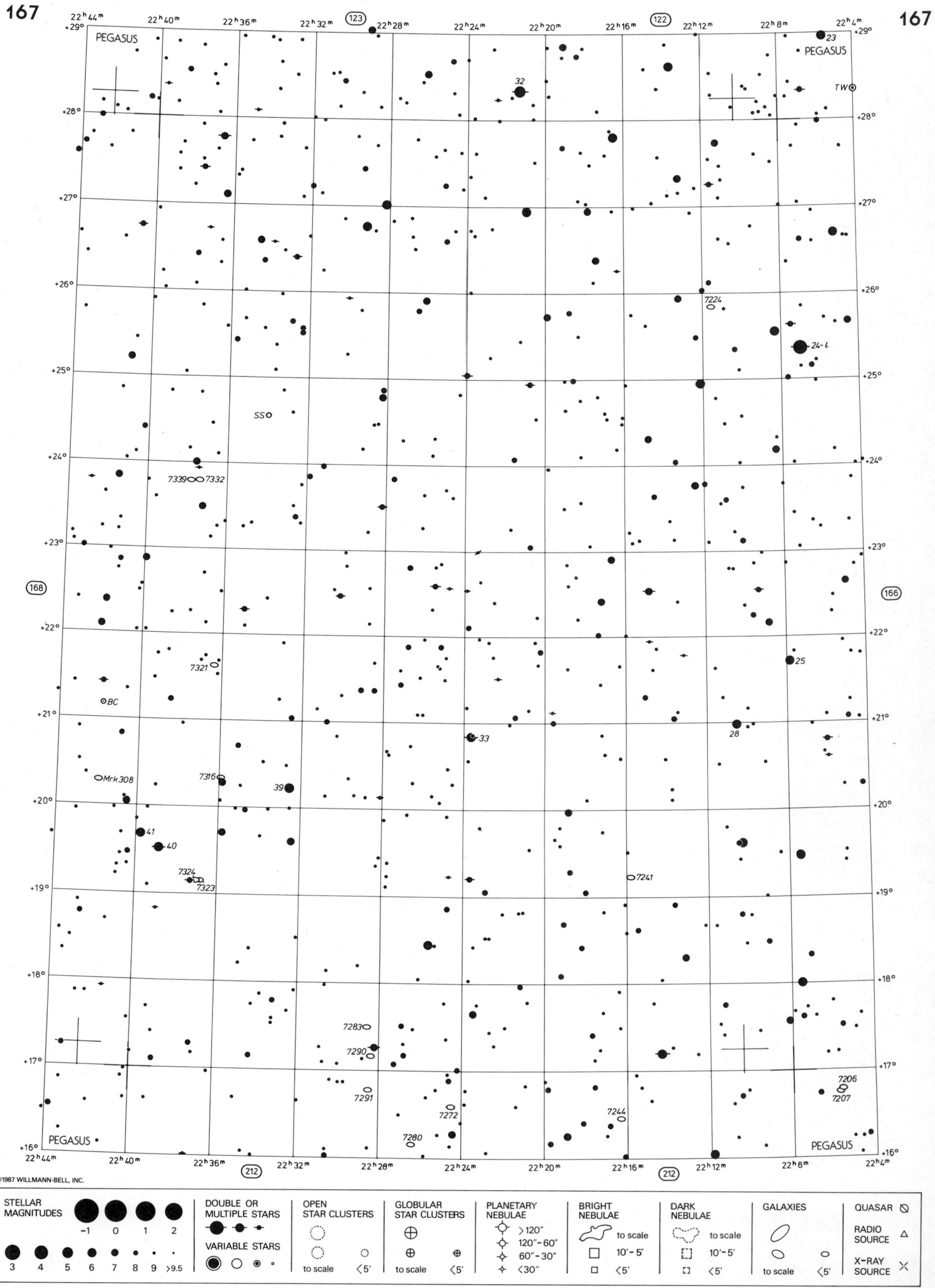

Barry Rappaport & Wil Tirion

Barry Rappaport & Wil Tirion

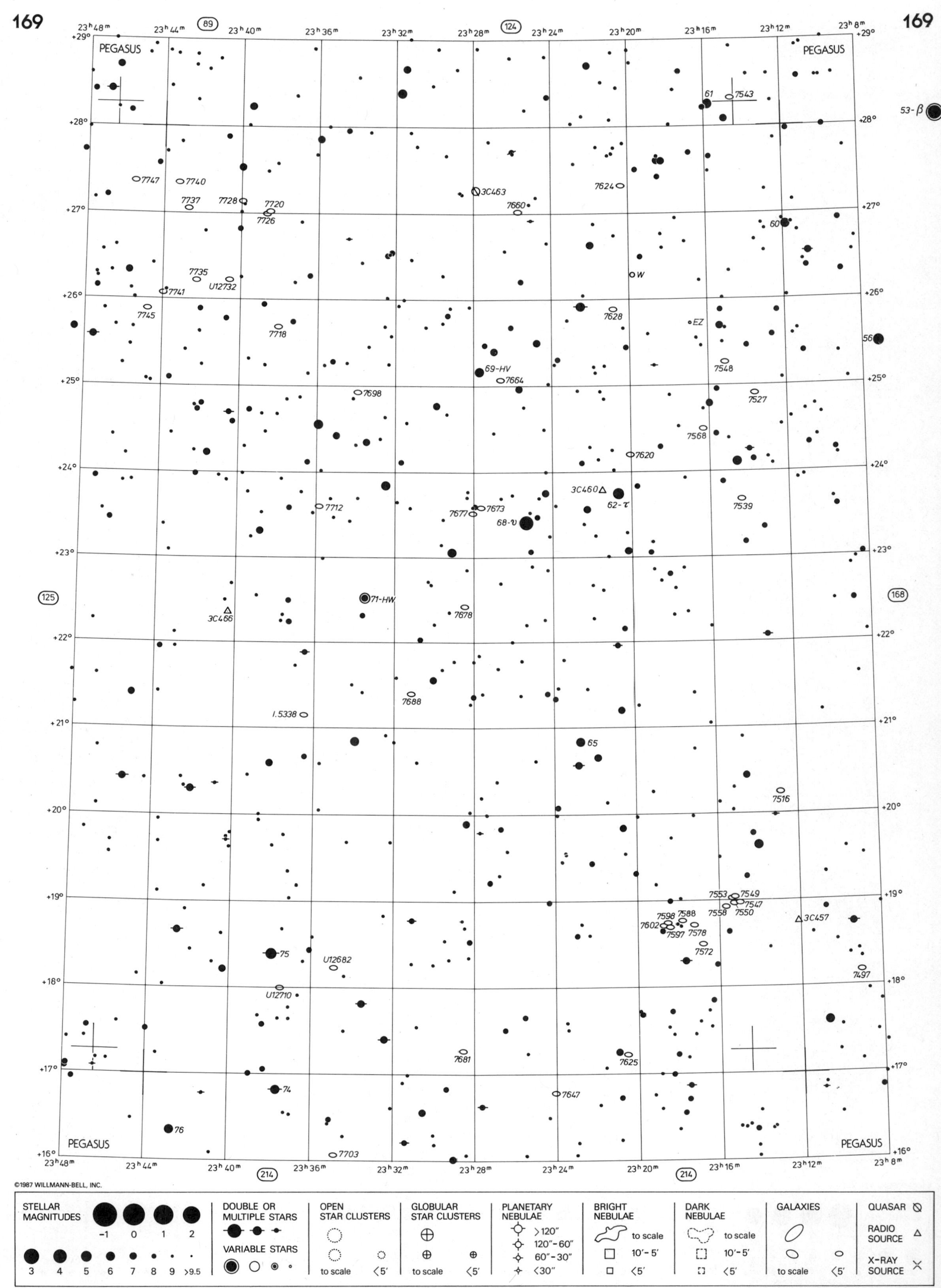

Barry Rappaport & Wil Tirion

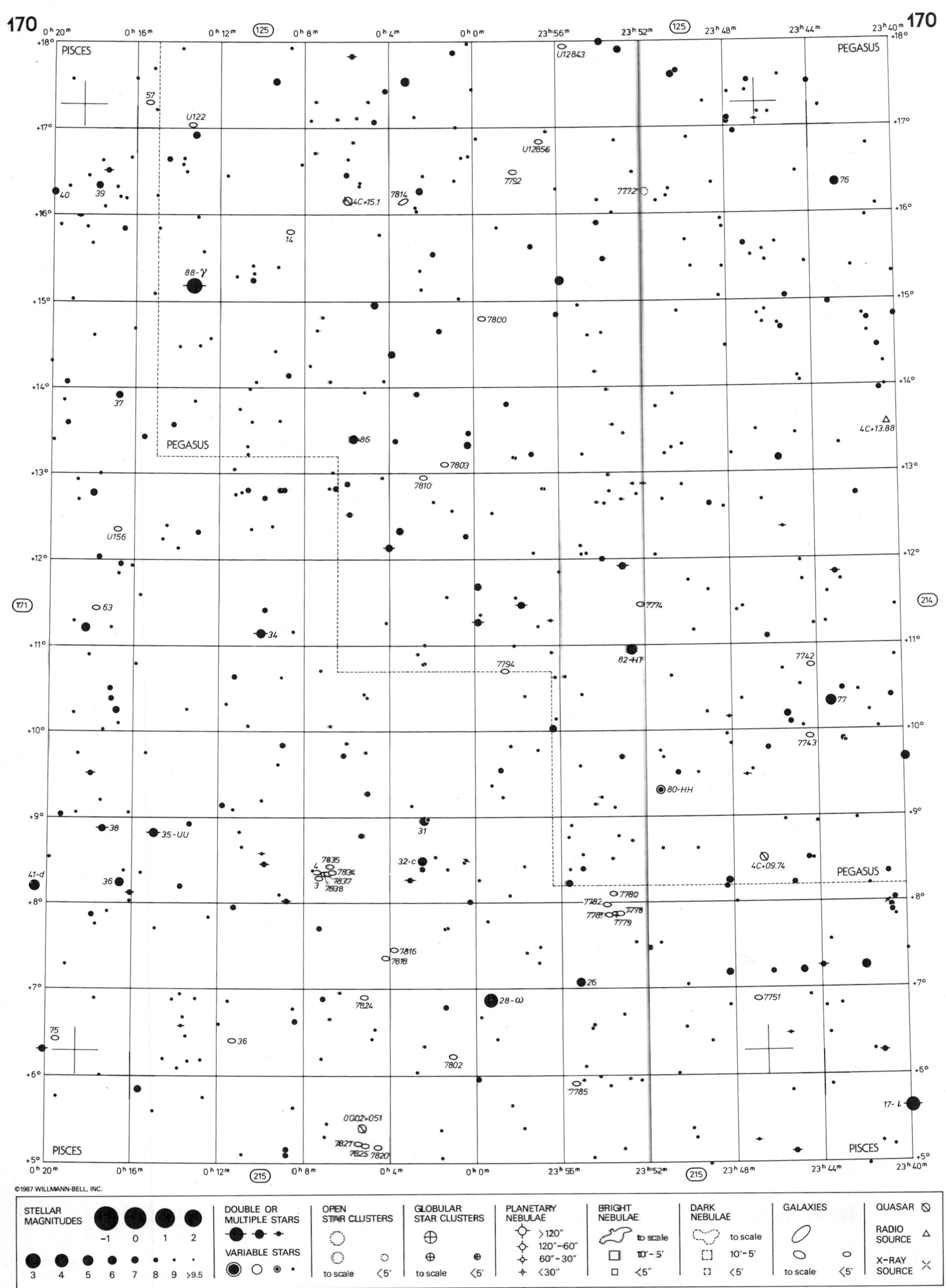

170
$0^h 20^m$
$0^h 16^m$
$0^h 12^m$
125
$0^h 8^m$
$0^h 4^m$
$0^h 0^m$
$23^h 56^m$
$23^h 52^m$
125
$23^h 48^m$
$23^h 44^m$
$23^h 40^m$
170
+18°
+17°
+16°
+15°
+14°
+13°
+12°
+11°
+10°
+9°
+8°
+7°
+6°
+5°
PISCES
PEGASUS
PEGASUS
PEGASUS
PISCES
PISCES
57
U122
U12843
U12856
7792
7772
7814
4C+15.1
40
39
76
14
88-γ
7800
37
4C+13.88
86
7803
7810
U156
171
214
63
7774
34
82-HT
7794
7742
77
7743
80-HH
38
35-UU
31
41-d
36
7835
7834
7837
7838
32-c
4C+09.74
7780
7782
7781
7778
7779
7816
7818
26
7824
28-ω
7751
75
36
7802
7785
17-ι
0002+051
7827
7825
7820
$0^h 20^m$
$0^h 16^m$
$0^h 12^m$
215
$0^h 8^m$
$0^h 4^m$
$0^h 0^m$
$23^h 56^m$
$23^h 52^m$
215
$23^h 48^m$
$23^h 44^m$
$23^h 40^m$
©1987 WILLMANN-BELL, INC.
STELLAR MAGNITUDES
-1 0 1 2
3 4 5 6 7 8 9 >9.5
DOUBLE OR MULTIPLE STARS
VARIABLE STARS
OPEN STAR CLUSTERS
to scale
<5'
GLOBULAR STAR CLUSTERS
to scale
<5'
PLANETARY NEBULAE
>120"
120"-60"
60"-30"
<30"
BRIGHT NEBULAE
to scale
10'-5'
<5'
DARK NEBULAE
to scale
10'-5'
<5'
GALAXIES
to scale
<5'
QUASAR
RADIO SOURCE
X-RAY SOURCE
Barry Rappaport & Wil Tirion

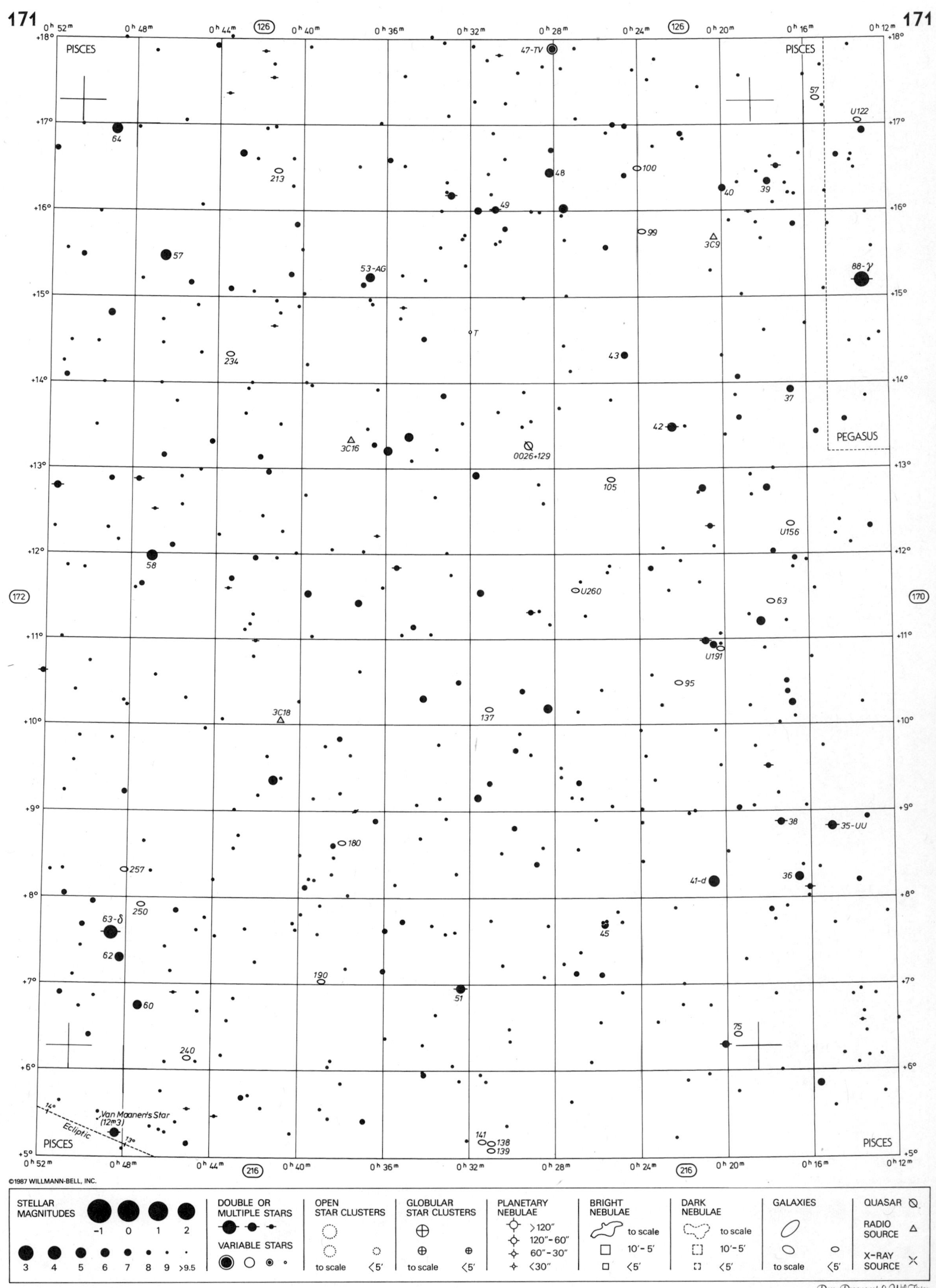
0h 52m
0h 48m
0h 44m
126
0h 40m
0h 36m
0h 32m
0h 28m
0h 24m
0h 20m
0h 16m
0h 12m
+18°
+17°
+16°
+15°
+14°
+13°
+12°
+11°
+10°
+9°
+8°
+7°
+6°
+5°
PISCES
PEGASUS
172
170
216
47-TV
57
U122
64
213
100
48
49
40
39
99
3C9
57
53-AG
88-γ
T
234
43
37
3C16
0026+129
42
105
U156
58
U260
63
U191
95
3C18
137
38
35-UU
180
257
41-d
36
250
63-δ
45
62
190
51
60
75
240
Van Maanen's Star
(12m3)
Ecliptic
14°
13°
141
138
139
©1987 WILLMANN-BELL, INC.
STELLAR MAGNITUDES
-1 0 1 2
3 4 5 6 7 8 9 >9.5
DOUBLE OR MULTIPLE STARS
VARIABLE STARS
OPEN STAR CLUSTERS
to scale
<5'
GLOBULAR STAR CLUSTERS
to scale
<5'
PLANETARY NEBULAE
>120"
120"-60"
60"-30"
<30"
BRIGHT NEBULAE
to scale
10'-5'
<5'
DARK NEBULAE
to scale
10'-5'
<5'
GALAXIES
to scale
<5'
QUASAR
RADIO SOURCE
X-RAY SOURCE
Barry Rappaport & Wil Tirion

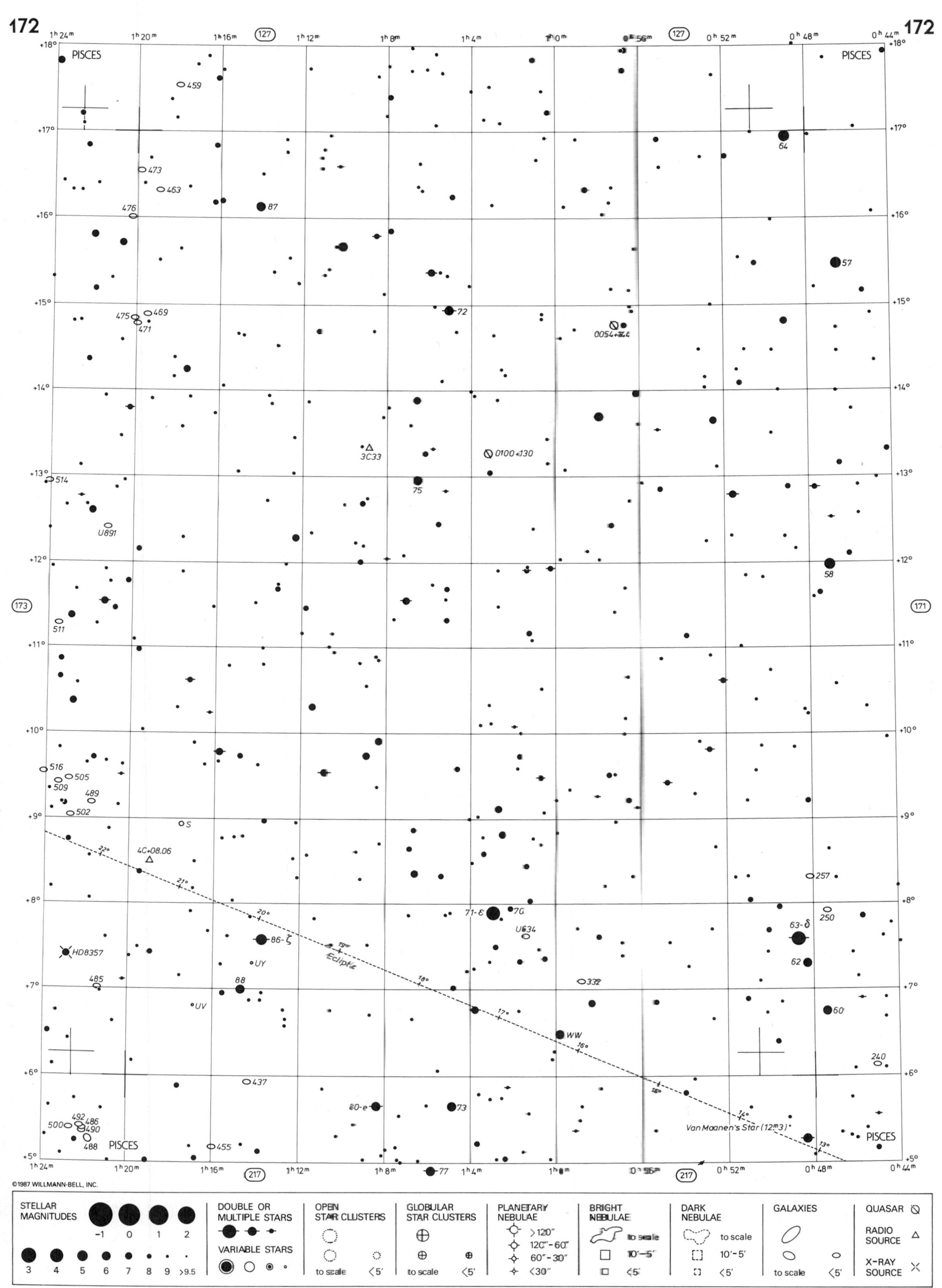
PISCES
87
64
57
72
0054+144
3C33
0100+130
75
U891
58
511
86-ζ
71-ε
70
U634
63-δ
62
HD8357
UY
88
UV
60
WW
Ecliptic
Van Maanen's Star (12m3)
80-e
73
77
STELLAR MAGNITUDES
DOUBLE OR MULTIPLE STARS
VARIABLE STARS
OPEN STAR CLUSTERS
GLOBULAR STAR CLUSTERS
PLANETARY NEBULAE
BRIGHT NEBULAE
DARK NEBULAE
GALAXIES
QUASAR
RADIO SOURCE
X-RAY SOURCE
to scale
10'-5'
<5'
>120"
120"-60"
60"-30"
<30"

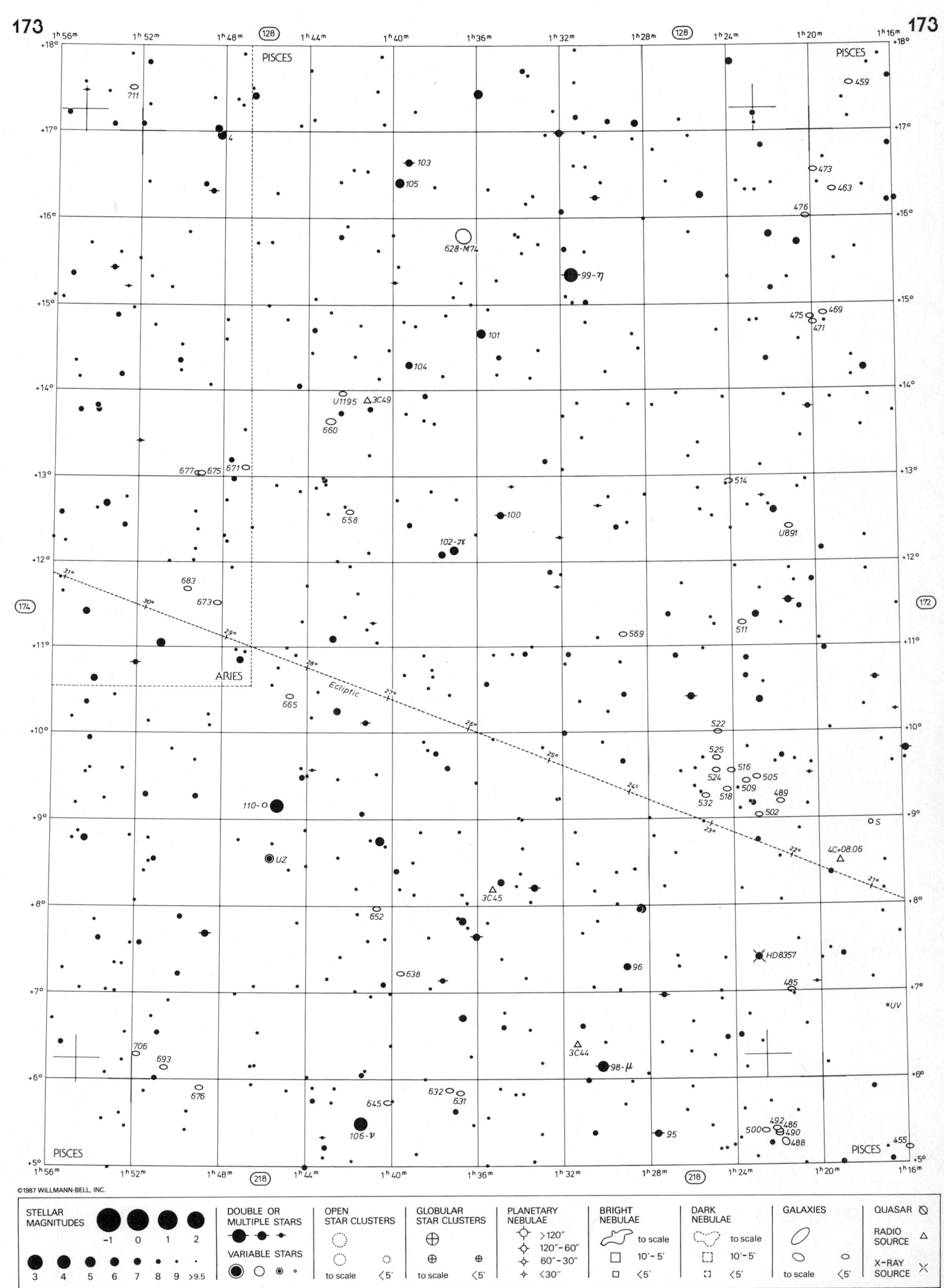

Barry Rappaport & Wil Tirion

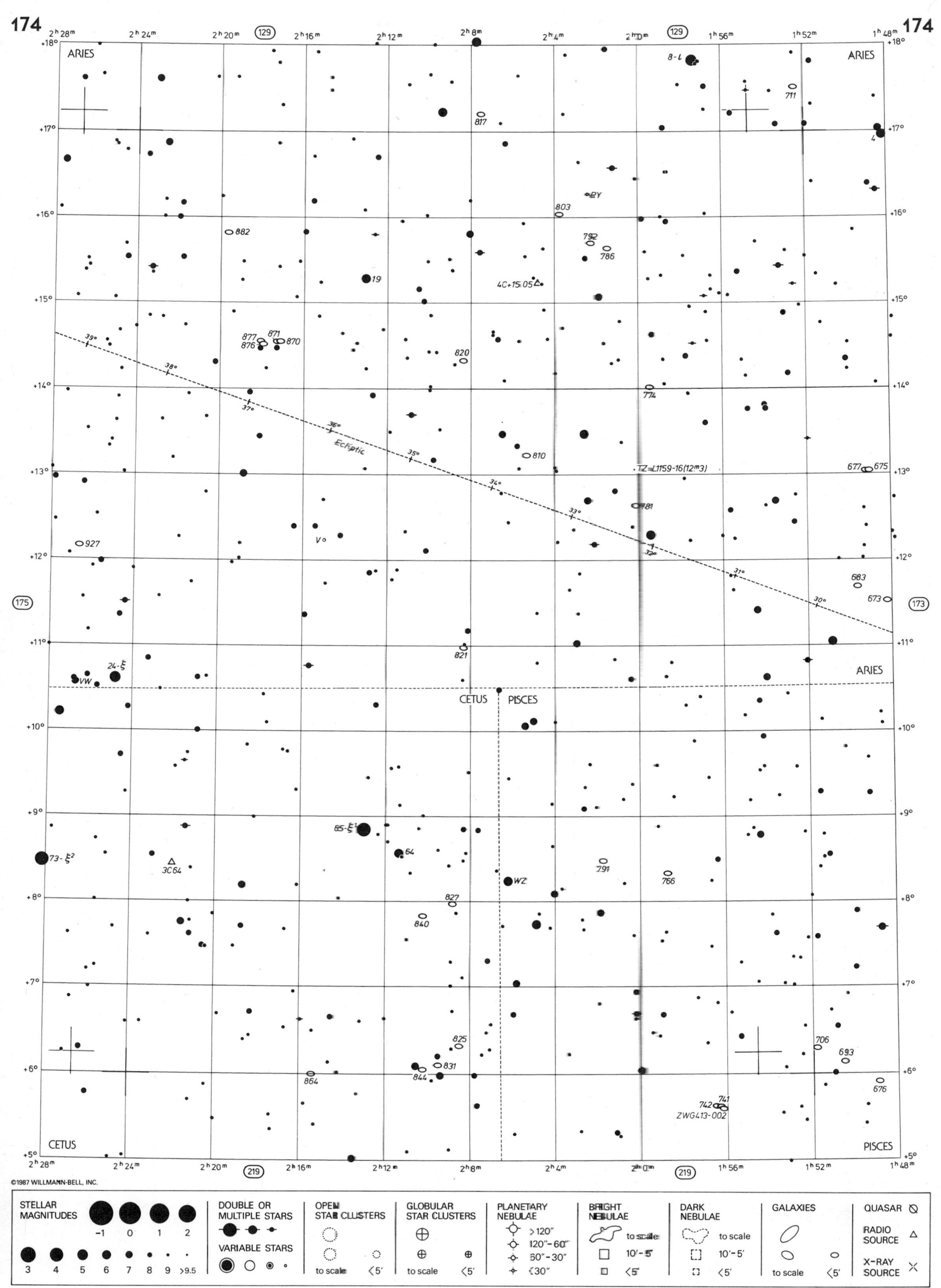

Barry Rappaport & Wil Tirion

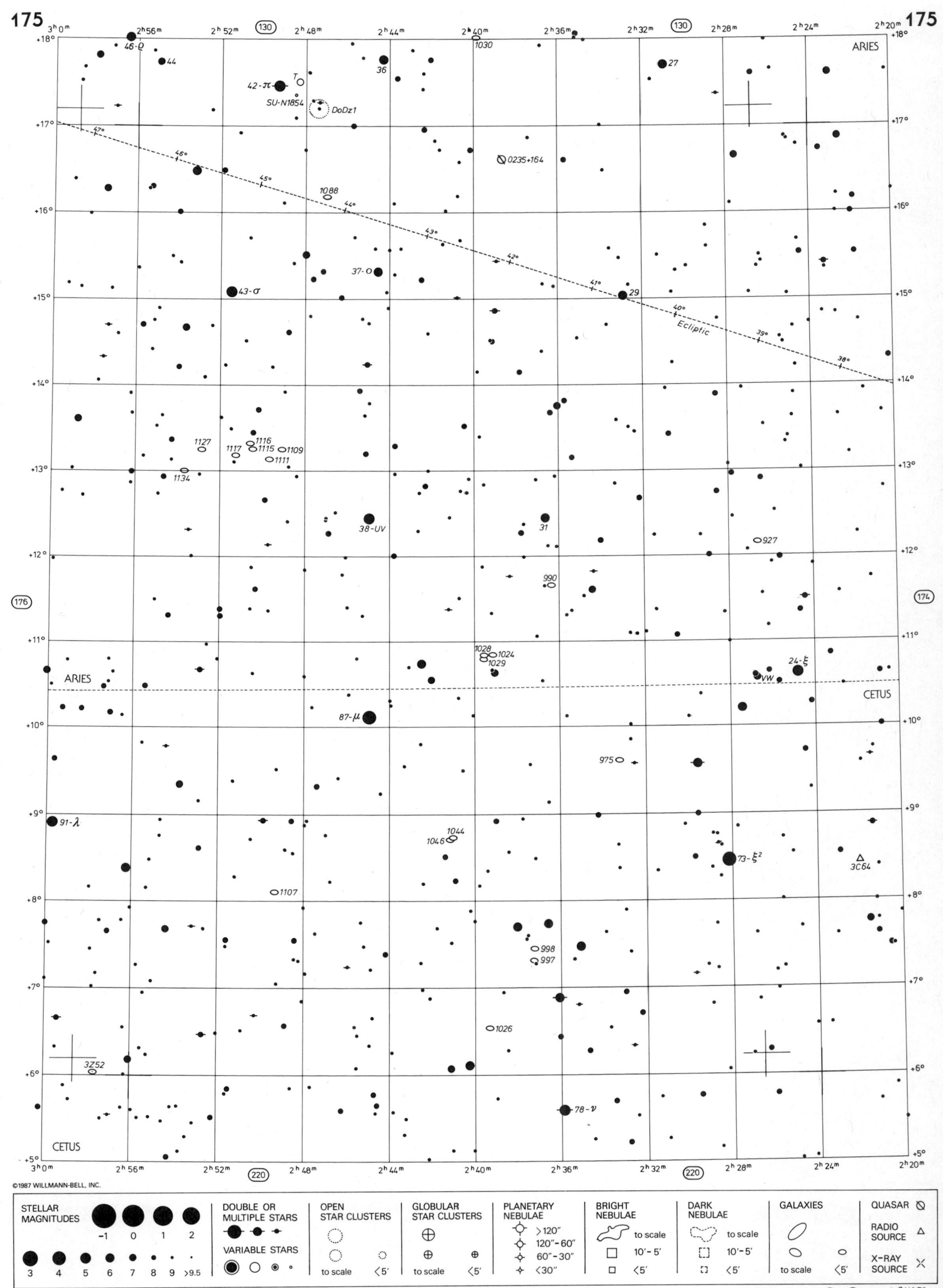

175
ARIES
CETUS
Ecliptic
46-ρ
44
42-π
T
SU-N1854
DoDz1
36
27
1030
0235+164
1088
37-ο
43-σ
29
1127
1116
1117
1115
1109
1111
1134
38-UV
31
927
990
1028
1024
1029
24-ξ
VW
87-μ
975
91-λ
1044
1046
73-ξ2
3C64
1107
998
997
1026
3Z52
78-ν
©1987 WILLMANN-BELL, INC.
STELLAR MAGNITUDES
DOUBLE OR MULTIPLE STARS
VARIABLE STARS
OPEN STAR CLUSTERS
to scale
GLOBULAR STAR CLUSTERS
PLANETARY NEBULAE
>120"
120"-60"
60"-30"
<30"
BRIGHT NEBULAE
DARK NEBULAE
GALAXIES
QUASAR
RADIO SOURCE
X-RAY SOURCE
Barry Rappaport & Wil Tirion

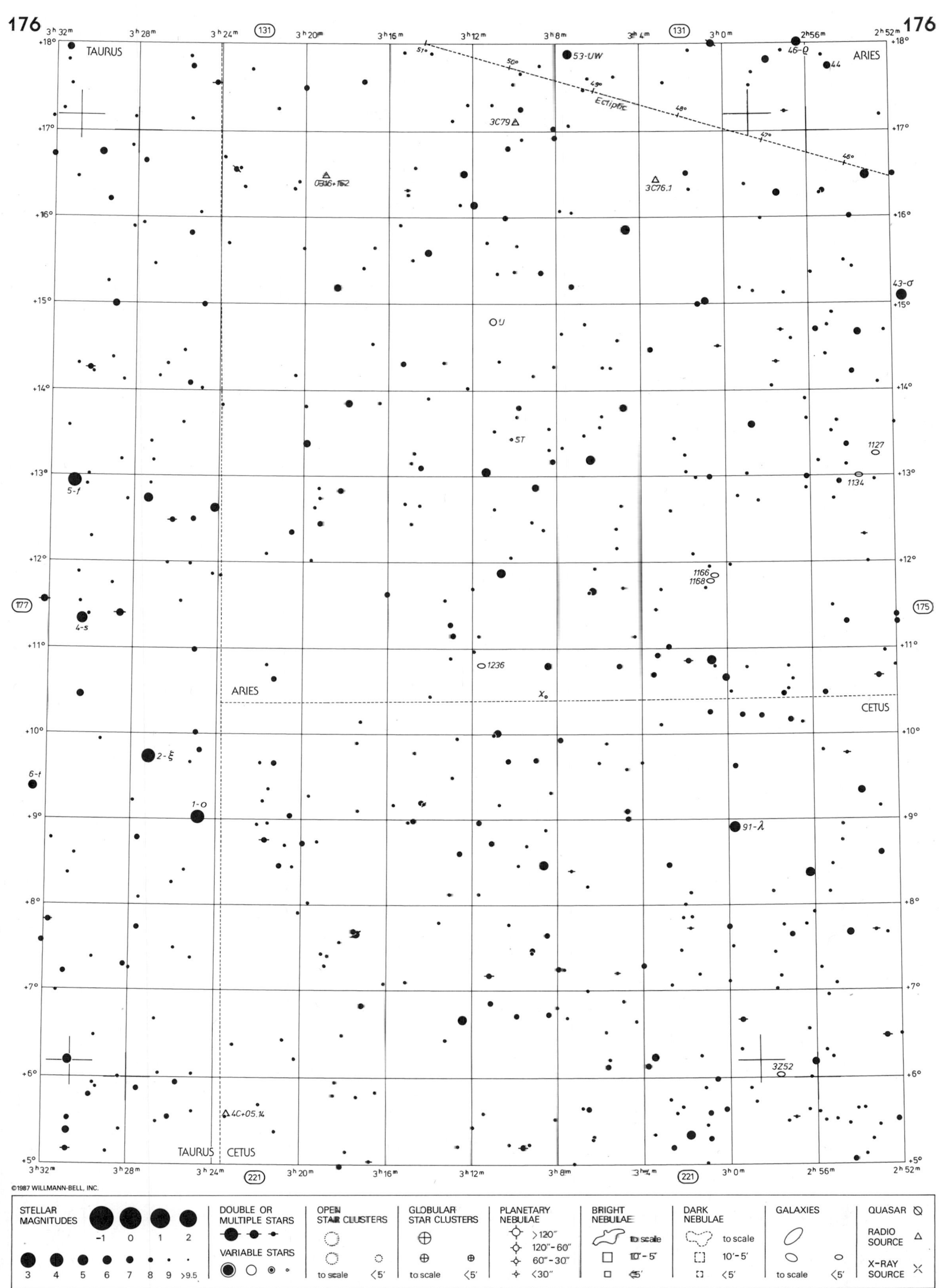

Barry Rappaport & Wil Tirion

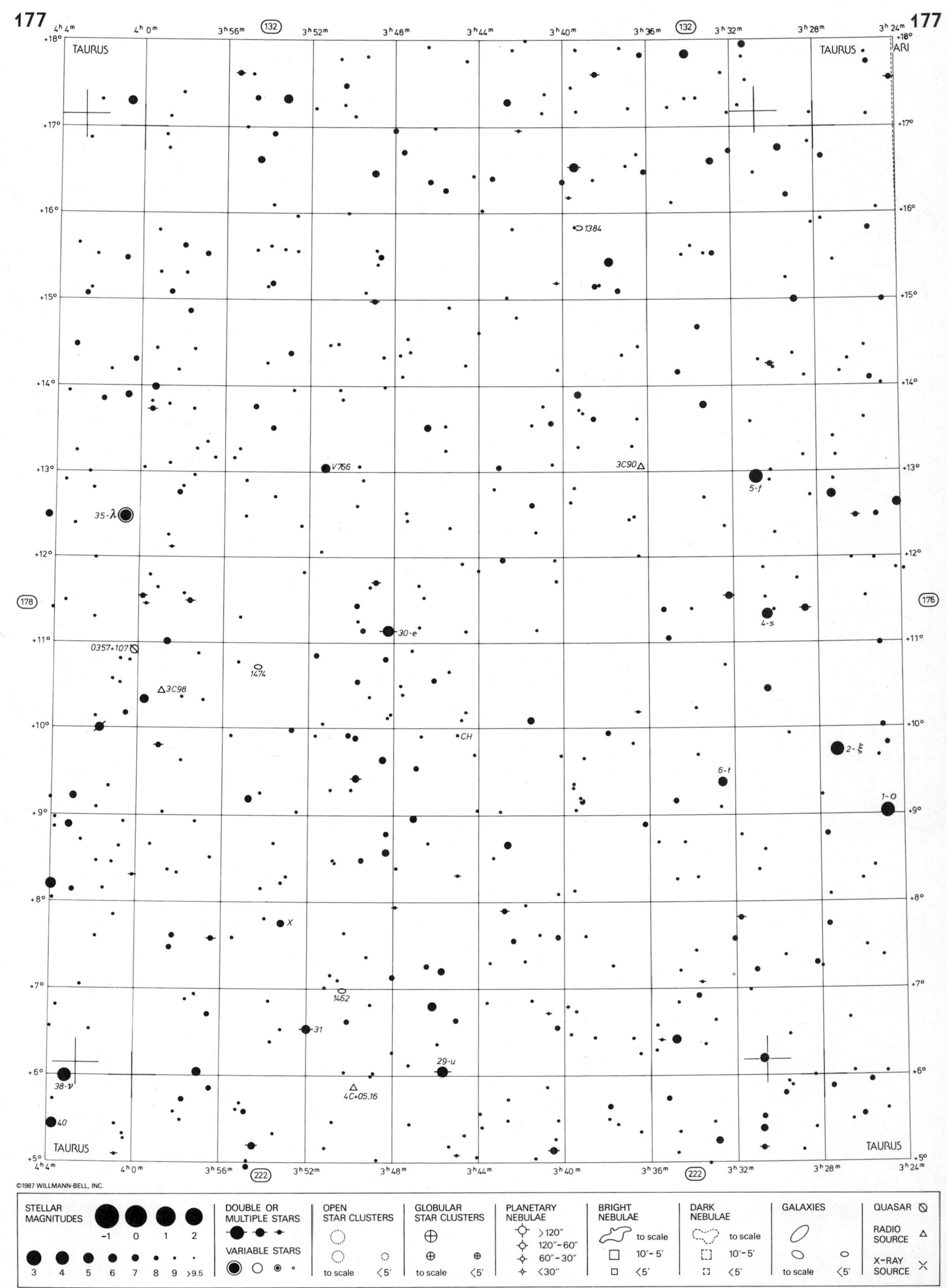
TAURUS
ARI
+18°
+17°
+16°
+15°
+14°
+13°
+12°
+11°
+10°
+9°
+8°
+7°
+6°
+5°
4h 4m
4h 0m
3h 56m
3h 52m
3h 48m
3h 44m
3h 40m
3h 36m
3h 32m
3h 28m
3h 24m
132
222
178
176
1384
V766
3C90
5-f
35-λ
30-e
4-s
0357+107
1474
3C98
CH
2-ξ
6-t
1-o
X
1462
31
29-u
38-ν
4C+05.16
40
©1987 WILLMANN-BELL, INC.
STELLAR MAGNITUDES
-1 0 1 2
3 4 5 6 7 8 9 >9.5
DOUBLE OR MULTIPLE STARS
VARIABLE STARS
OPEN STAR CLUSTERS
to scale
<5'
GLOBULAR STAR CLUSTERS
to scale
<5'
PLANETARY NEBULAE
>120"
120"-60"
60"-30"
<30"
BRIGHT NEBULAE
to scale
10'-5'
<5'
DARK NEBULAE
to scale
10'-5'
<5'
GALAXIES
to scale
<5'
QUASAR
RADIO SOURCE
X-RAY SOURCE
Barry Rappaport & Wil Tirion

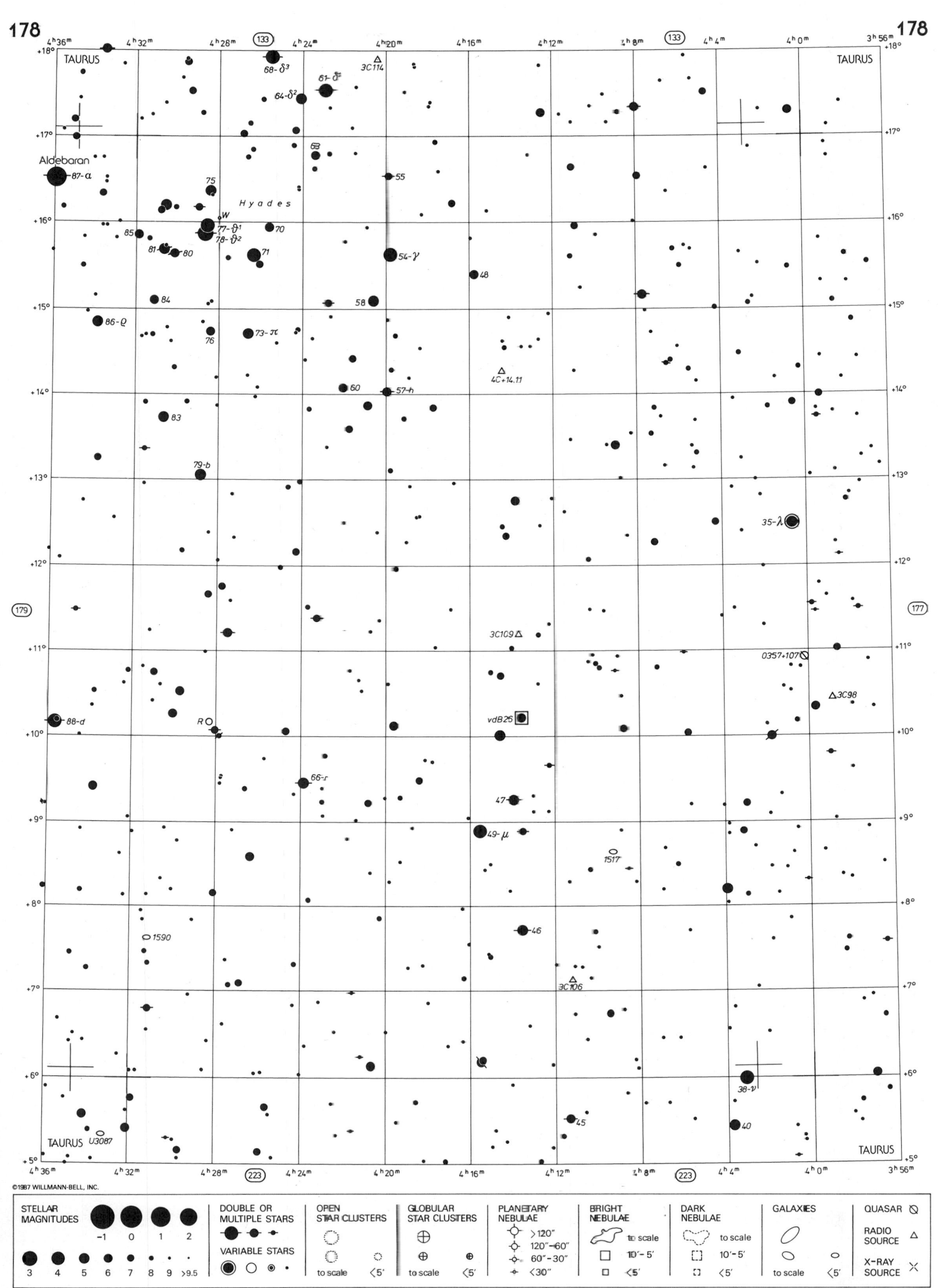
TAURUS
Aldebaran
87-α
Hyades
68-δ3
64-δ2
61-δ1
3C114
55
54-γ
48
77-θ1
78-θ2
71
70
75
85
81
80
84
58
86-ρ
76
73-π
4C+14.11
60
57-h
83
79-b
35-λ
3C109
0357+107
3C98
88-d
R
vdB26
66-r
47
49-μ
1517
46
1590
3C106
38-ν
45
40
U3087
4h36m
4h32m
4h28m
4h24m
4h20m
4h16m
4h12m
4h8m
4h4m
4h0m
3h56m
+18°
+17°
+16°
+15°
+14°
+13°
+12°
+11°
+10°
+9°
+8°
+7°
+6°
+5°
133
179
177
223
©1987 WILLMANN-BELL, INC.
STELLAR MAGNITUDES
-1 0 1 2
3 4 5 6 7 8 9 >9.5
DOUBLE OR MULTIPLE STARS
VARIABLE STARS
OPEN STAR CLUSTERS
to scale <5'
GLOBULAR STAR CLUSTERS
to scale <5'
PLANETARY NEBULAE
>120"
120"-60"
60"-30"
<30"
BRIGHT NEBULAE
to scale
10'-5'
<5'
DARK NEBULAE
to scale
10'-5'
<5'
GALAXIES
to scale <5'
QUASAR
RADIO SOURCE
X-RAY SOURCE
Barry Rappaport & Wil Tirion

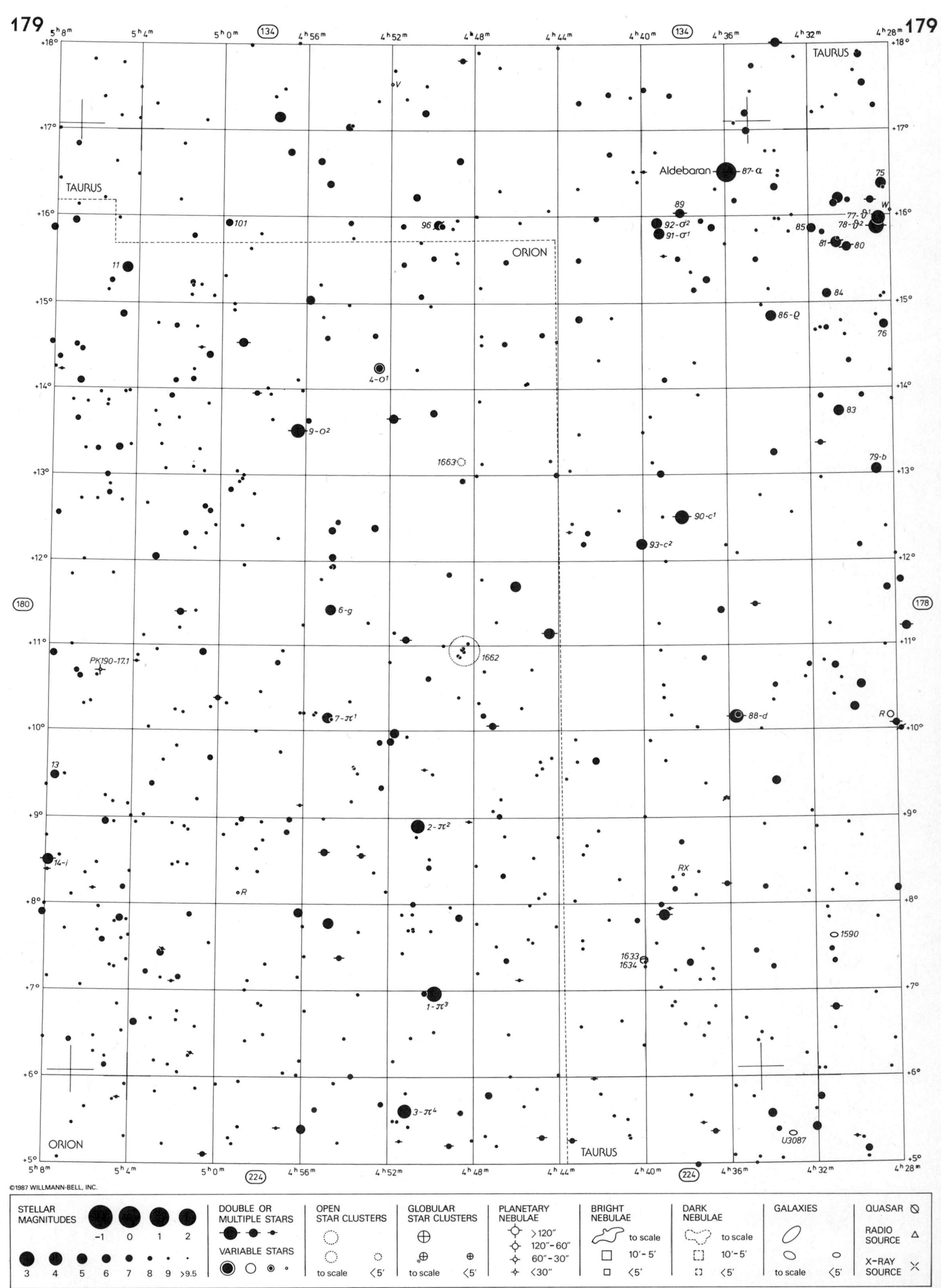

Barry Rappaport & Wil Tirion

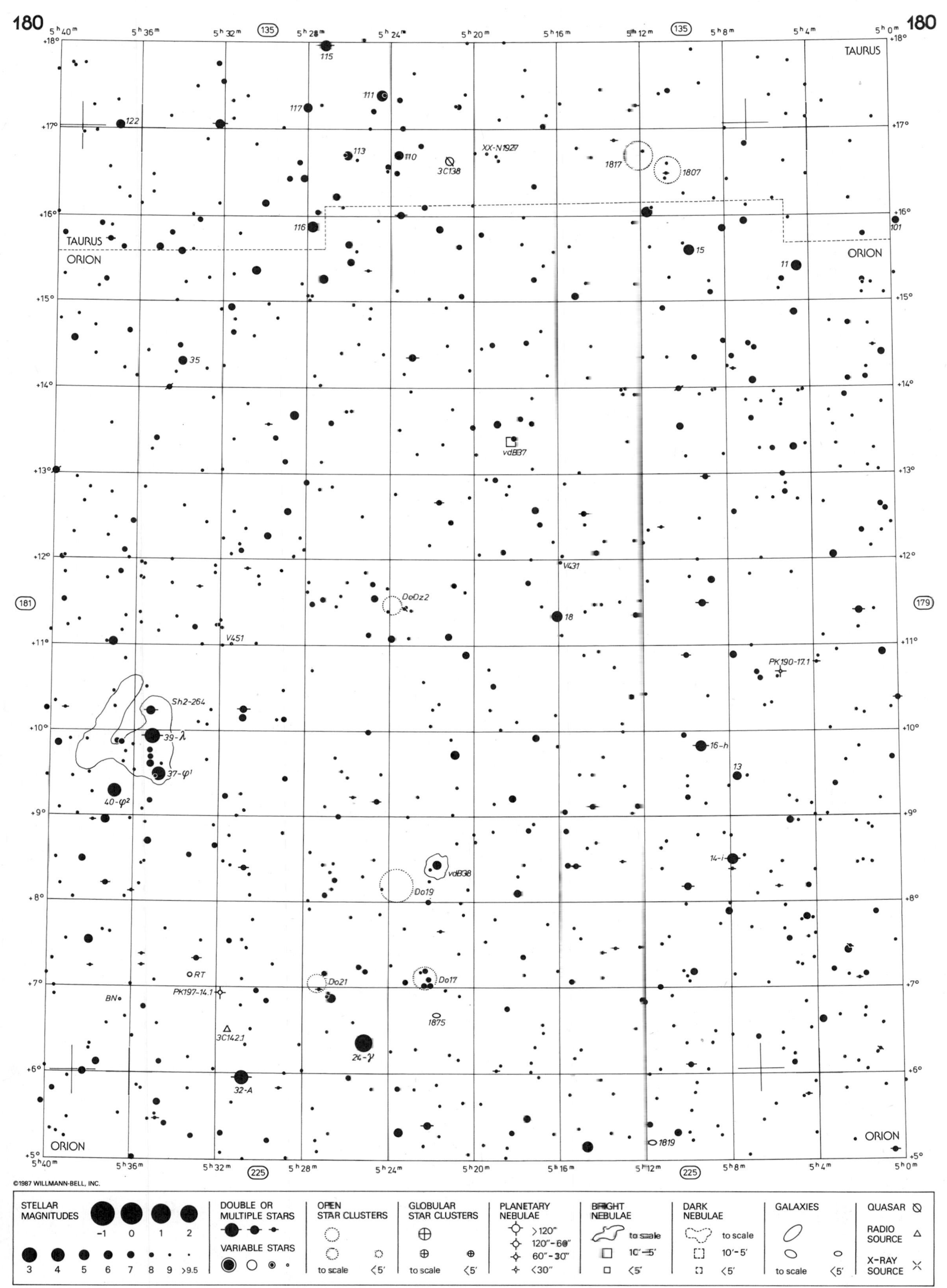

STELLAR MAGNITUDES	DOUBLE OR MULTIPLE STARS	OPEN STAR CLUSTERS	GLOBULAR STAR CLUSTERS	PLANETARY NEBULAE	BRIGHT NEBULAE	DARK NEBULAE	GALAXIES	QUASAR
−1 0 1 2	VARIABLE STARS	to scale ‹5′	to scale ‹5′	>120″ 120″–60″ 60″–30″ ‹30″	to scale 10′–5′ ‹5′	to scale 10′–5′ ‹5′	to scale ‹5′	RADIO SOURCE
3 4 5 6 7 8 9 >9.5								X-RAY SOURCE

Barry Rappaport & Wil Tirion

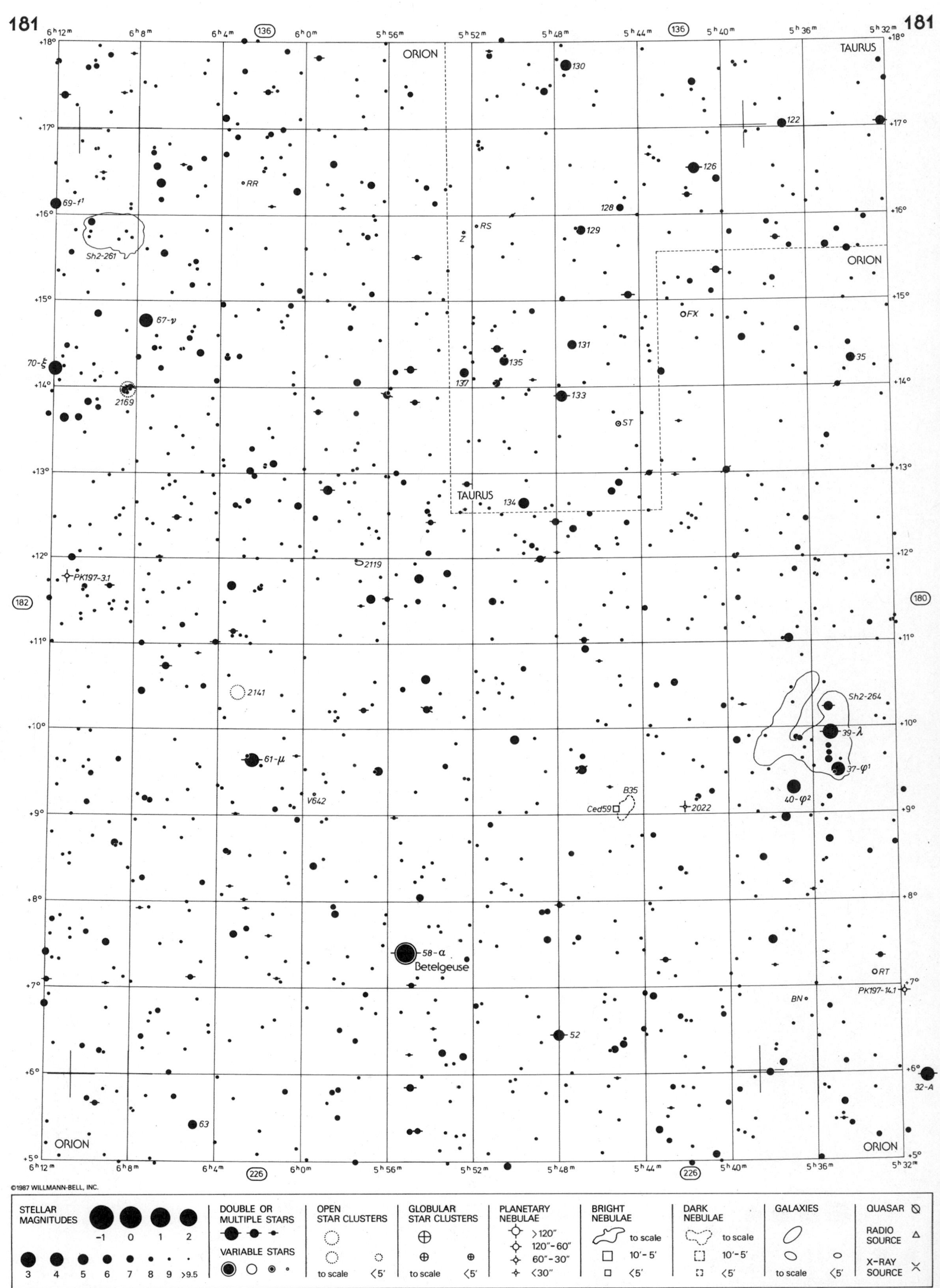
181
6h12m
6h8m
6h4m
6h0m
5h56m
5h52m
5h48m
5h44m
5h40m
5h36m
5h32m
136
226
182
180
+18°
+17°
+16°
+15°
+14°
+13°
+12°
+11°
+10°
+9°
+8°
+7°
+6°
+5°
ORION
TAURUS
130
122
126
128
129
RR
RS
Z
69-f¹
Sh2-261
67-ν
70-ξ
2169
FX
131
135
137
133
35
ST
TAURUS
134
2119
PK197-3.1
2141
Sh2-264
39-λ
37-φ¹
40-φ²
61-μ
V642
B35
Ced59
2022
58-α
Betelgeuse
RT
PK197-14.1
BN
52
32-A
63
©1987 WILLMANN-BELL, INC.
STELLAR MAGNITUDES
-1 0 1 2
3 4 5 6 7 8 9 >9.5
DOUBLE OR MULTIPLE STARS
VARIABLE STARS
OPEN STAR CLUSTERS
to scale
<5′
GLOBULAR STAR CLUSTERS
to scale
<5′
PLANETARY NEBULAE
>120″
120″-60″
60″-30″
<30″
BRIGHT NEBULAE
to scale
10′-5′
<5′
DARK NEBULAE
to scale
10′-5′
<5′
GALAXIES
to scale
<5′
QUASAR
RADIO SOURCE
X-RAY SOURCE
Barry Rappaport & Wil Tirion

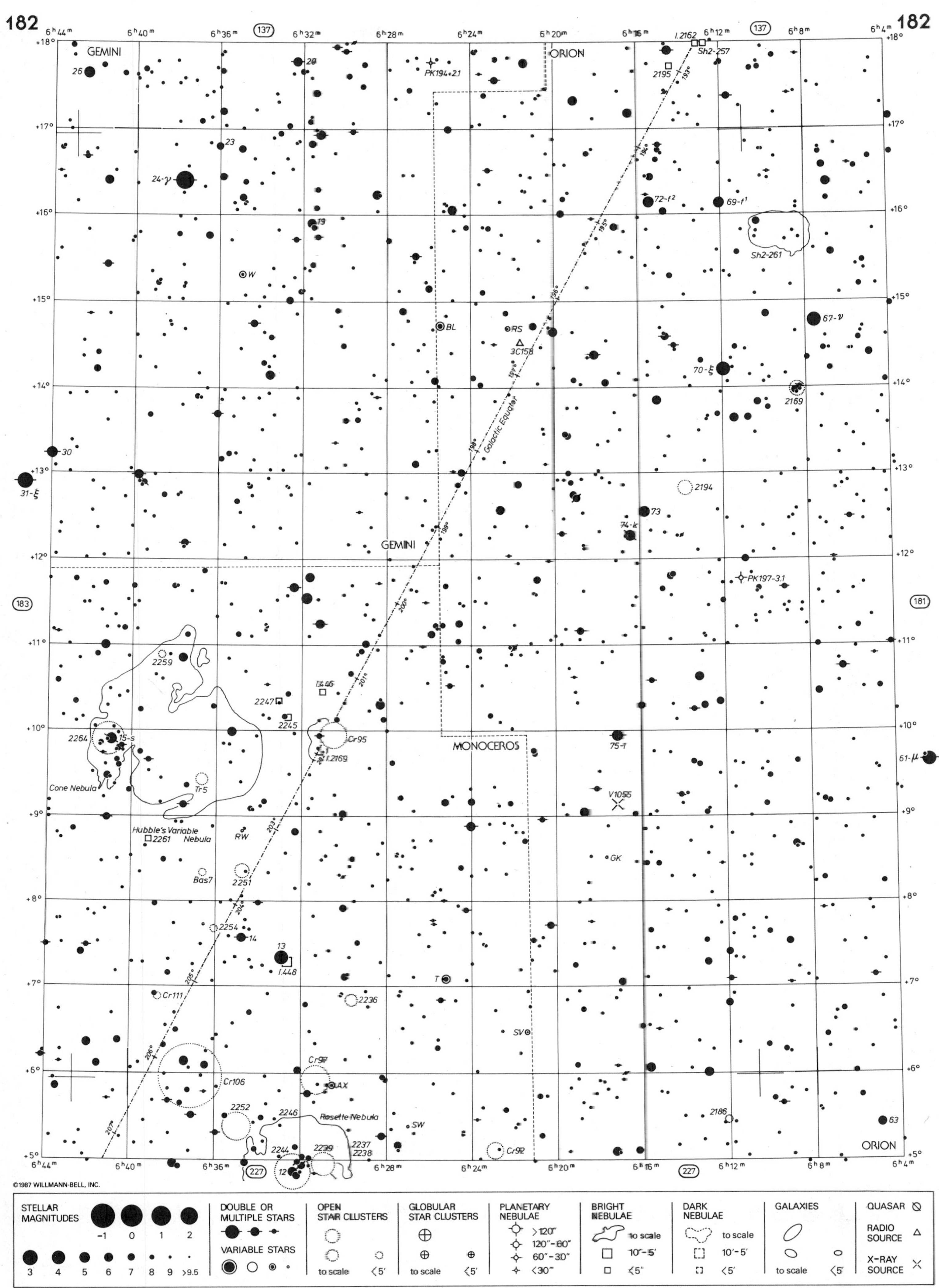

GEMINI
ORION
MONOCEROS
6h44m 6h40m 6h36m 6h32m 6h28m 6h24m 6h20m 6h16m 6h12m 6h8m 6h4m
+18° +17° +16° +15° +14° +13° +12° +11° +10° +9° +8° +7° +6° +5°
137
183
181
227
26
28
PK194+2.1
I.2162
Sh2-257
2195
23
24-γ
72-f2
69-f1
13
Sh2-261
W
BL
RS
3C158
67-ν
70-ξ
2169
Galactic Equator
30
31-ξ
2194
73
74-k
PK197-3.1
2259
I.446
2247
2245
2264
15-S
Cr95
I.2169
75-l
61-μ
Cone Nebula
Tr5
V1055
Hubble's Variable Nebula
2261
RW
GK
Bas7
2251
2254
14
13
I.448
T
Cr111
2236
SV
Cr106
Cr97
AX
2252
2246
Rosette Nebula
SW
2186
63
2244
2239
2237
2238
Cr92
12
©1987 WILLMANN-BELL, INC.
STELLAR MAGNITUDES
-1 0 1 2
3 4 5 6 7 8 9 >9.5
DOUBLE OR MULTIPLE STARS
VARIABLE STARS
OPEN STAR CLUSTERS
to scale
<5'
GLOBULAR STAR CLUSTERS
to scale
<5'
PLANETARY NEBULAE
>120"
120"-60"
60"-30"
<30"
BRIGHT NEBULAE
to scale
10'-5'
<5'
DARK NEBULAE
to scale
10'-5'
<5'
GALAXIES
to scale
<5'
QUASAR
RADIO SOURCE
X-RAY SOURCE
Barry Rappaport & Wil Tirion

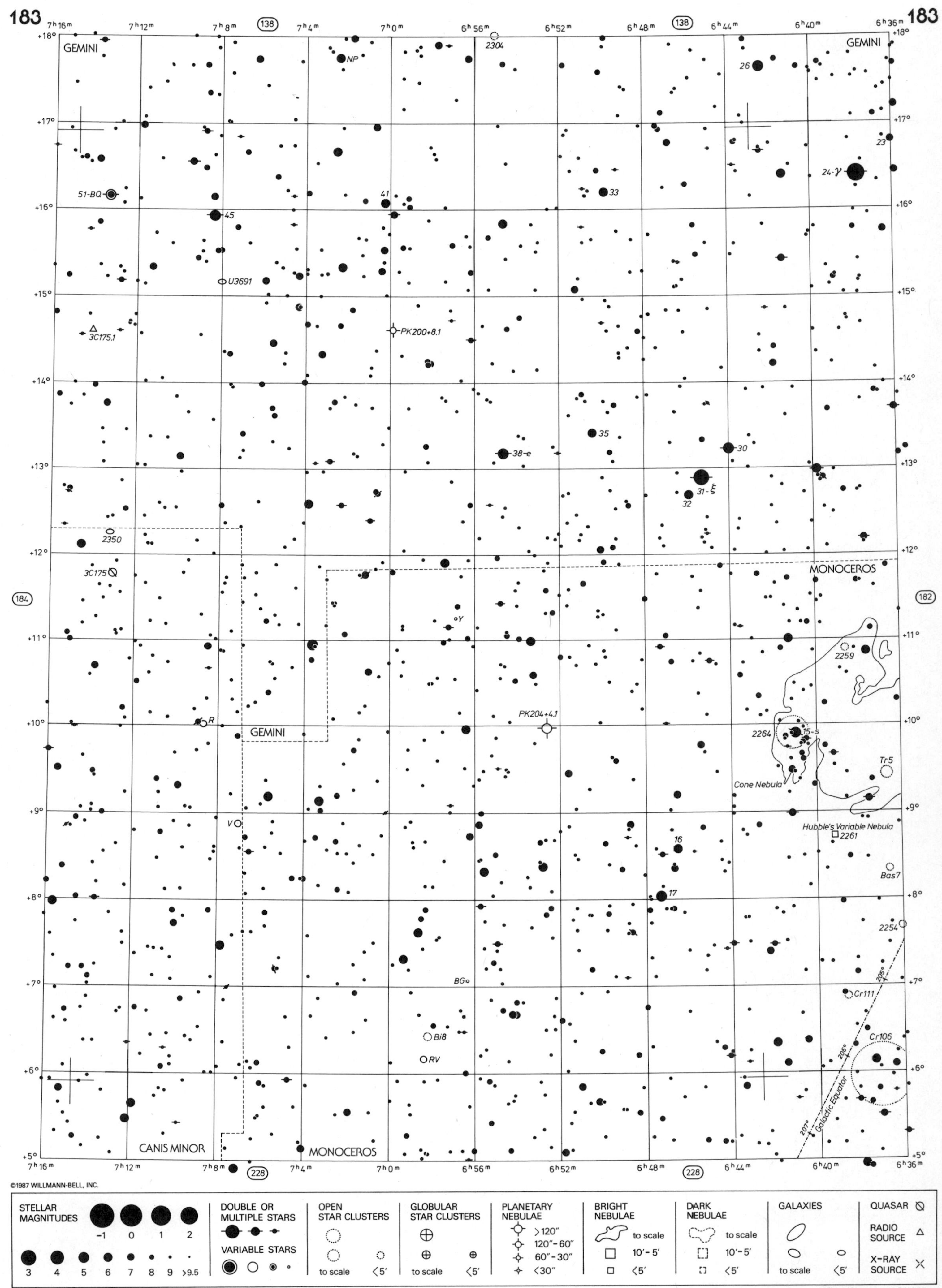

Barry Rappaport & Wil Tirion

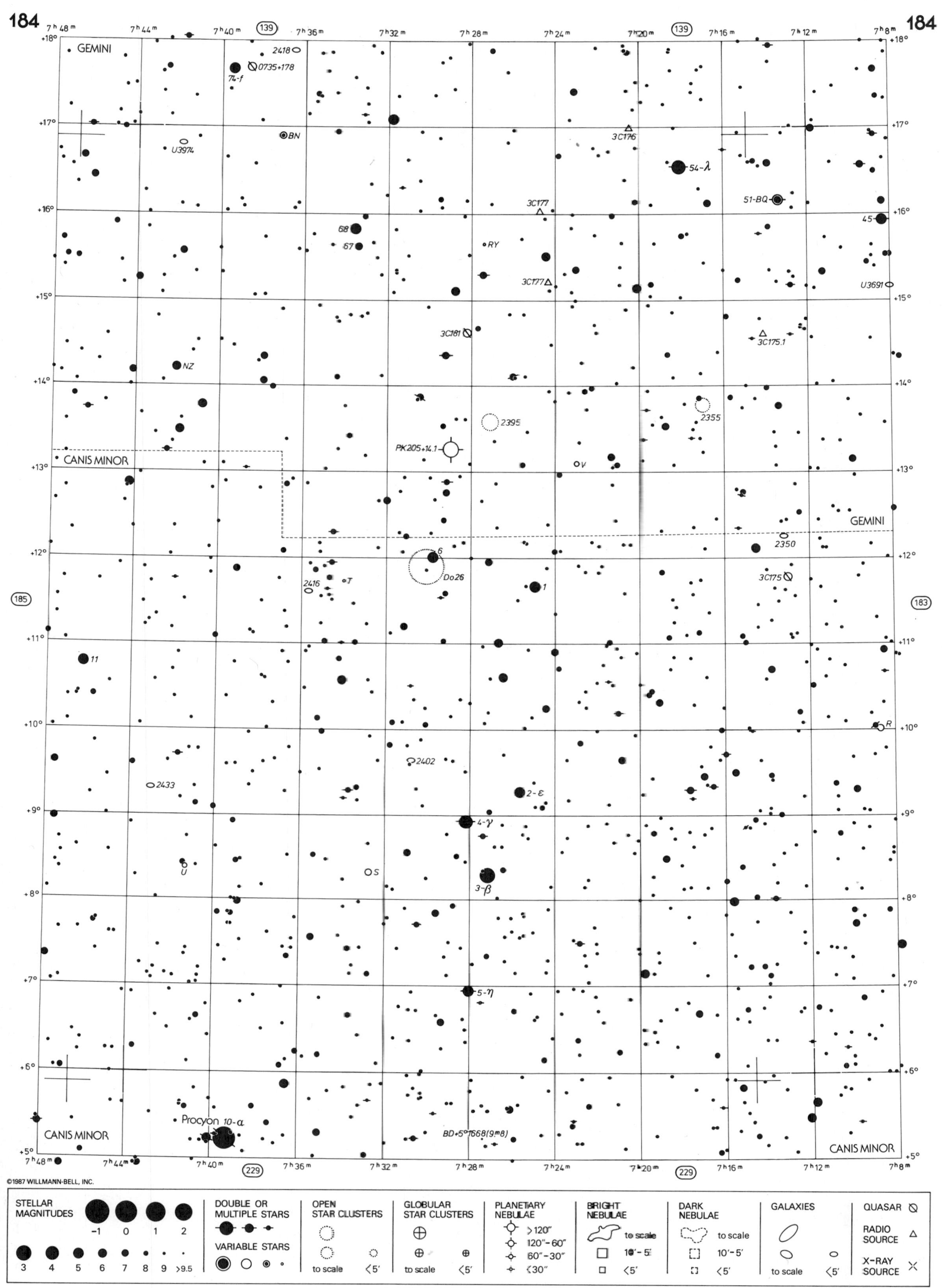
GEMINI
CANIS MINOR
GEMINI
CANIS MINOR
CANIS MINOR
Procyon 10-α
54-λ
51-BQ
45
4-γ
3-β
2-ε
5-η
0735+178
3C176
3C177
3C177
3C181
3C175.1
3C175
U3974
U3691
2418
2395
2355
2350
2416
2402
2433
PK205+14.1
Do26
BD+5°1668(9m8)
©1987 WILLMANN-BELL, INC.
STELLAR MAGNITUDES
-1 0 1 2
3 4 5 6 7 8 9 >9.5
DOUBLE OR MULTIPLE STARS
VARIABLE STARS
OPEN STAR CLUSTERS
to scale <5′
GLOBULAR STAR CLUSTERS
to scale <5′
PLANETARY NEBULAE
>120″
120″-60″
60″-30″
<30″
BRIGHT NEBULAE
to scale
10′-5′
<5′
DARK NEBULAE
to scale
10′-5′
<5′
GALAXIES
to scale <5′
QUASAR
RADIO SOURCE
X-RAY SOURCE
Barry Rappaport & Wil Tirion

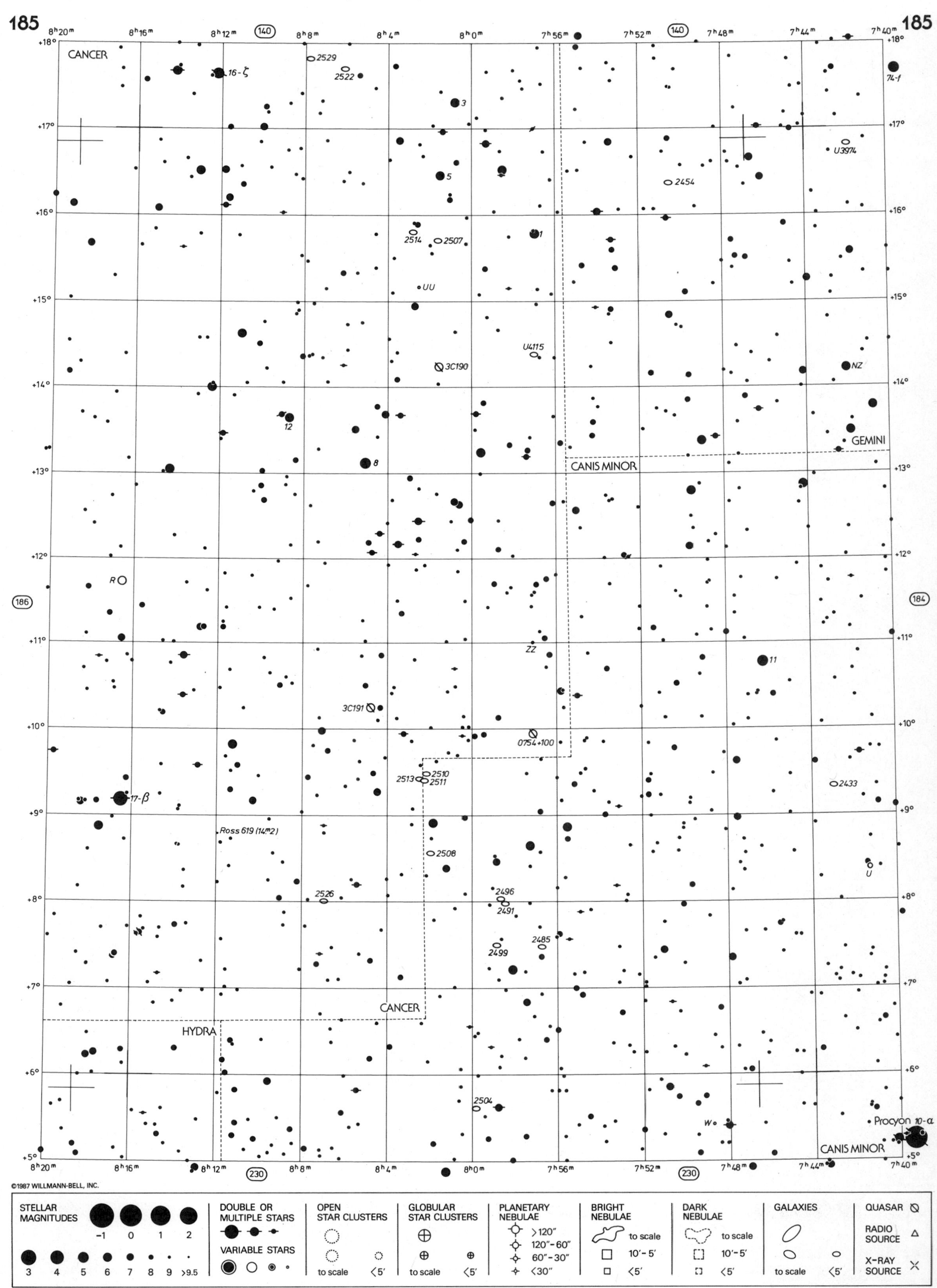

Barry Rappaport & Wil Tirion

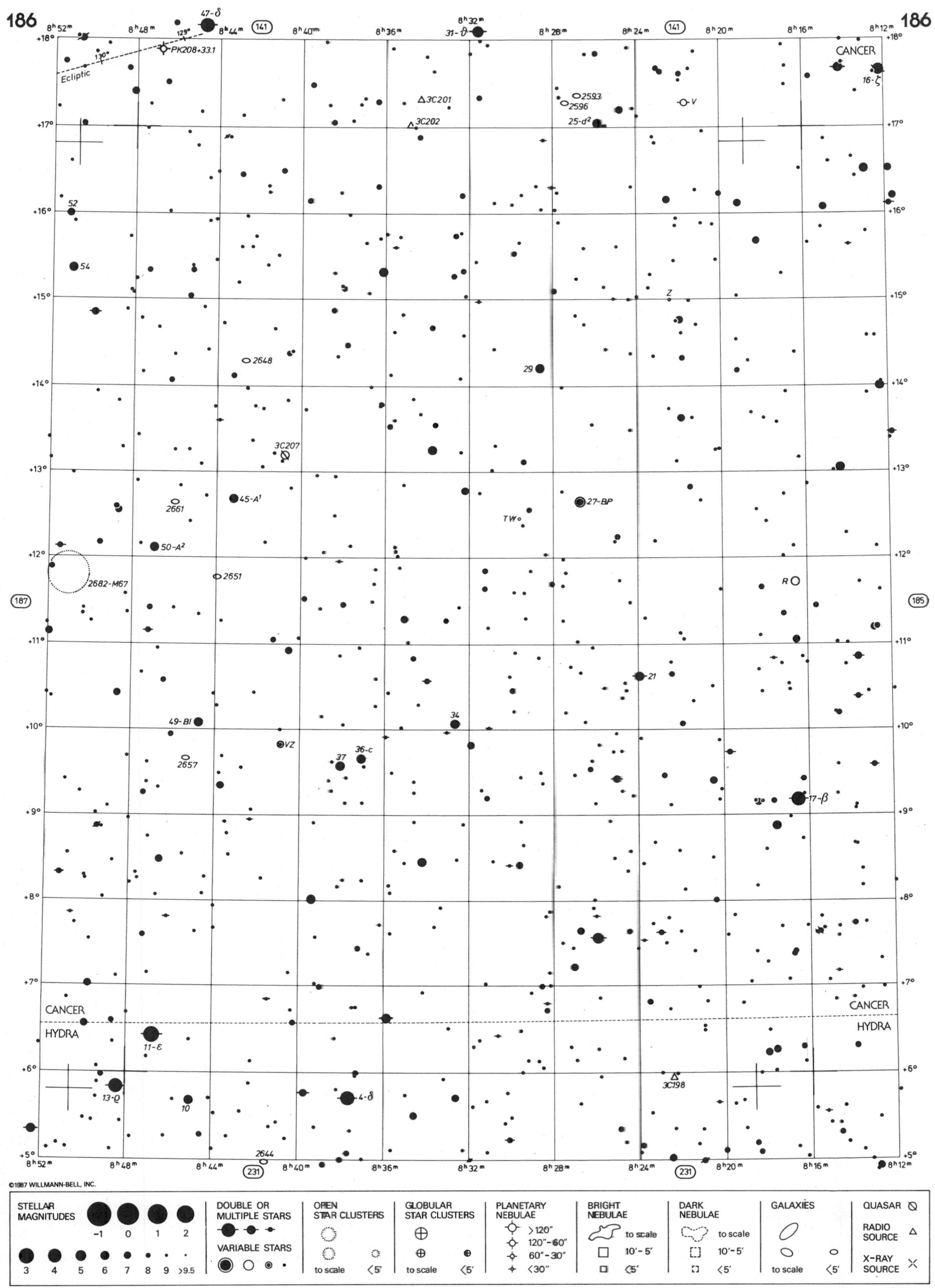

STELLAR MAGNITUDES	DOUBLE OR MULTIPLE STARS	OPEN STAR CLUSTERS	GLOBULAR STAR CLUSTERS	PLANETARY NEBULAE	BRIGHT NEBULAE	DARK NEBULAE	GALAXIES	QUASAR
-1 0 1 2	VARIABLE STARS	to scale <5′	to scale <5′	>120″ 120″-60″ 60″-30″ <30″	to scale 10′-5′ <5′	to scale 10′-5′ <5′	to scale <5′	RADIO SOURCE
3 4 5 6 7 8 9 >9.5								X-RAY SOURCE

Barry Rappaport & Wil Tirion

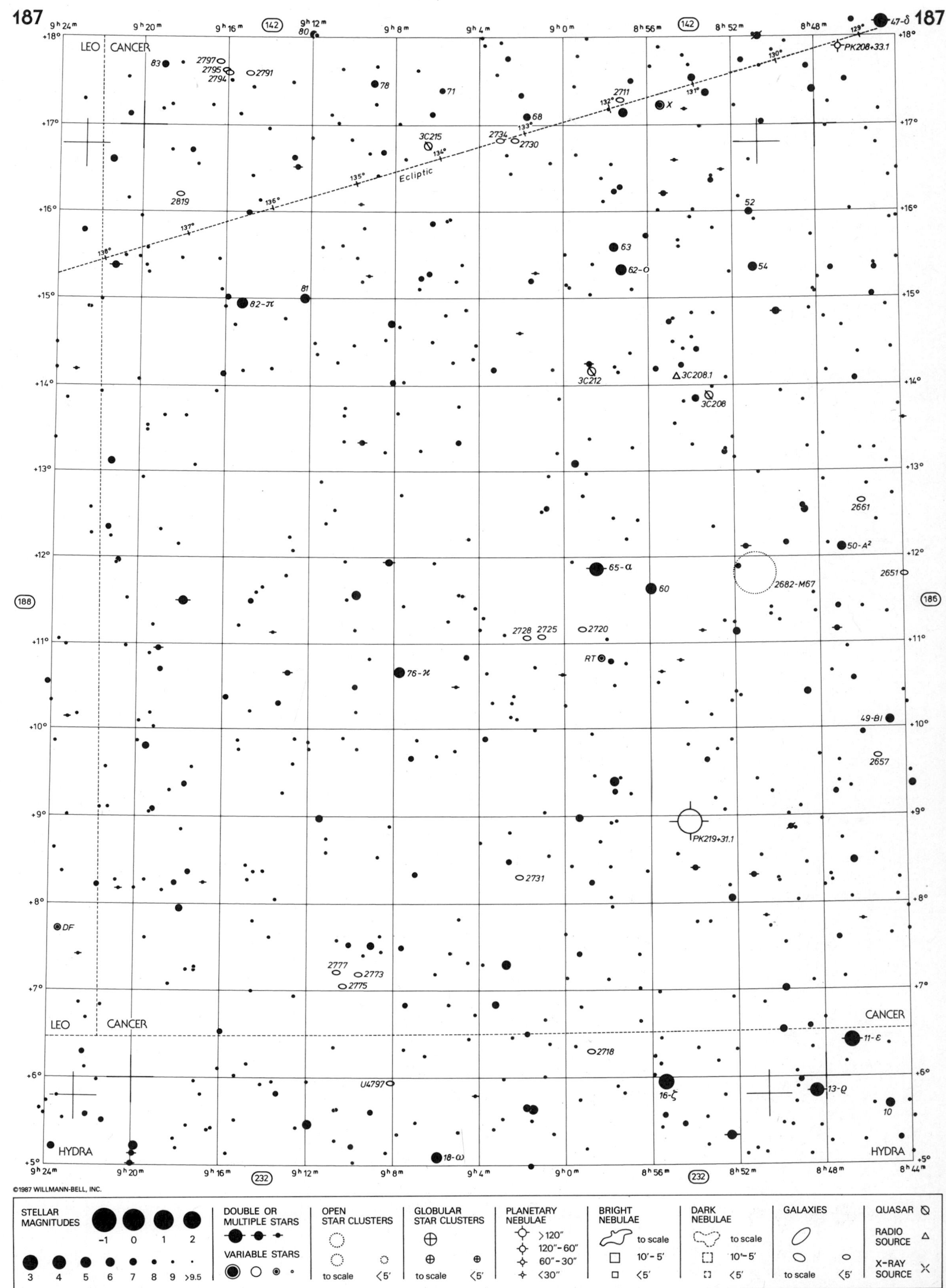

187
9h 24m
9h 20m
9h 16m
9h 12m
9h 8m
9h 4m
9h 0m
8h 56m
8h 52m
8h 48m
8h 44m
142
232
188
186
+18°
+17°
+16°
+15°
+14°
+13°
+12°
+11°
+10°
+9°
+8°
+7°
+6°
+5°
LEO
CANCER
HYDRA
Ecliptic
47-δ
PK208+33.1
3C215
3C212
3C208.1
3C208
2682-M67
65-α
76-ϰ
PK219+31.1
16-ζ
13-ρ
11-ε
18-ω
82-π
62-o
50-A²
49-BI
U4797
2797
2795
2794
2791
2819
2734
2730
2711
2728
2725
2720
2731
2718
2777
2773
2775
2661
2651
2657
RT
DF
X
©1987 WILLMANN-BELL, INC.
STELLAR MAGNITUDES
DOUBLE OR MULTIPLE STARS
VARIABLE STARS
OPEN STAR CLUSTERS
GLOBULAR STAR CLUSTERS
PLANETARY NEBULAE
BRIGHT NEBULAE
DARK NEBULAE
GALAXIES
QUASAR
RADIO SOURCE
X-RAY SOURCE
to scale
<5′
>120″
120″-60″
60″-30″
<30″
10′-5′
Barry Rappaport & Wil Tirion

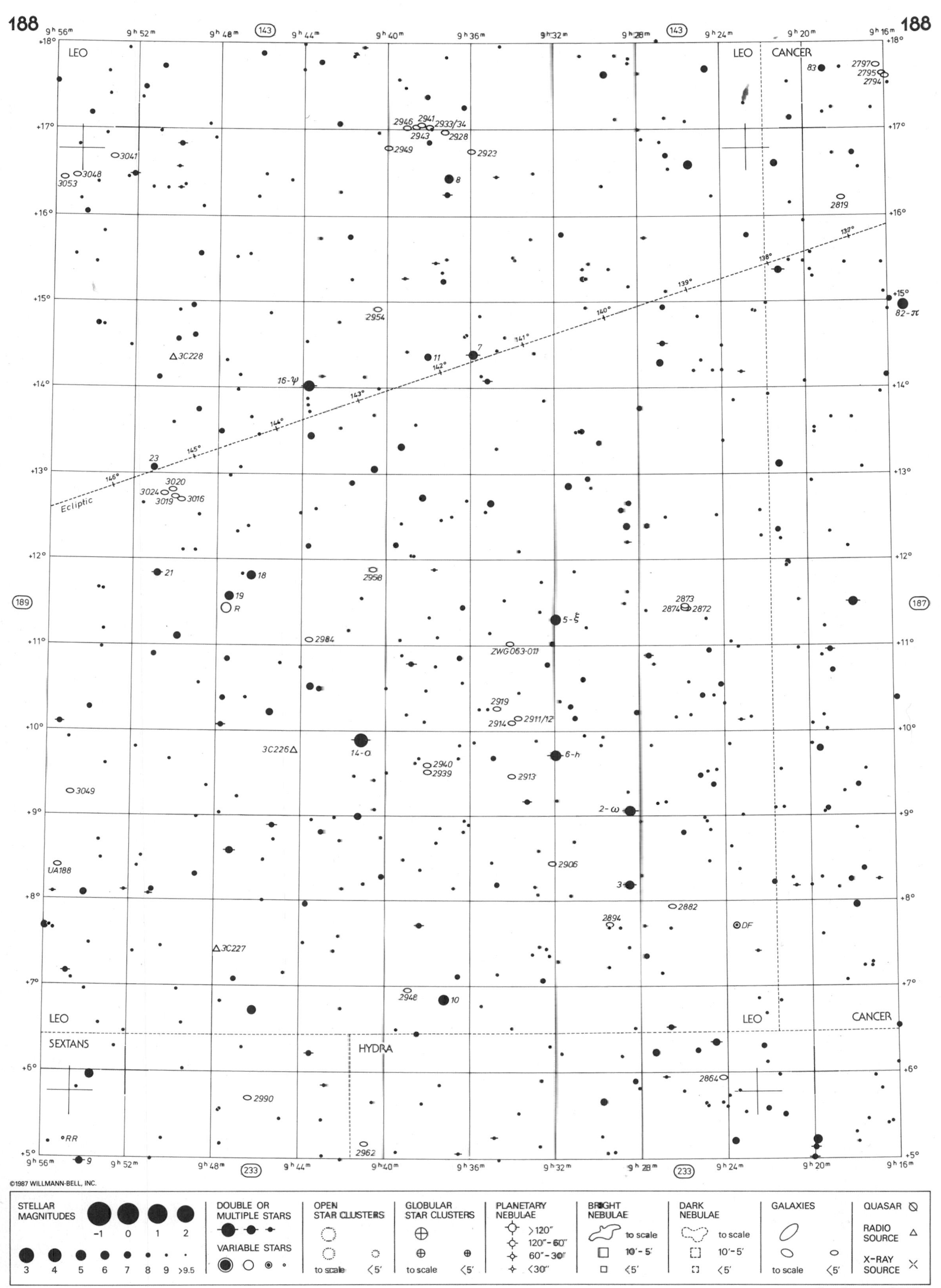

Barry Rappaport & Wil Tirion

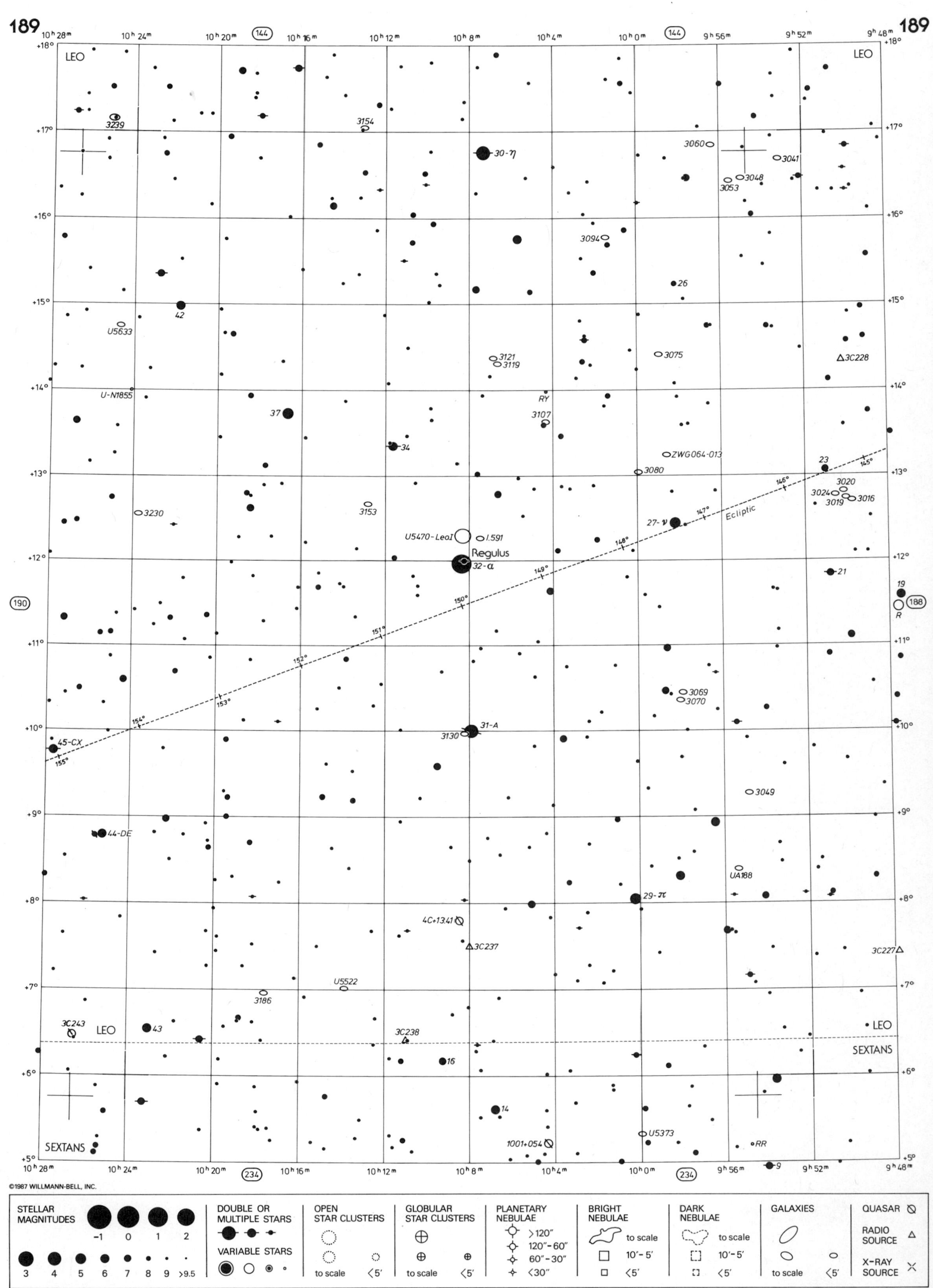

Barry Rappaport & Wil Tirion

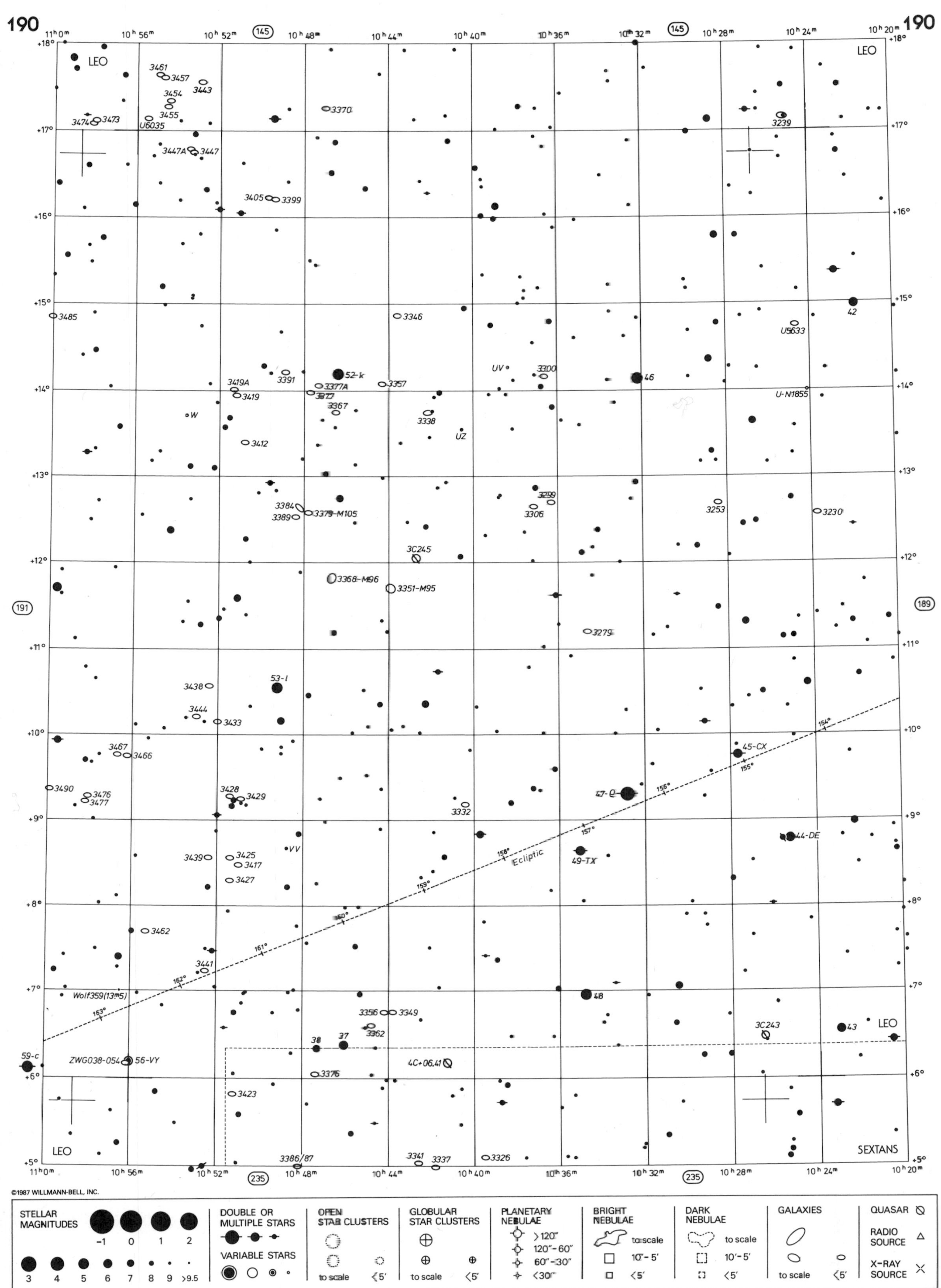
LEO
SEXTANS
Ecliptic
11h 0m
10h 56m
10h 52m
10h 48m
10h 44m
10h 40m
10h 36m
10h 32m
10h 28m
10h 24m
10h 20m
+18°
+17°
+16°
+15°
+14°
+13°
+12°
+11°
+10°
+9°
+8°
+7°
+6°
+5°
145
191
189
235
3461
3457
3443
3454
3455
U6035
3474
3473
3447A
3447
3405
3399
3370
3239
3485
3346
U5633
42
3391
52-k
3419A
3419
3377A
3377
3367
3357
UV
3300
46
U-N1855
W
3338
UZ
3412
3384
3389
3379-M105
3299
3306
3253
3230
3C245
3368-M96
3351-M95
3279
53-l
3438
3444
3433
3467
3466
45-CX
3490
3476
3477
3428
3429
47-ρ
3332
44-DE
VV
3439
3425
3417
3427
49-TX
3462
3441
Wolf359(13m5)
48
3356
3349
3362
3C243
43
38
37
59-c
ZWG038-054
56-VY
4C+06.41
3376
3423
3386/87
3341
3337
3326
©1987 WILLMANN-BELL, INC.
STELLAR MAGNITUDES
-1
0
1
2
3
4
5
6
7
8
9
>9.5
DOUBLE OR MULTIPLE STARS
VARIABLE STARS
OPEN STAR CLUSTERS
to scale
<5'
GLOBULAR STAR CLUSTERS
to scale
<5'
PLANETARY NEBULAE
>120"
120"-60"
60"-30"
<30"
BRIGHT NEBULAE
to scale
10'-5'
<5'
DARK NEBULAE
to scale
10'-5'
<5'
GALAXIES
to scale
<5'
QUASAR
RADIO SOURCE
X-RAY SOURCE
Barry Rappaport & Wil Tirion

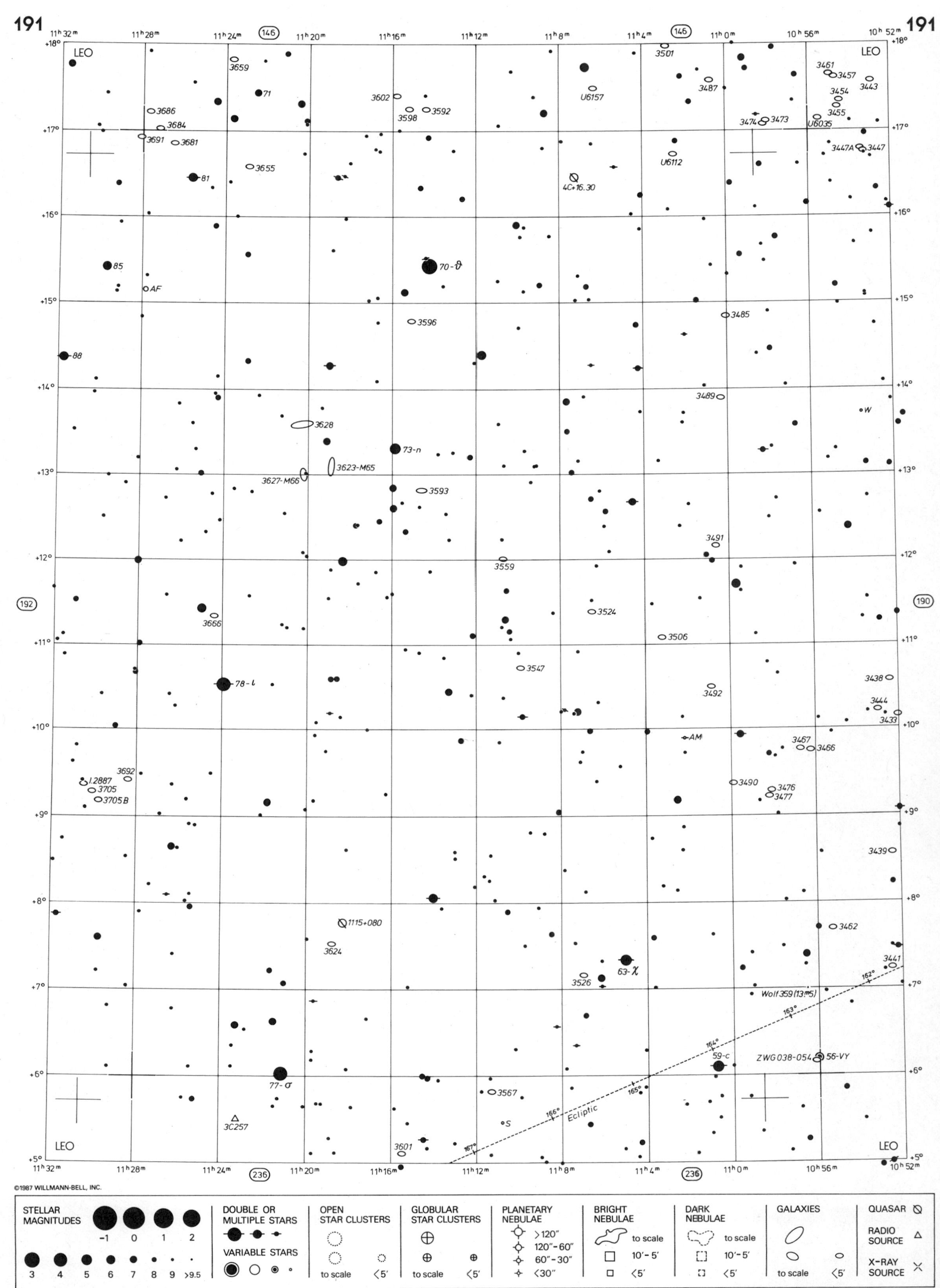

Barry Rappaport & Wil Tirion

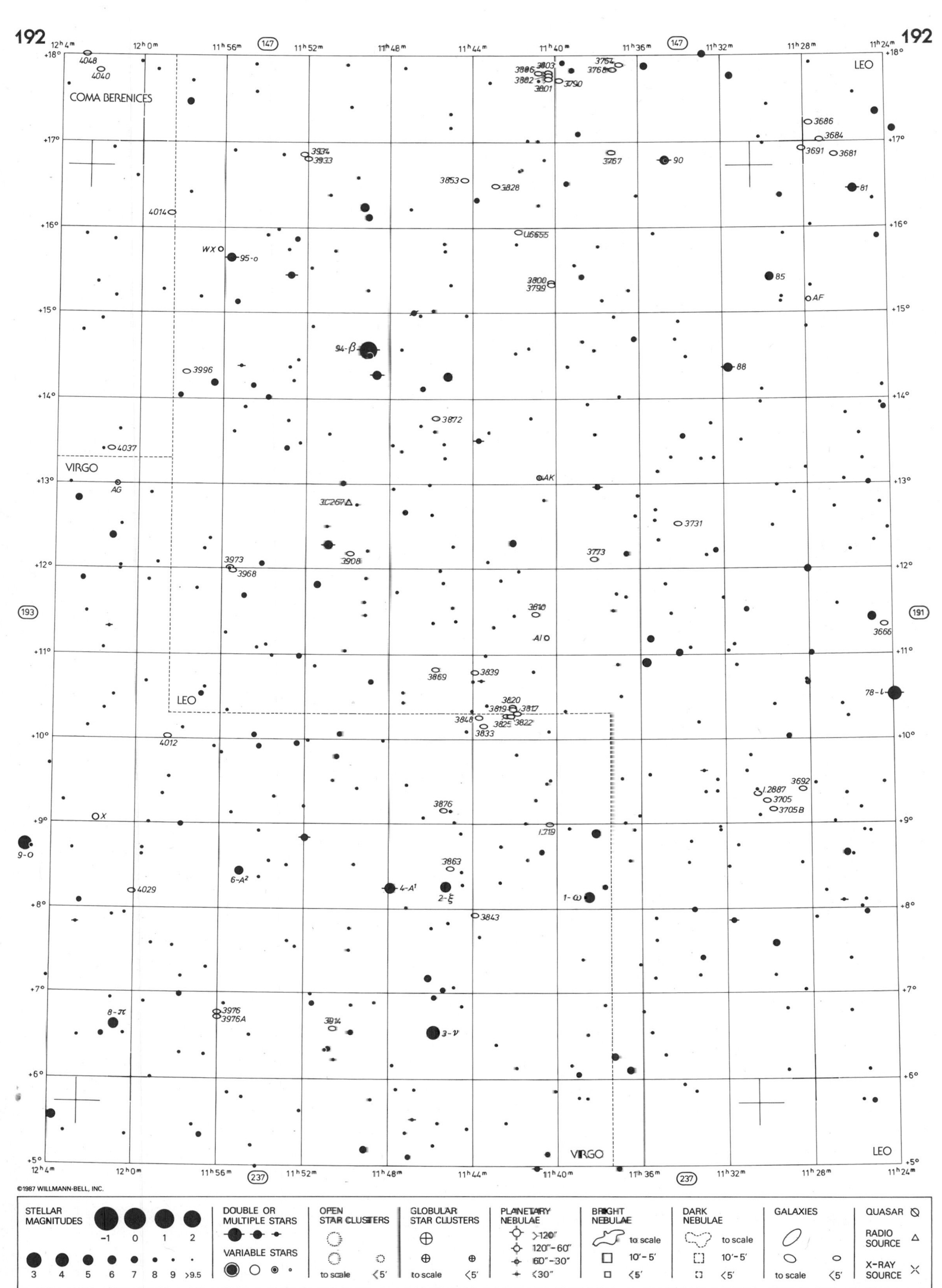

Barry Rappaport & Wil Tirion

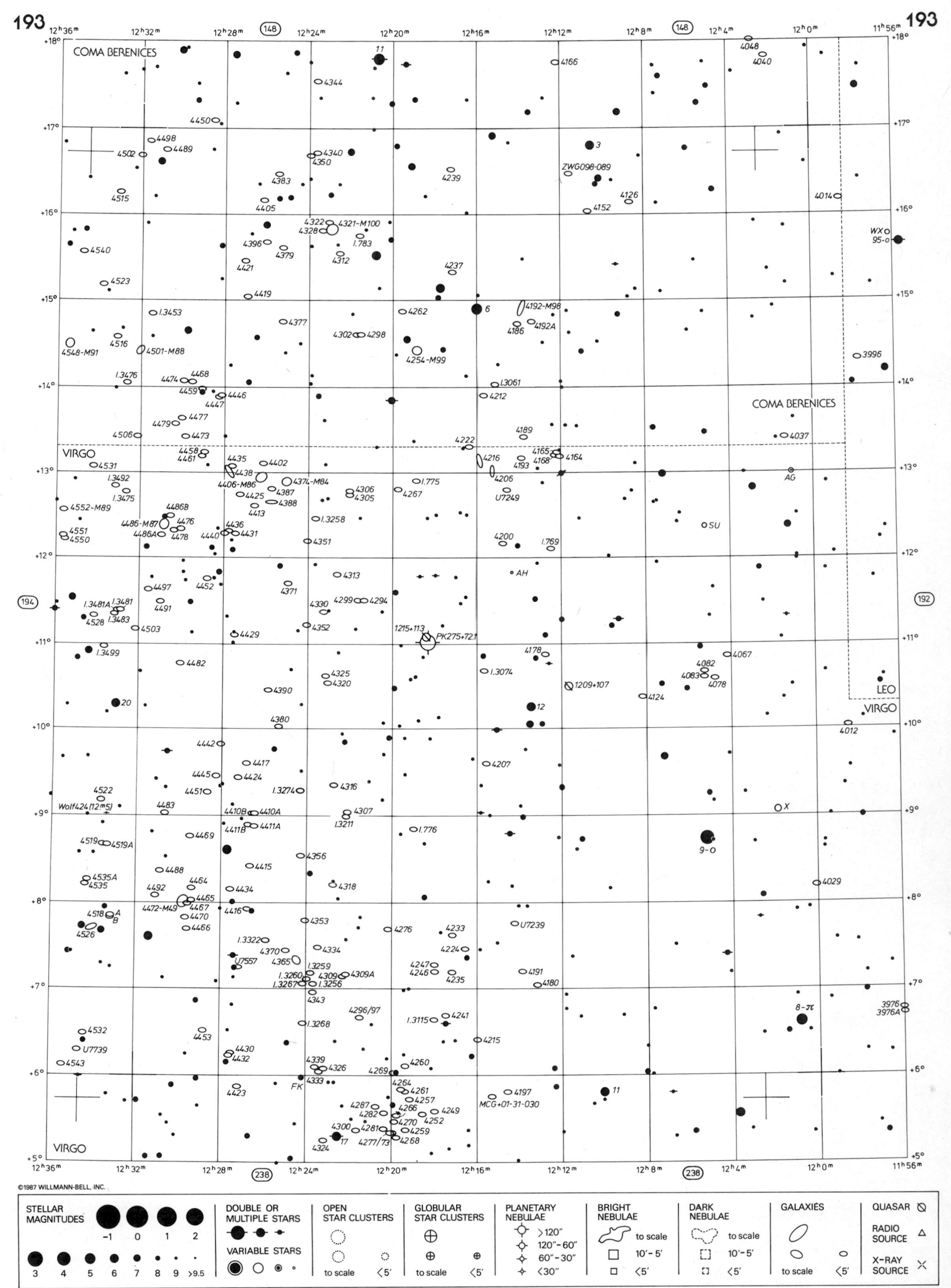

Barry Rappaport & Wil Tirion

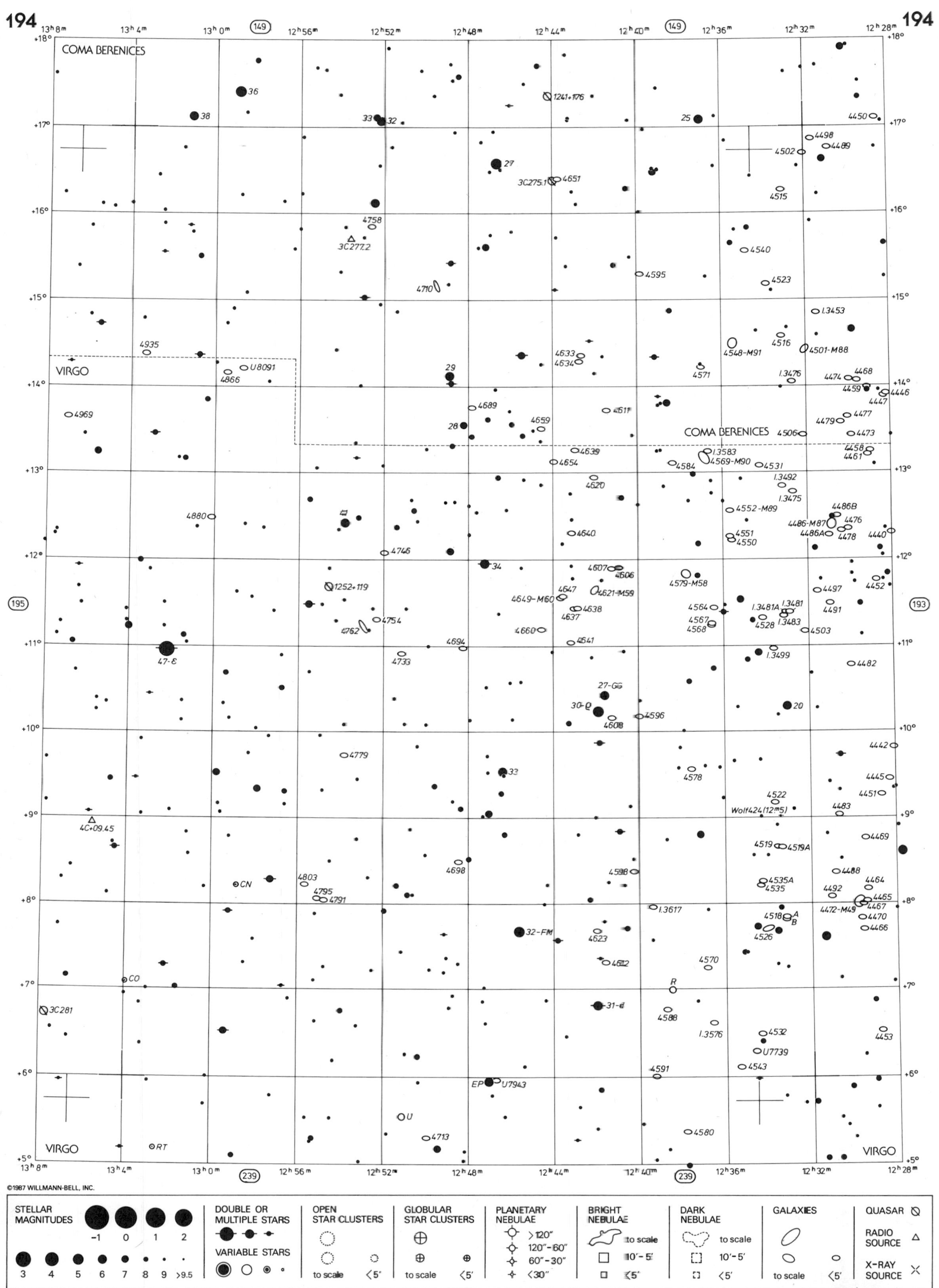

Barry Rappaport & Wil Tirion

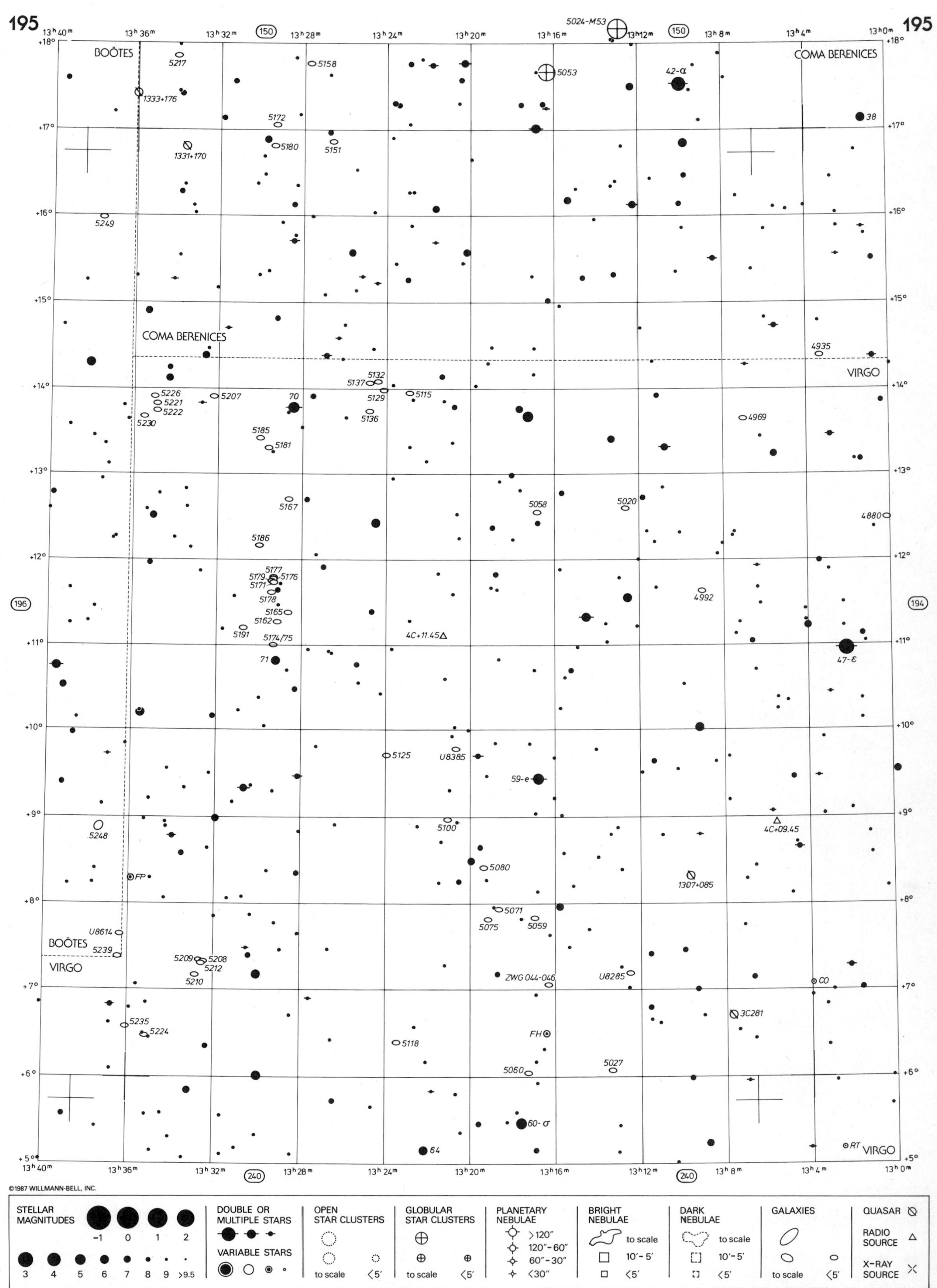

Barry Rappaport & Wil Tirion

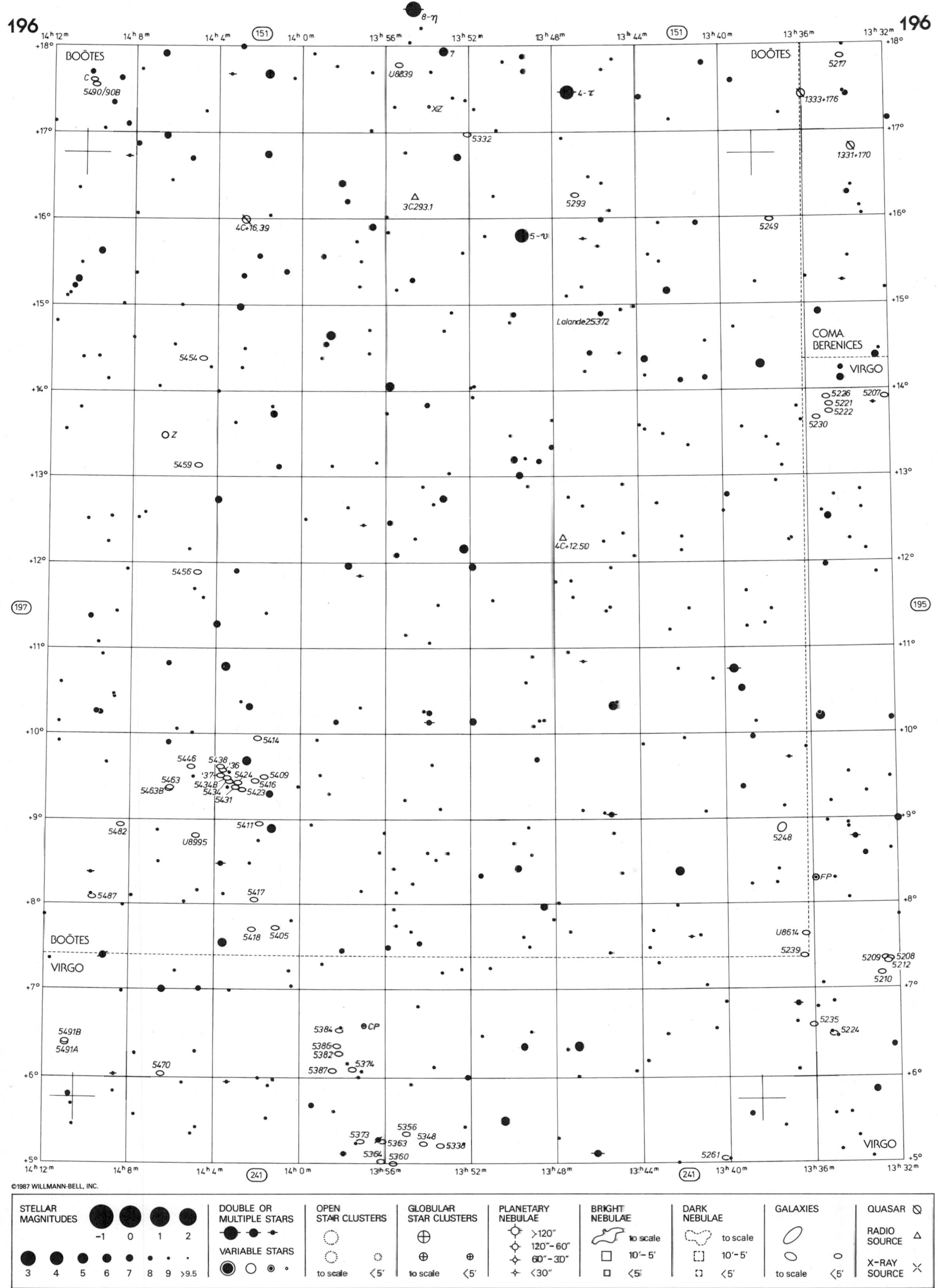

Barry Rappaport & Wil Tirion

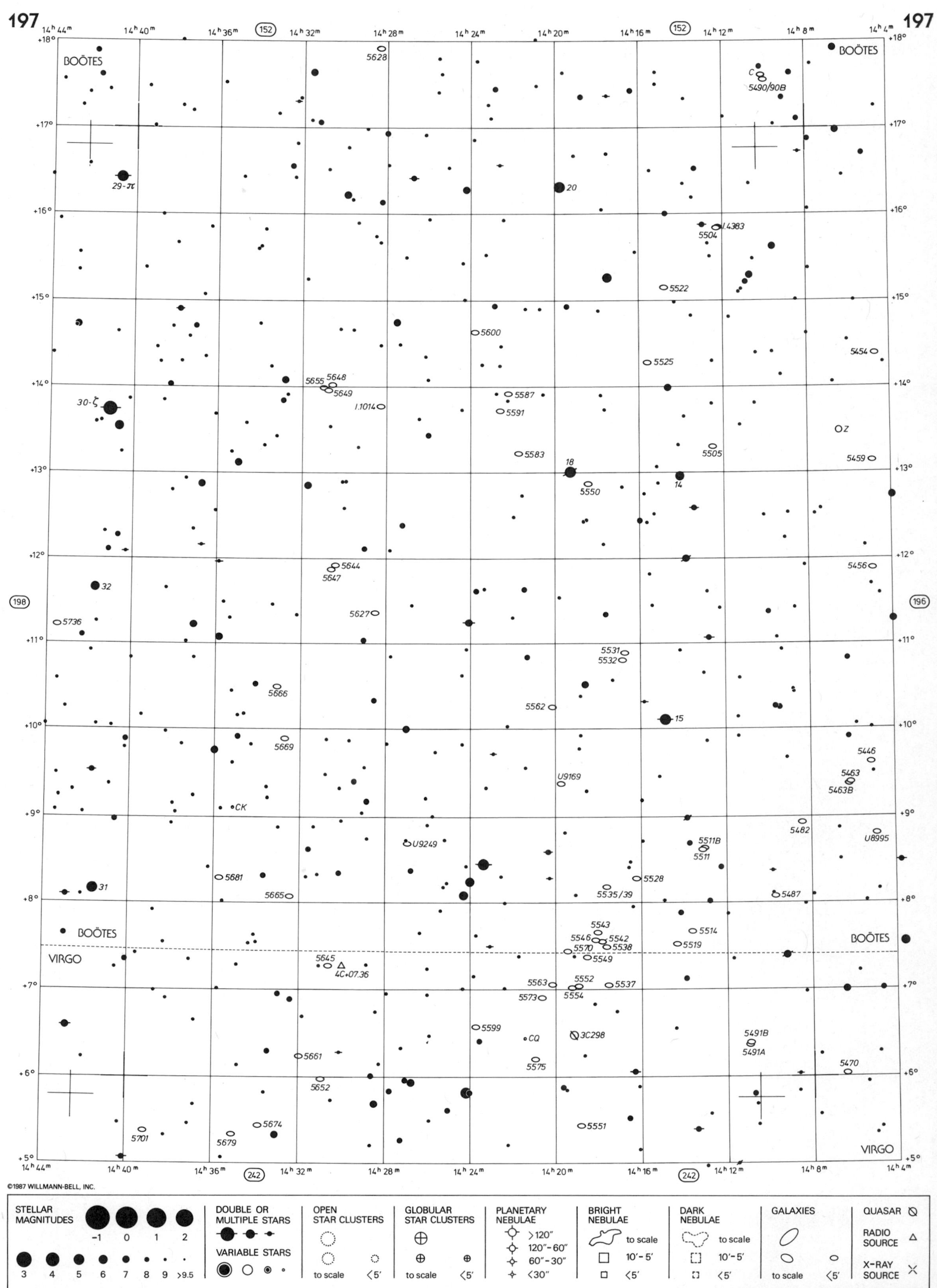

Barry Rappaport & Wil Tirion

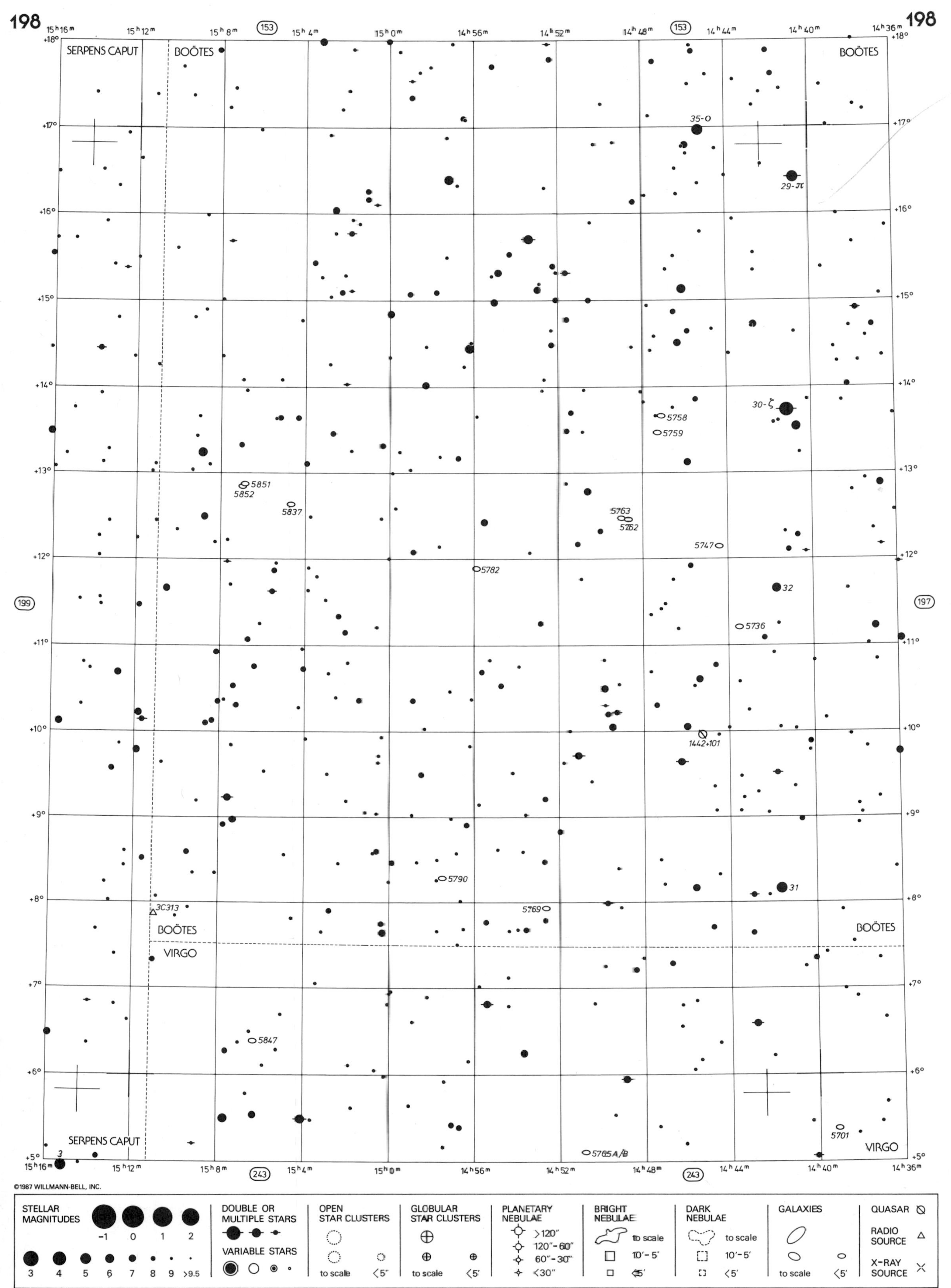

Barry Rappaport & Wil Tirion

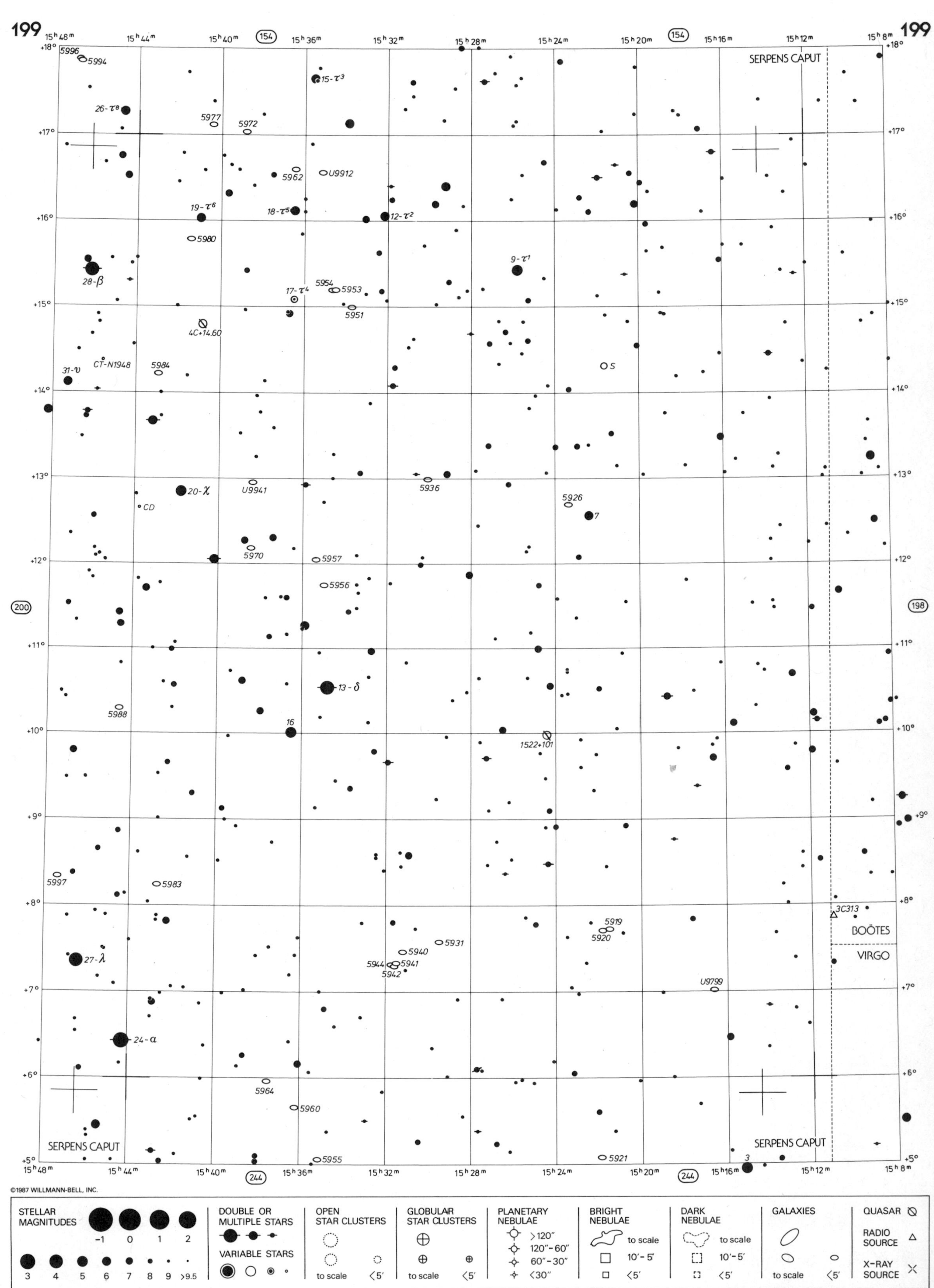

Barry Rappaport & Wil Tirion

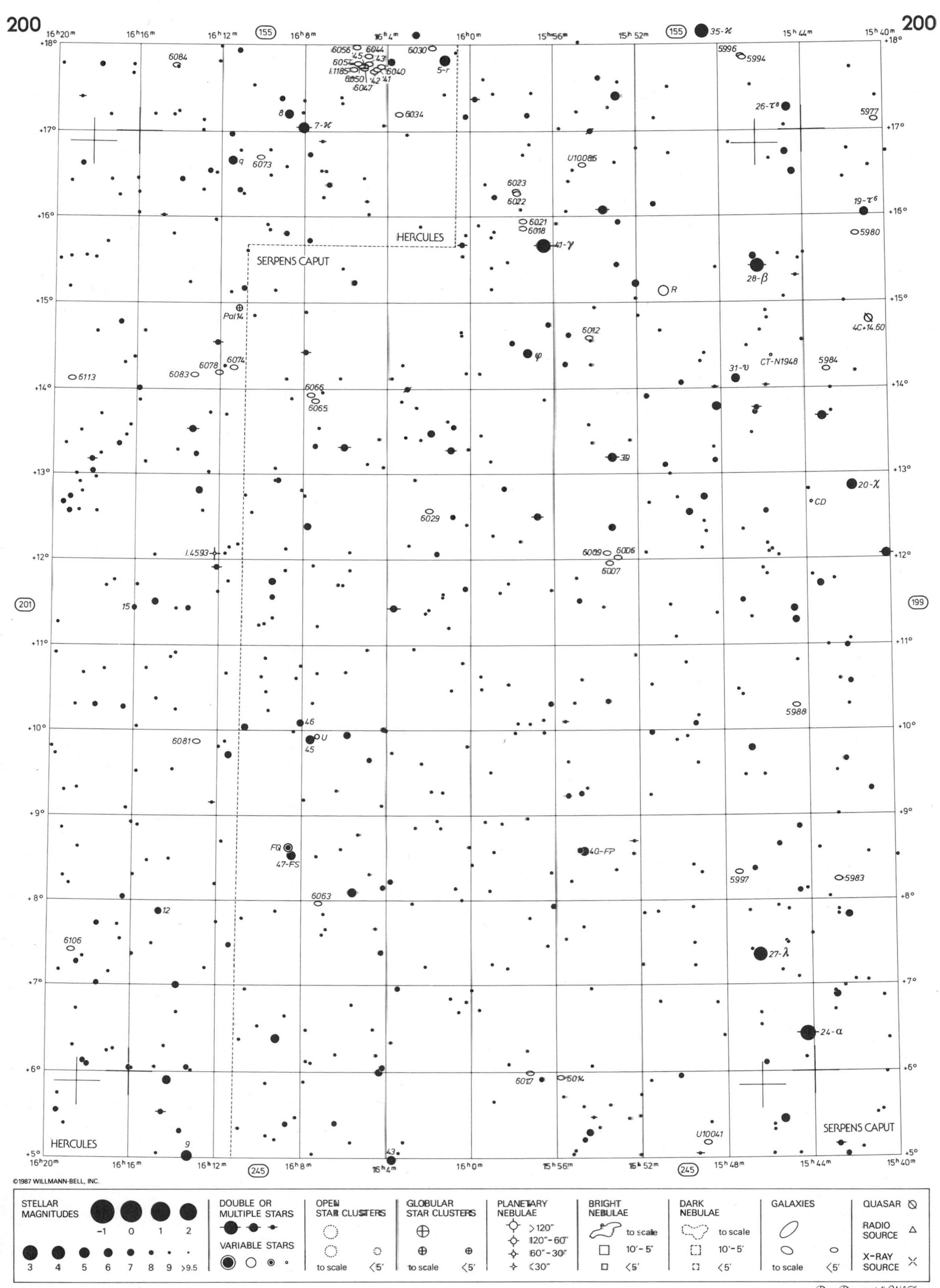

Barry Rappaport & Wil Tirion

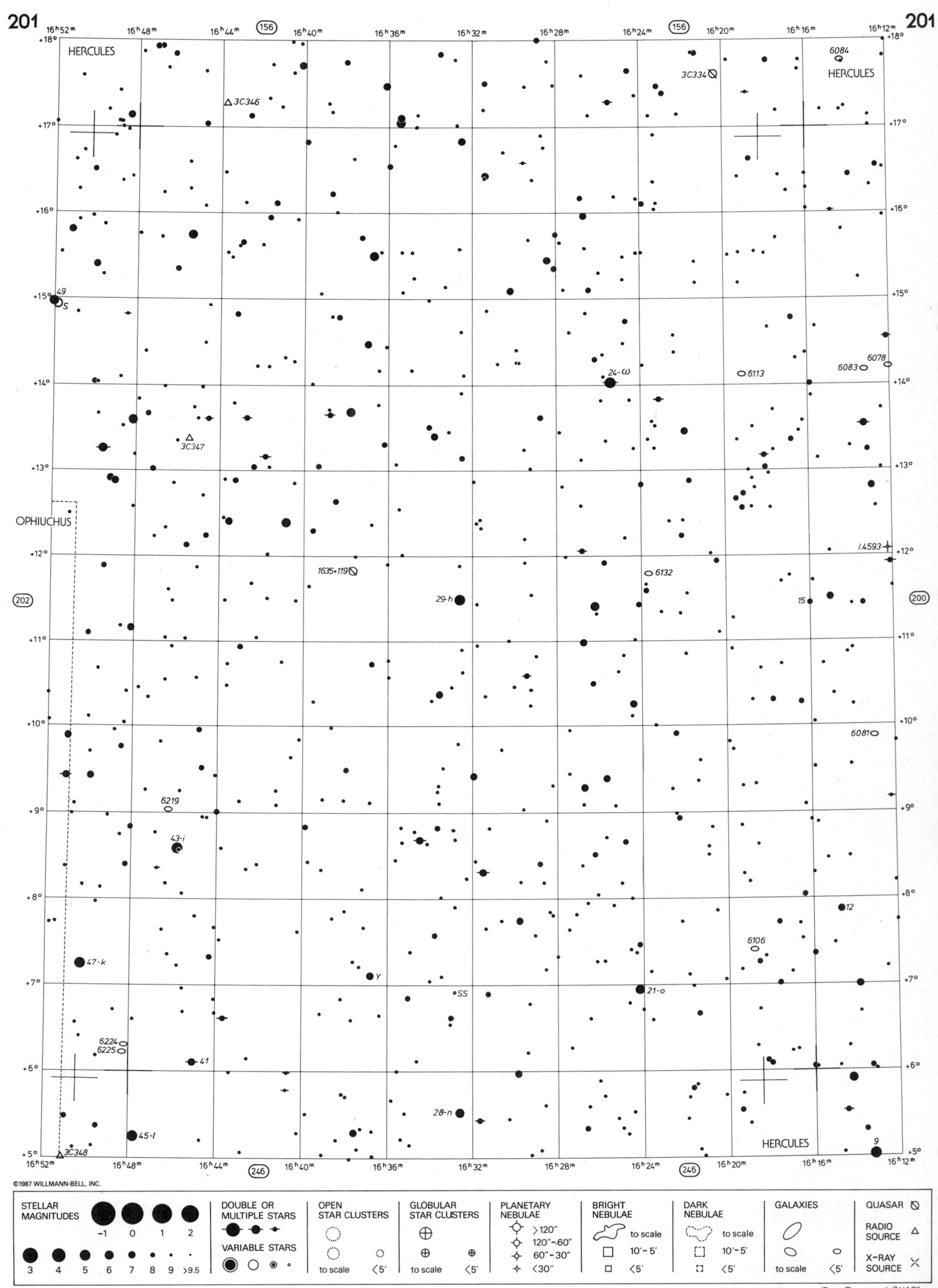

Barry Rappaport & Wil Tirion

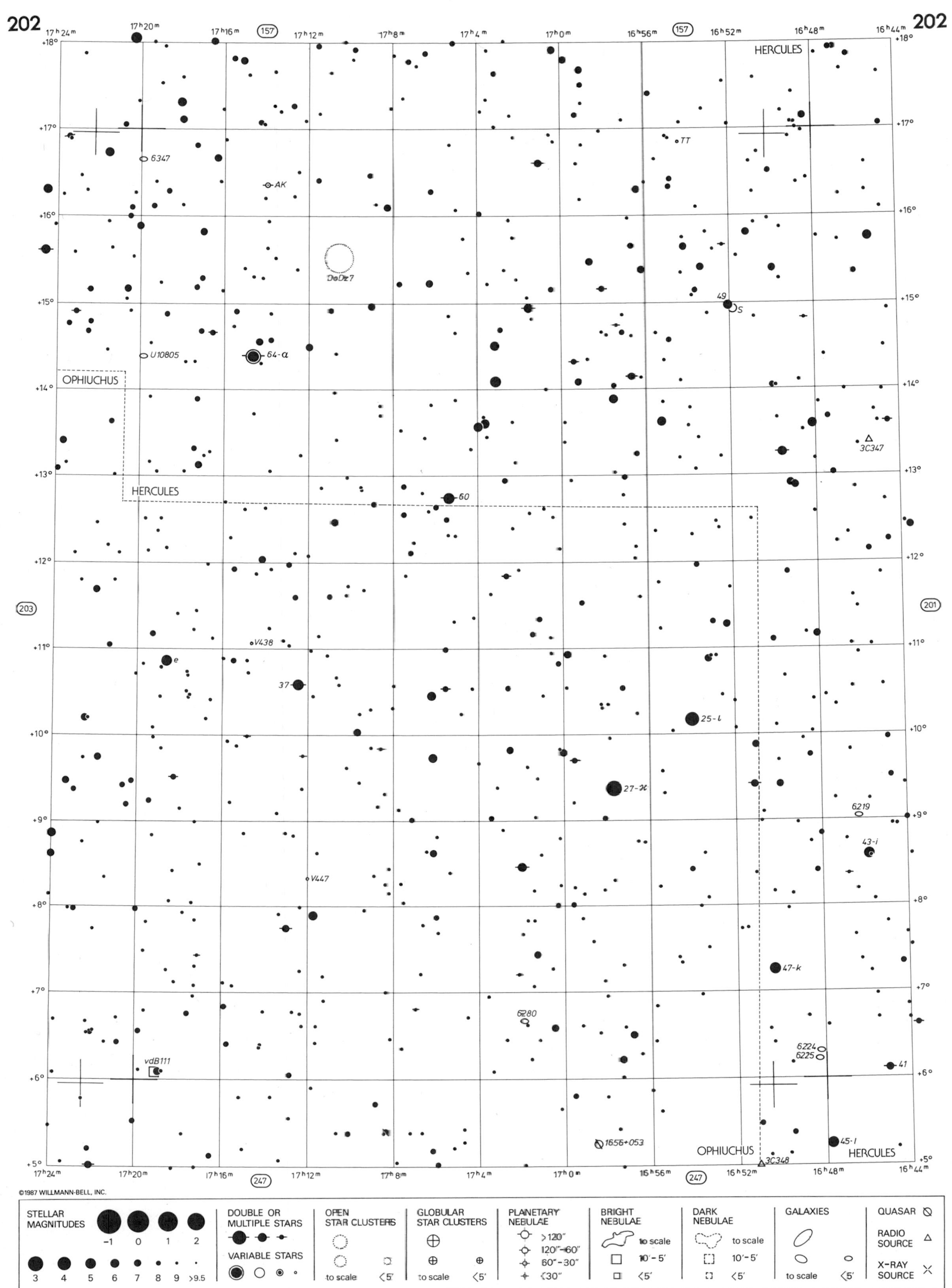

Barry Rappaport & Wil Tirion

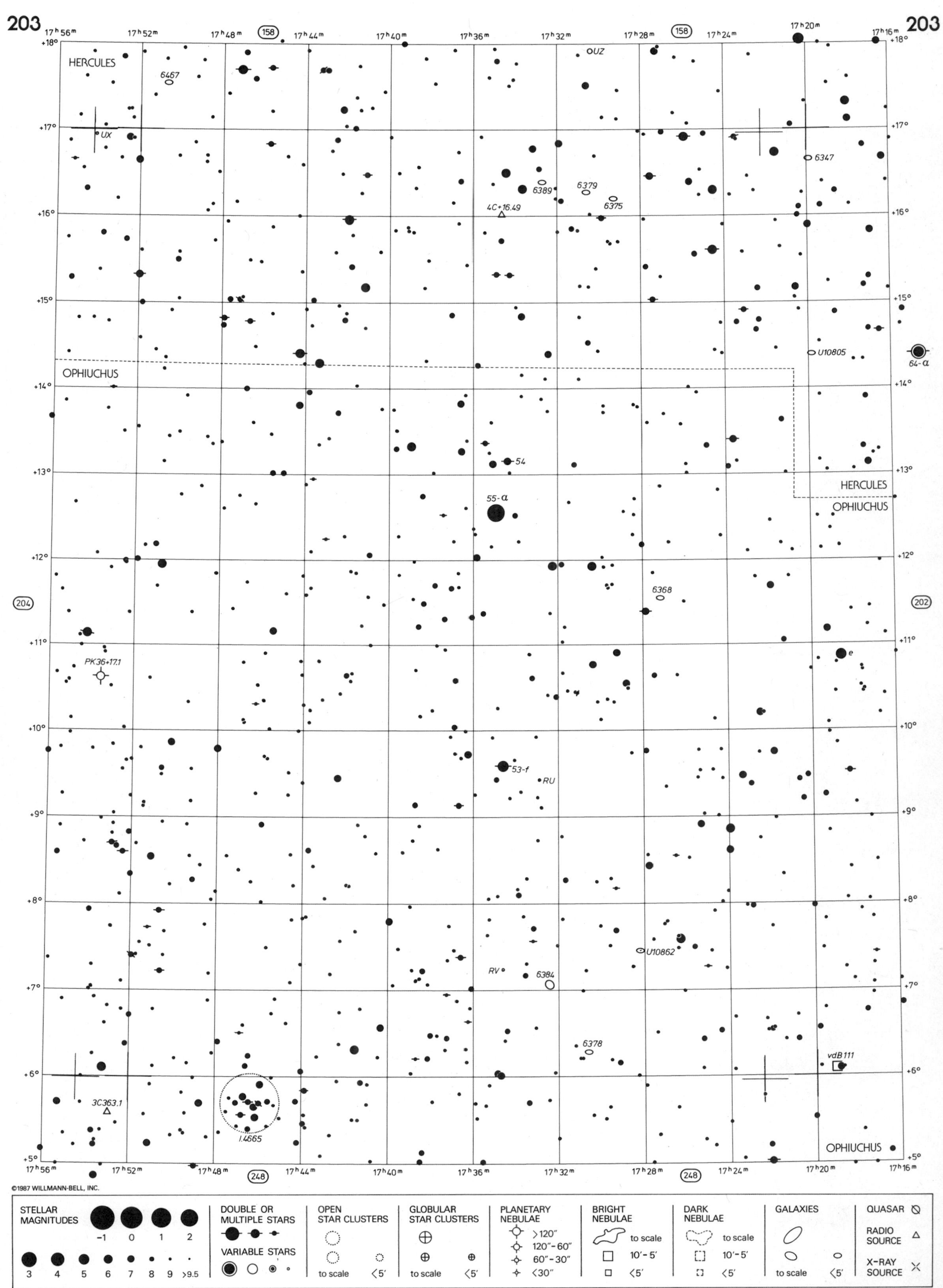

Barry Rappaport & Wil Tirion

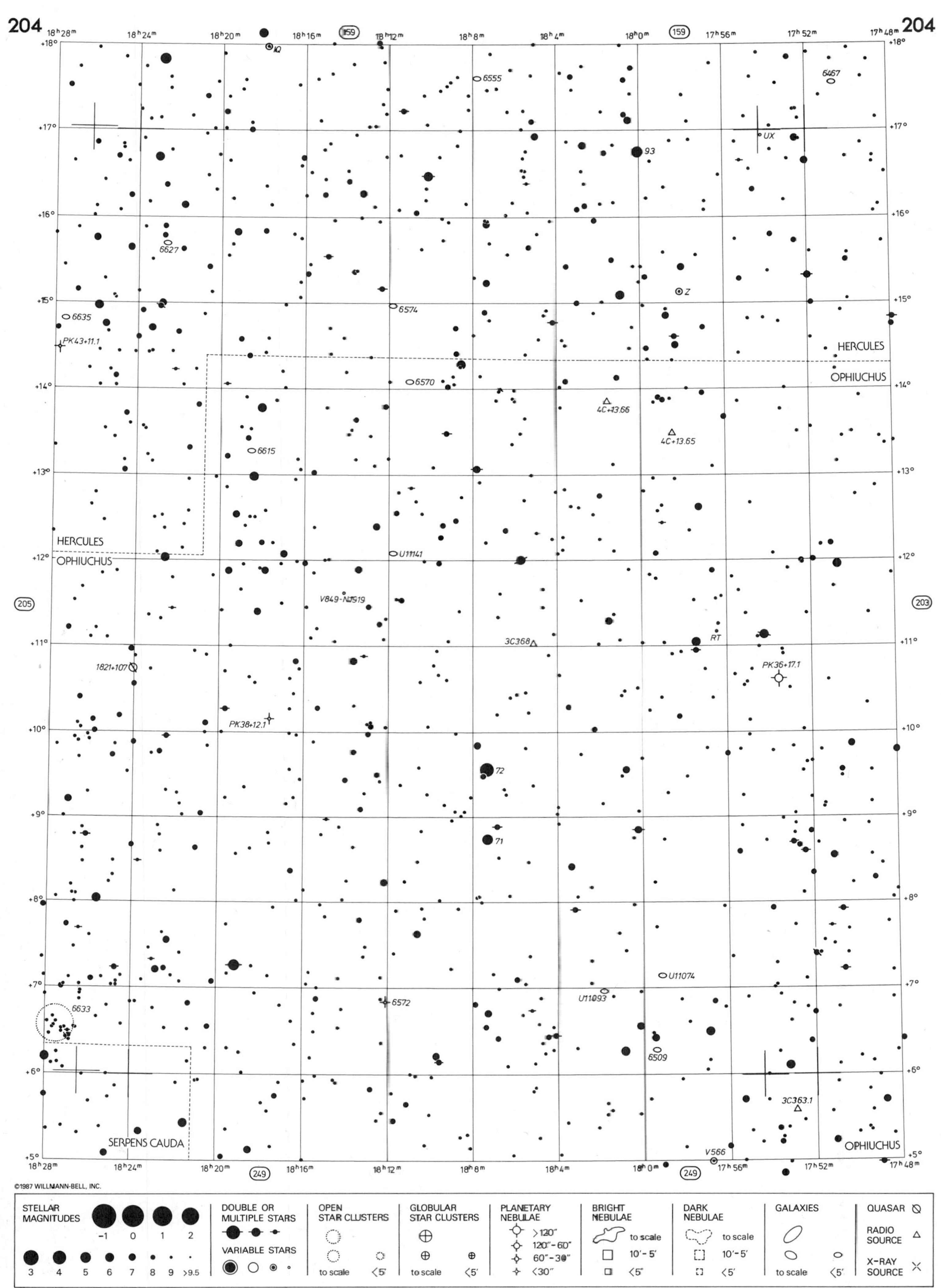

STELLAR MAGNITUDES: -1 0 1 2 3 4 5 6 7 8 9 >9.5
DOUBLE OR MULTIPLE STARS
VARIABLE STARS
OPEN STAR CLUSTERS: to scale, <5'
GLOBULAR STAR CLUSTERS: to scale, <5'
PLANETARY NEBULAE: >120", 120"-60", 60"-30", <30"
BRIGHT NEBULAE: to scale, 10'-5', <5'
DARK NEBULAE: to scale, 10'-5', <5'
GALAXIES: to scale, <5'
QUASAR
RADIO SOURCE
X-RAY SOURCE

Barry Rappaport & Wil Tirion

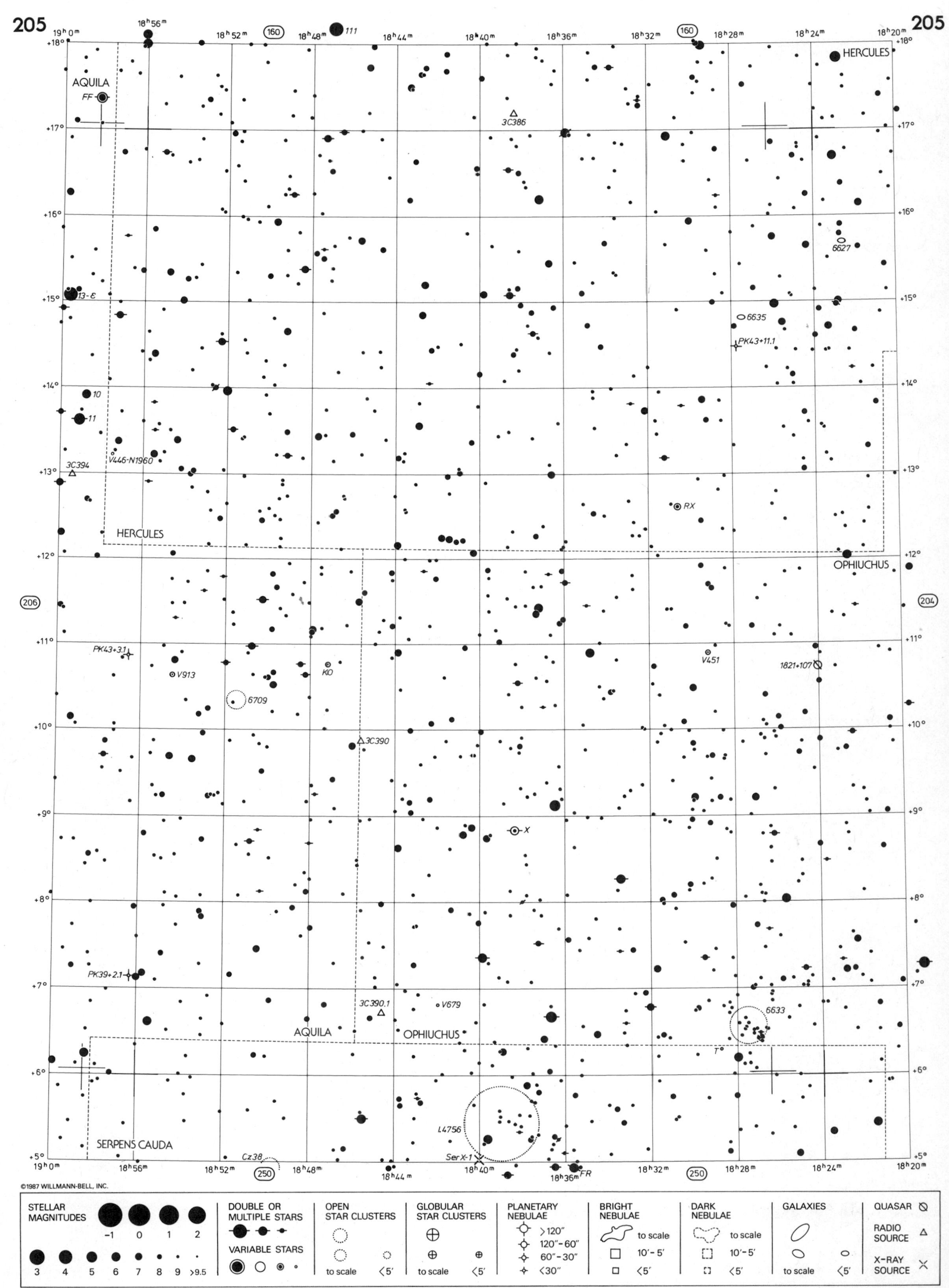

Barry Rappaport & Wil Tirion

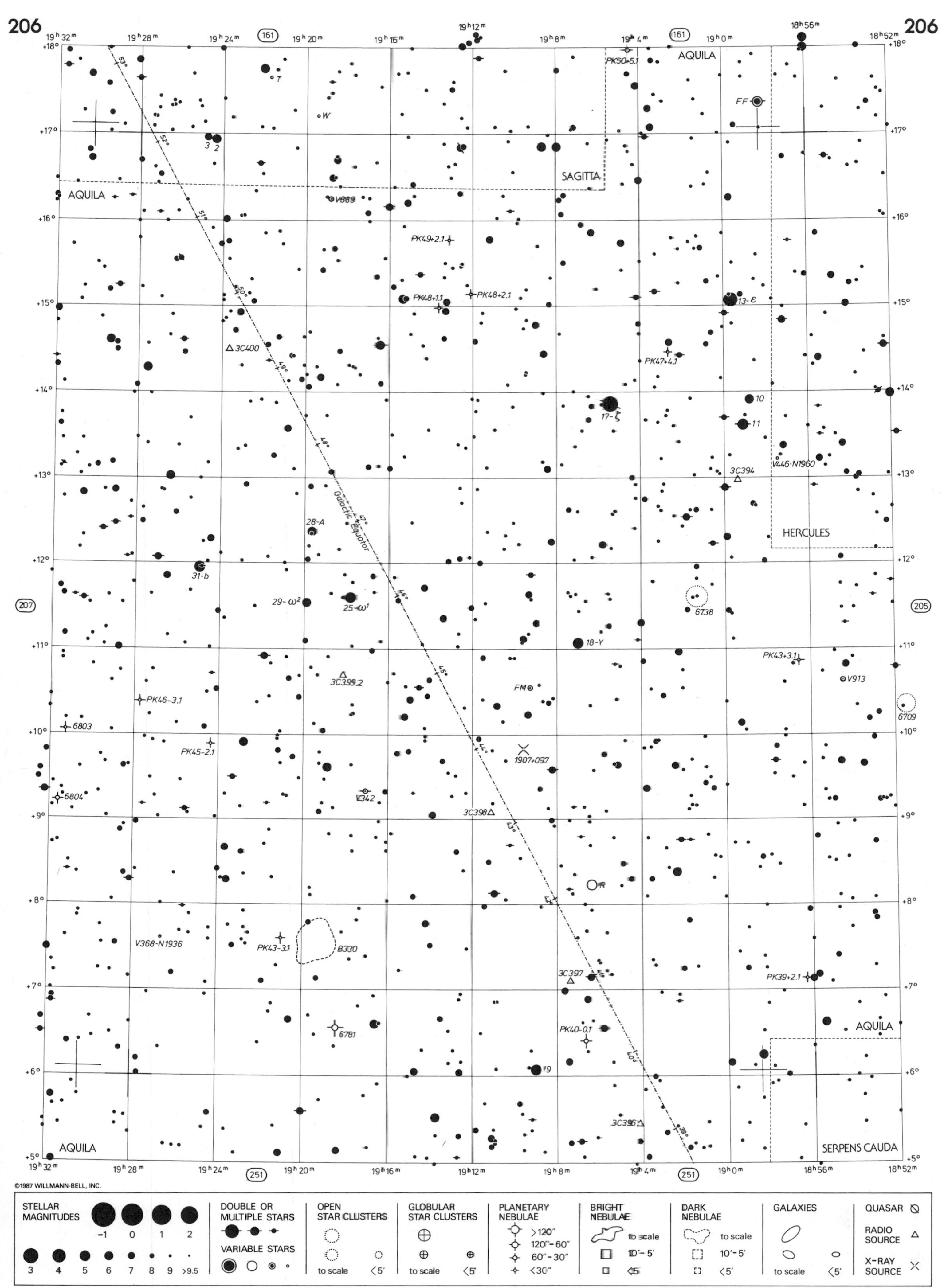

Barry Rappaport & Wil Tirion

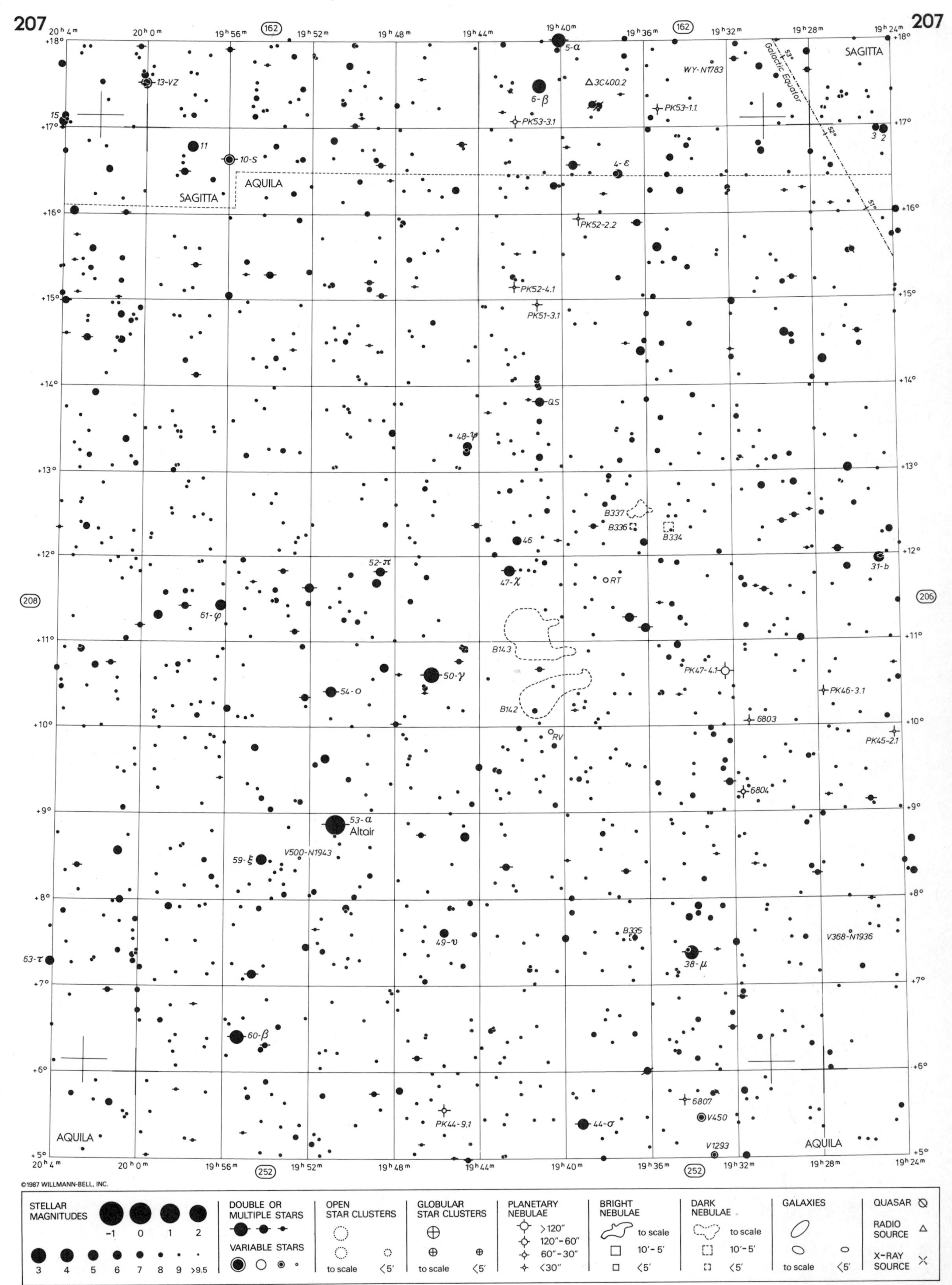

Barry Rappaport & Wil Tirion

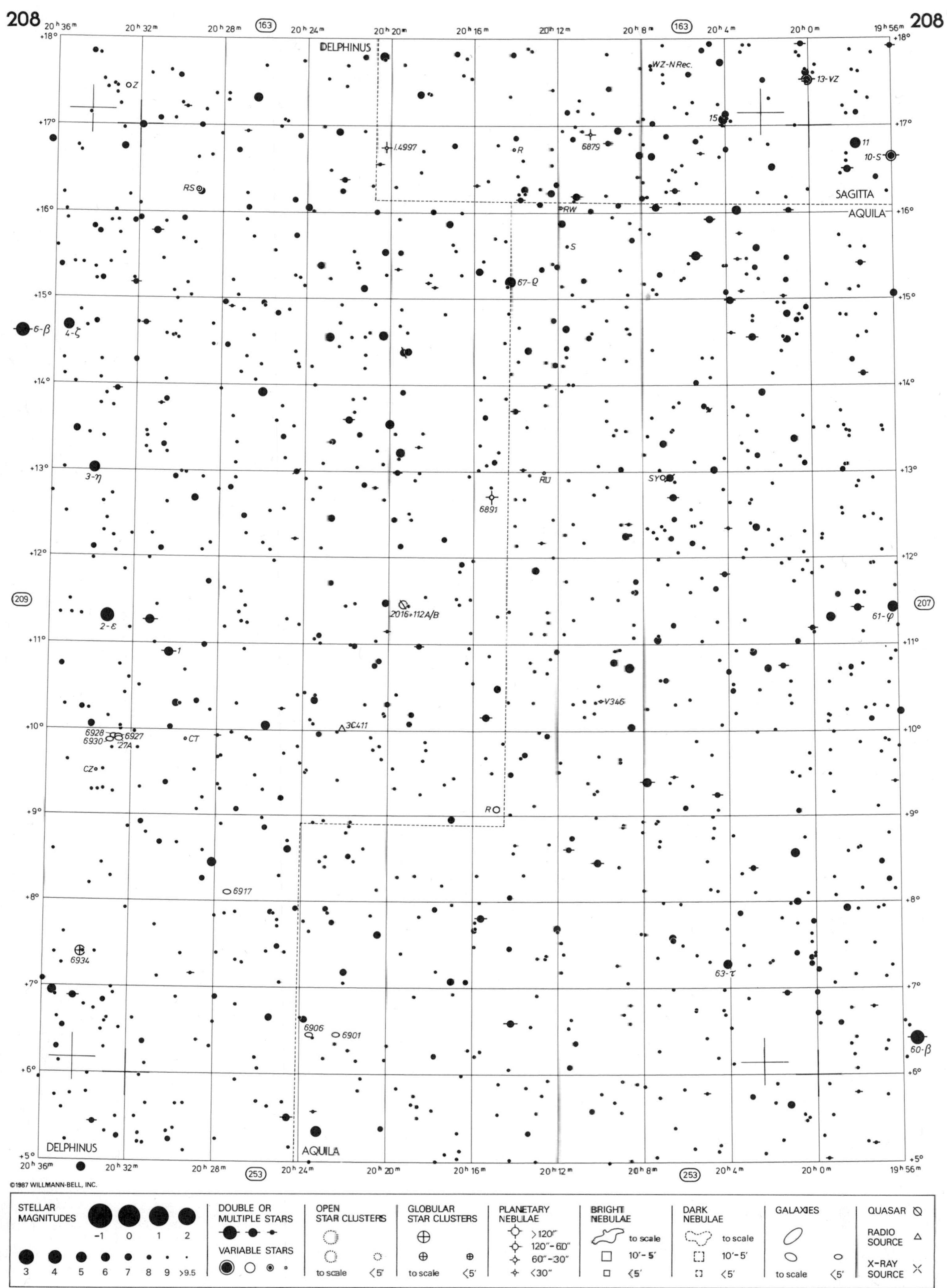

Barry Rappaport & Wil Tirion

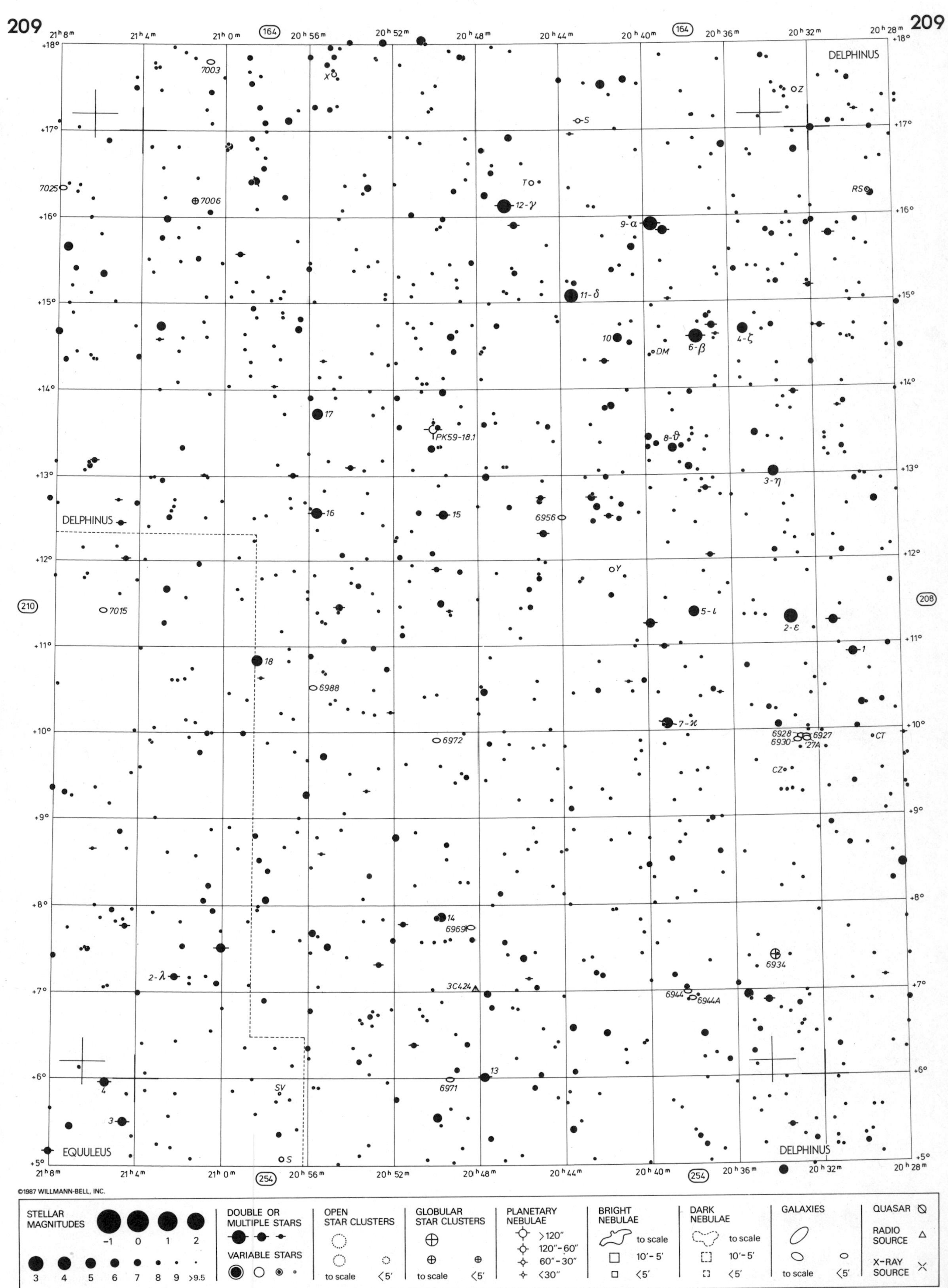

Barry Rappaport & Wil Tirion

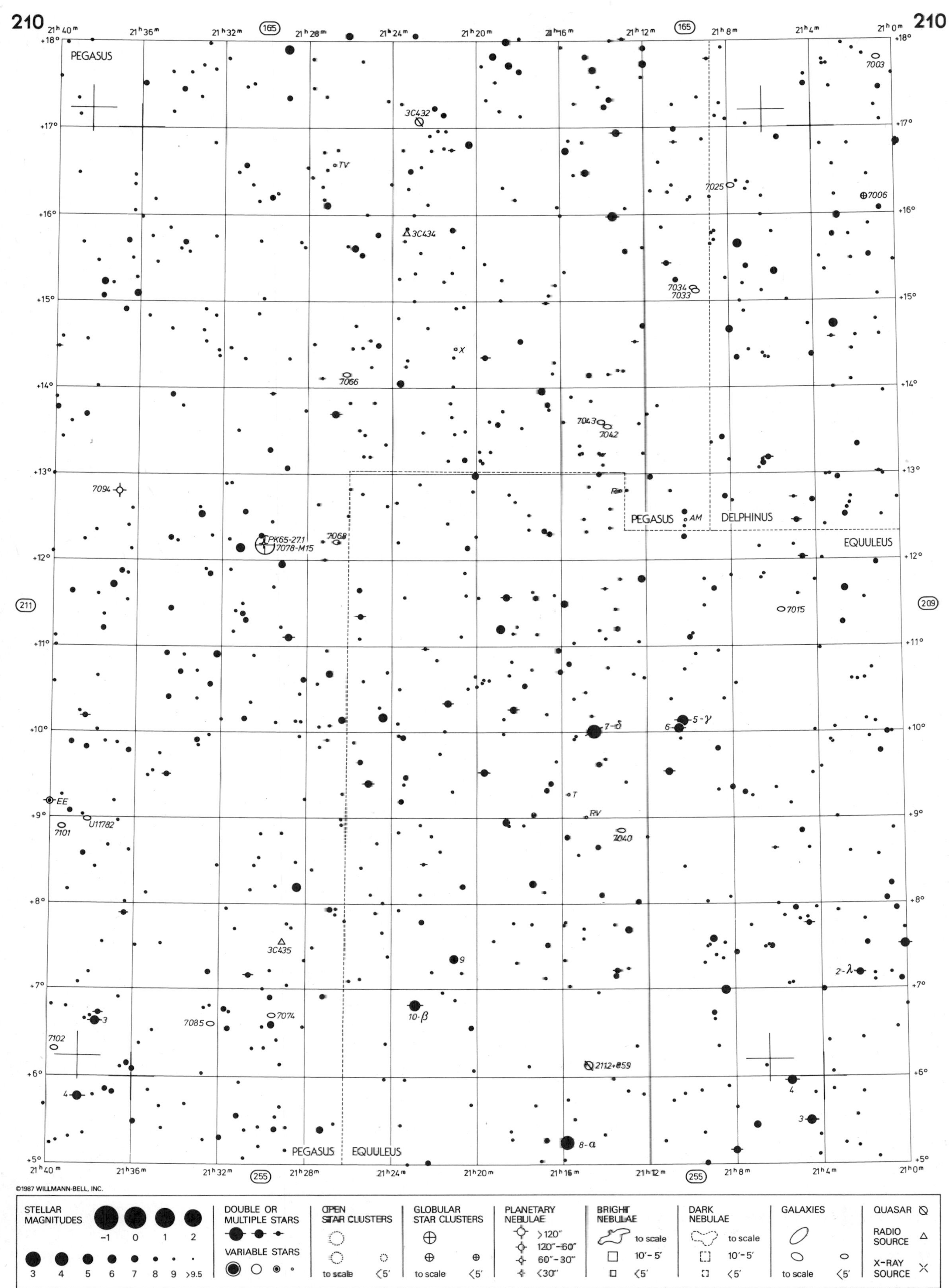

STELLAR MAGNITUDES −1 0 1 2 3 4 5 6 7 8 9 >9.5
DOUBLE OR MULTIPLE STARS
VARIABLE STARS
OPEN STAR CLUSTERS to scale <5′
GLOBULAR STAR CLUSTERS to scale <5′
PLANETARY NEBULAE >120″ 120″–60″ 60″–30″ <30″
BRIGHT NEBULAE to scale 10′–5′ <5′
DARK NEBULAE to scale 10′–5′ <5′
GALAXIES to scale <5′
QUASAR
RADIO SOURCE
X-RAY SOURCE

Barry Rappaport & Wil Tirion

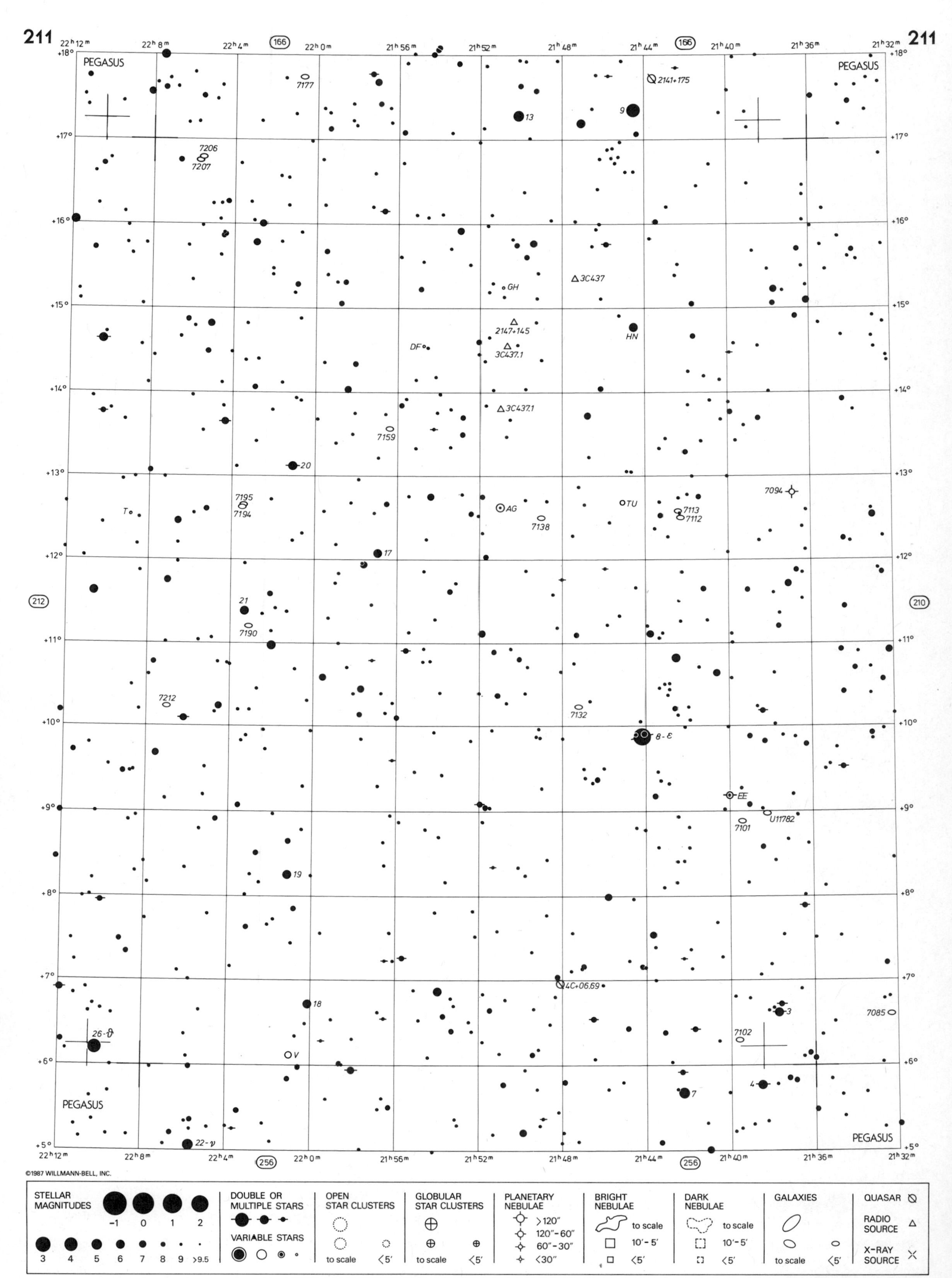
PEGASUS
22h12m
22h8m
22h4m
22h0m
21h56m
21h52m
21h48m
21h44m
21h40m
21h36m
21h32m
+18°
+17°
+16°
+15°
+14°
+13°
+12°
+11°
+10°
+9°
+8°
+7°
+6°
+5°
166
256
212
210
7177
2141+175
13
9
7206
7207
3C437
GH
2147+145
HN
DF
3C437.1
3C437.1
7159
20
7094
7195
7194
T
AG
7138
TU
7113
7112
17
21
7190
7212
7132
8-ε
EE
U11782
7101
19
4C+06.69
18
3
7085
26-θ
7102
V
4
7
22-ν
©1987 WILLMANN-BELL, INC.
STELLAR MAGNITUDES
-1 0 1 2
3 4 5 6 7 8 9 >9.5
DOUBLE OR MULTIPLE STARS
VARIABLE STARS
OPEN STAR CLUSTERS
to scale <5'
GLOBULAR STAR CLUSTERS
to scale <5'
PLANETARY NEBULAE
>120"
120"-60"
60"-30"
<30"
BRIGHT NEBULAE
to scale
10'-5'
<5'
DARK NEBULAE
to scale
10'-5'
<5'
GALAXIES
to scale <5'
QUASAR
RADIO SOURCE
X-RAY SOURCE
Barry Rappaport & Wil Tirion

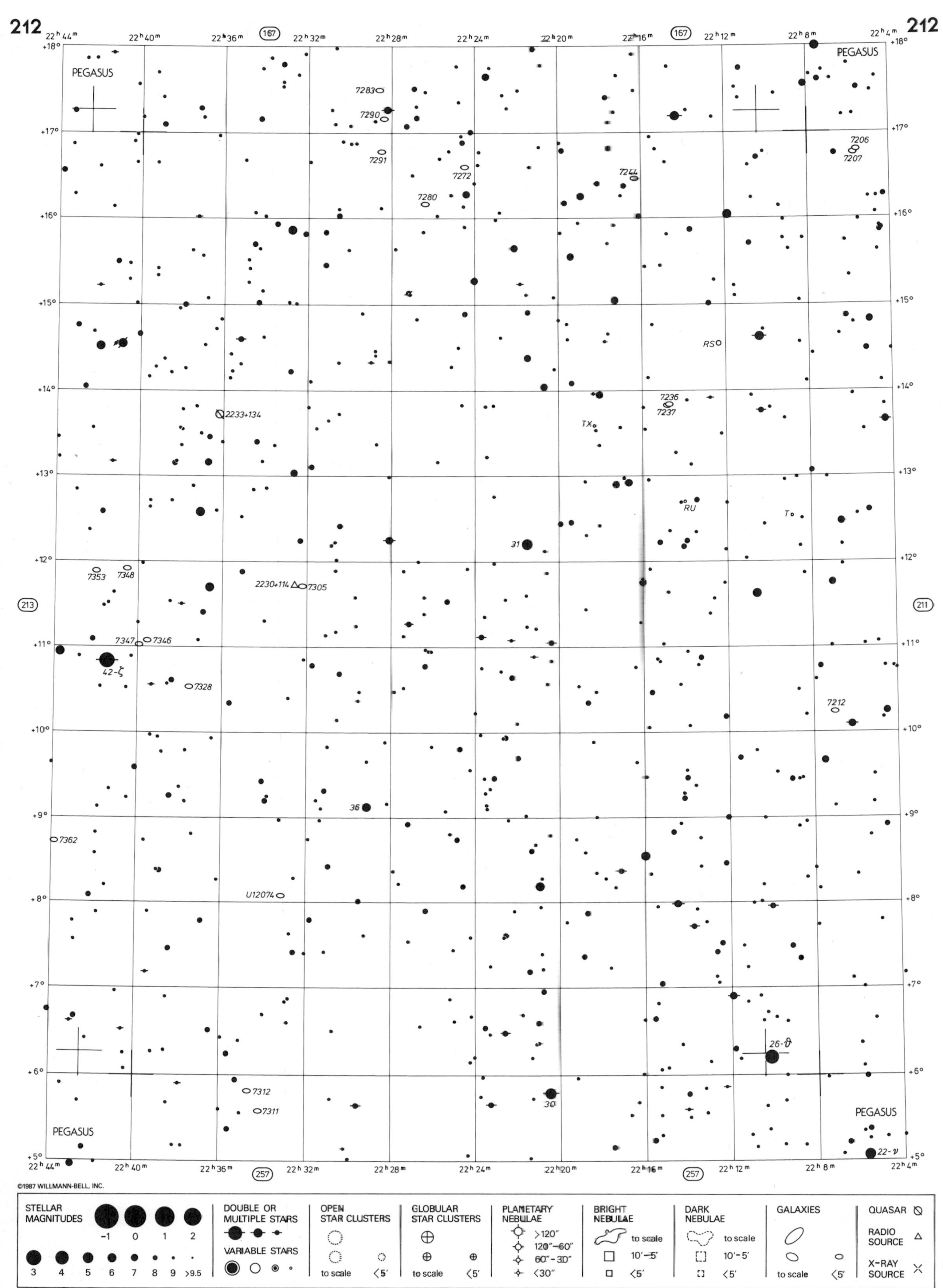
PEGASUS
22h 44m
22h 40m
22h 36m
22h 32m
22h 28m
22h 24m
22h 20m
22h 16m
22h 12m
22h 8m
22h 4m
+18°
+17°
+16°
+15°
+14°
+13°
+12°
+11°
+10°
+9°
+8°
+7°
+6°
+5°
167
257
213
211
7283
7290
7291
7272
7280
7244
7206
7207
RS
7236
7237
TX
2233+134
RU
T
31
7353
7348
2230+114
7305
7347
7346
42-ζ
7328
7212
36
7362
U12074
26-θ
7312
7311
30
22-ν
©1987 WILLMANN-BELL, INC.
STELLAR MAGNITUDES
-1
0
1
2
3
4
5
6
7
8
9
>9.5
DOUBLE OR MULTIPLE STARS
VARIABLE STARS
OPEN STAR CLUSTERS
to scale
<5'
GLOBULAR STAR CLUSTERS
to scale
<5'
PLANETARY NEBULAE
>120"
120"–60"
60"–30"
<30"
BRIGHT NEBULAE
to scale
10'–5'
<5'
DARK NEBULAE
to scale
10'–5'
<5'
GALAXIES
to scale
<5'
QUASAR
RADIO SOURCE
X-RAY SOURCE
Barry Rappaport & Wil Tirion

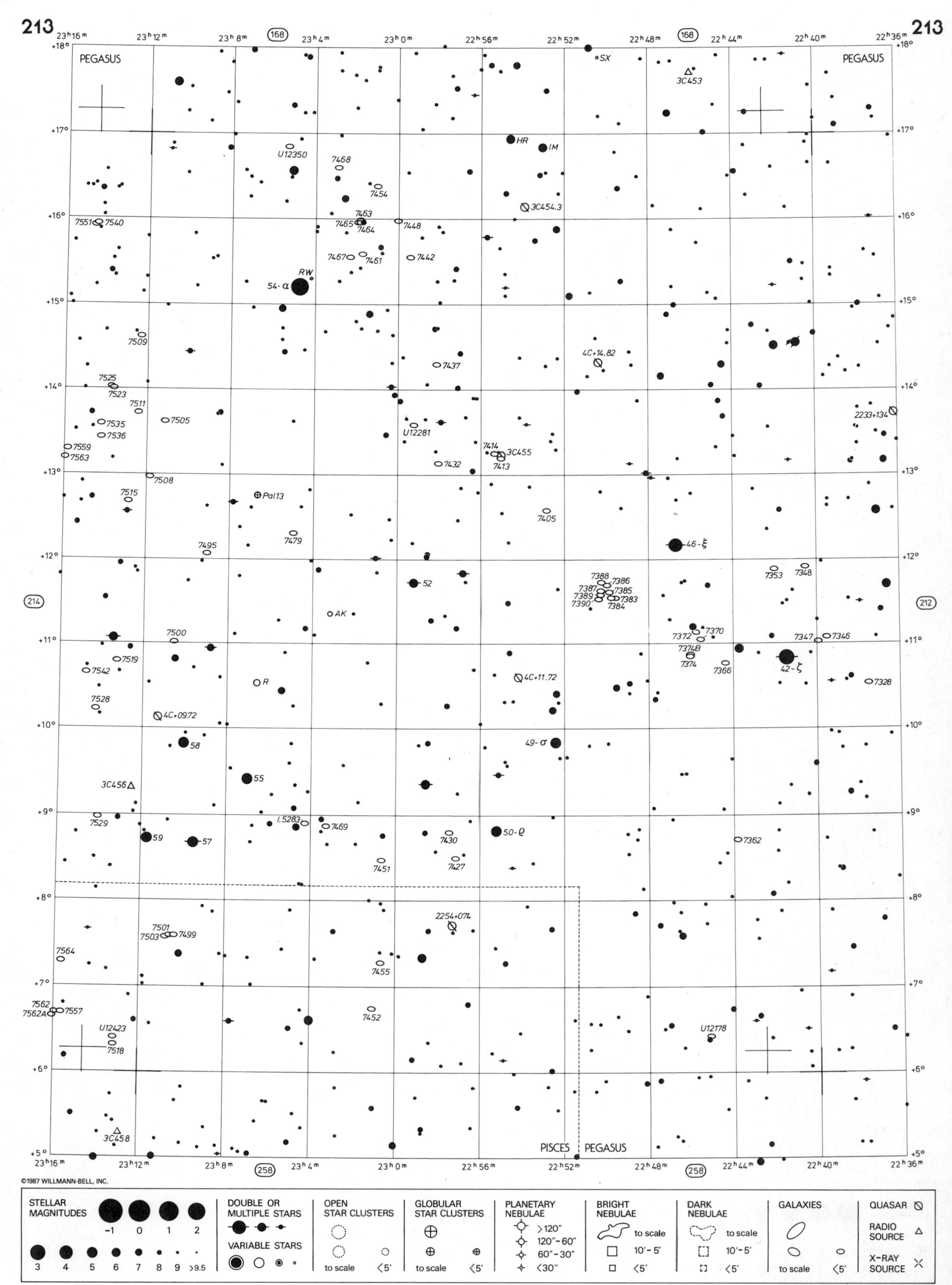

213
213
23h16m
23h12m
23h8m
23h4m
23h0m
22h56m
22h52m
22h48m
22h44m
22h40m
22h36m
168
168
258
258
214
212
+18°
+17°
+16°
+15°
+14°
+13°
+12°
+11°
+10°
+9°
+8°
+7°
+6°
+5°
PEGASUS
PEGASUS
PISCES
PEGASUS
SX
3C453
HR
IM
U12350
7468
7454
3C454.3
7551
7540
7463
7465
7464
7448
7467
7461
7442
RW
54-α
7509
4C+14.82
7437
7525
7523
7511
7505
7535
7536
2233+134
U12281
7559
7563
7414
3C455
7413
7432
7508
7515
Pal13
7405
7495
7479
46-ξ
7353
7348
52
7388
7386
7387
7385
7389
7383
7390
7384
AK
7500
7372
7370
7347
7346
7374B
7519
7374
7366
42-ζ
7542
R
4C+11.72
7328
7528
4C+09.72
58
49-σ
55
3C456
7529
I.5283
7469
59
57
7430
50-ρ
7362
7451
7427
2254+074
7501
7503
7499
7564
7455
7562
7562A
7557
7452
U12423
U12178
7518
3C458
©1987 WILLMANN-BELL, INC.
STELLAR MAGNITUDES
-1
0
1
2
3
4
5
6
7
8
9
>9.5
DOUBLE OR MULTIPLE STARS
VARIABLE STARS
OPEN STAR CLUSTERS
to scale
<5'
GLOBULAR STAR CLUSTERS
to scale
<5'
PLANETARY NEBULAE
>120"
120"-60"
60"-30"
<30"
BRIGHT NEBULAE
to scale
10'-5'
<5'
DARK NEBULAE
to scale
10'-5'
<5'
GALAXIES
to scale
<5'
QUASAR
RADIO SOURCE
X-RAY SOURCE
Barry Rappaport & Wil Tirion

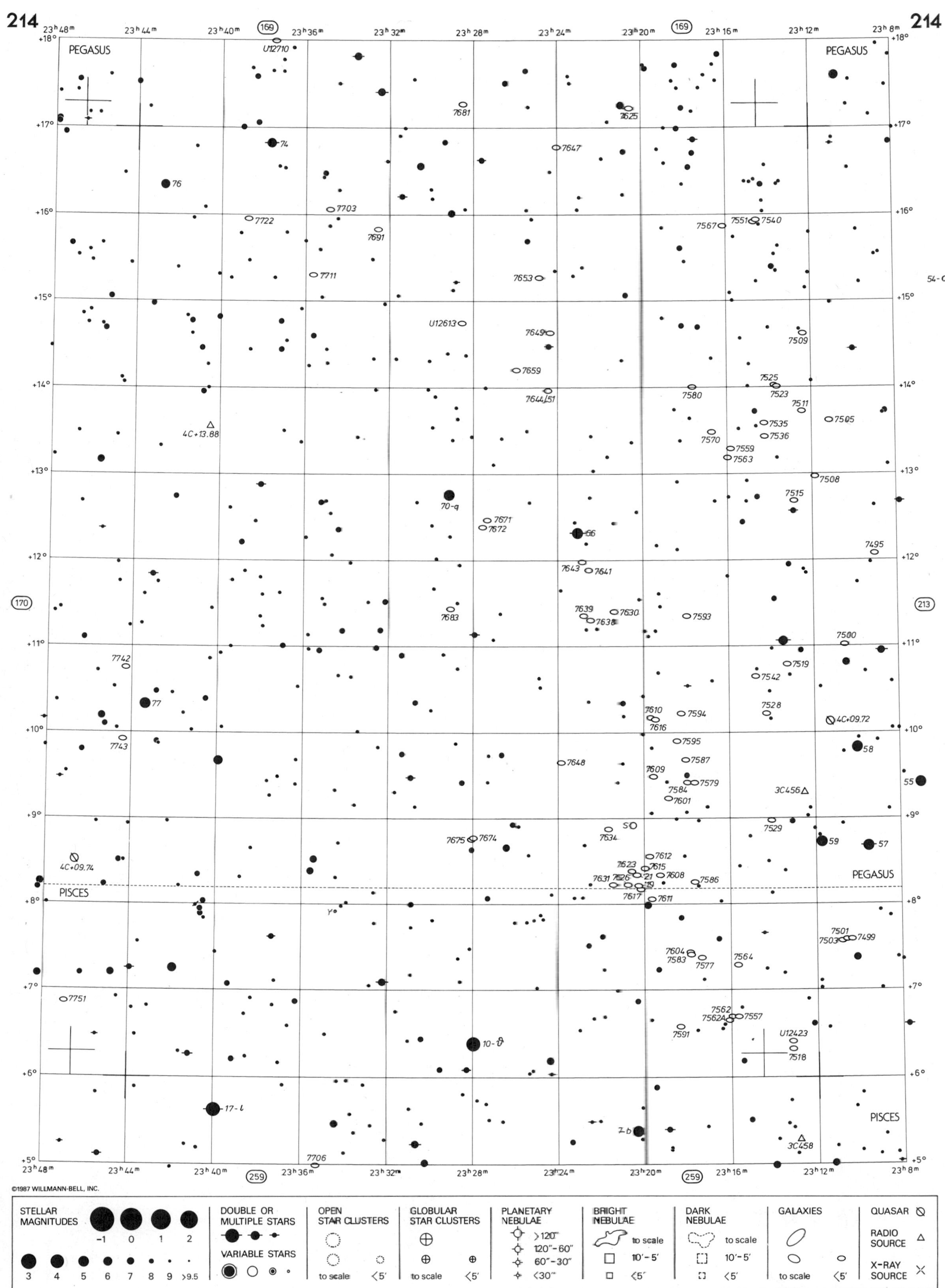
PEGASUS
PISCES
23h48m
23h44m
23h40m
23h36m
23h32m
23h28m
23h24m
23h20m
23h16m
23h12m
23h8m
+18°
+17°
+16°
+15°
+14°
+13°
+12°
+11°
+10°
+9°
+8°
+7°
+6°
+5°
169
170
213
259
54-α
70-q
10-ϑ
17-ι
©1987 WILLMANN-BELL, INC.
STELLAR MAGNITUDES
-1 0 1 2
3 4 5 6 7 8 9 >9.5
DOUBLE OR MULTIPLE STARS
VARIABLE STARS
OPEN STAR CLUSTERS
to scale
<5'
GLOBULAR STAR CLUSTERS
to scale
<5'
PLANETARY NEBULAE
>120"
120"-60"
60"-30"
<30"
BRIGHT NEBULAE
to scale
10'-5'
<5'
DARK NEBULAE
to scale
10'-5'
<5'
GALAXIES
to scale
<5'
QUASAR
RADIO SOURCE
X-RAY SOURCE
Barry Rappaport & Wil Tirion

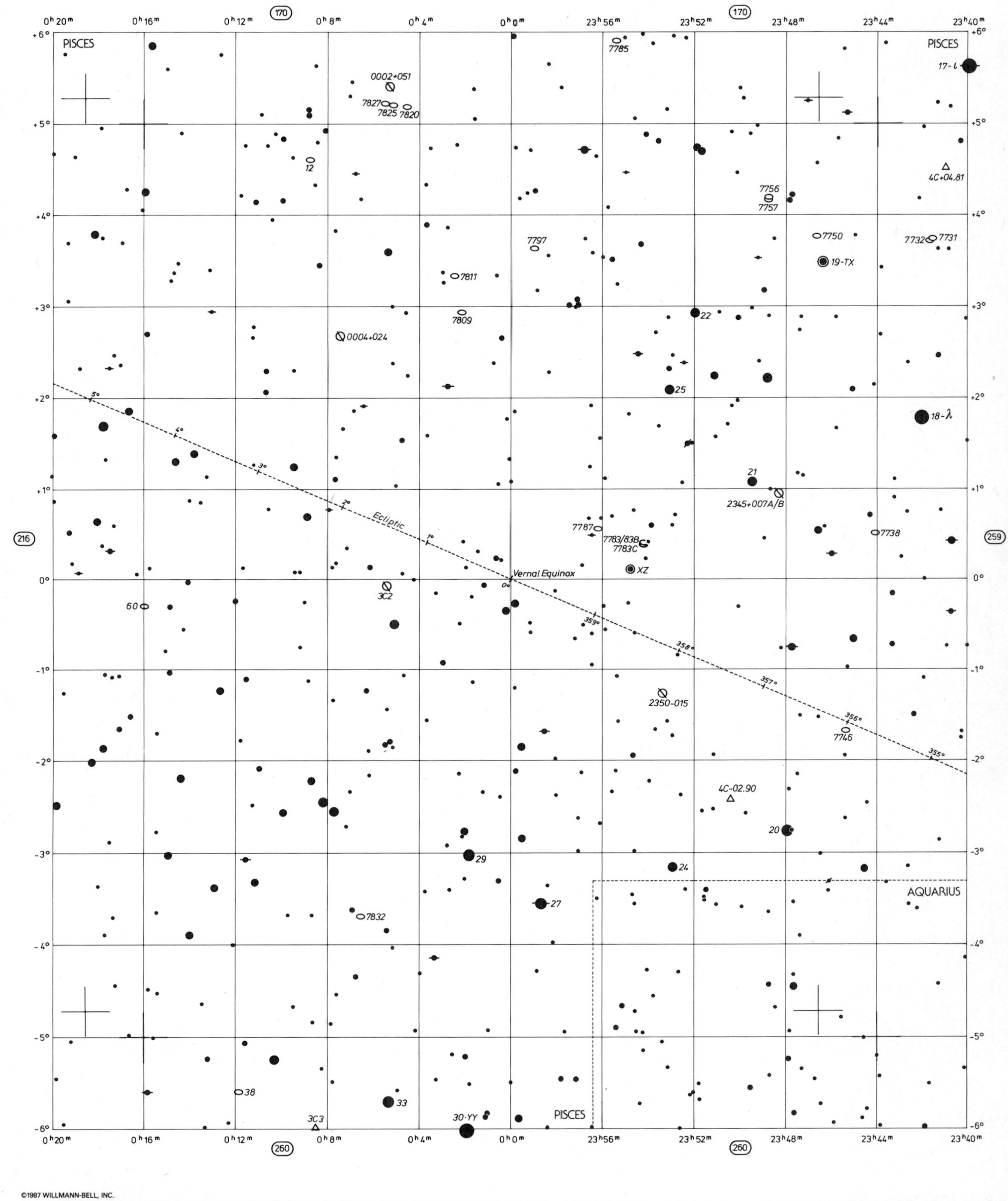

STELLAR MAGNITUDES	DOUBLE OR MULTIPLE STARS	OPEN STAR CLUSTERS	GLOBULAR STAR CLUSTERS	PLANETARY NEBULAE	BRIGHT NEBULAE	DARK NEBULAE	GALAXIES	QUASAR
-1 0 1 2	VARIABLE STARS	to scale <5′	to scale <5′	>120″ 120″-60″ 60″-30″ <30″	to scale 10′-5′ <5′	to scale 10′-5′ <5′	to scale <5′	RADIO SOURCE
3 4 5 6 7 8 9 >9.5								X-RAY SOURCE

Barry Rappaport & Wil Tirion

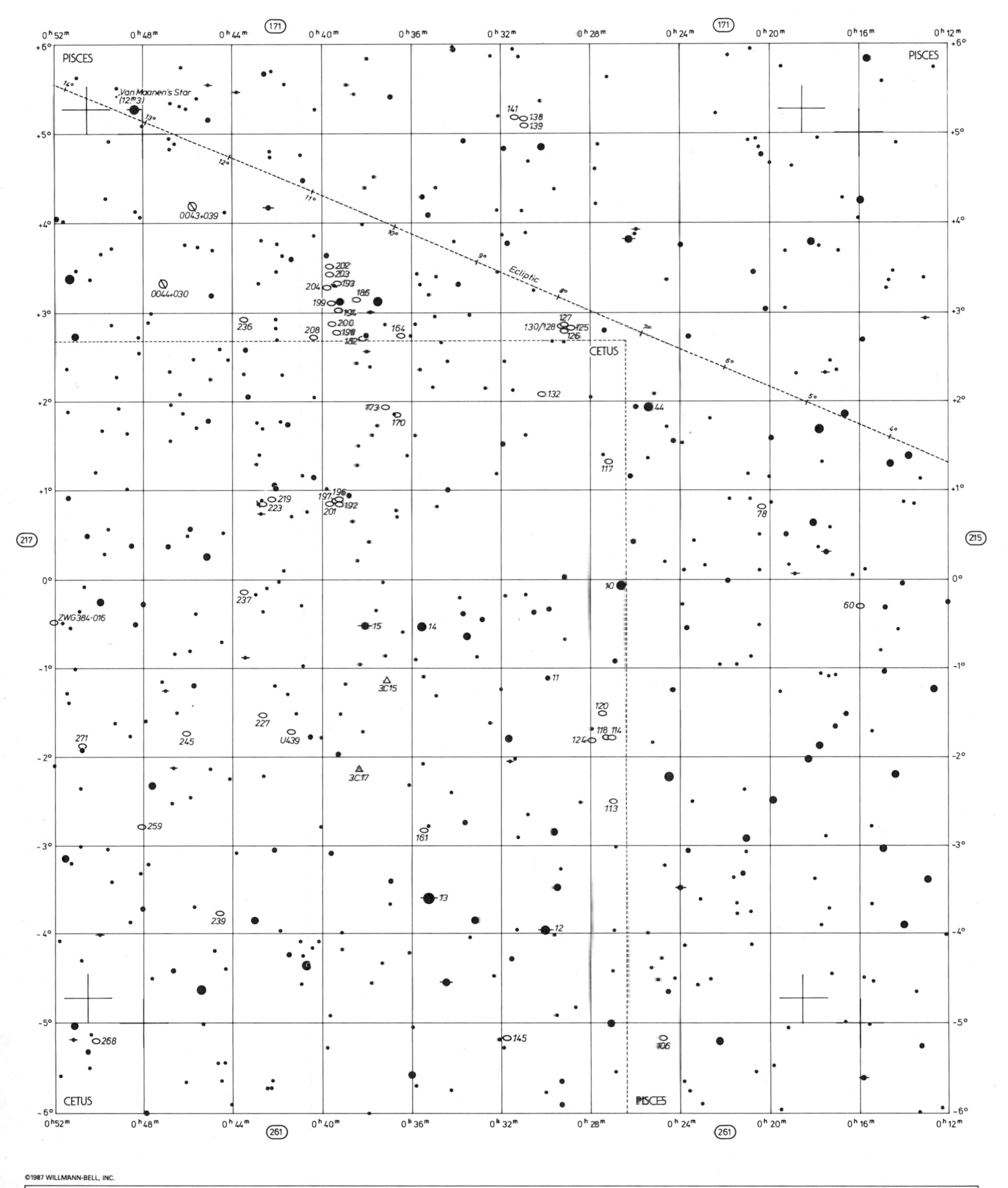

STELLAR MAGNITUDES	DOUBLE OR MULTIPLE STARS	OPEN STAR CLUSTERS	GLOBULAR STAR CLUSTERS	PLANETARY NEBULAE	BRIGHT NEBULAE	DARK NEBULAE	GALAXIES	QUASAR
-1 0 1 2	VARIABLE STARS	to scale <5'	to scale <5'	>120" 120"-60" 60"-30" <30"	to scale 10'-5' <5'	to scale 10'-5' <5'	to scale <5'	RADIO SOURCE
3 4 5 6 7 8 9 >9.5								X-RAY SOURCE

Barry Rappaport & Wil Tirion

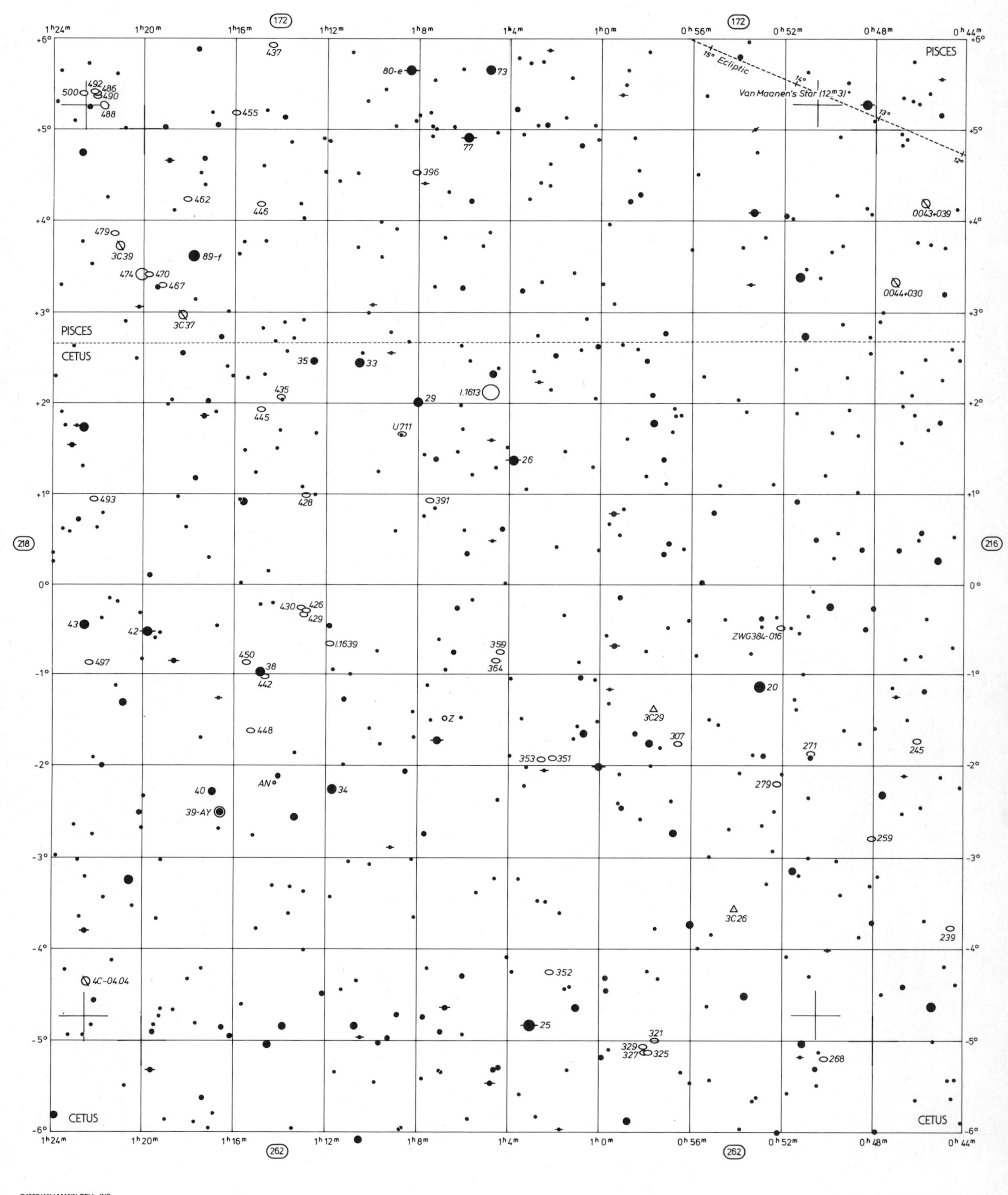

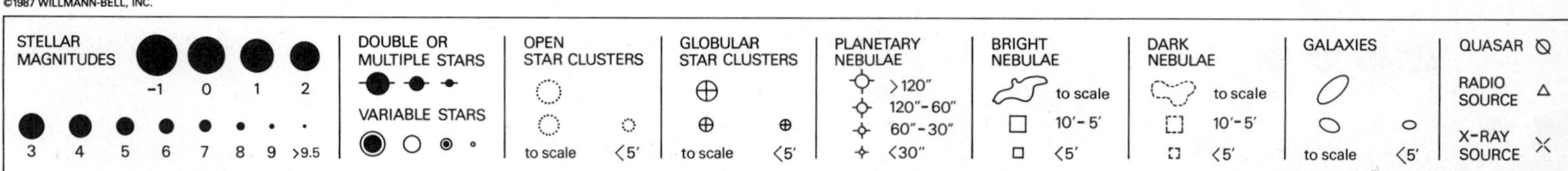

Barry Rappaport & Wil Tirion

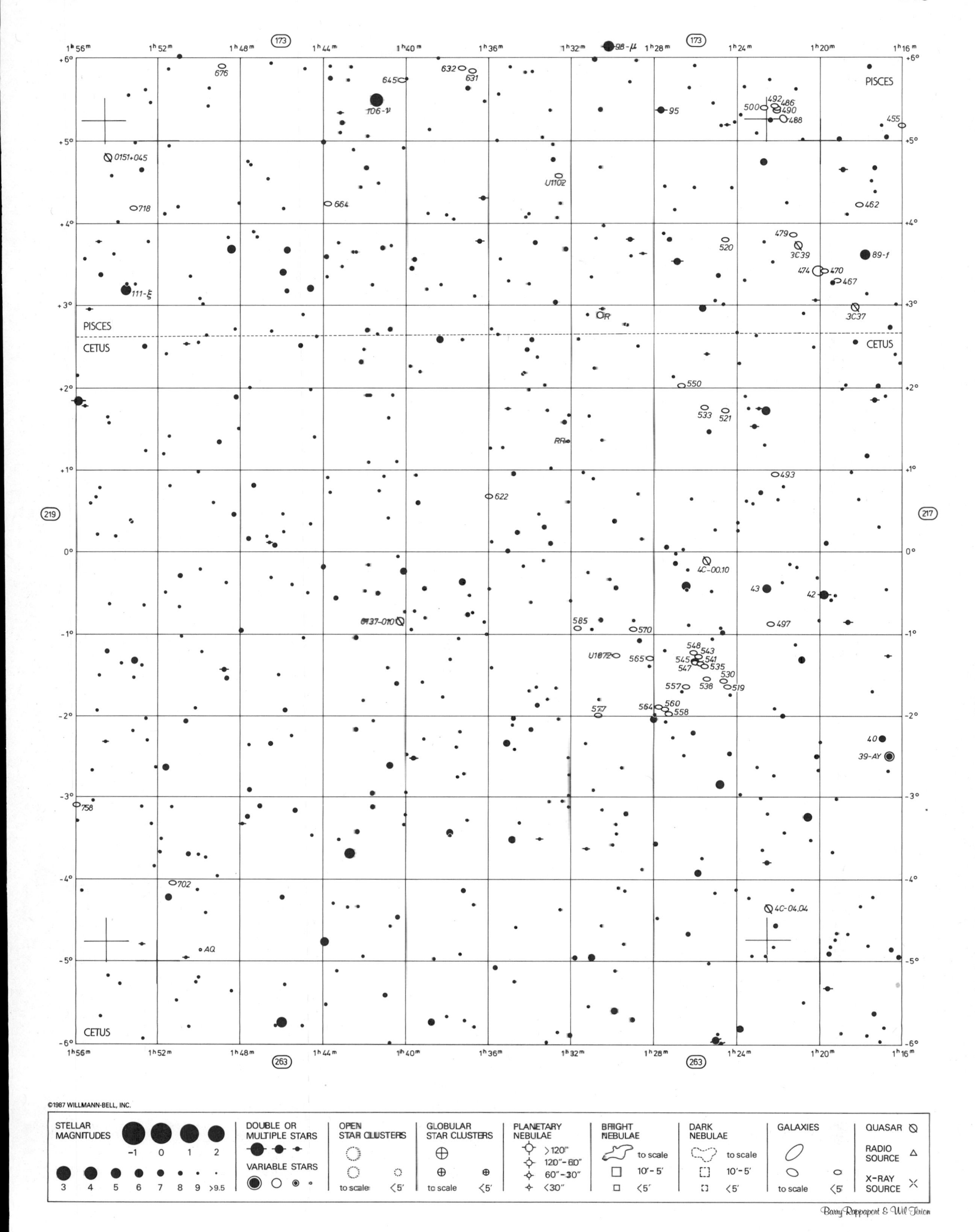

Barry Rappaport & Wil Tirion

219

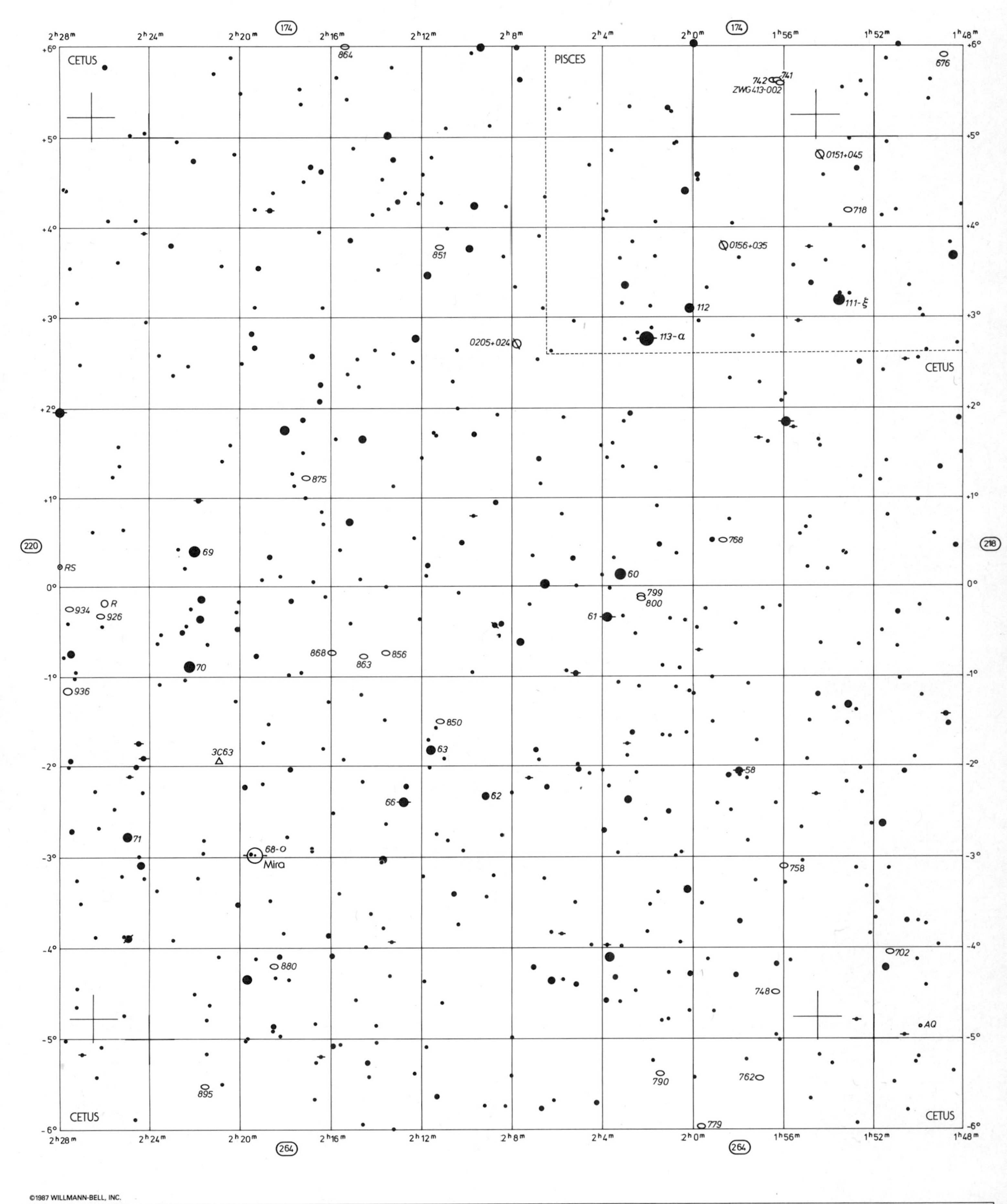

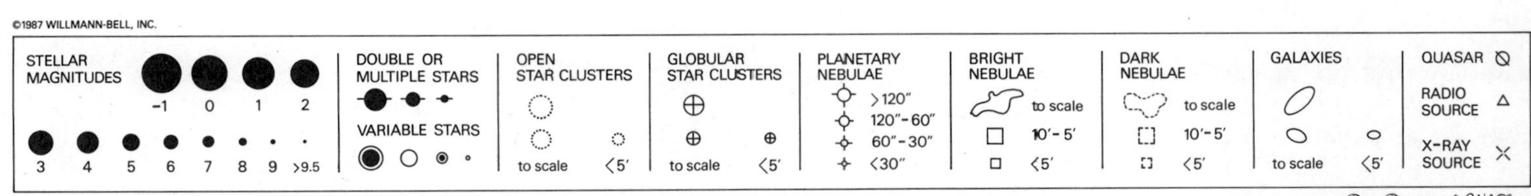

Barry Rappaport & Wil Tirion

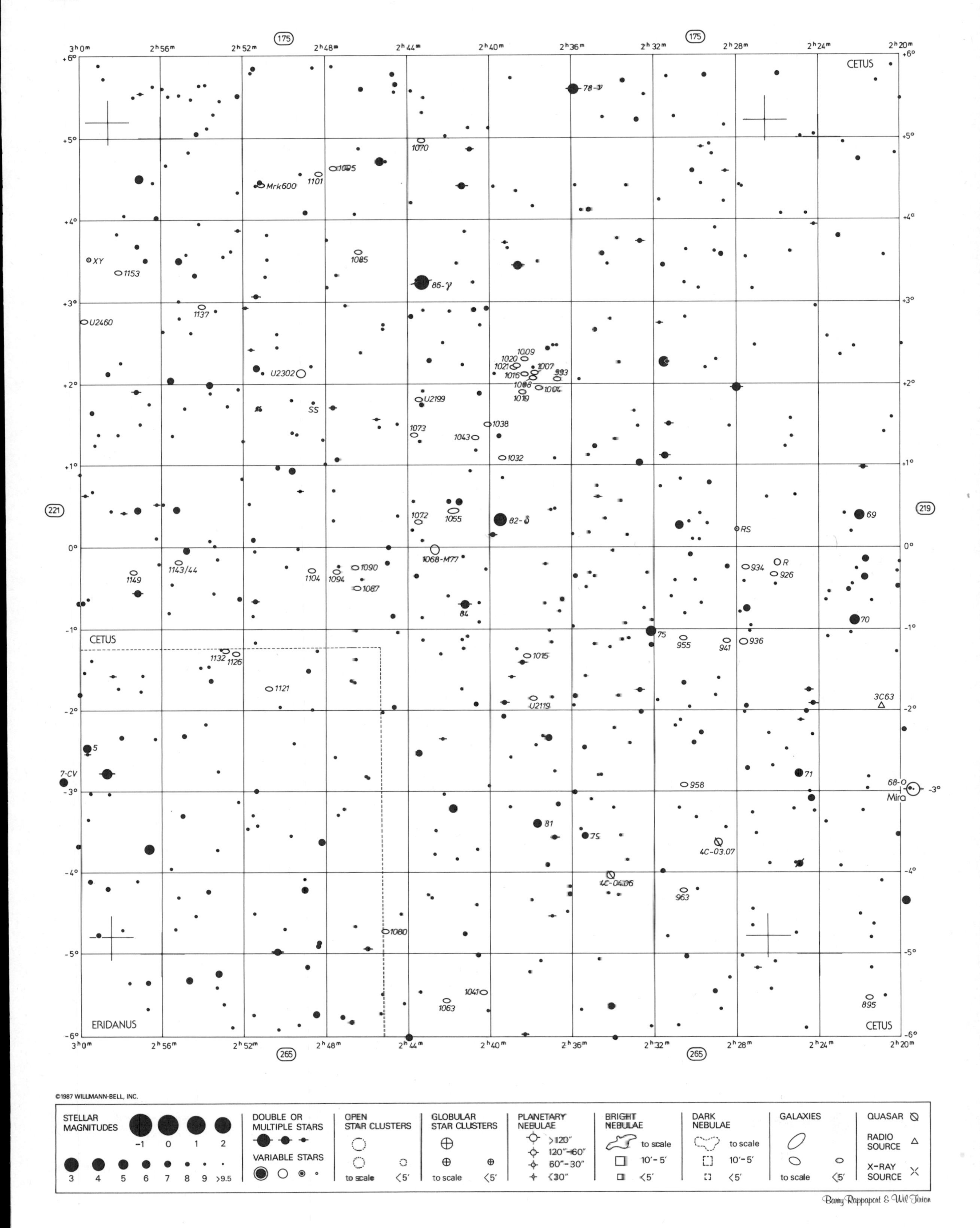

STELLAR MAGNITUDES -1 0 1 2 3 4 5 6 7 8 9 >9.5 | DOUBLE OR MULTIPLE STARS | VARIABLE STARS | OPEN STAR CLUSTERS to scale <5' | GLOBULAR STAR CLUSTERS to scale <5' | PLANETARY NEBULAE >120" 120"-60" 60"-30" <30" | BRIGHT NEBULAE to scale 10'-5' <5' | DARK NEBULAE to scale 10'-5' <5' | GALAXIES to scale <5' | QUASAR | RADIO SOURCE | X-RAY SOURCE

Barry Rappaport & Wil Tirion

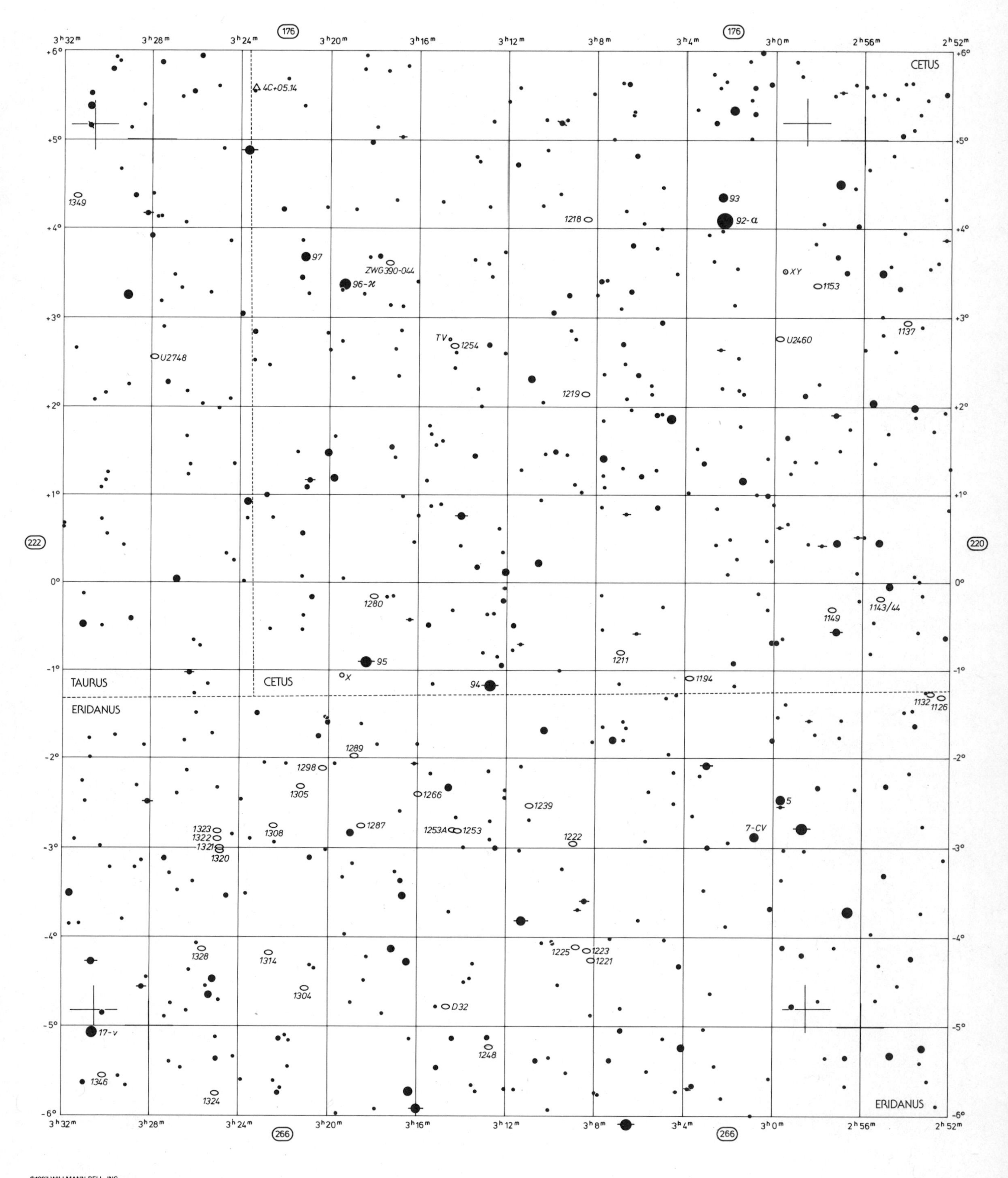

STELLAR MAGNITUDES: -1, 0, 1, 2, 3, 4, 5, 6, 7, 8, 9, >9.5

DOUBLE OR MULTIPLE STARS

VARIABLE STARS

OPEN STAR CLUSTERS: to scale, <5'

GLOBULAR STAR CLUSTERS: to scale, <5'

PLANETARY NEBULAE: >120", 120"-60", 60"-30", <30"

BRIGHT NEBULAE: to scale, 10'-5', <5'

DARK NEBULAE: to scale, 10'-5', <5'

GALAXIES: to scale, <5'

QUASAR

RADIO SOURCE

X-RAY SOURCE

Barry Rappaport & Wil Tirion

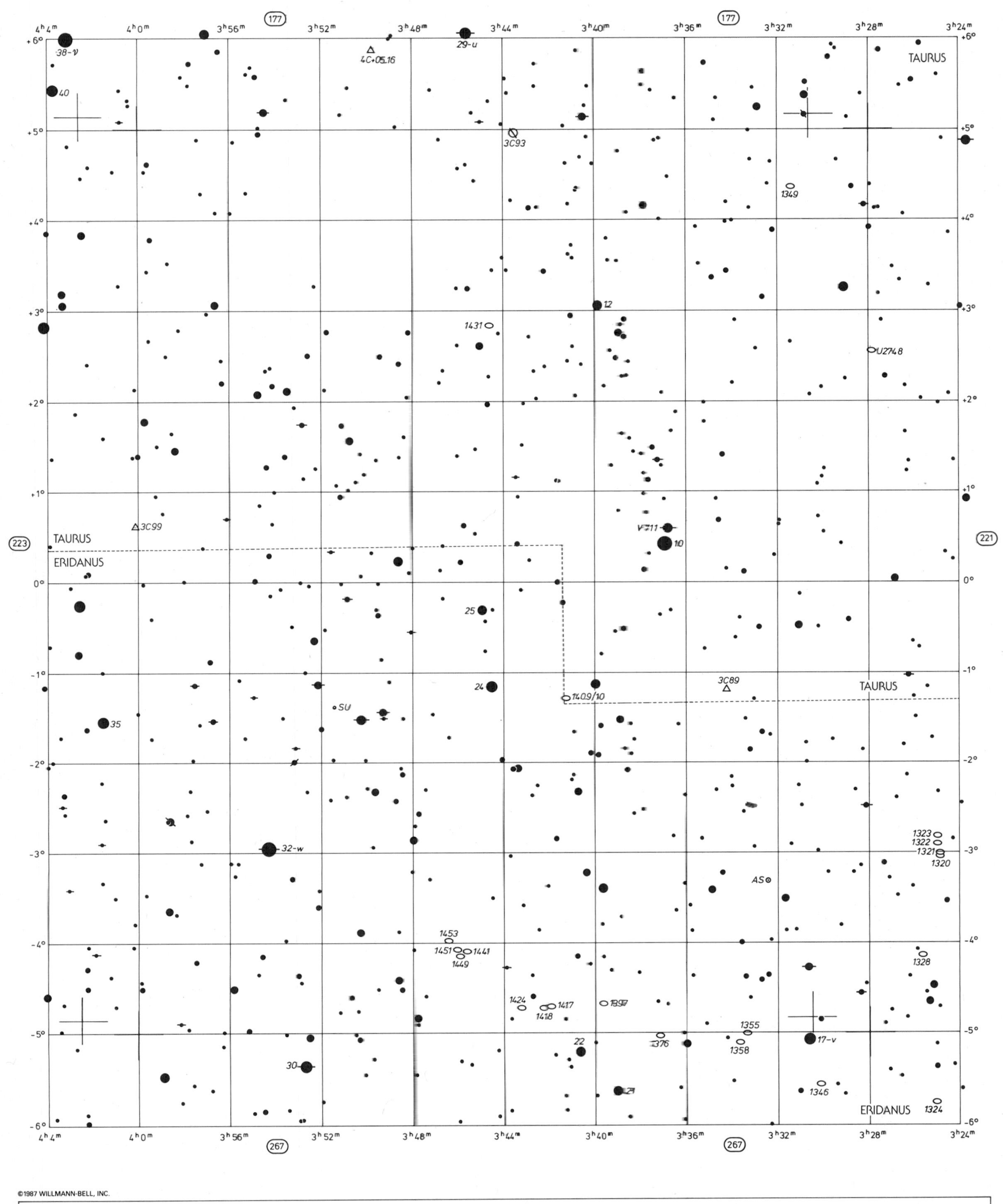

STELLAR MAGNITUDES -1 0 1 2 3 4 5 6 7 8 9 >9.5
DOUBLE OR MULTIPLE STARS
VARIABLE STARS
OPEN STAR CLUSTERS to scale <5'
GLOBULAR STAR CLUSTERS to scale <5'
PLANETARY NEBULAE >120" 120"-60" 60"-30" <30"
BRIGHT NEBULAE to scale 10'-5' <5'
DARK NEBULAE to scale 10'-5' <5'
GALAXIES to scale <5'
QUASAR
RADIO SOURCE
X-RAY SOURCE

Barry Rappaport & Wil Tirion

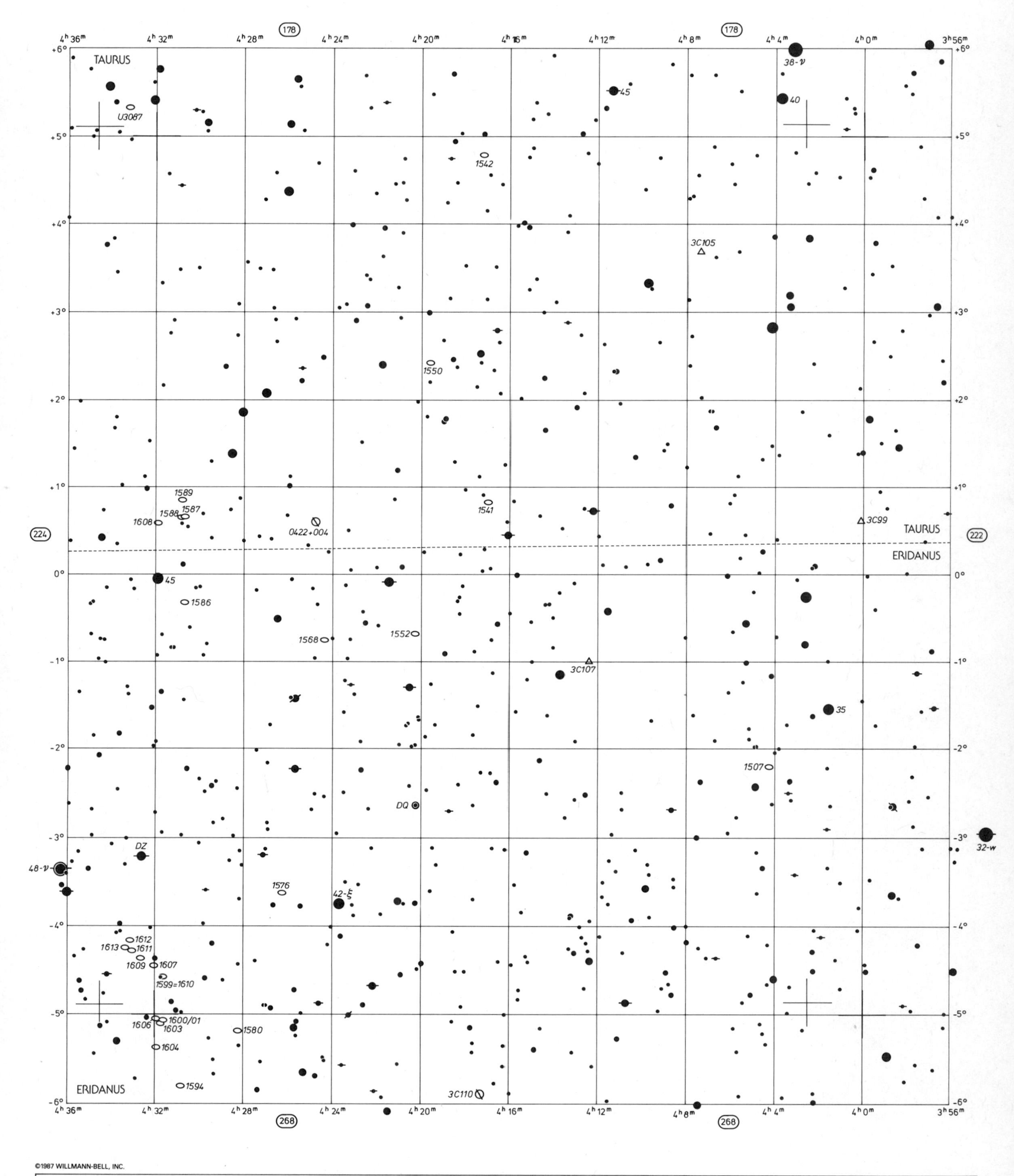

STELLAR MAGNITUDES: -1, 0, 1, 2, 3, 4, 5, 6, 7, 8, 9, >9.5

DOUBLE OR MULTIPLE STARS

VARIABLE STARS

OPEN STAR CLUSTERS: to scale, <5′

GLOBULAR STAR CLUSTERS: to scale, <5′

PLANETARY NEBULAE: >120″, 120″–60″, 60″–30″, <30″

BRIGHT NEBULAE: to scale, 10′–5′, <5′

DARK NEBULAE: to scale, 10′–5′, <5′

GALAXIES: to scale, <5′

QUASAR

RADIO SOURCE

X-RAY SOURCE

Barry Rappaport & Wil Tirion

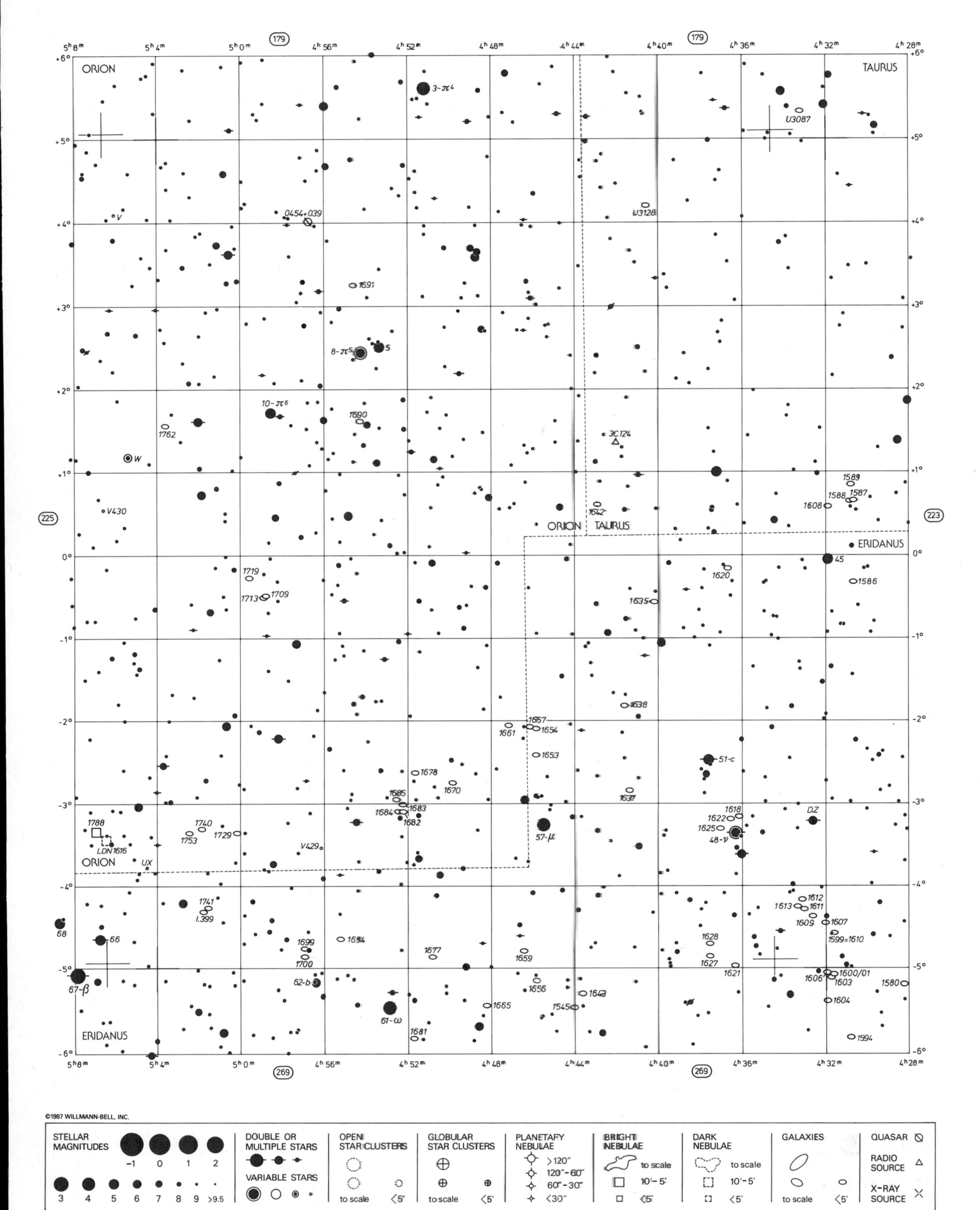

STELLAR MAGNITUDES -1 0 1 2 3 4 5 6 7 8 9 >9.5 | DOUBLE OR MULTIPLE STARS | VARIABLE STARS | OPEN STAR CLUSTERS to scale <5' | GLOBULAR STAR CLUSTERS to scale <5' | PLANETARY NEBULAE >120" 120"-60" 60"-30" <30" | BRIGHT NEBULAE to scale 10'-5' <5' | DARK NEBULAE to scale 10'-5' <5' | GALAXIES to scale <5' | QUASAR | RADIO SOURCE | X-RAY SOURCE

Barry Rappaport & Wil Tirion

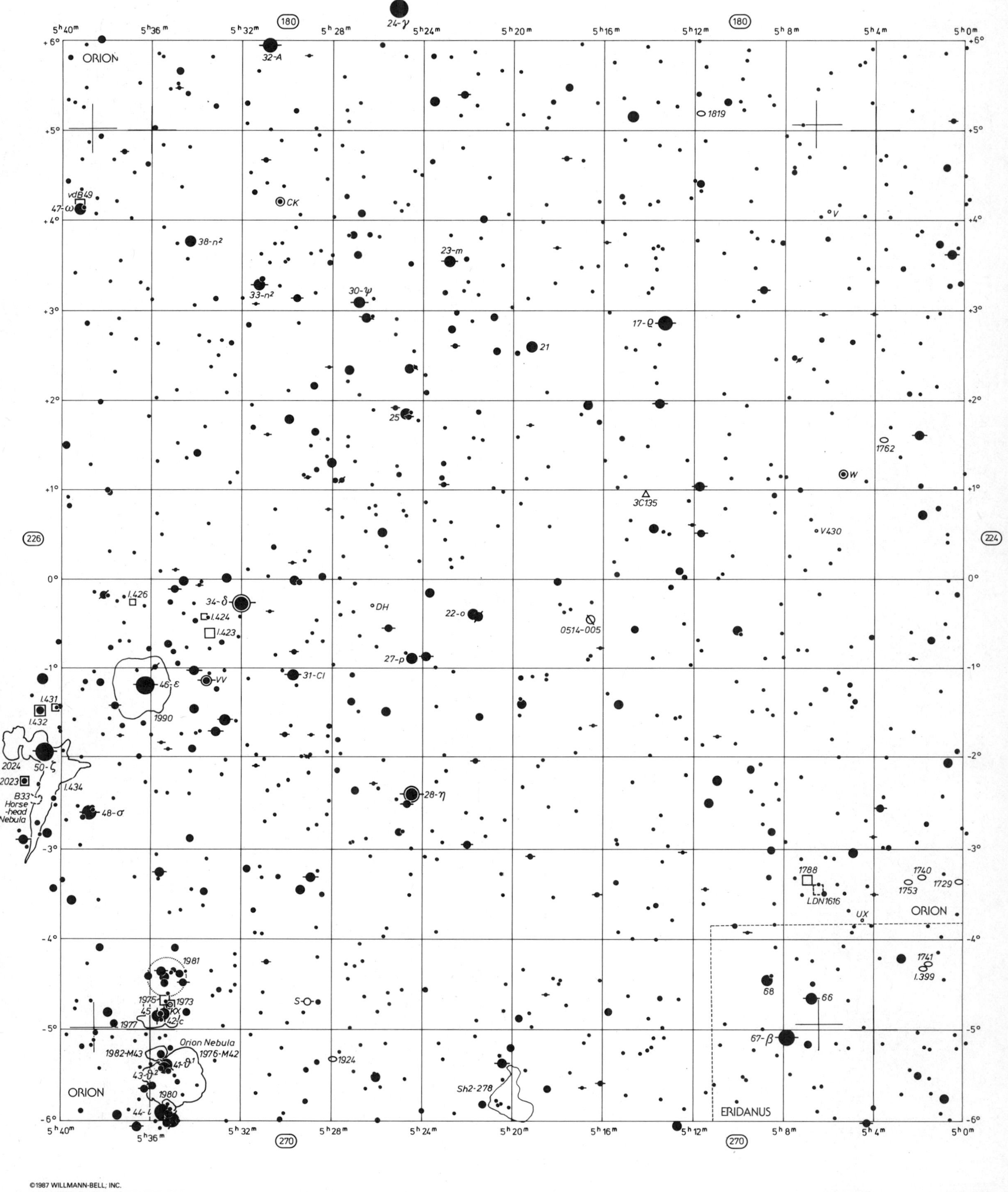

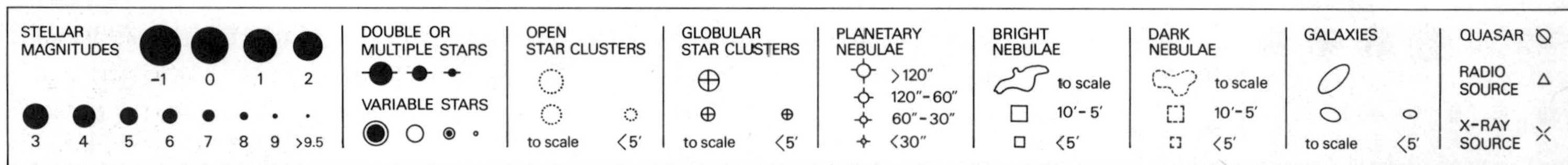

Barry Rappaport & Wil Tirion

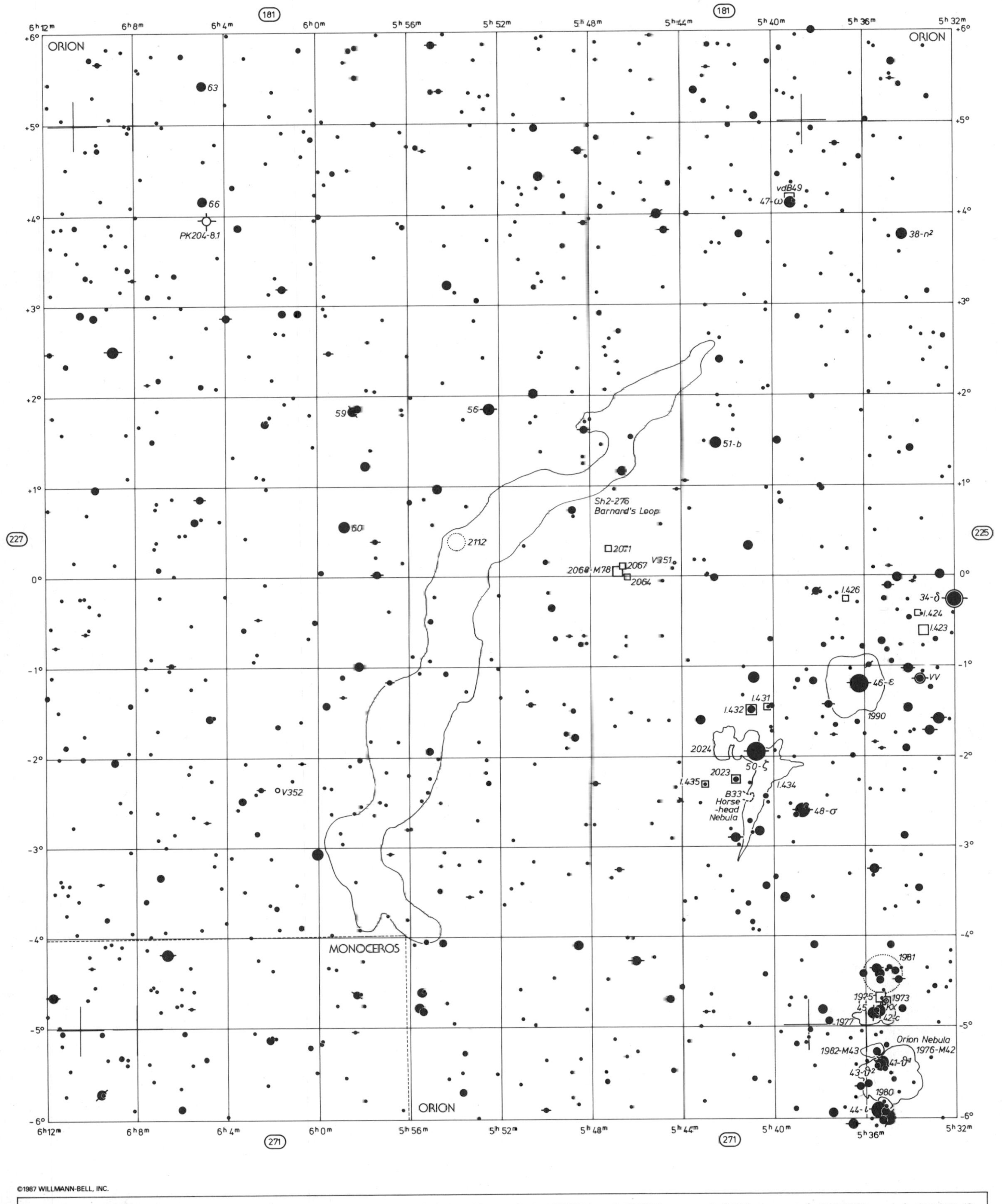

STELLAR MAGNITUDES -1 0 1 2 3 4 5 6 7 8 9 >9.5
DOUBLE OR MULTIPLE STARS
VARIABLE STARS
OPEN STAR CLUSTERS to scale <5′
GLOBULAR STAR CLUSTERS to scale <5′
PLANETARY NEBULAE >120″ 120″-60″ 60″-30″ <30″
BRIGHT NEBULAE to scale 10′-5′ <5′
DARK NEBULAE to scale 10′-5′ <5′
GALAXIES to scale <5′
QUASAR
RADIO SOURCE
X-RAY SOURCE

Barry Rappaport & Wil Tirion

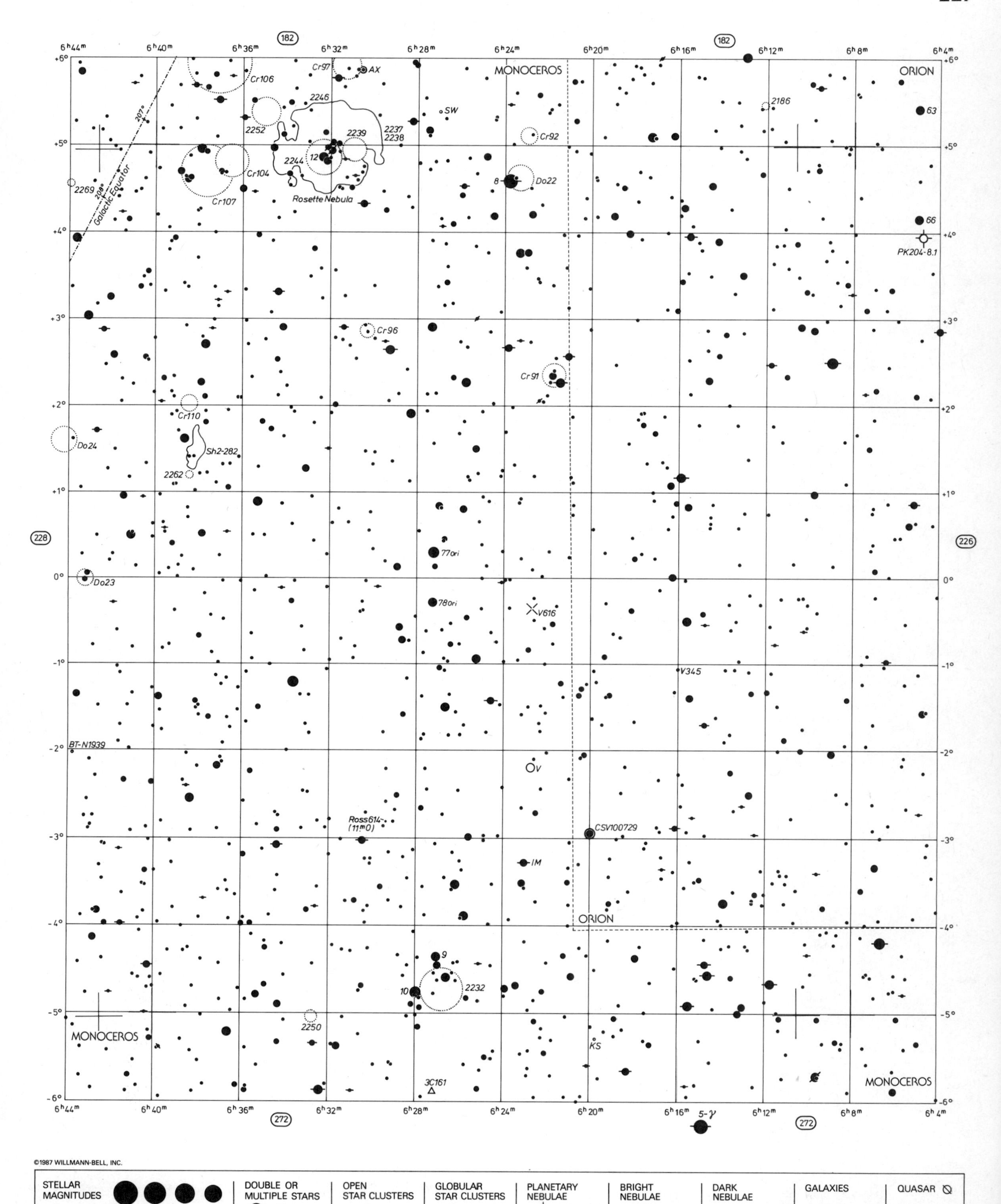

STELLAR MAGNITUDES	DOUBLE OR MULTIPLE STARS	OPEN STAR CLUSTERS	GLOBULAR STAR CLUSTERS	PLANETARY NEBULAE	BRIGHT NEBULAE	DARK NEBULAE	GALAXIES	QUASAR / RADIO SOURCE / X-RAY SOURCE
-1 0 1 2	VARIABLE STARS	to scale	to scale	>120"	to scale	to scale	to scale	QUASAR
3 4 5 6 7 8 9 >9.5		<5'	<5'	120"-60"	10'-5'	10'-5'	<5'	RADIO SOURCE
				60"-30"	<5'	<5'		X-RAY SOURCE
				<30"				

Barry Rappaport & Wil Tirion

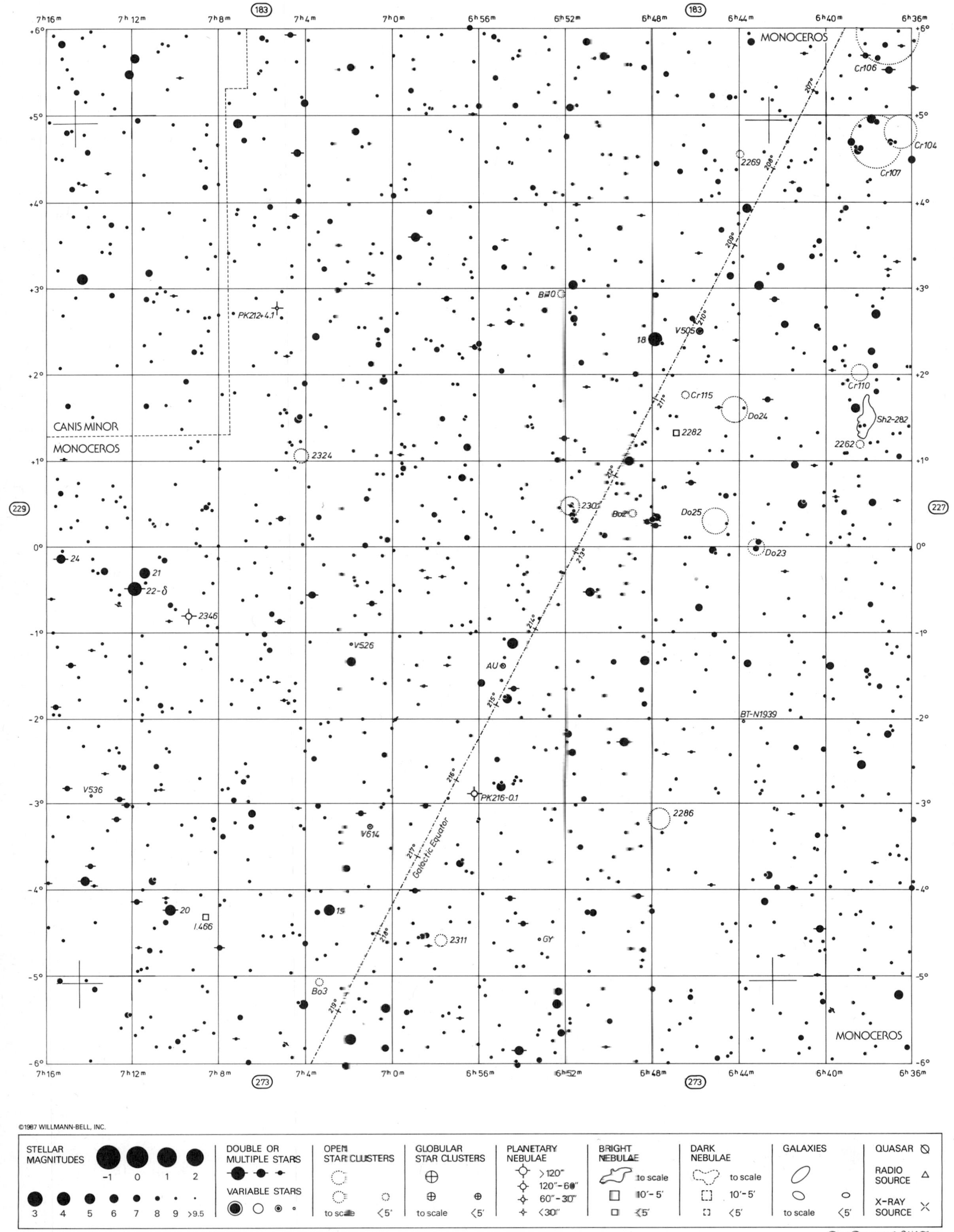
MONOCEROS
CANIS MINOR
MONOCEROS
MONOCEROS
Galactic Equator
Cr106
Cr104
Cr107
2269
V505
18
Cr115
Cr110
Do24
Sh2-282
2282
2262
2324
2301
Do25
Do23
24
21
22-δ
2346
V526
AU
BT-N1939
V536
PK216-0.1
2286
V614
20
I.466
2311
GY
Bo3
PK212+4.1
STELLAR MAGNITUDES
DOUBLE OR MULTIPLE STARS
VARIABLE STARS
OPEN STAR CLUSTERS
GLOBULAR STAR CLUSTERS
PLANETARY NEBULAE
BRIGHT NEBULAE
DARK NEBULAE
GALAXIES
QUASAR
RADIO SOURCE
X-RAY SOURCE
to scale
Barry Rappaport & Wil Tirion

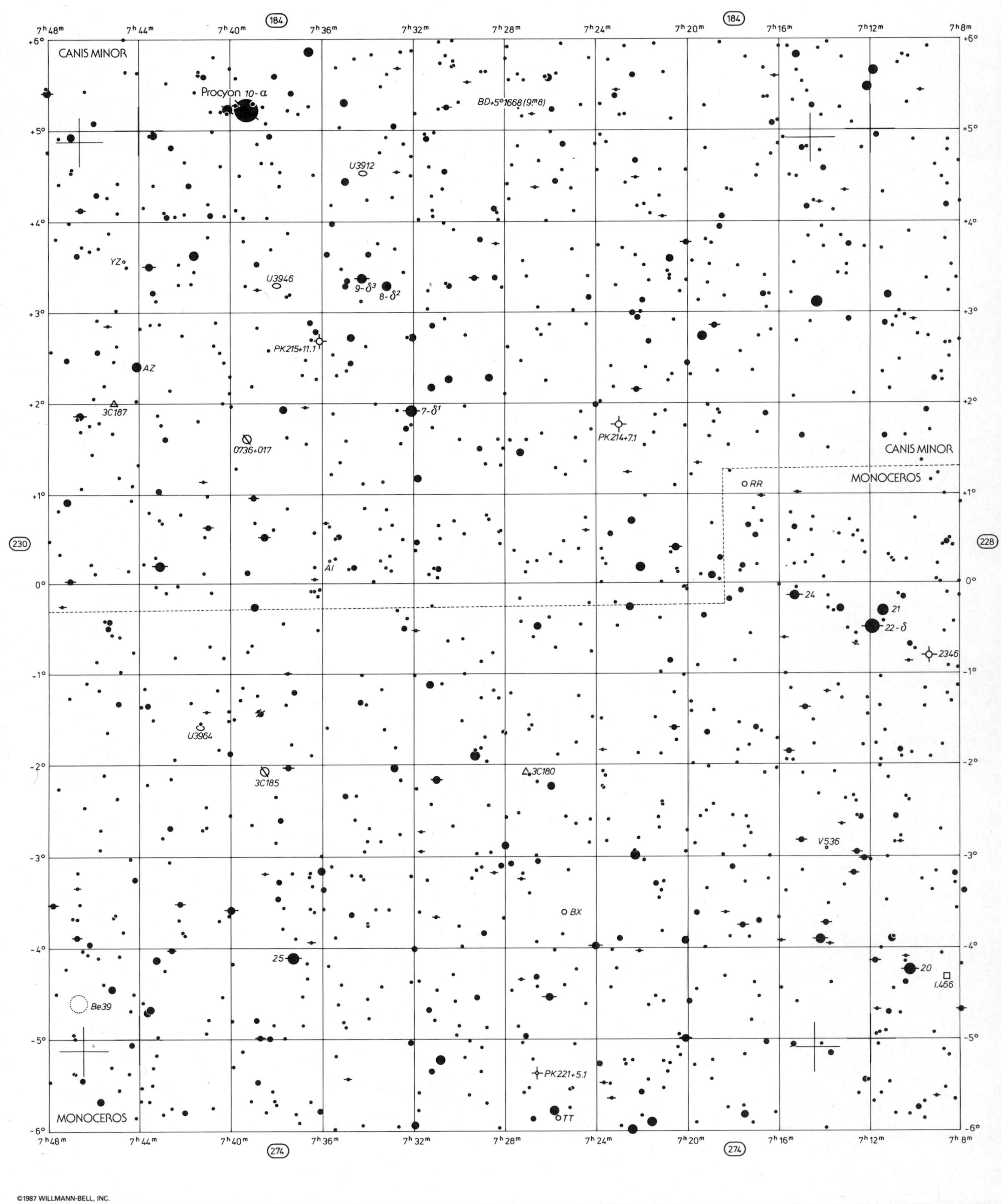

STELLAR MAGNITUDES
-1 0 1 2
3 4 5 6 7 8 9 >9.5

DOUBLE OR MULTIPLE STARS

VARIABLE STARS

OPEN STAR CLUSTERS
to scale <5′

GLOBULAR STAR CLUSTERS
to scale <5′

PLANETARY NEBULAE
>120″
120″–60″
60″–30″
<30″

BRIGHT NEBULAE
to scale
10′–5′
<5′

DARK NEBULAE
to scale
10′–5′
<5′

GALAXIES
to scale <5′

QUASAR

RADIO SOURCE

X-RAY SOURCE

Barry Rappaport & Wil Tirion

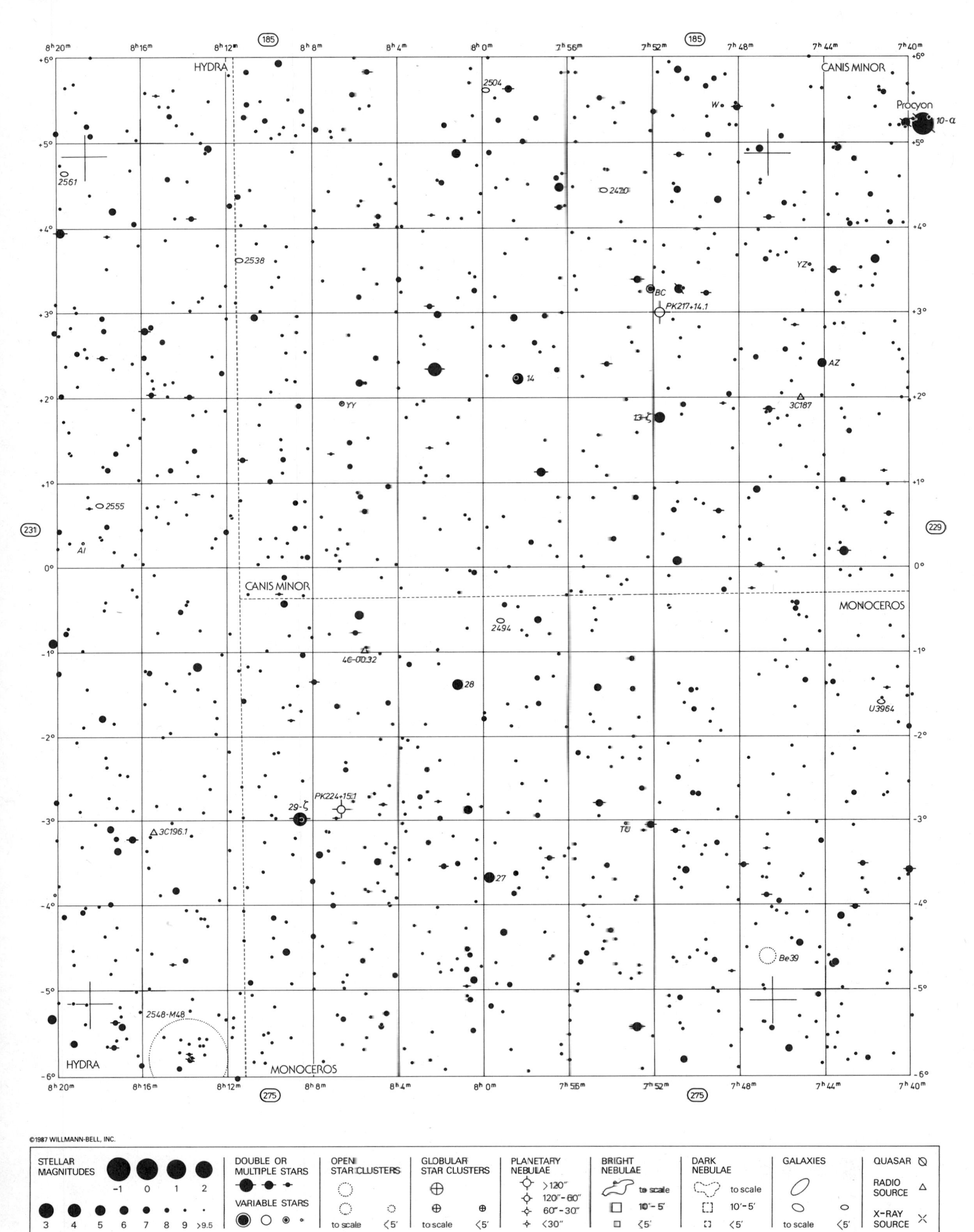

STELLAR MAGNITUDES	DOUBLE OR MULTIPLE STARS	OPEN STAR CLUSTERS	GLOBULAR STAR CLUSTERS	PLANETARY NEBULAE	BRIGHT NEBULAE	DARK NEBULAE	GALAXIES	QUASAR / RADIO SOURCE / X-RAY SOURCE
-1 0 1 2	VARIABLE STARS	to scale <5′	to scale <5′	>120″ 120″-60″ 60″-30″ <30″	to scale 10′-5′ <5′	to scale 10′-5′ <5′	to scale <5′	
3 4 5 6 7 8 9 >9.5								

Barry Rappaport & Wil Tirion

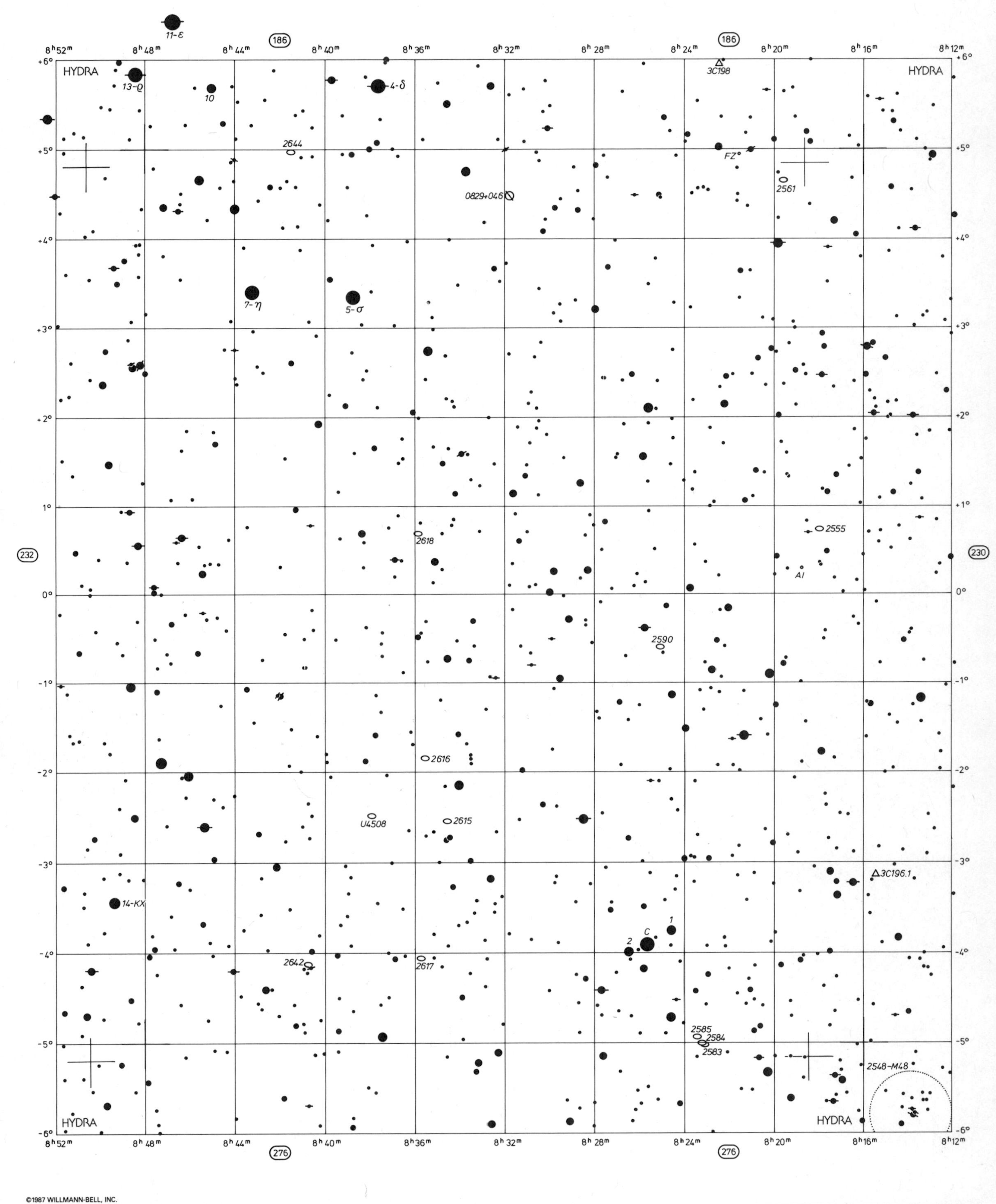

STELLAR MAGNITUDES	DOUBLE OR MULTIPLE STARS	OPEN STAR CLUSTERS	GLOBULAR STAR CLUSTERS	PLANETARY NEBULAE	BRIGHT NEBULAE	DARK NEBULAE	GALAXIES	QUASAR
-1 0 1 2	VARIABLE STARS	to scale <5′	to scale <5′	>120″ 120″–60″ 60″–30″ <30″	to scale 10′–5′ <5′	to scale 10′–5′ <5′	to scale <5′	RADIO SOURCE
3 4 5 6 7 8 9 >9.5								X-RAY SOURCE

Barry Rappaport & Wil Tirion

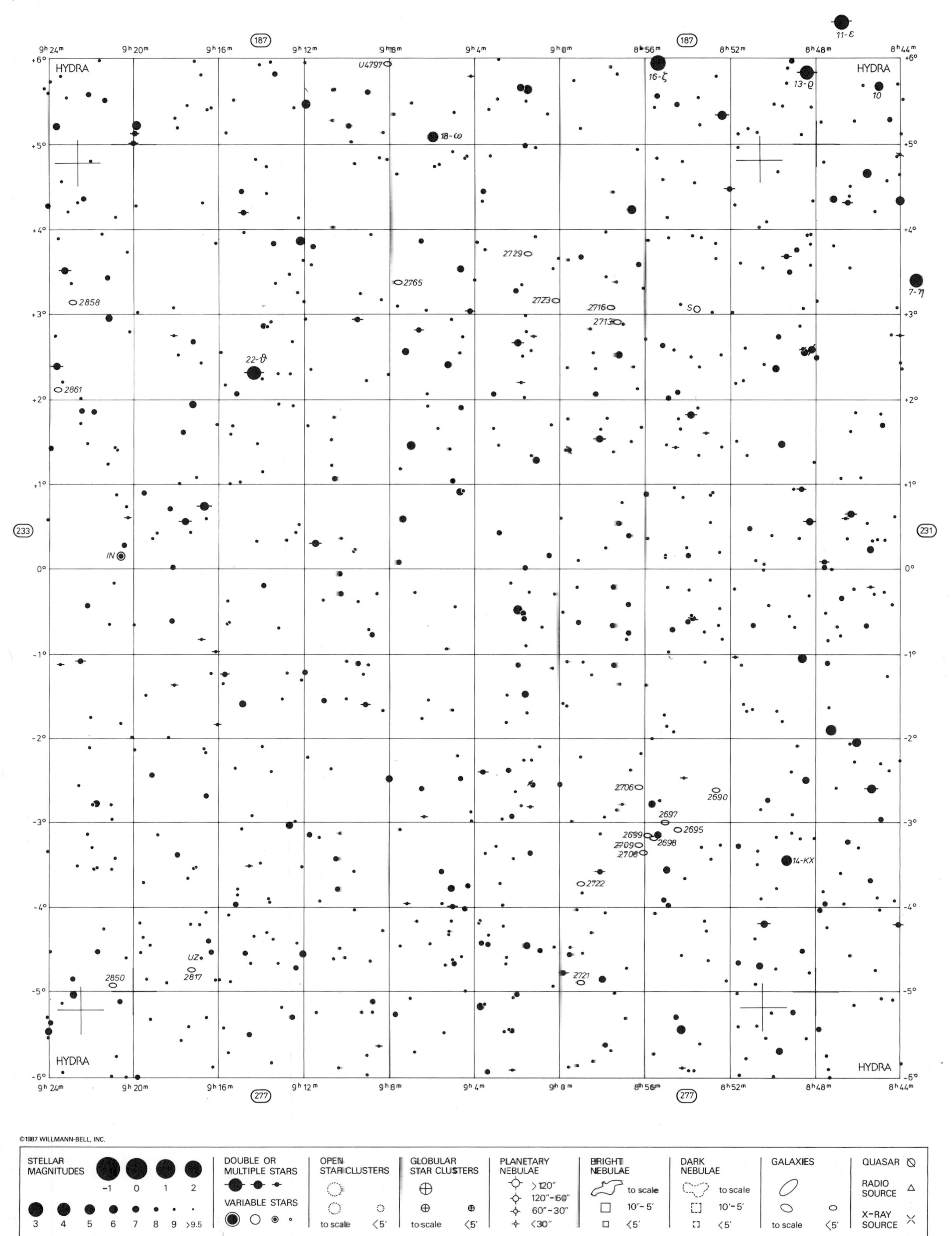

STELLAR MAGNITUDES: -1 0 1 2 3 4 5 6 7 8 9 >9.5 | DOUBLE OR MULTIPLE STARS | VARIABLE STARS | OPEN STAR CLUSTERS: to scale, <5′ | GLOBULAR STAR CLUSTERS: to scale, <5′ | PLANETARY NEBULAE: >120″, 120″–60″, 60″–30″, <30″ | BRIGHT NEBULAE: to scale, 10′–5′, <5′ | DARK NEBULAE: to scale, 10′–5′, <5′ | GALAXIES: to scale, <5′ | QUASAR | RADIO SOURCE | X-RAY SOURCE

Barry Rappaport & Wil Tirion

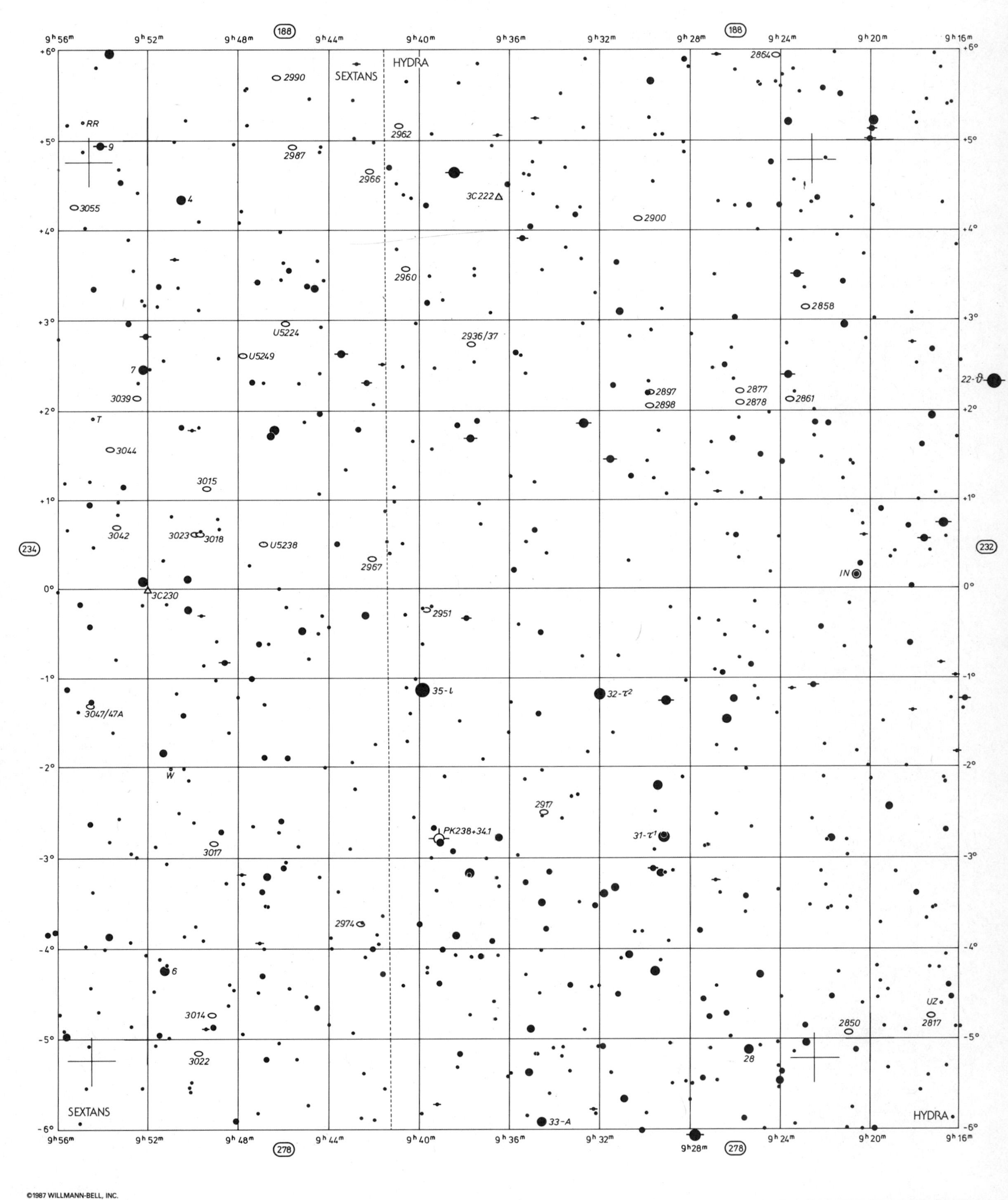

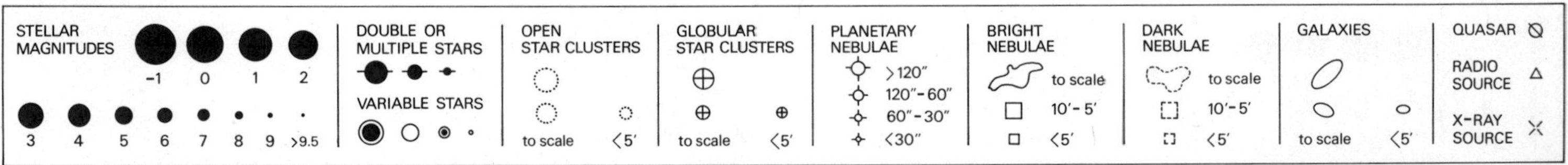

Barry Rappaport & Wil Tirion

234

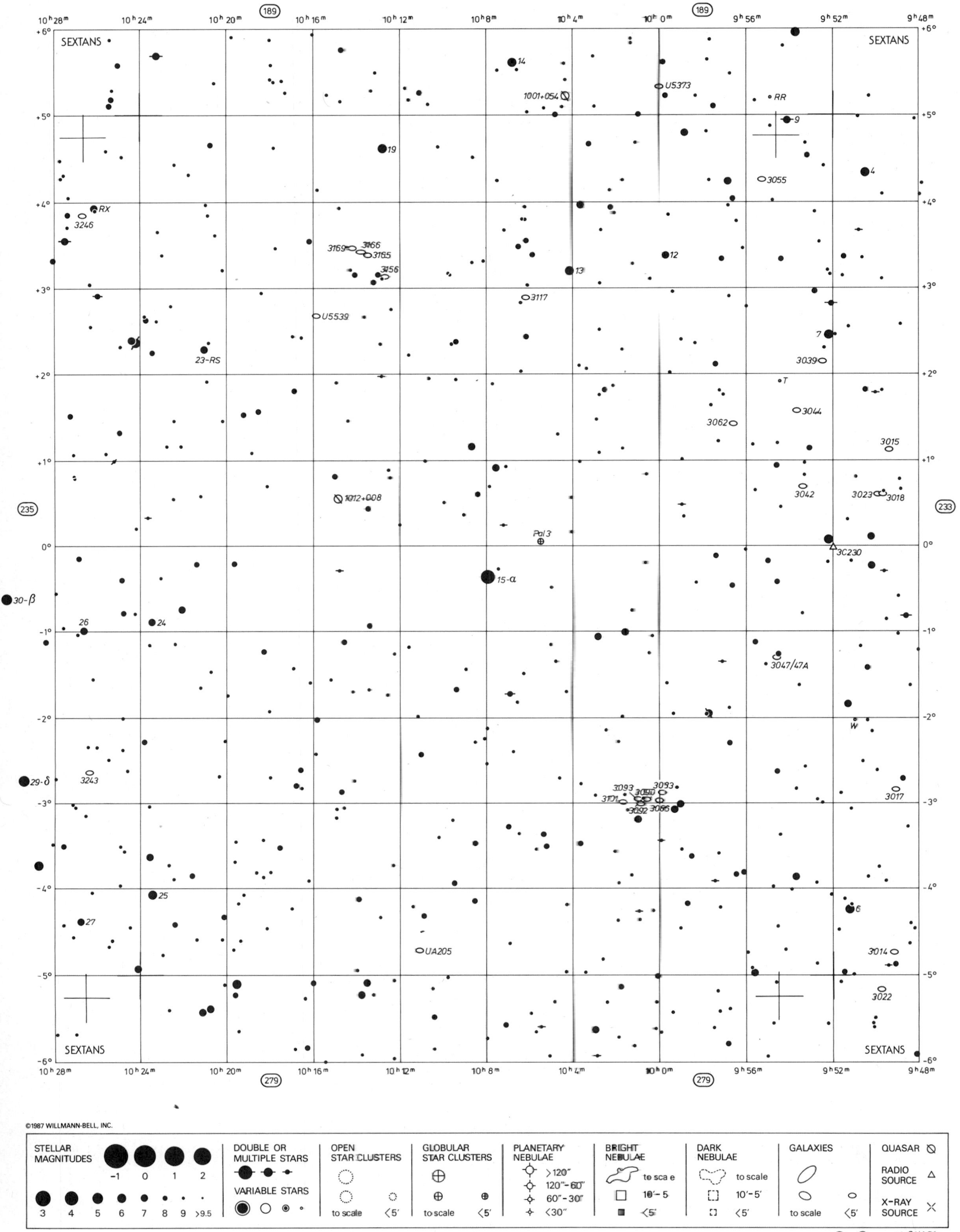

STELLAR MAGNITUDES	DOUBLE OR MULTIPLE STARS	OPEN STAR CLUSTERS	GLOBULAR STAR CLUSTERS	PLANETARY NEBULAE	BRIGHT NEBULAE	DARK NEBULAE	GALAXIES	QUASAR
-1 0 1 2	VARIABLE STARS	to scale <5′	to scale <5′	>120″ 120″–60″ 60″–30″ <30″	to scale 10′–5′ <5′	to scale 10′–5′ <5′	to scale <5′	RADIO SOURCE
3 4 5 6 7 8 9 >9.5								X-RAY SOURCE

Barry Rappaport & Wil Tirion

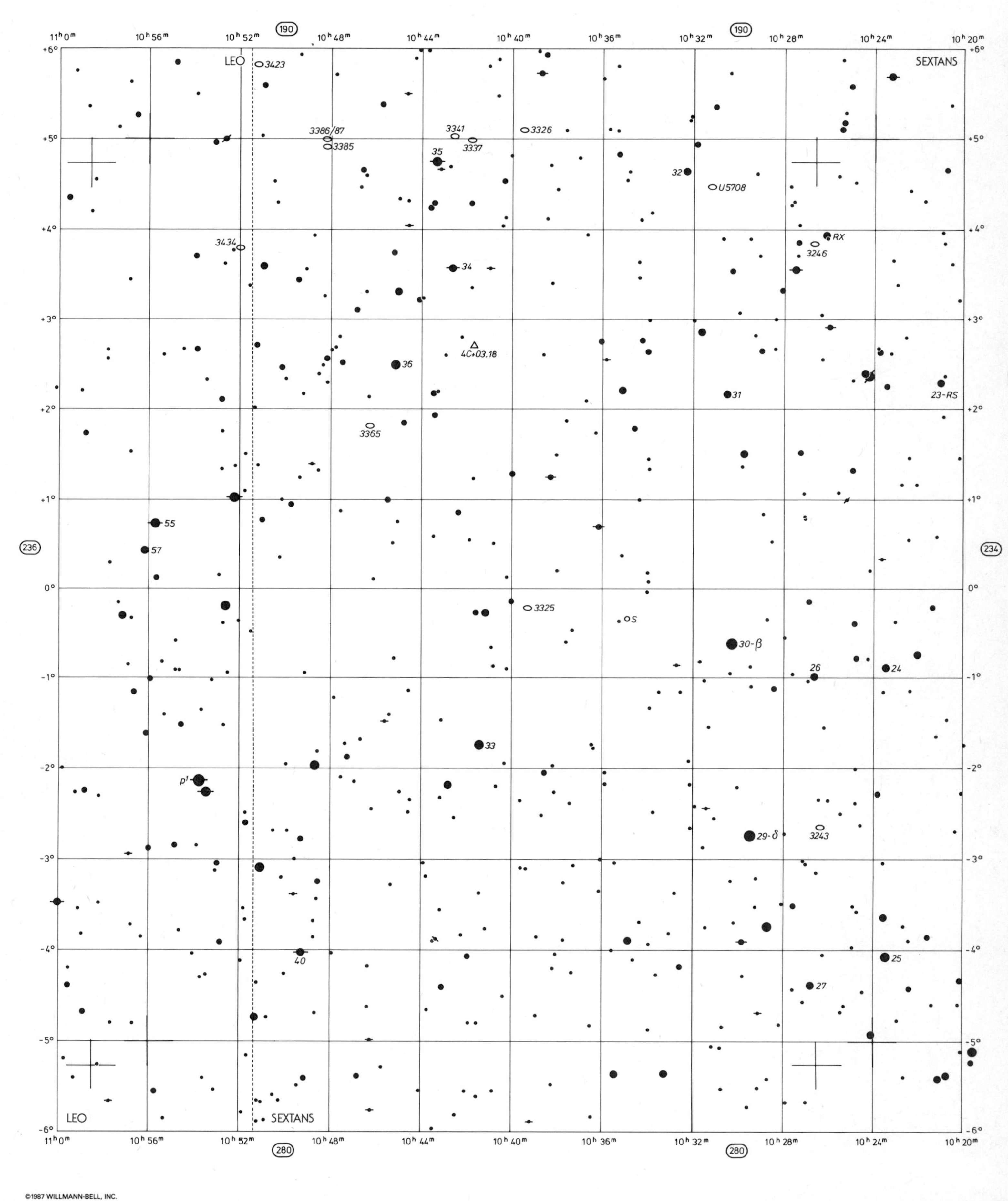

STELLAR MAGNITUDES	DOUBLE OR MULTIPLE STARS	OPEN STAR CLUSTERS	GLOBULAR STAR CLUSTERS	PLANETARY NEBULAE	BRIGHT NEBULAE	DARK NEBULAE	GALAXIES	QUASAR
-1 0 1 2	VARIABLE STARS	to scale <5′	to scale <5′	>120″ 120″-60″ 60″-30″ <30″	to scale 10′-5′ <5′	to scale 10′-5′ <5′	to scale <5′	RADIO SOURCE
3 4 5 6 7 8 9 >9.5								X-RAY SOURCE

Barry Rappaport & Wil Tirion

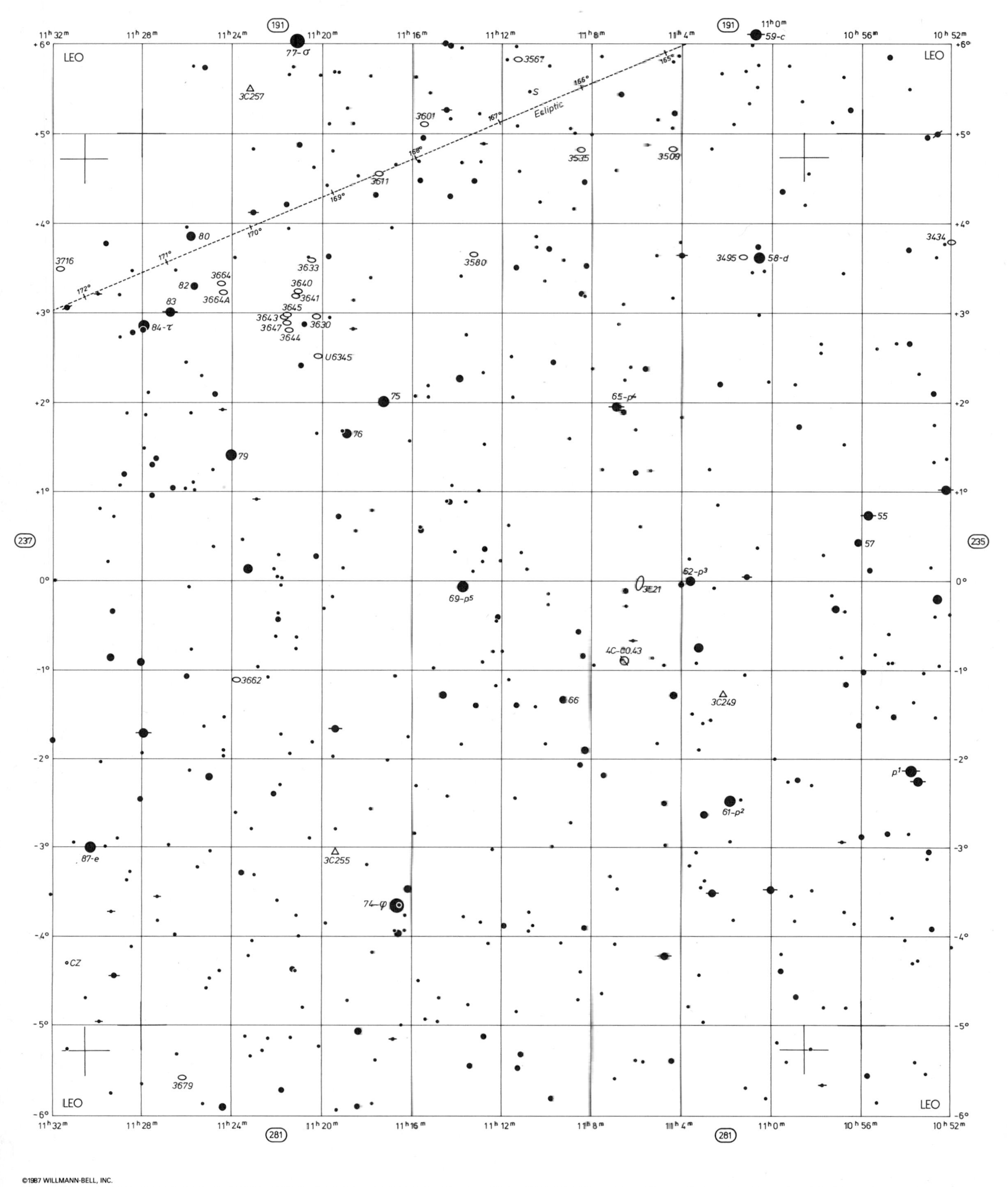

STELLAR MAGNITUDES -1 0 1 2 3 4 5 6 7 8 9 >9.5

DOUBLE OR MULTIPLE STARS

VARIABLE STARS

OPEN STAR CLUSTERS to scale <5'

GLOBULAR STAR CLUSTERS to scale <5'

PLANETARY NEBULAE >120" 120"-60" 60"-30" <30"

BRIGHT NEBULAE to scale 10'-5' <5'

DARK NEBULAE to scale 10'-5' <5'

GALAXIES to scale <5'

QUASAR

RADIO SOURCE

X-RAY SOURCE

Barry Rappaport & Wil Tirion

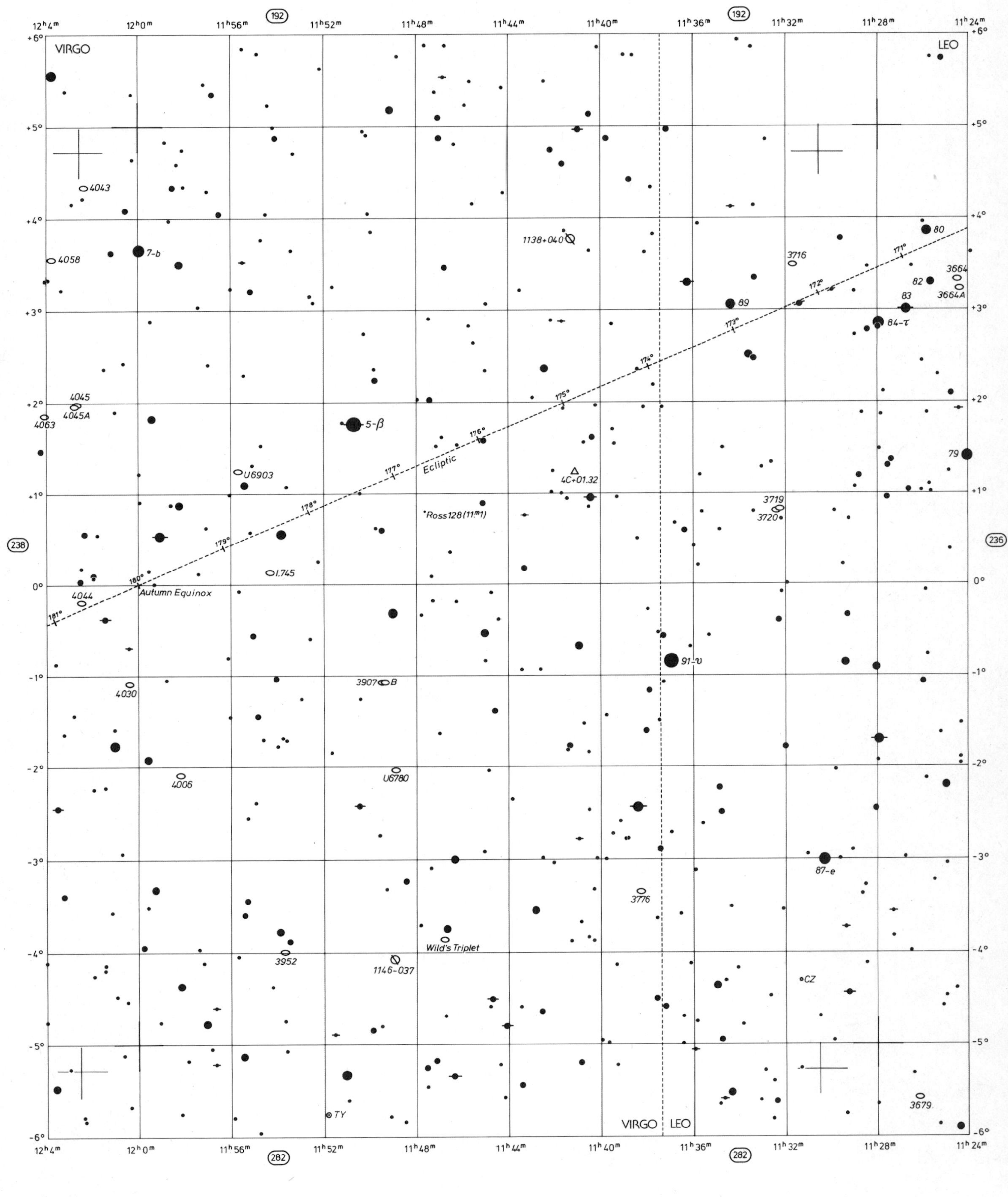

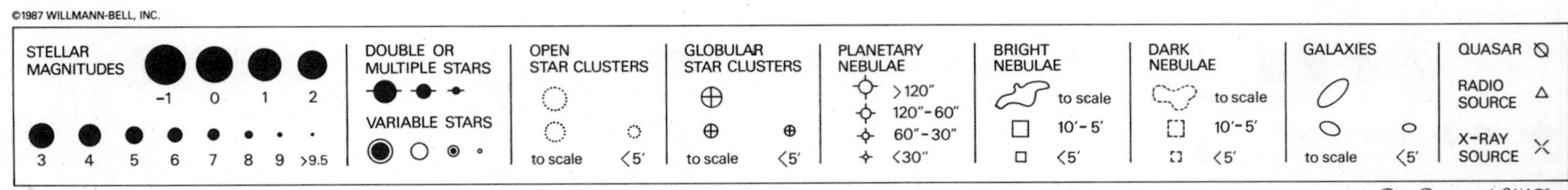

Barry Rappaport & Wil Tirion

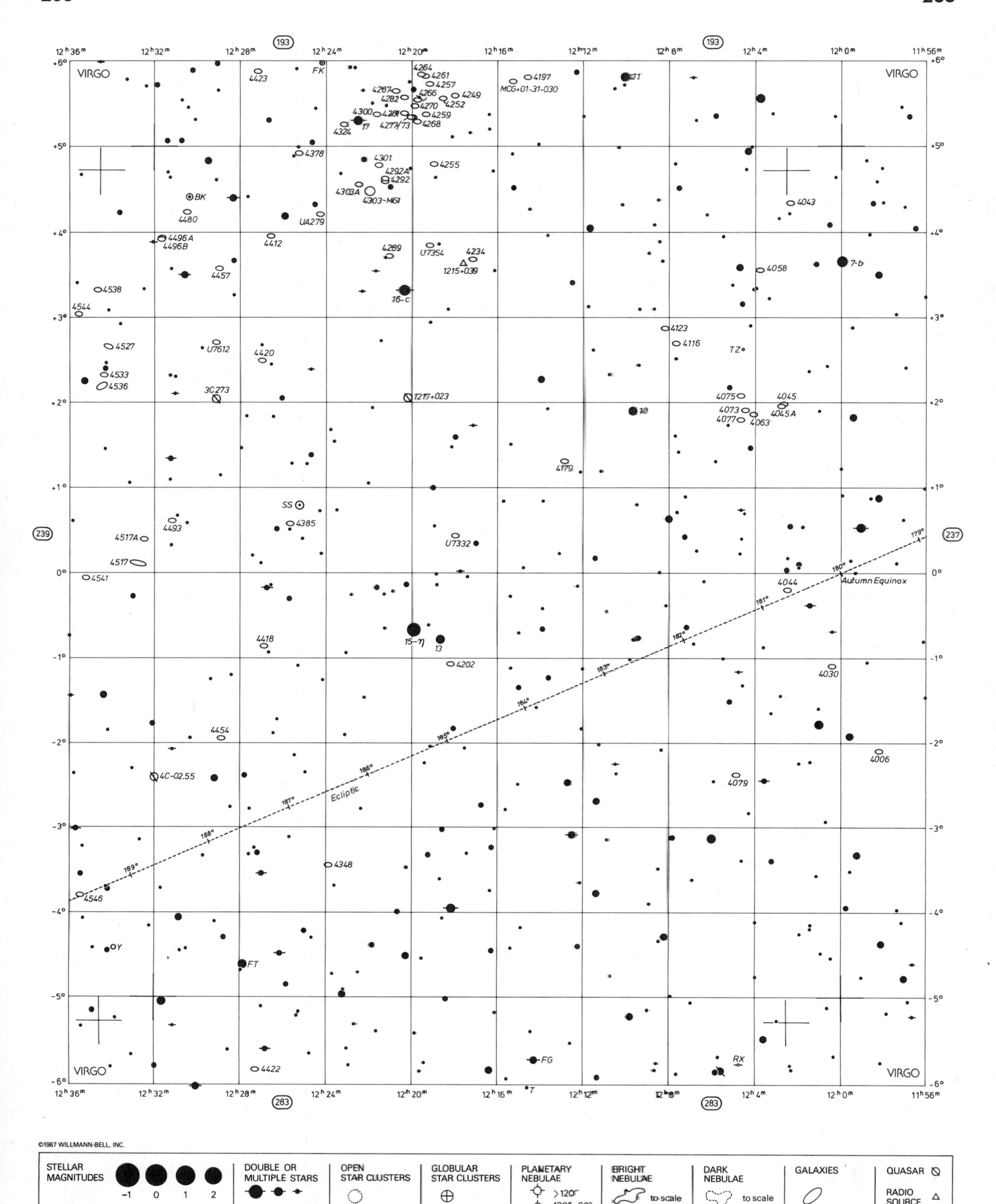

STELLAR MAGNITUDES	DOUBLE OR MULTIPLE STARS	OPEN STAR CLUSTERS	GLOBULAR STAR CLUSTERS	PLANETARY NEBULAE	BRIGHT NEBULAE	DARK NEBULAE	GALAXIES	QUASAR
-1 0 1 2	VARIABLE STARS	to scale	to scale	>120"	to scale	to scale	to scale	RADIO SOURCE
3 4 5 6 7 8 9 >9.5		<5'	<5'	120"–60"	10'–5'	10'–5'	<5'	X-RAY SOURCE
				60"–30"	<5'	<5'		
				<30"				

Barry Rappaport & Wil Tirion

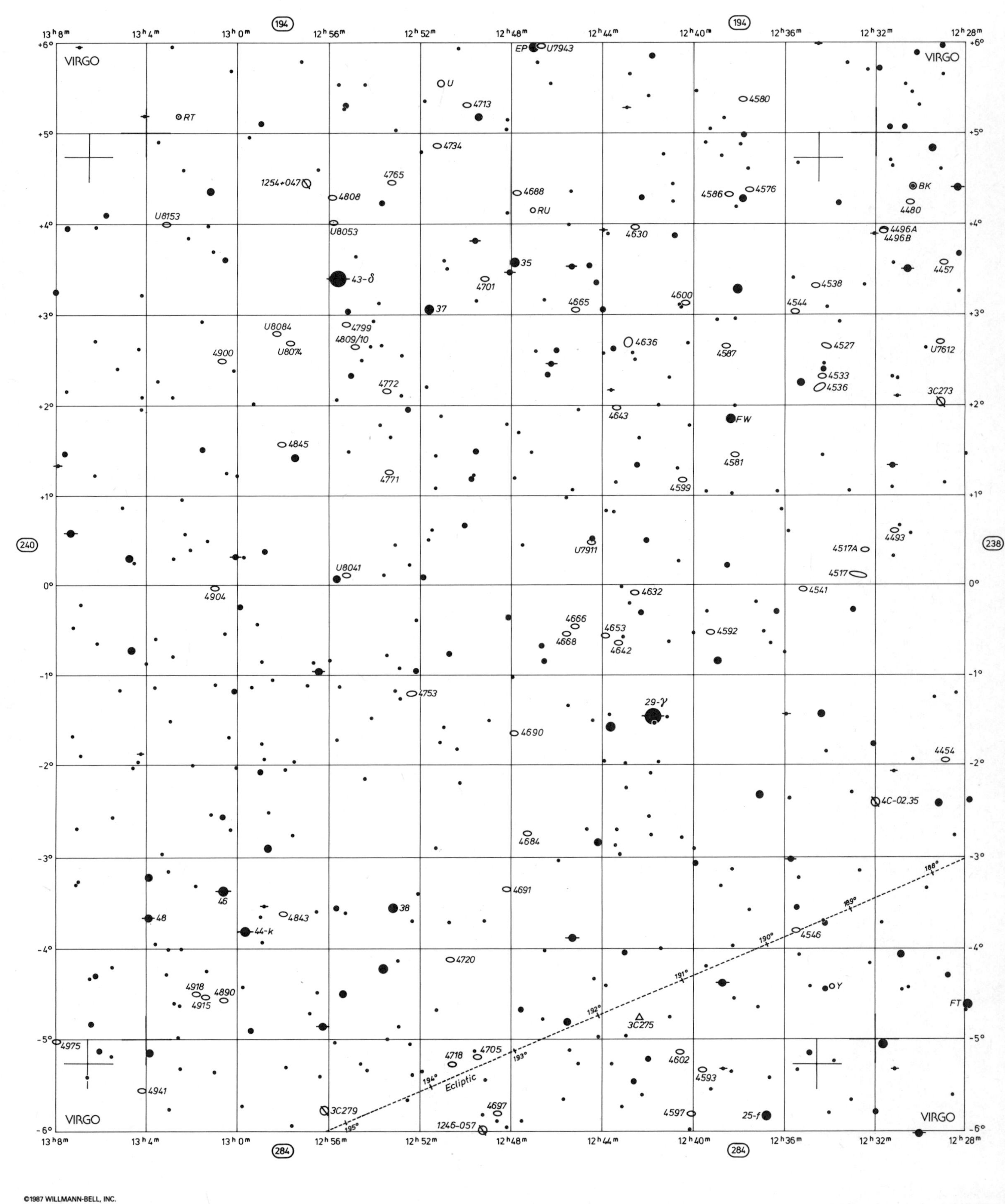

STELLAR MAGNITUDES	DOUBLE OR MULTIPLE STARS	OPEN STAR CLUSTERS	GLOBULAR STAR CLUSTERS	PLANETARY NEBULAE	BRIGHT NEBULAE	DARK NEBULAE	GALAXIES	QUASAR
-1 0 1 2	VARIABLE STARS	to scale ‹5′	to scale ‹5′	›120″ 120″-60″ 60″-30″ ‹30″	to scale 10′-5′ ‹5′	to scale 10′-5′ ‹5′	to scale ‹5′	RADIO SOURCE
3 4 5 6 7 8 9 ›9.5								X-RAY SOURCE

Barry Rappaport & Wil Tirion

240

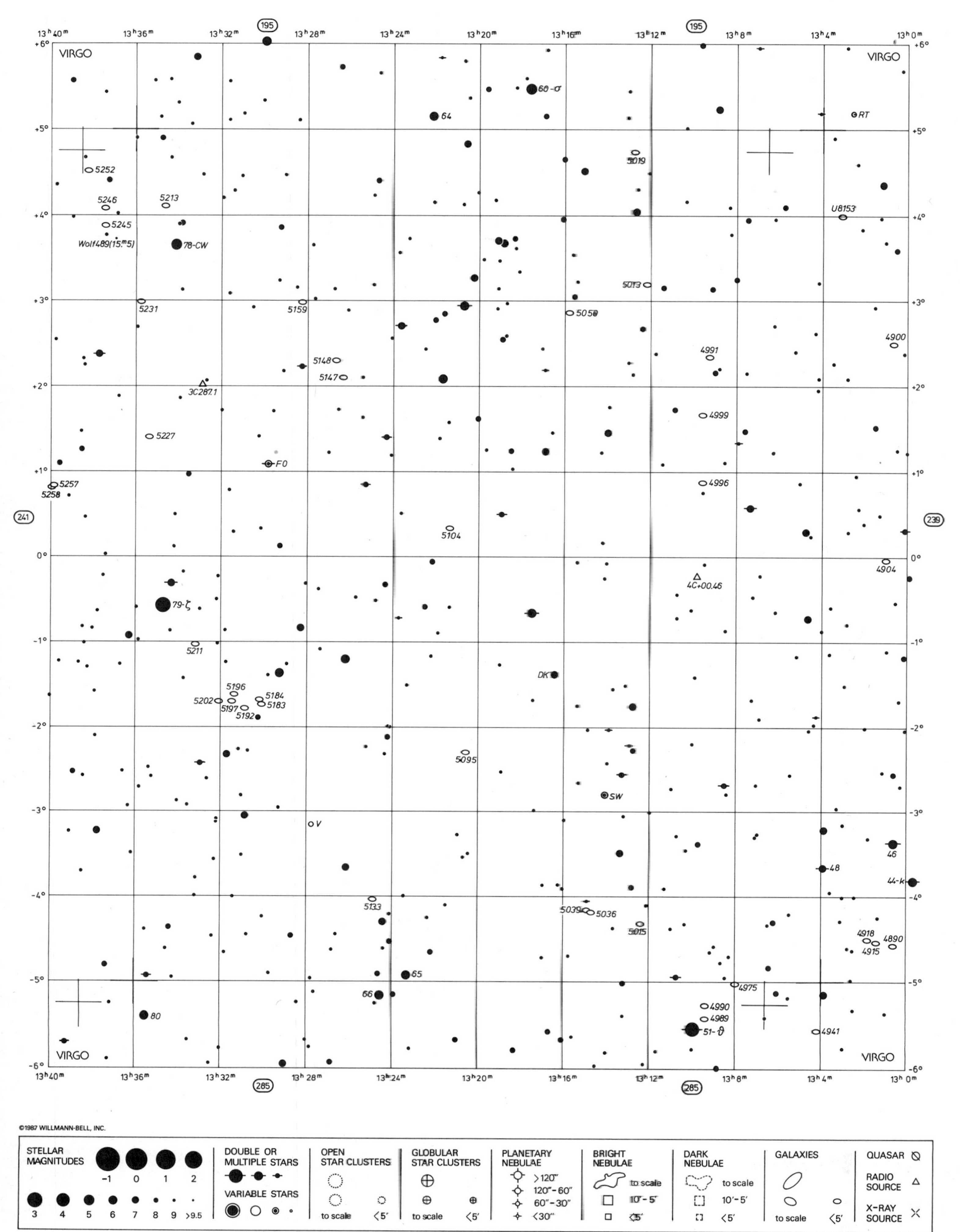

STELLAR MAGNITUDES	DOUBLE OR MULTIPLE STARS	OPEN STAR CLUSTERS	GLOBULAR STAR CLUSTERS	PLANETARY NEBULAE	BRIGHT NEBULAE	DARK NEBULAE	GALAXIES	
-1 0 1 2	VARIABLE STARS	to scale <5′	to scale <5′	>120″ 120″-60″ 60″-30″ <30″	to scale 10′-5′ <5′	to scale 10′-5′ <5′	to scale <5′	QUASAR RADIO SOURCE X-RAY SOURCE
3 4 5 6 7 8 9 >9.5								

Barry Rappaport & Wil Tirion

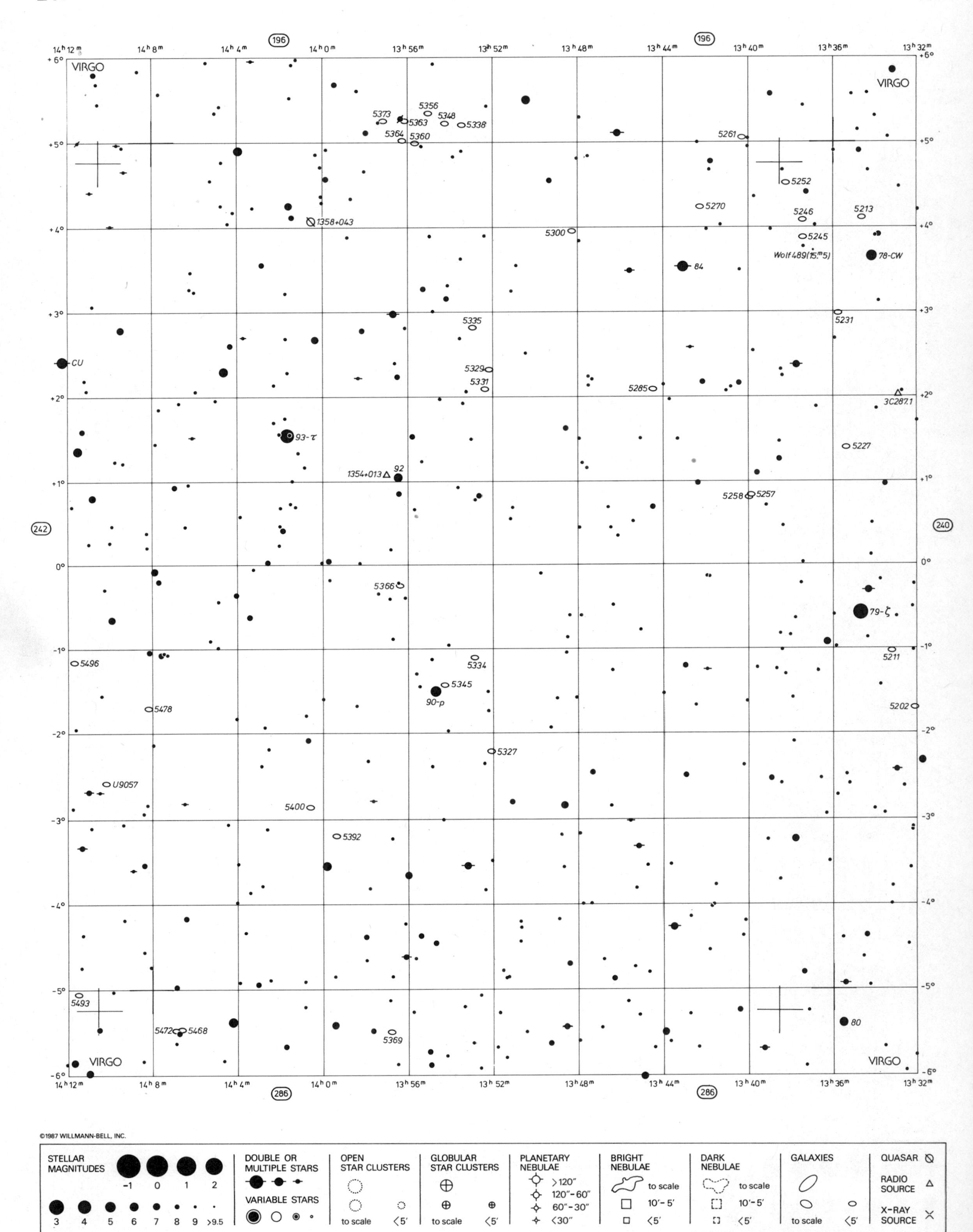

STELLAR MAGNITUDES: -1, 0, 1, 2, 3, 4, 5, 6, 7, 8, 9, >9.5
DOUBLE OR MULTIPLE STARS
VARIABLE STARS
OPEN STAR CLUSTERS: to scale, <5'
GLOBULAR STAR CLUSTERS: to scale, <5'
PLANETARY NEBULAE: >120", 120"–60", 60"–30", <30"
BRIGHT NEBULAE: to scale, 10'–5', <5'
DARK NEBULAE: to scale, 10'–5', <5'
GALAXIES: to scale, <5'
QUASAR
RADIO SOURCE
X-RAY SOURCE

Barry Rappaport & Wil Tirion

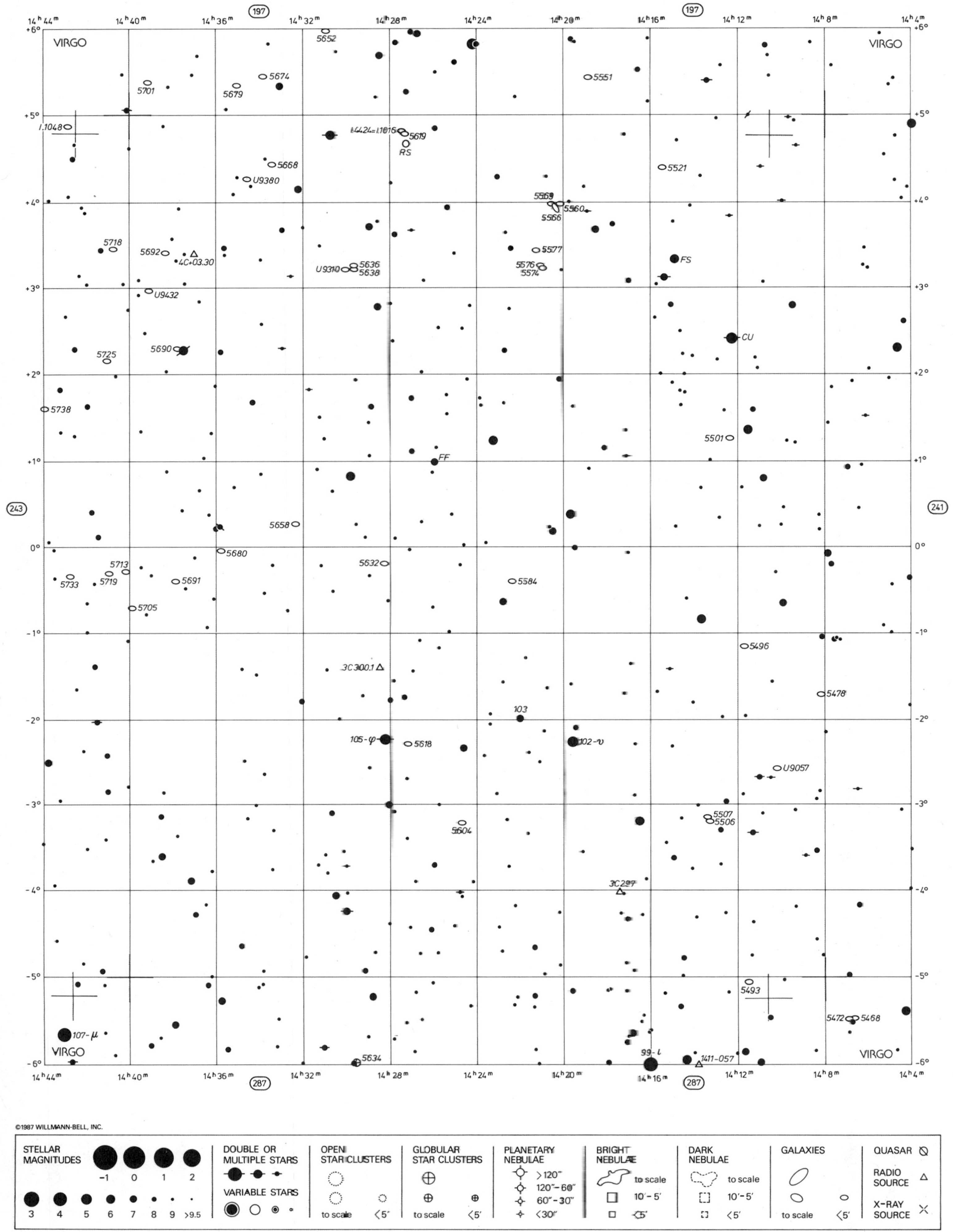
VIRGO
VIRGO
VIRGO
VIRGO
14h 44m
14h 40m
14h 36m
14h 32m
14h 28m
14h 24m
14h 20m
14h 16m
14h 12m
14h 8m
14h 4m
+6°
+5°
+4°
+3°
+2°
+1°
0°
-1°
-2°
-3°
-4°
-5°
-6°
197
287
243
241
5652
5674
5701
5679
5551
I.1048
I.4424=I.1016
5619
RS
5668
5521
U9380
5569
5560
5566
5718
5692
4C+03.30
5577
5636
5638
U9310
5576
5574
FS
U9432
CU
5725
5690
5738
5501
FF
5658
5680
5632
5713
5733
5719
5691
5584
5705
5496
3C 300.1
5478
103
105-φ
5618
102-υ
U9057
5507
5506
5604
3C 297
5493
5472
5468
107-μ
5634
99-ι
1411-057
©1987 WILLMANN-BELL, INC.
STELLAR MAGNITUDES
-1 0 1 2
3 4 5 6 7 8 9 >9.5
DOUBLE OR MULTIPLE STARS
VARIABLE STARS
OPEN STAR CLUSTERS
to scale
<5'
GLOBULAR STAR CLUSTERS
to scale
<5'
PLANETARY NEBULAE
>120"
120"-60"
60"-30"
<30"
BRIGHT NEBULAE
to scale
10'-5'
<5'
DARK NEBULAE
to scale
10'-5'
<5'
GALAXIES
to scale
<5'
QUASAR
RADIO SOURCE
X-RAY SOURCE
Barry Rappaport & Wil Tirion

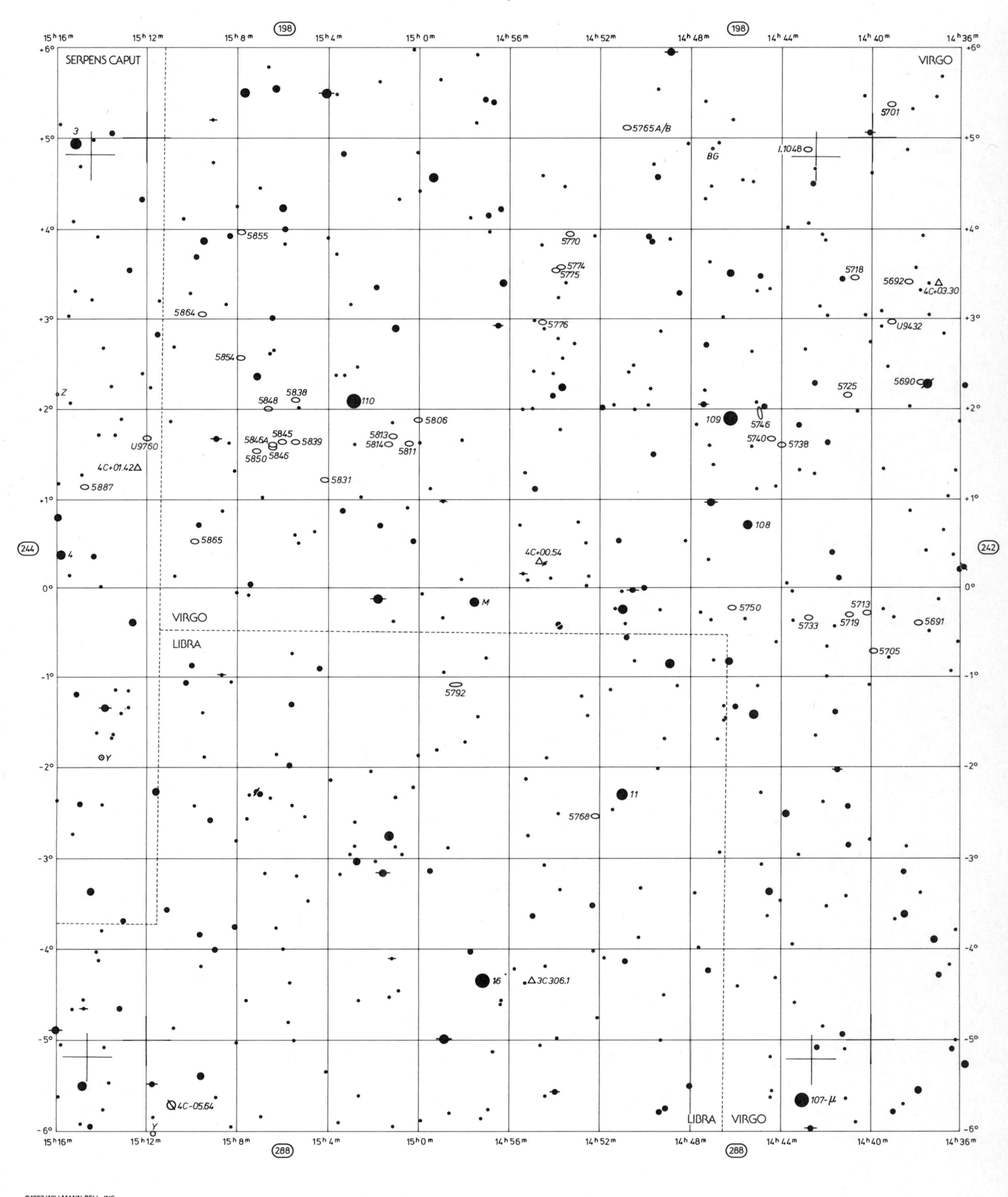

STELLAR MAGNITUDES -1 0 1 2 3 4 5 6 7 8 9 >9.5 | DOUBLE OR MULTIPLE STARS | VARIABLE STARS | OPEN STAR CLUSTERS to scale <5' | GLOBULAR STAR CLUSTERS to scale <5' | PLANETARY NEBULAE >120" 120"–60" 60"–30" <30" | BRIGHT NEBULAE to scale 10'–5' <5' | DARK NEBULAE to scale 10'–5' <5' | GALAXIES to scale <5' | QUASAR | RADIO SOURCE | X-RAY SOURCE

Barry Rappaport & Wil Tirion

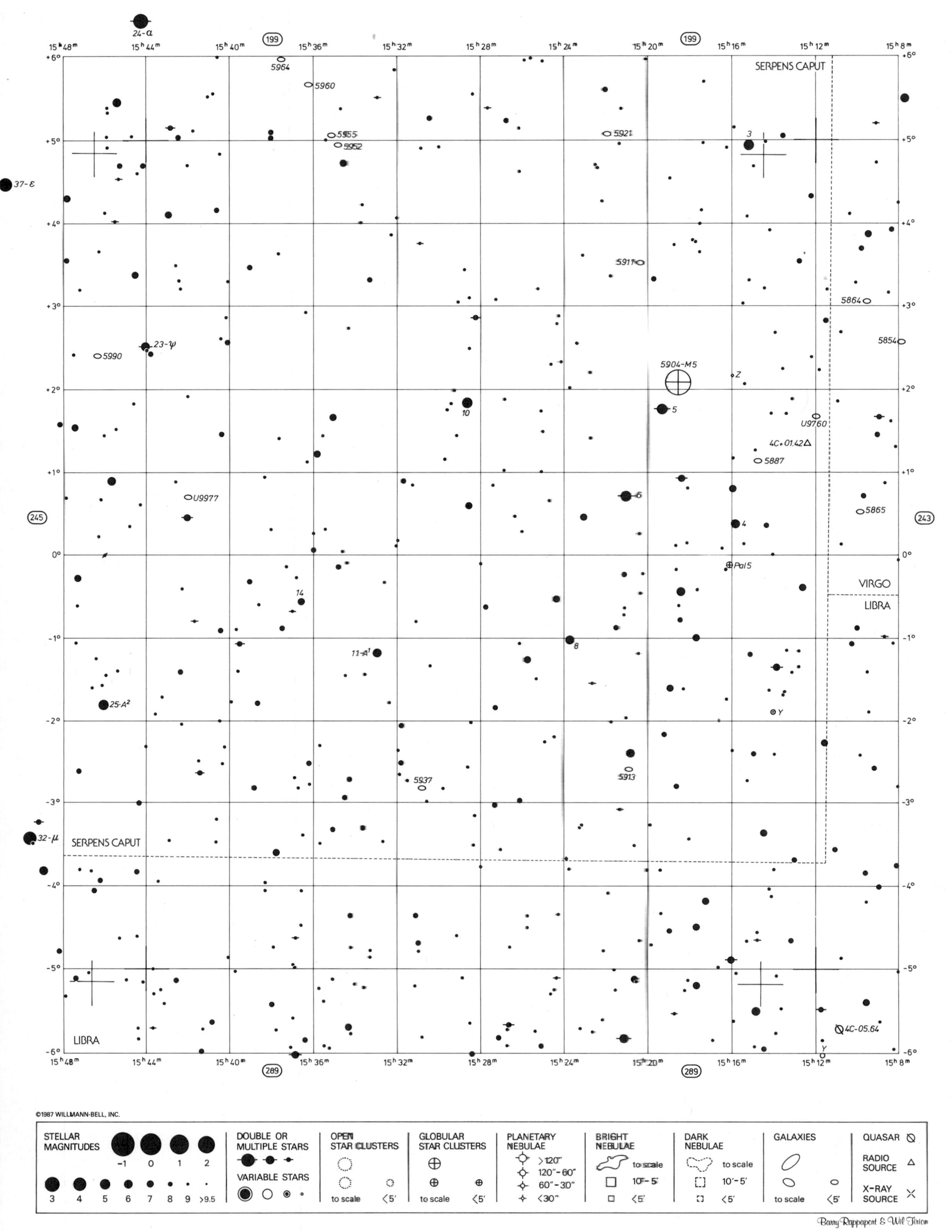

SERPENS CAPUT
VIRGO
LIBRA
24-α
37-ε
23-ψ
25-A²
32-μ
11-A¹
5904-M5
5964
5960
5955
5952
5921
5911
5864
5854
5990
U9977
U9760
4C+01.42
5887
5865
Pal5
5937
5913
4C-05.64
STELLAR MAGNITUDES
DOUBLE OR MULTIPLE STARS
VARIABLE STARS
OPEN STAR CLUSTERS
GLOBULAR STAR CLUSTERS
PLANETARY NEBULAE
BRIGHT NEBULAE
DARK NEBULAE
GALAXIES
QUASAR
RADIO SOURCE
X-RAY SOURCE
to scale
Barry Rappaport & Wil Tirion

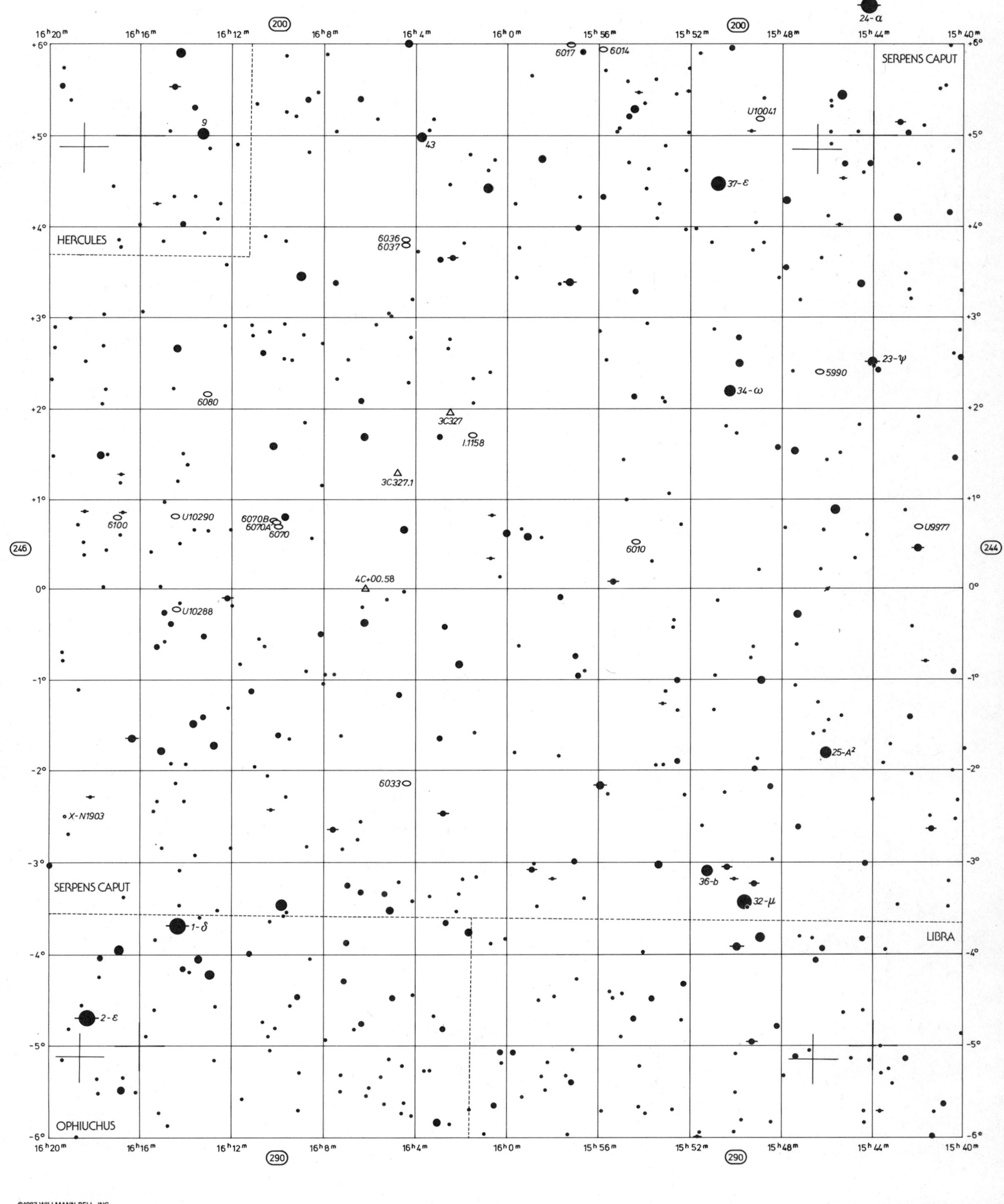

STELLAR MAGNITUDES	DOUBLE OR MULTIPLE STARS	OPEN STAR CLUSTERS	GLOBULAR STAR CLUSTERS	PLANETARY NEBULAE	BRIGHT NEBULAE	DARK NEBULAE	GALAXIES	QUASAR
-1 0 1 2	VARIABLE STARS	to scale <5'	to scale <5'	>120" 120"-60" 60"-30" <30"	to scale 10'-5' <5'	to scale 10'-5' <5'	to scale <5'	RADIO SOURCE
3 4 5 6 7 8 9 >9.5								X-RAY SOURCE

Barry Rappaport & Wil Tirion

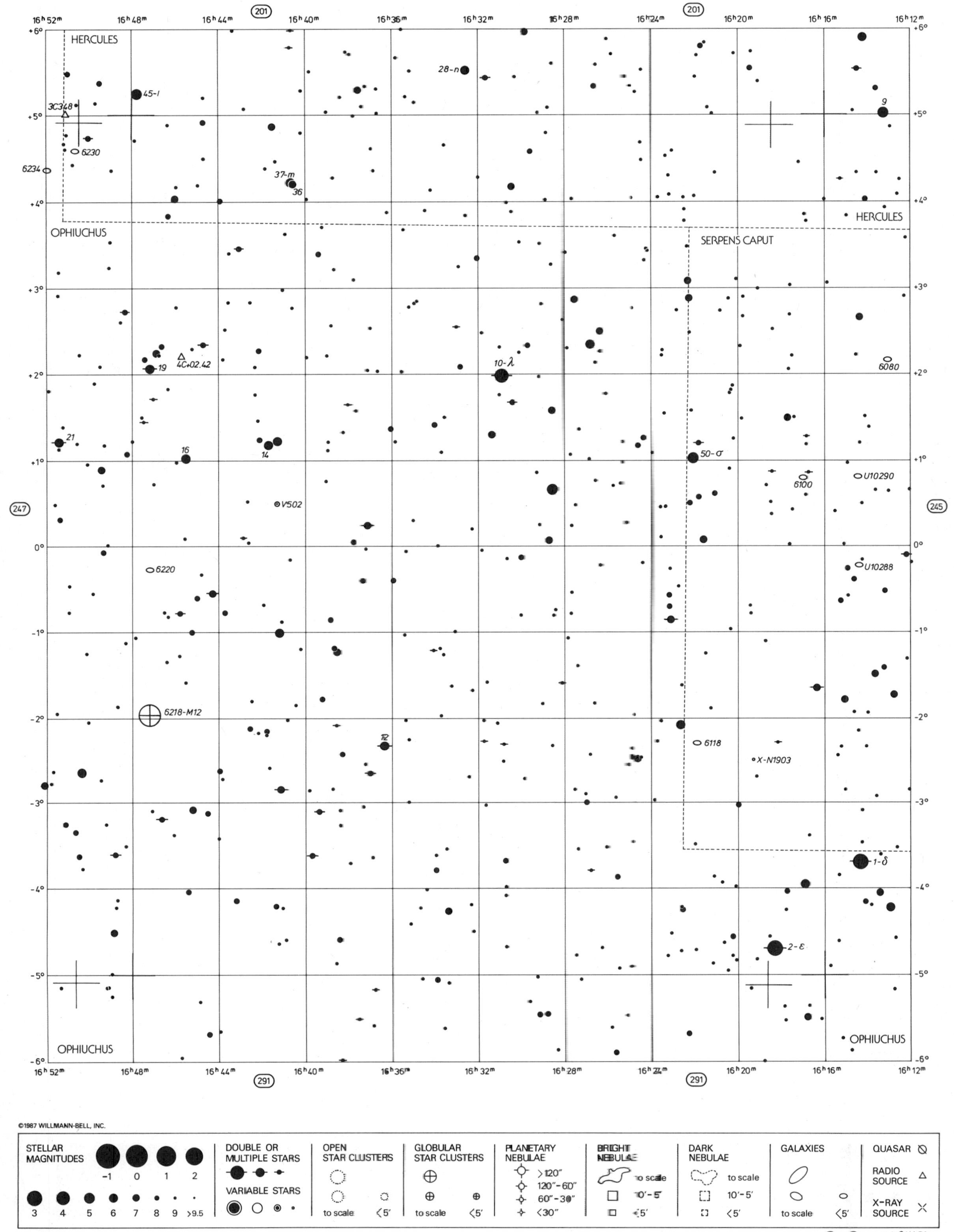

STELLAR MAGNITUDES	DOUBLE OR MULTIPLE STARS	OPEN STAR CLUSTERS	GLOBULAR STAR CLUSTERS	PLANETARY NEBULAE	BRIGHT NEBULAE	DARK NEBULAE	GALAXIES	QUASAR
-1 0 1 2	VARIABLE STARS	to scale <5′	to scale <5′	>120″ 120″-60″ 60″-30″ <30″	to scale 10′-5′ <5′	to scale 10′-5′ <5′	to scale <5′	RADIO SOURCE
3 4 5 6 7 8 9 >9.5								X-RAY SOURCE

Barry Rappaport & Wil Tirion

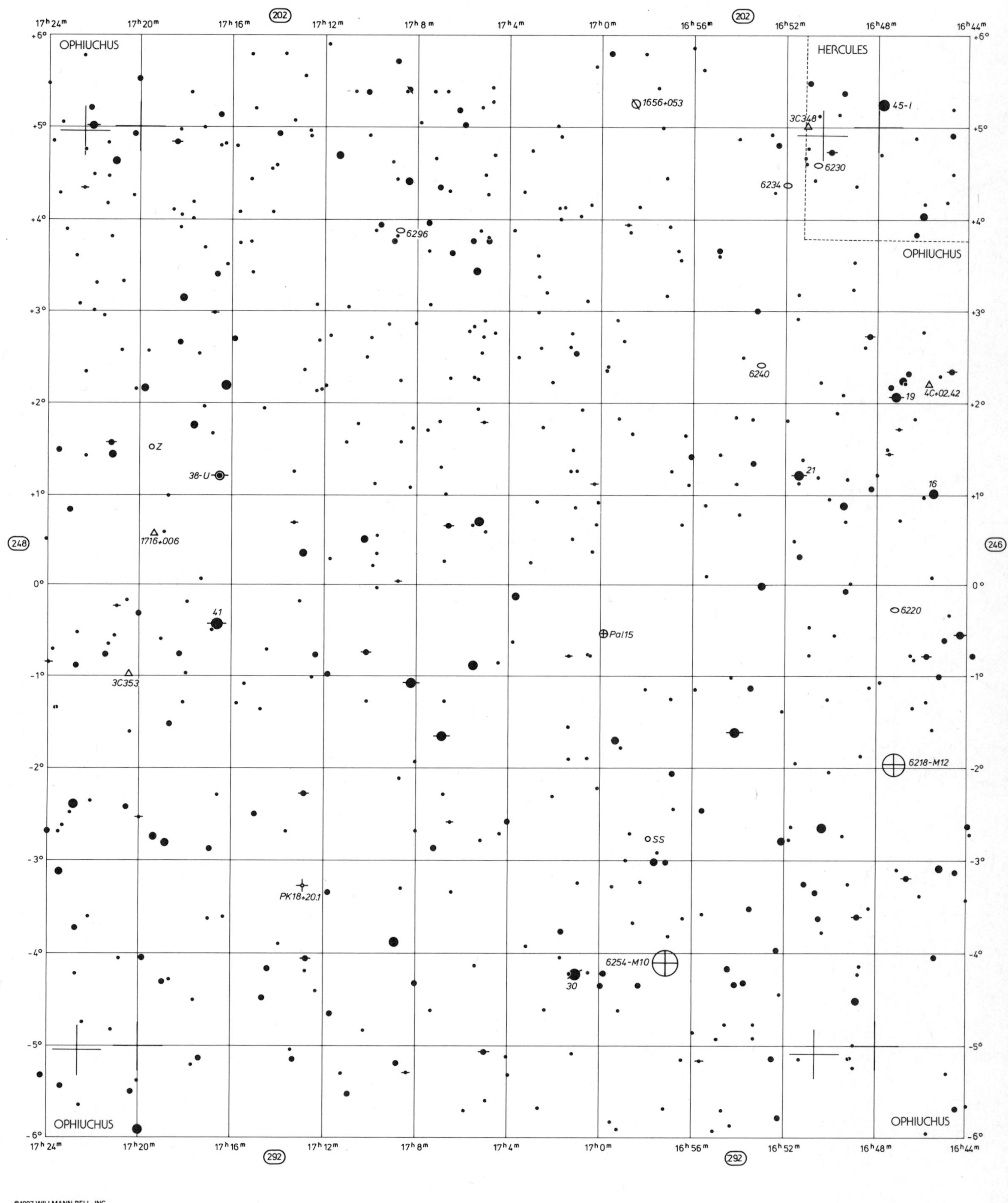

STELLAR MAGNITUDES	DOUBLE OR MULTIPLE STARS	OPEN STAR CLUSTERS	GLOBULAR STAR CLUSTERS	PLANETARY NEBULAE	BRIGHT NEBULAE	DARK NEBULAE	GALAXIES	QUASAR / RADIO SOURCE / X-RAY SOURCE
-1 0 1 2	VARIABLE STARS	to scale <5′	to scale <5′	>120″ 120″–60″ 60″–30″ <30″	to scale 10′–5′ <5′	to scale 10′–5′ <5′	to scale <5′	QUASAR RADIO SOURCE X-RAY SOURCE
3 4 5 6 7 8 9 >9.5								

Barry Rappaport & Wil Tirion

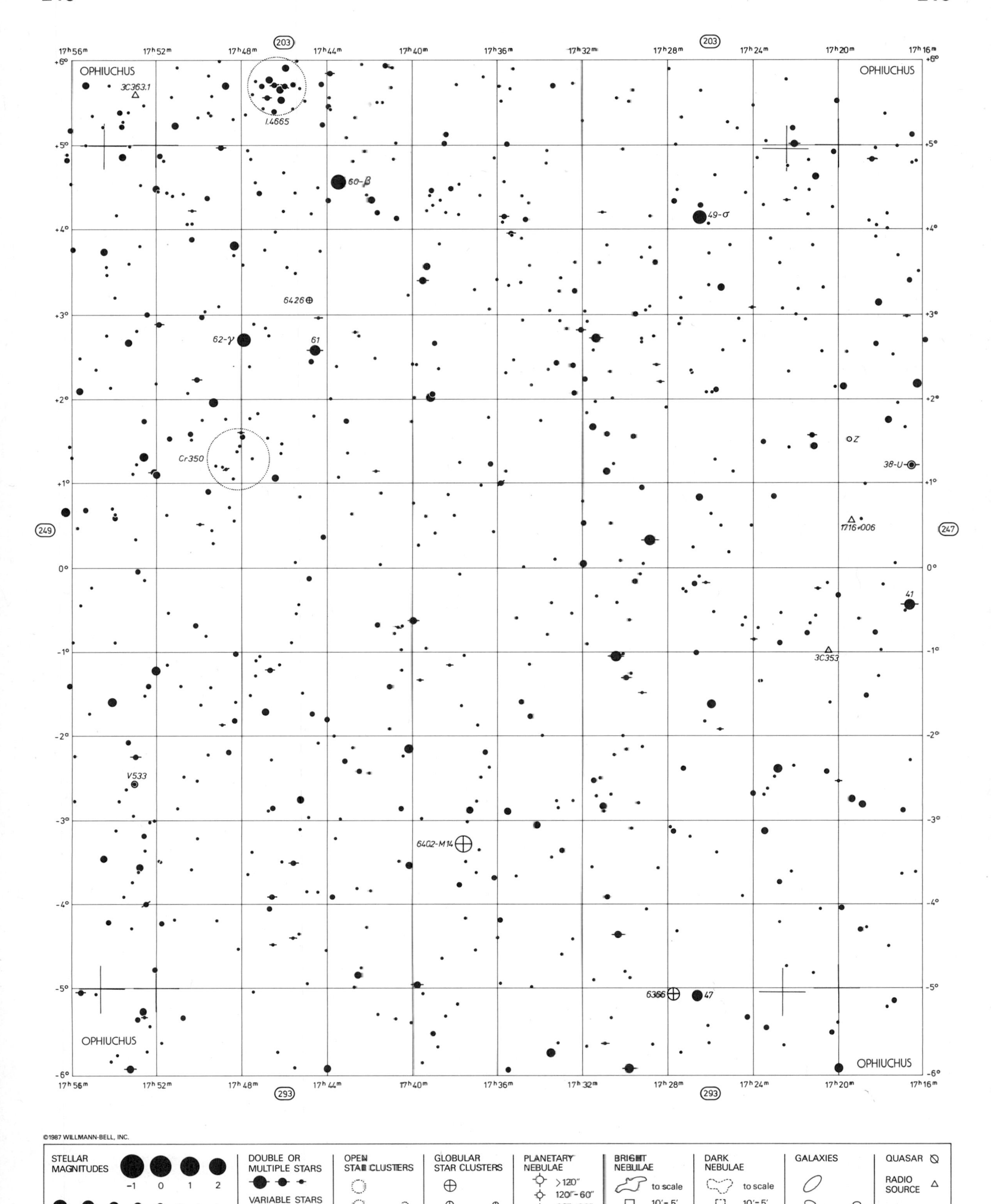

STELLAR MAGNITUDES −1 0 1 2 3 4 5 6 7 8 9 >9.5
DOUBLE OR MULTIPLE STARS
VARIABLE STARS
OPEN STAR CLUSTERS to scale <5′
GLOBULAR STAR CLUSTERS to scale <5′
PLANETARY NEBULAE >120″ 120″–60″ 60″–30″ <30″
BRIGHT NEBULAE to scale 10′–5′ <5′
DARK NEBULAE to scale 10′–5′ <5′
GALAXIES to scale <5′
QUASAR
RADIO SOURCE
X-RAY SOURCE

Barry Rappaport & Wil Tirion

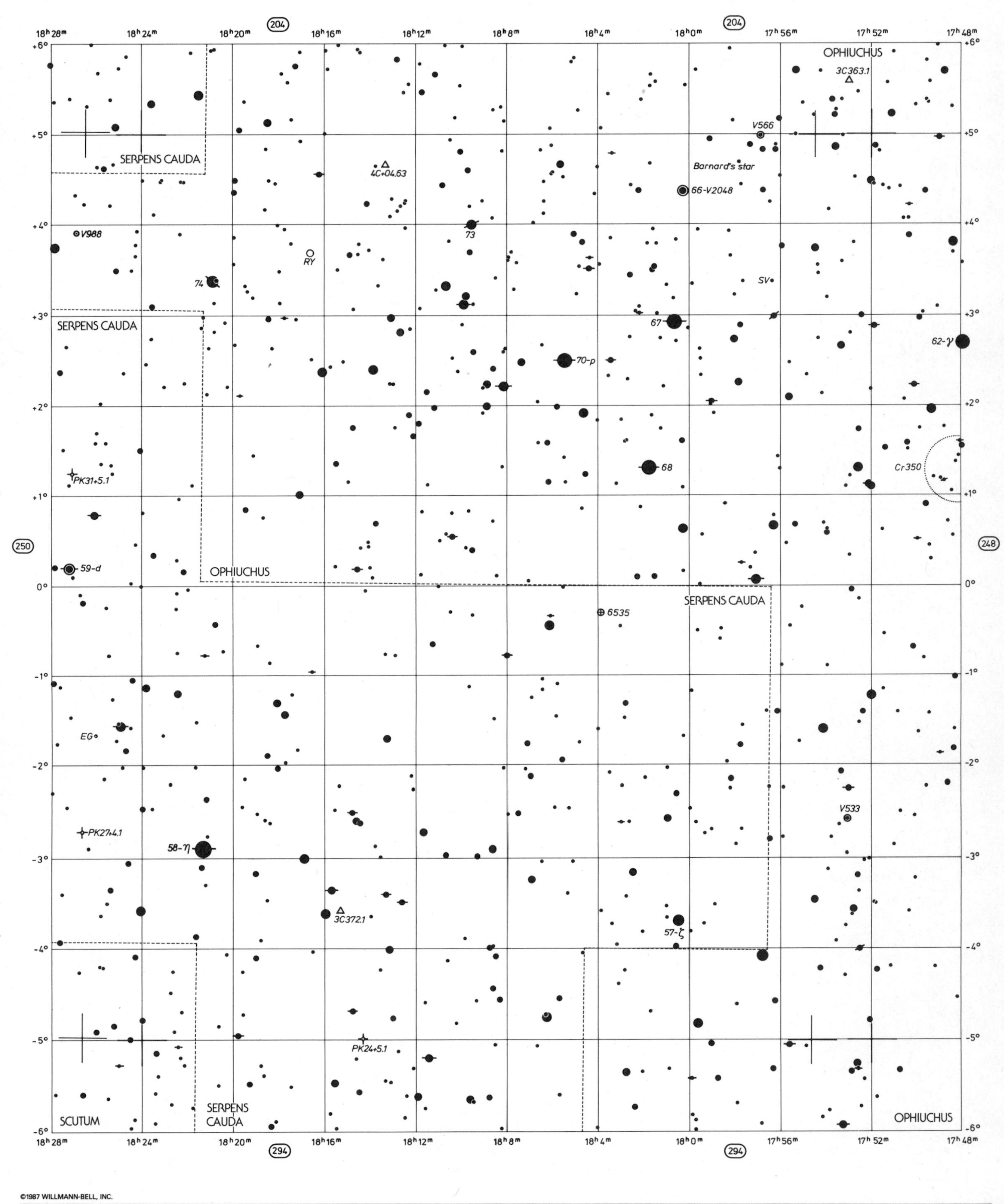

STELLAR MAGNITUDES: -1, 0, 1, 2, 3, 4, 5, 6, 7, 8, 9, >9.5

DOUBLE OR MULTIPLE STARS

VARIABLE STARS

OPEN STAR CLUSTERS: to scale, <5'

GLOBULAR STAR CLUSTERS: to scale, <5'

PLANETARY NEBULAE: >120", 120"-60", 60"-30", <30"

BRIGHT NEBULAE: to scale, 10'-5', <5'

DARK NEBULAE: to scale, 10'-5', <5'

GALAXIES: to scale, <5'

QUASAR

RADIO SOURCE

X-RAY SOURCE

Barry Rappaport & Wil Tirion

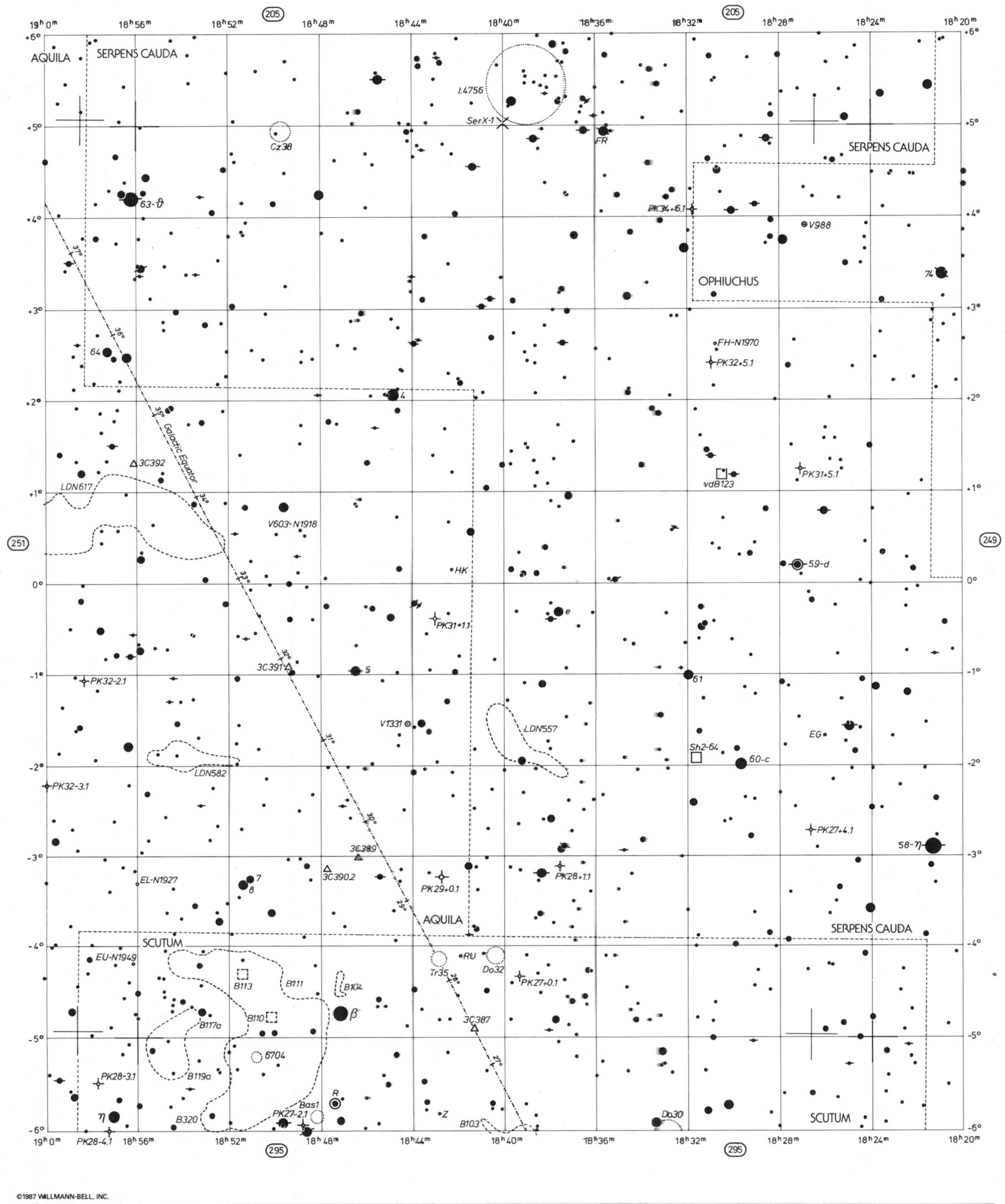

STELLAR MAGNITUDES: -1, 0, 1, 2, 3, 4, 5, 6, 7, 8, 9, >9.5

DOUBLE OR MULTIPLE STARS

VARIABLE STARS

OPEN STAR CLUSTERS: to scale, <5'

GLOBULAR STAR CLUSTERS: to scale, <5'

PLANETARY NEBULAE: >120", 120"-60", 60"-30", <30"

BRIGHT NEBULAE: to scale, 10'-5', <5'

DARK NEBULAE: to scale, 10'-5', <5'

GALAXIES: to scale, <5'

QUASAR

RADIO SOURCE

X-RAY SOURCE

Barry Rappaport & Wil Tirion

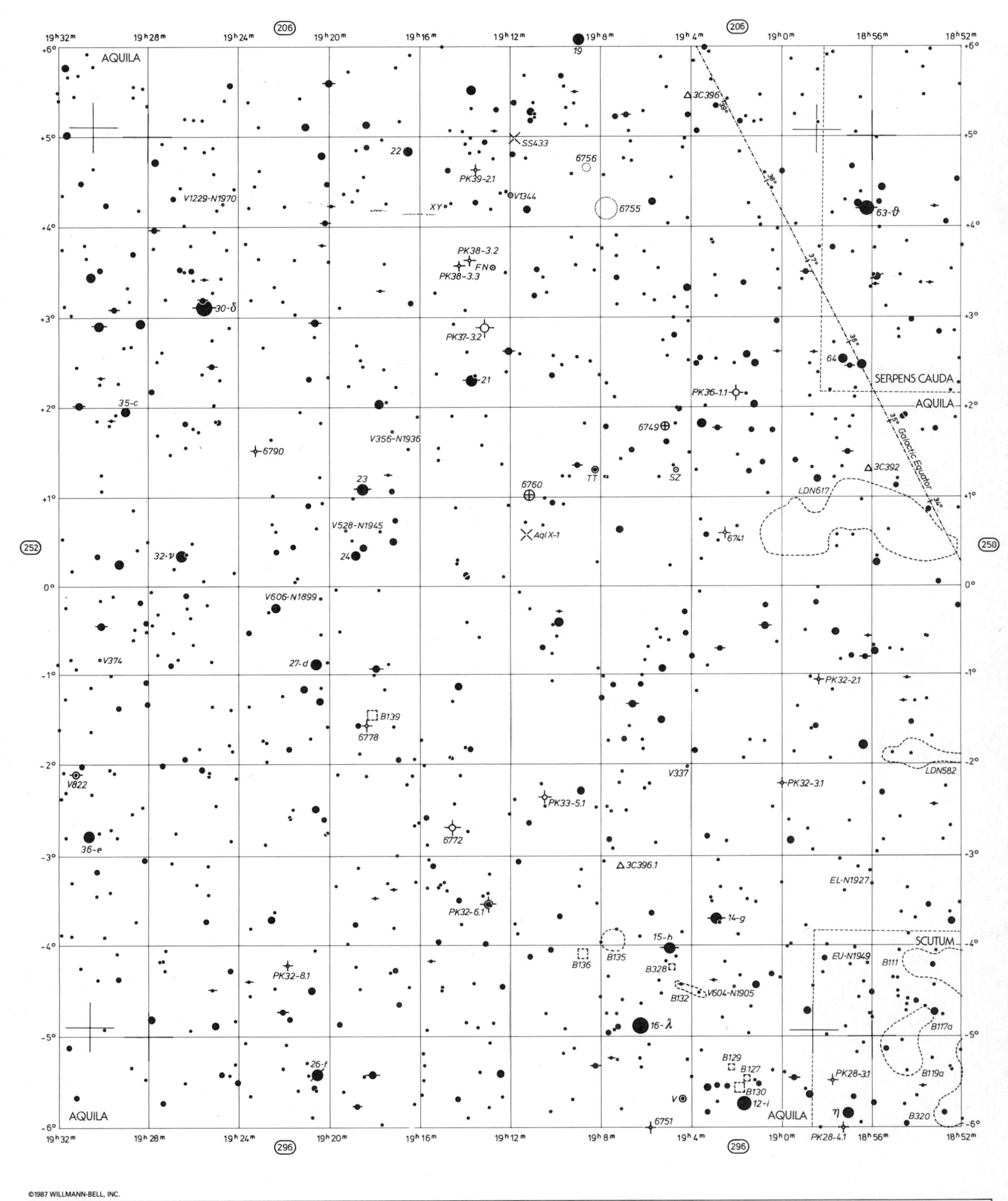

STELLAR MAGNITUDES	DOUBLE OR MULTIPLE STARS	OPEN STAR CLUSTERS	GLOBULAR STAR CLUSTERS	PLANETARY NEBULAE	BRIGHT NEBULAE	DARK NEBULAE	GALAXIES	
-1 0 1 2	VARIABLE STARS	to scale	to scale	>120"	to scale	to scale	to scale	QUASAR
3 4 5 6 7 8 9 >9.5		<5'	<5'	120"–60"	10'–5'	10'–5'	<5'	RADIO SOURCE
				60"–30"	<5'	<5'		X-RAY SOURCE
				<30"				

Barry Rappaport & Wil Tirion

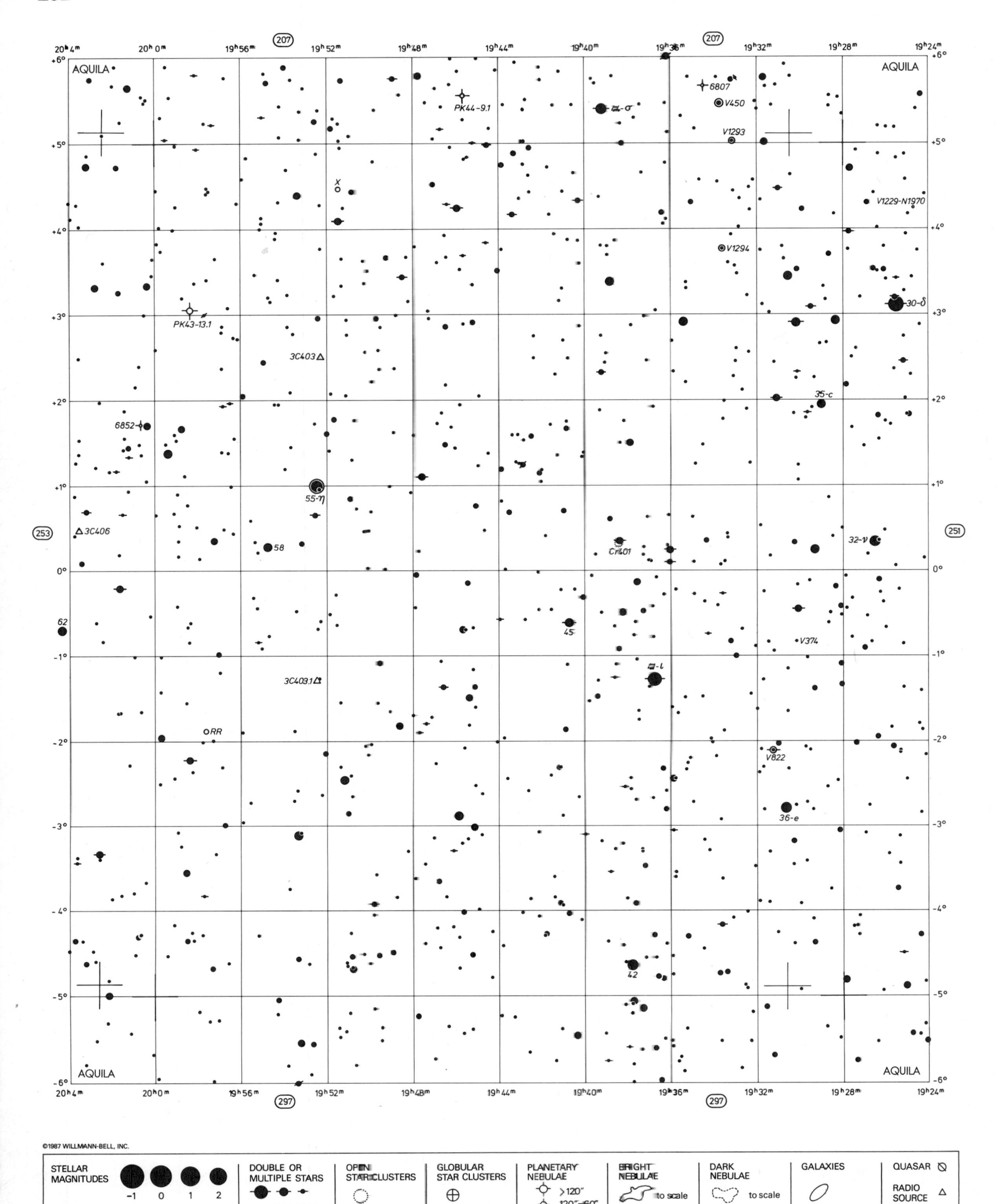

Barry Rappaport & Wil Tirion

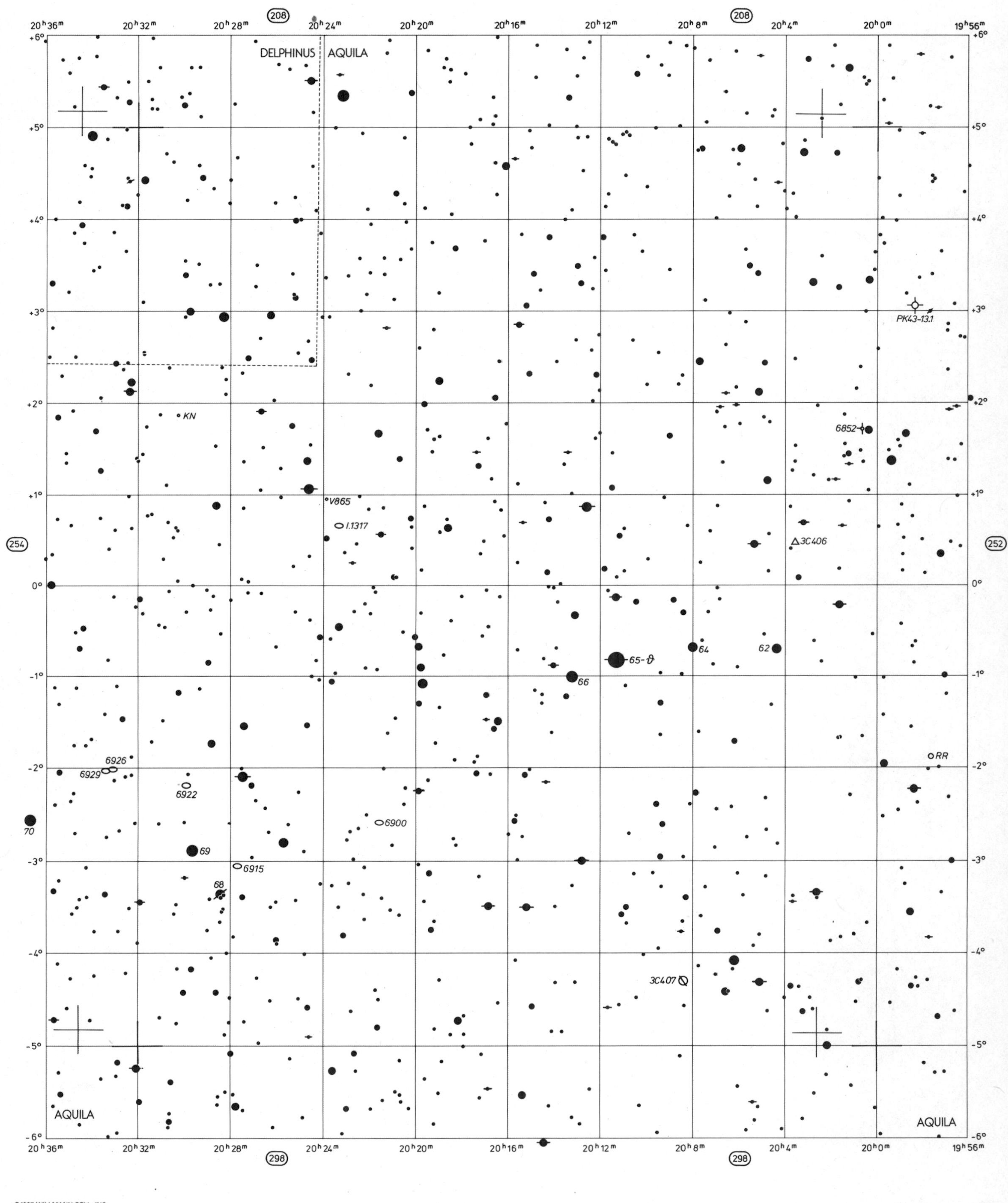

STELLAR MAGNITUDES	DOUBLE OR MULTIPLE STARS	OPEN STAR CLUSTERS	GLOBULAR STAR CLUSTERS	PLANETARY NEBULAE	BRIGHT NEBULAE	DARK NEBULAE	GALAXIES	QUASAR
-1 0 1 2	VARIABLE STARS			>120"	to scale	to scale		RADIO SOURCE
3 4 5 6 7 8 9 >9.5		to scale <5'	to scale <5'	120"-60" 60"-30" <30"	10'-5' <5'	10'-5' <5'	to scale <5'	X-RAY SOURCE

Barry Rappaport & Wil Tirion

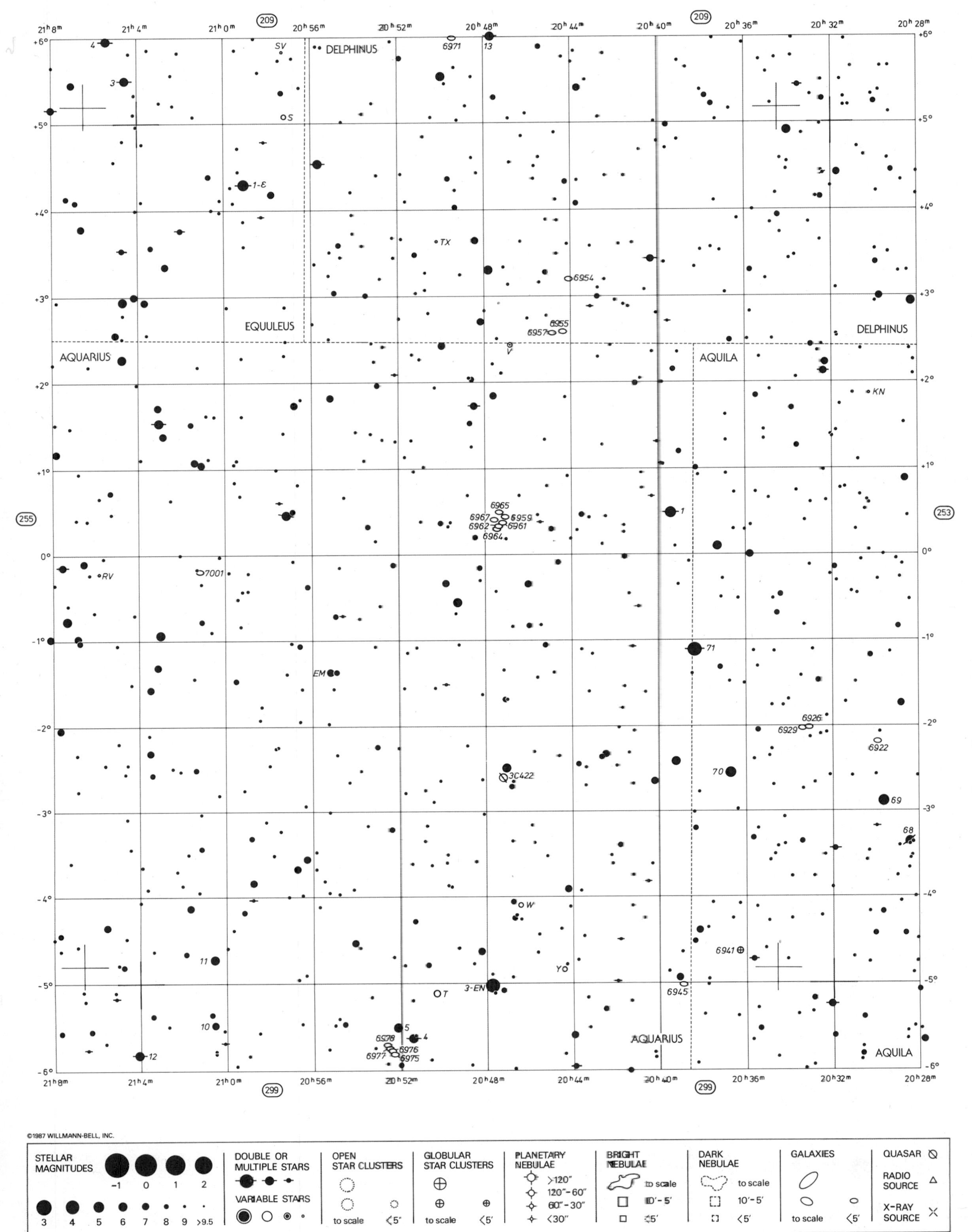

STELLAR MAGNITUDES	DOUBLE OR MULTIPLE STARS	OPEN STAR CLUSTERS	GLOBULAR STAR CLUSTERS	PLANETARY NEBULAE	BRIGHT NEBULAE	DARK NEBULAE	GALAXIES	QUASAR
-1 0 1 2	VARIABLE STARS	to scale <5′	to scale <5′	>120″ 120″–60″ 60″–30″ <30″	to scale 10′–5′ <5′	to scale 10′–5′ <5′	to scale <5′	RADIO SOURCE
3 4 5 6 7 8 9 >9.5								X-RAY SOURCE

Barry Rappaport & Wil Tirion

255

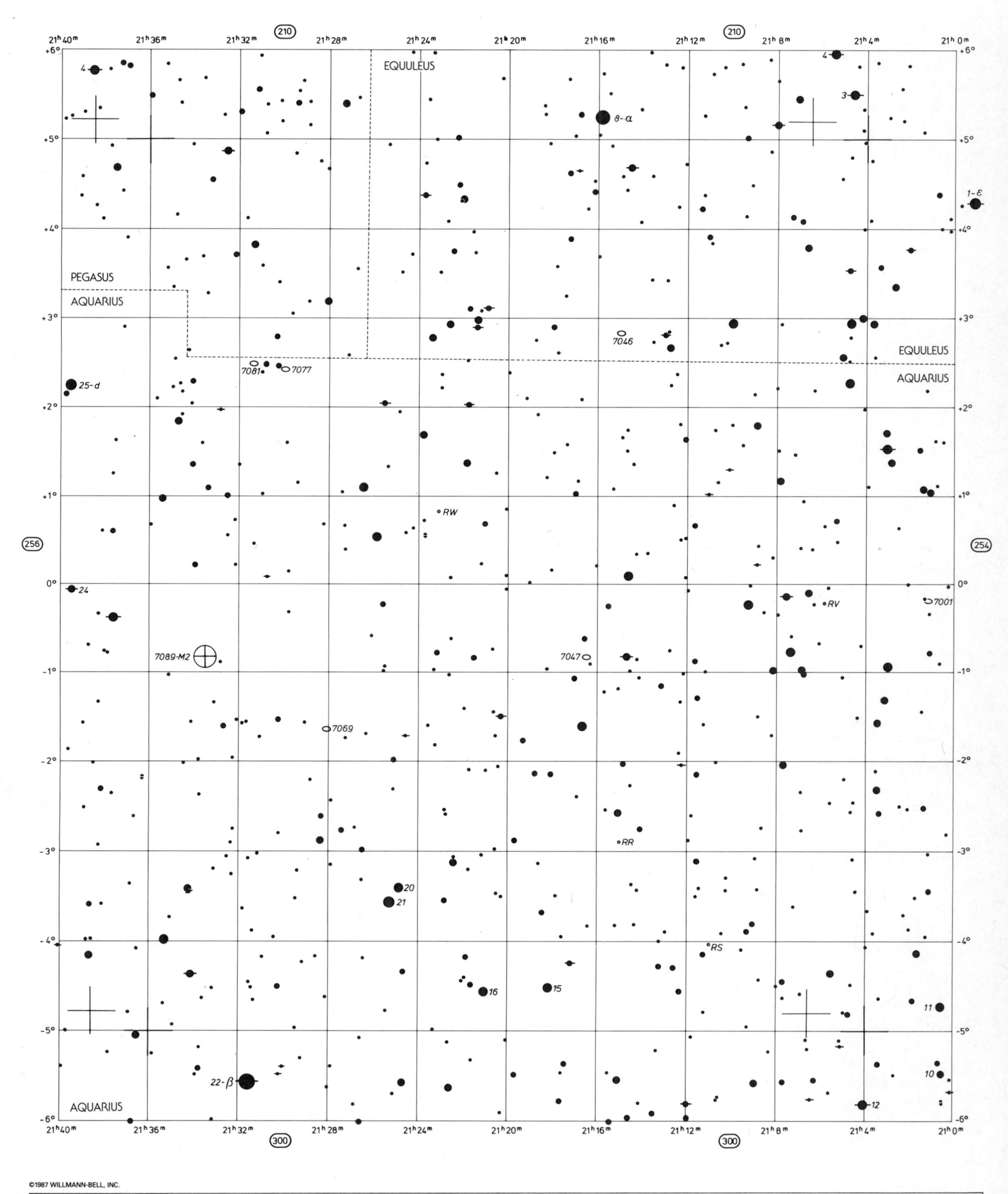

STELLAR MAGNITUDES	DOUBLE OR MULTIPLE STARS	OPEN STAR CLUSTERS	GLOBULAR STAR CLUSTERS	PLANETARY NEBULAE	BRIGHT NEBULAE	DARK NEBULAE	GALAXIES	QUASAR
-1 0 1 2	VARIABLE STARS			>120"	to scale	to scale		RADIO SOURCE
3 4 5 6 7 8 9 >9.5		to scale <5'	to scale <5'	120"-60" 60"-30" <30"	10'-5' <5'	10'-5' <5'	to scale <5'	X-RAY SOURCE

Barry Rappaport & Wil Tirion

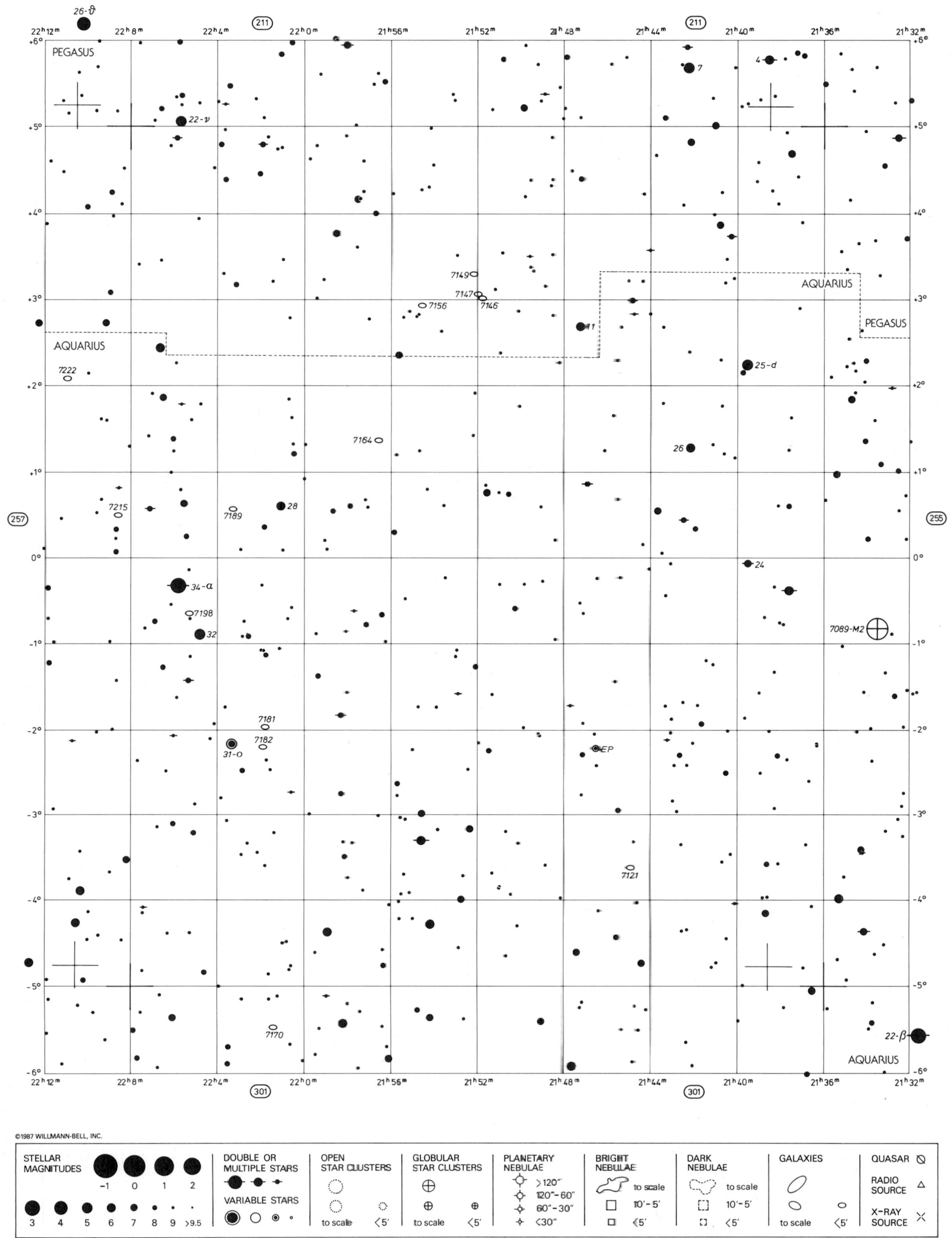

STELLAR MAGNITUDES −1 0 1 2 3 4 5 6 7 8 9 >9.5 | DOUBLE OR MULTIPLE STARS | VARIABLE STARS | OPEN STAR CLUSTERS to scale <5′ | GLOBULAR STAR CLUSTERS to scale <5′ | PLANETARY NEBULAE >120″ 120″–60″ 60″–30″ <30″ | BRIGHT NEBULAE to scale 10′–5′ <5′ | DARK NEBULAE to scale 10′–5′ <5′ | GALAXIES to scale <5′ | QUASAR | RADIO SOURCE | X-RAY SOURCE

Barry Rappaport & Wil Tirion

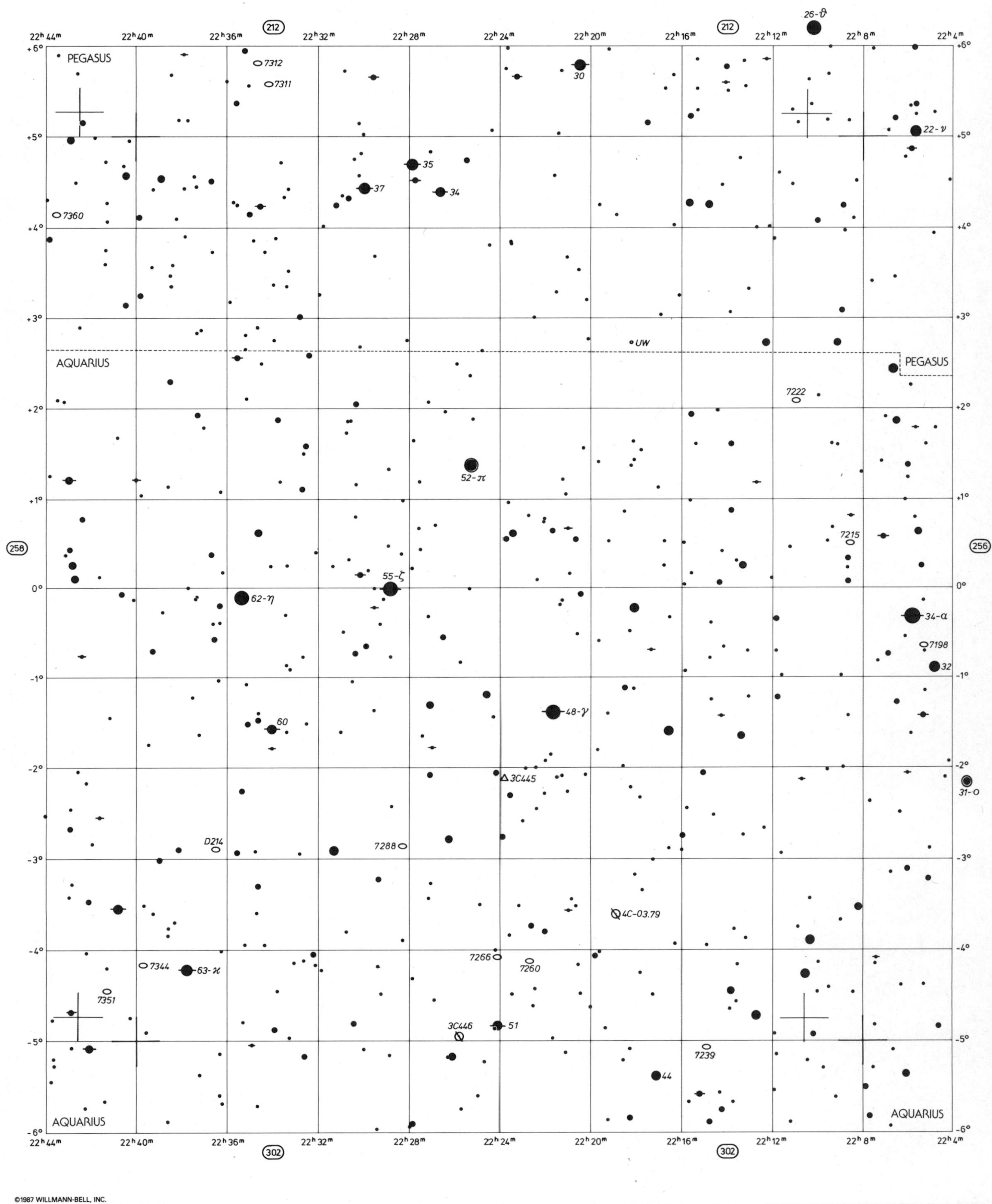

STELLAR MAGNITUDES	DOUBLE OR MULTIPLE STARS	OPEN STAR CLUSTERS	GLOBULAR STAR CLUSTERS	PLANETARY NEBULAE	BRIGHT NEBULAE	DARK NEBULAE	GALAXIES	
-1 0 1 2	VARIABLE STARS	to scale	to scale	>120″	to scale	to scale	to scale	QUASAR
3 4 5 6 7 8 9 >9.5		<5′	<5′	120″–60″	10′–5′	10′–5′	<5′	RADIO SOURCE
				60″–30″	<5′	<5′		X-RAY SOURCE
				<30″				

Barry Rappaport & Wil Tirion

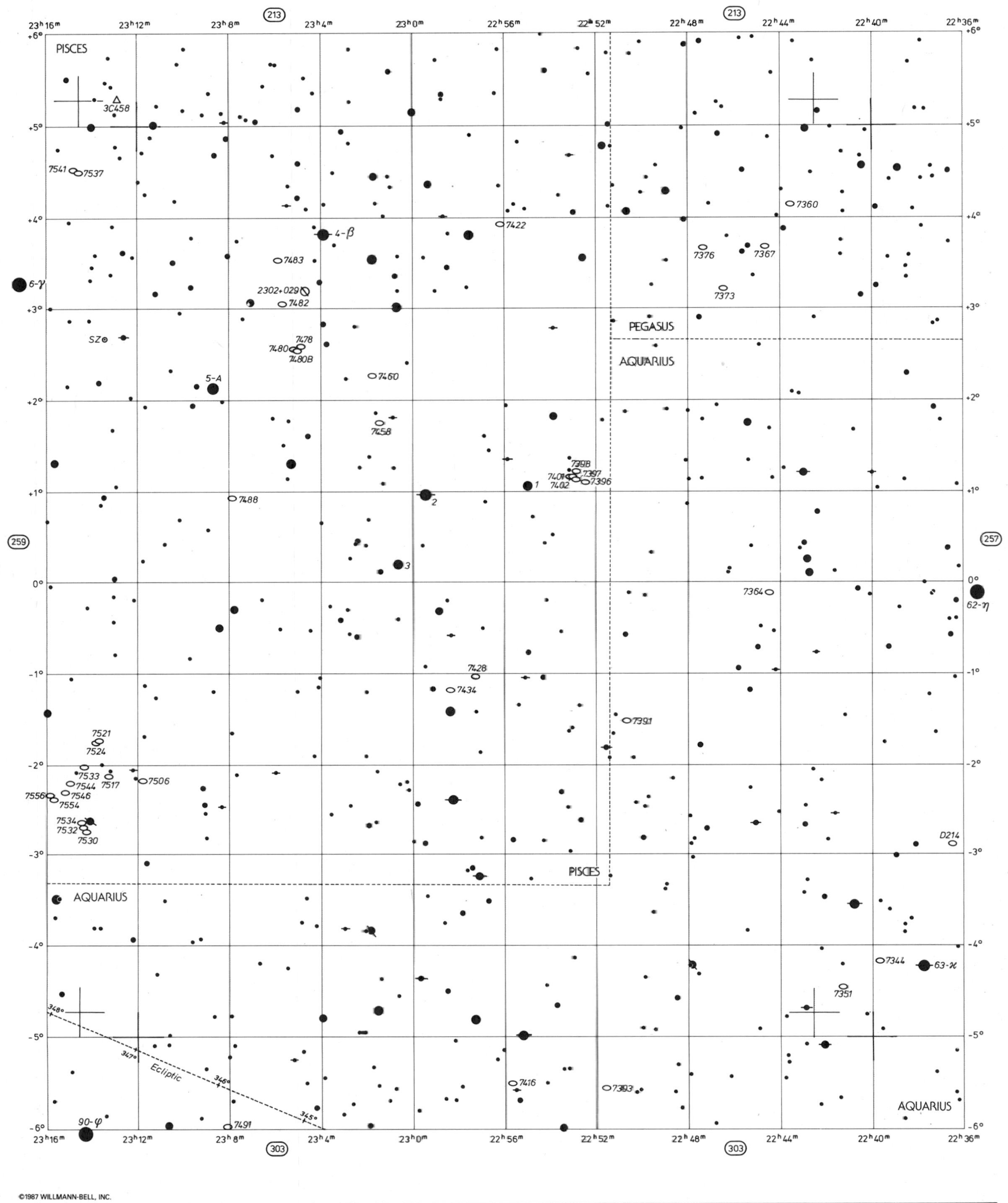

STELLAR MAGNITUDES -1 0 1 2 3 4 5 6 7 8 9 >9.5 | DOUBLE OR MULTIPLE STARS | VARIABLE STARS | OPEN STAR CLUSTERS to scale <5′ | GLOBULAR STAR CLUSTERS to scale <5′ | PLANETARY NEBULAE >120″ 120″-60″ 60″-30″ <30″ | BRIGHT NEBULAE to scale 10′-5′ <5′ | DARK NEBULAE to scale 10′-5′ <5′ | GALAXIES to scale <5′ | QUASAR | RADIO SOURCE | X-RAY SOURCE

Barry Rappaport & Wil Tirion

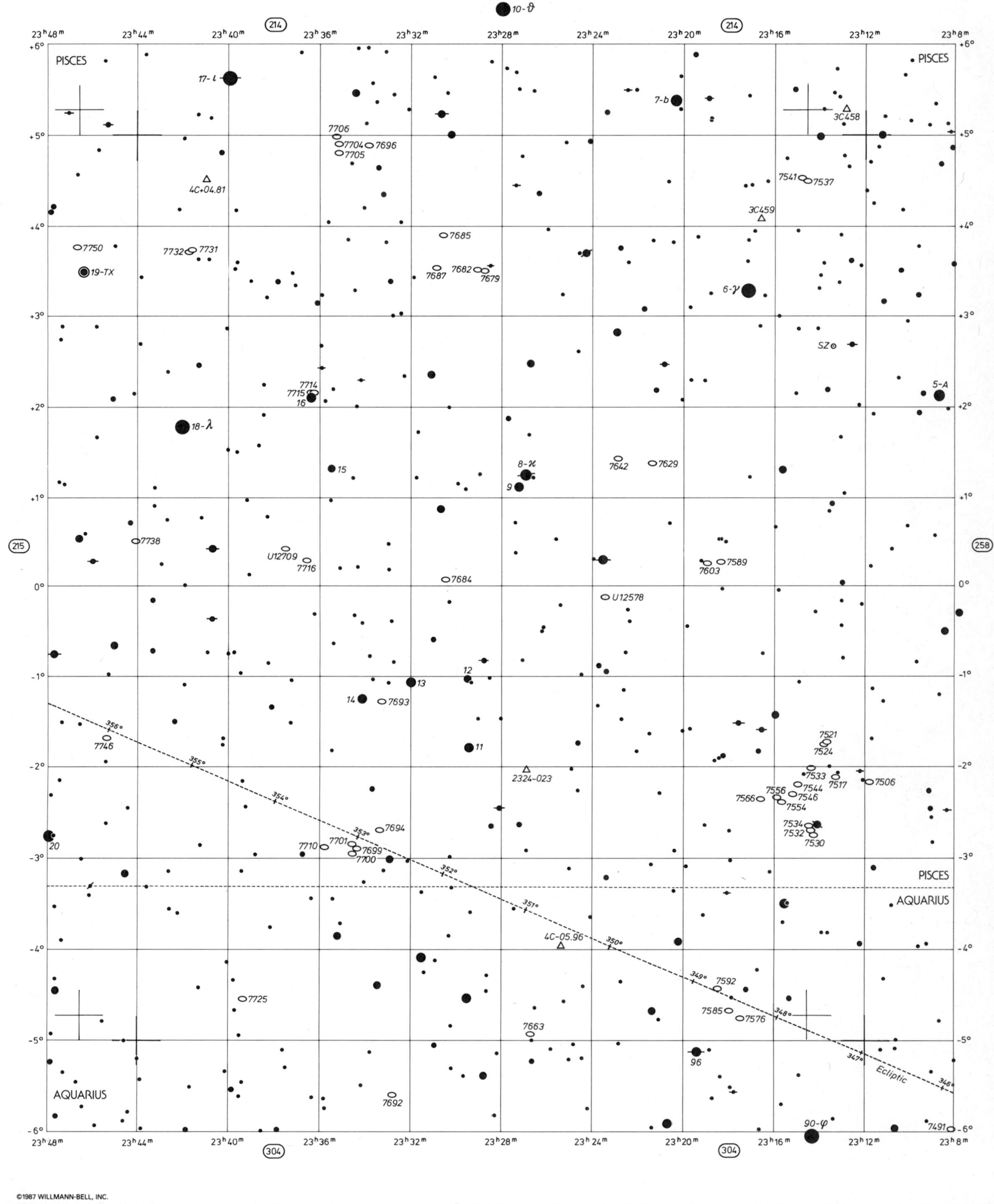

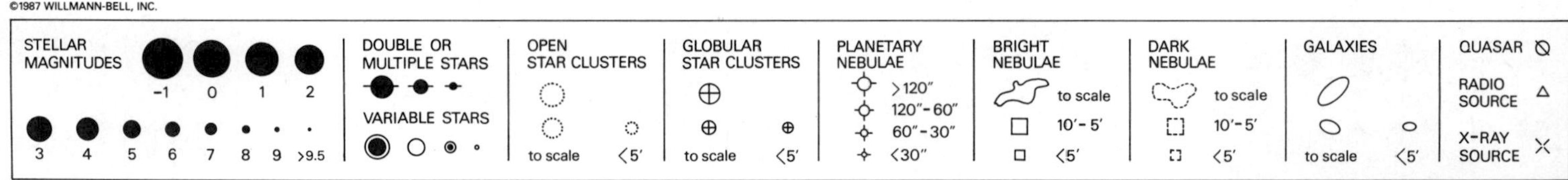

Barry Rappaport & Wil Tirion

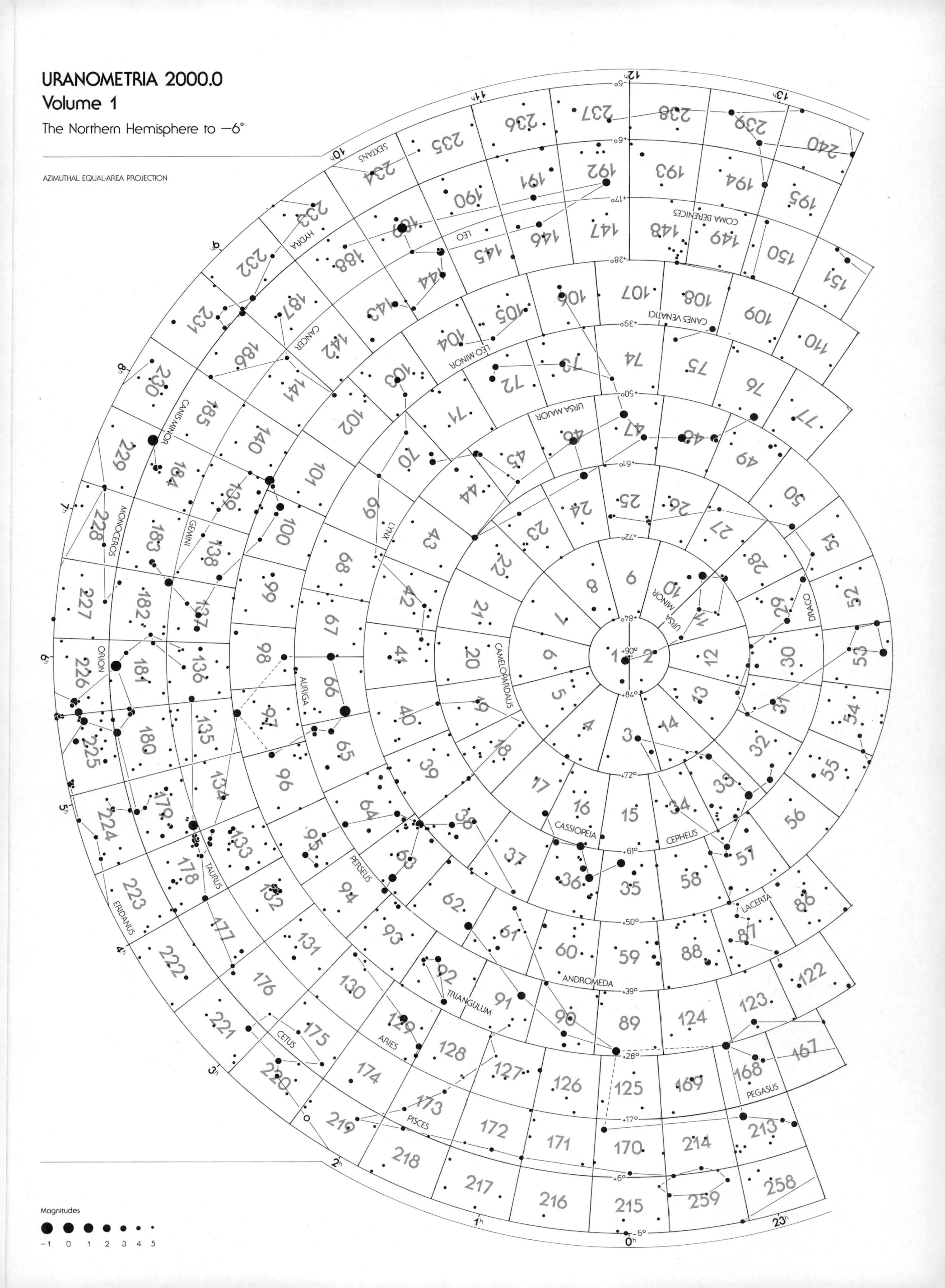

URANOMETRIA 2000.0
Volume 1
The Northern Hemisphere to −6°
AZIMUTHAL EQUAL-AREA PROJECTION
URSA MINOR
CAMELOPARDALIS
DRACO
CASSIOPEIA
CEPHEUS
URSA MAJOR
LYNX
AURIGA
PERSEUS
LACERTA
ANDROMEDA
TRIANGULUM
LEO MINOR
CANES VENATICI
COMA BERENICES
LEO
SEXTANS
HYDRA
CANCER
CANIS MINOR
GEMINI
MONOCEROS
ORION
TAURUS
ERIDANUS
CETUS
ARIES
PISCES
PEGASUS
Magnitudes
−1 0 1 2 3 4 5

INDEX

URSA MINOR
DRACO
CAMELOPARDALIS
CASSIOPEIA
CEPHEUS
URSA MAJOR
LEO MINOR
LEO
CANES VENATICI
COMA BERENICES
VIRGO
BOOTES
SERPENS CAPUT
CORONA BOREALIS
HERCULES
OPHIUCHUS
LYRA
SERPENS CAUDA
SCUTUM
VULPECULA
CYGNUS
SAGITTA
AQUILA
LACERTA
DELPHINUS
ANDROMEDA
TRIANGULUM
PEGASUS
EQUULEUS
AQUARIUS

0h 1h 11h 12h 13h 14h 15h 16h 17h 18h 19h 20h 21h 22h 23h

+90° +84° +78° +72° +61° +50° +39° +28° +17° +6° −6°

Wil Tirion

NOTES

NOTES

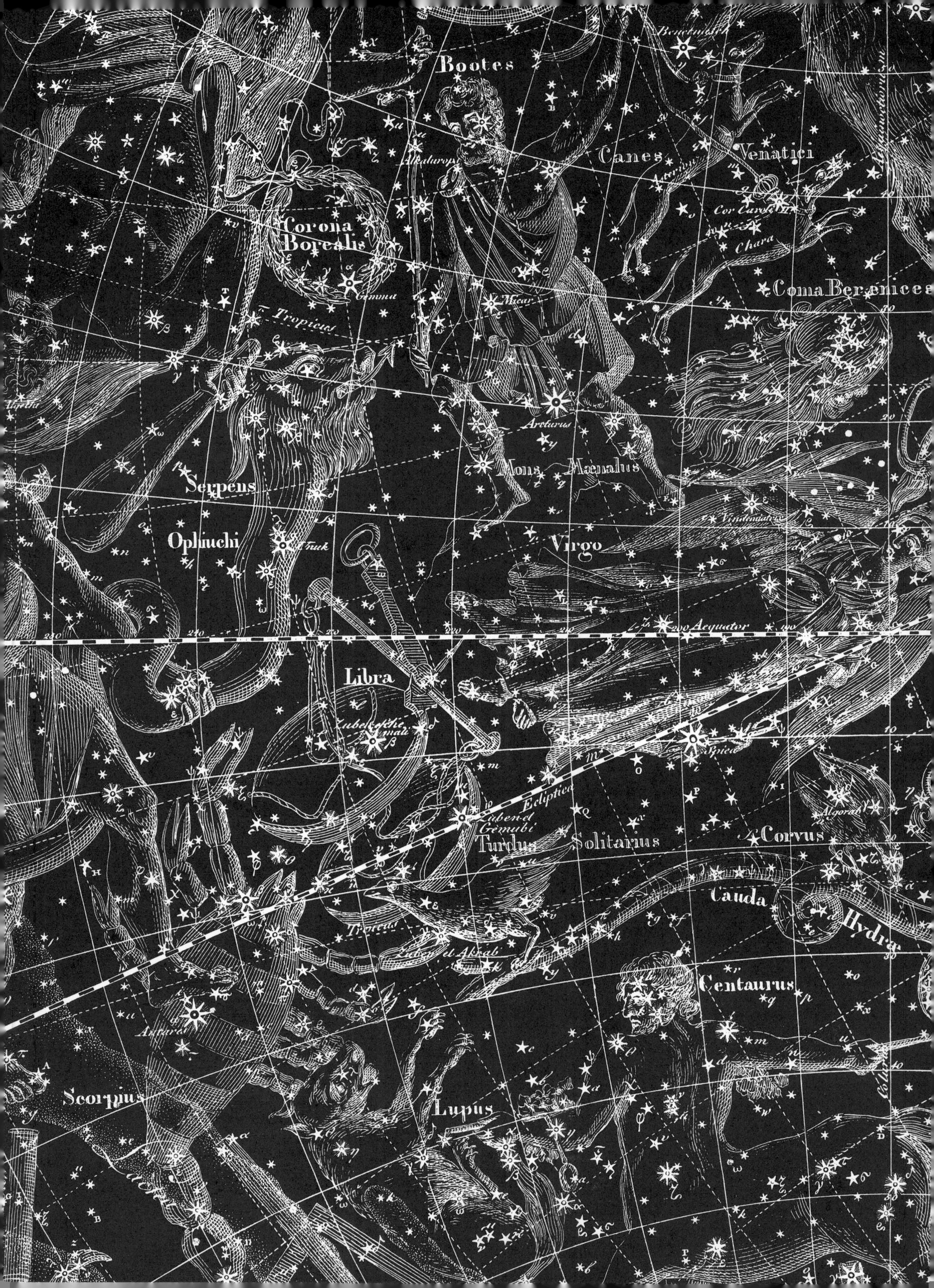

Bootes
Canes
Venatici
Cor Caroli
Chara
Coma Berenices
Corona
Borealis
Gemma
Tropicus
Arcturus
Mons Mænalus
Serpens
Ophiuchi
Virgo
Aequator
Libra
Spica
Ecliptica
Turdus
Solitarius
Corvus
Cauda
Hydra
Centaurus
Antares
Scorpius
Lupus